THE STUDENTS
ELEMENTS OF GEOLOGY
BY SIR CHARLES LYELL

PREFACE.

The LAST or sixth EDITION of my "Elements of Geology" was already out of print before the end of 1868, in which year I brought out the tenth edition of my "Principles of Geology."

In writing the last-mentioned work I had been called upon to pass in review almost all the leading points of speculation and controversy to which the rapid advance of the science had given rise, and when I proposed to bring out a new edition of the "Elements" I was strongly urged by my friends not to repeat these theoretical discussions, but to confine myself in the new treatise to those parts of the "Elements" which were most indispensable to a beginner. This was to revert, to a certain extent, to the original plan of the first edition; but I found, after omitting a great number of subjects, that the necessity of bringing up to the day those which remained, and adverting, however briefly, to new discoveries, made it most difficult to confine the proposed abridgment within moderate limits. Some chapters had to be entirely recast, some additional illustrations to be introduced, and figures of some organic remains to be replaced by new ones from specimens more perfect than those which had been at my command on former occasions. By these changes the work assumed a form so different from the sixth edition of the "Elements," that I resolved to give it a new title and call it the "Student's Elements of Geology."

In executing this task I have found it very difficult to meet the requirements of those who are entirely ignorant of the science. It is only the adept who has already overcome the first steps as an observer, and is familiar with many of the technical terms, who can profit by a brief and concise manual. Beginners wish for a short and cheap book in which they may find a full explanation of the leading facts and principles of Geology. Their wants, I fear, somewhat resemble those of the old woman in New England, who asked a bookseller to supply her with "the cheapest Bible in the largest possible print."

But notwithstanding the difficulty of reconciling brevity with the copiousness of illustration demanded by those who have not yet mastered the rudiments of the science, I have endeavoured to abridge the work in the manner above hinted at, so as to place it within the reach of many to whom it was before inaccessible.

CHARLES LYELL.
73 Harley Street, London, December, 1870.

STUDENT'S ELEMENTS OF GEOLOGY.
CHAPTER I.
ON THE DIFFERENT CLASSES OF ROCKS.

Geology defined.
Successive Formation of the Earth's Crust.
Classification of Rocks according to their Origin and Age.
Aqueous Rocks.
Their Stratification and imbedded Fossils.
Volcanic Rocks, with and without Cones and Craters.
Plutonic Rocks, and their Relation to the Volcanic.
Metamorphic Rocks, and their probable Origin.
The term Primitive, why erroneously applied to the Crystalline Formations.
Leading Division of the Work.

Of what materials is the earth composed, and in what manner are these materials arranged? These are the first inquiries with which Geology is occupied, a science which derives its name from the Greek ge, the earth, and logos, a discourse. Previously to experience we might have imagined that investigations of this kind would relate exclusively to the mineral kingdom, and to the various rocks, soils, and metals, which occur upon the surface of the earth, or at various depths beneath it. But, in pursuing such researches, we soon find ourselves led on to consider the successive changes which have taken place in the former state of the earth's surface and interior, and the causes which have given rise to these changes; and, what is still more singular and unexpected, we soon become engaged in researches into the history of the animate creation, or of the various tribes of animals and plants which have, at different periods of the past, inhabited the globe.

All are aware that the solid parts of the earth consist of distinct substances, such as clay, chalk, sand, limestone, coal, slate, granite, and the like; but previously to observation it is commonly imagined that all these had remained from the first in the state in which we now see them— that they were created in their present form, and in their present position. The geologist soon comes to a different conclusion, discovering proofs that the external parts of the earth were not all produced in the beginning of things in the state in which we now behold them, nor in an instant of time. On the contrary, he can show that they have acquired their actual configuration and condition gradually, under a great variety of circumstances, and at successive periods, during each of which distinct races of living beings have flourished on the land and in the waters, the remains of these creatures still lying buried in the crust of the earth.

By the "earth's crust," is meant that small portion of the exterior of our planet which is accessible to human observation. It comprises not merely all of which the structure is laid open in mountain precipices, or in cliffs overhanging a river or the sea, or whatever the miner may reveal in artificial excavations; but the whole of that outer covering of the planet on which we are enabled to reason by observations made at or near the surface. These reasonings may extend to a depth of several miles, perhaps ten miles; and even then it may be said, that such a thickness is no more than 1/400 part of the distance from the surface to the centre. The remark is just: but although the dimensions of such a crust are, in truth, insignificant when compared to the entire globe, yet they are vast, and of magnificent extent in relation to man, and to the organic beings which people our globe. Referring to this standard of magnitude, the geologist may admire the ample limits of his domain, and admit, at the same time, that not only the exterior of the planet, but the entire earth, is but an atom in the midst of the countless worlds surveyed by the astronomer.

The materials of this crust are not thrown together confusedly; but distinct mineral masses, called rocks, are found to occupy definite spaces, and to exhibit a certain order of arrangement. The term ROCK is applied indifferently by geologists to all these substances, whether they be soft or stony, for clay and sand are included in the term, and some have even brought peat under this denomination. Our old writers endeavoured to avoid offering such violence to our language, by speaking of the component materials of the earth as consisting of rocks and SOILS. But there is often so insensible a passage from a soft and incoherent state to that of stone, that geologists of all countries have found it indispensable to have one technical term to include both, and in this sense we find ROCHE applied in French, ROCCA in Italian, and FELSART in German. The beginner, however, must constantly bear in mind that the term rock by no means implies that a mineral mass is in an indurated or stony condition.

The most natural and convenient mode of classifying the various rocks which compose the earth's crust, is to refer, in the first place, to their origin, and in the second to their relative age. I shall therefore begin by endeavouring briefly to explain to the student how all rocks may be divided into four great classes by reference to their different origin, or, in other words, by reference to the different circumstances and causes by which they have been produced.

The first two divisions, which will at once be understood as natural, are the aqueous and volcanic, or the products of watery and those of igneous action at or near the surface.

AQUEOUS ROCKS.

The aqueous rocks, sometimes called the sedimentary, or fossiliferous, cover a larger part of the earth's surface than any others. They consist chiefly of mechanical deposits (pebbles, sand, and mud), but are partly of chemical and some of them of organic origin, especially the limestones. These rocks are STRATIFIED, or divided into distinct layers, or strata. The term STRATUM means simply a bed, or any thing spread out or STREWED over a given surface; and we infer that these strata have been generally spread out by the action of water, from what we daily see taking place near the mouths of rivers, or on the land during temporary inundations. For, whenever a running stream charged with mud or sand, has its velocity checked, as when it enters a lake or sea, or overflows a plain, the sediment, previously held in suspension by the motion of the water, sinks, by its own gravity to the bottom. In this manner layers of mud and sand are thrown down one upon another.

If we drain a lake which has been fed by a small stream, we frequently find at the bottom a series of deposits, disposed with considerable regularity, one above the other; the uppermost, perhaps, may be a stratum of peat, next below a more dense and solid variety of the same material; still lower a bed of shell- marl, alternating with peat or sand, and then other beds of marl, divided by layers of clay. Now, if a second pit be sunk through the same continuous lacustrine FORMATION at some distance from the first, nearly the same series of beds is commonly met with, yet with slight variations; some, for example, of the layers of sand, clay, or marl, may be wanting, one or more of them having thinned out and given place to others, or sometimes one of the masses first examined is observed to increase in thickness to the exclusion of other beds.

The term "FORMATION," which I have used in the above explanation, expresses in geology any assemblage of rocks which have some character in common, whether of origin, age, or composition. Thus we speak of stratified and unstratified, fresh-water and marine, aqueous and volcanic, ancient and modern, metalliferous and non-metalliferous formations.

In the estuaries of large rivers, such as the Ganges and the Mississippi, we may observe, at low water, phenomena analogous to those of the drained lakes above mentioned, but on a grander scale, and extending over areas several hundred miles in length and breadth. When the periodical inundations subside, the river hollows out a channel to the depth of many yards through horizontal beds of clay and sand, the ends of which are seen exposed in perpendicular cliffs. These beds vary in their mineral composition, or colour, or in the fineness or coarseness of their particles, and some of them are occasionally characterised by containing drift-wood. At the junction of the river and the sea, especially in lagoons nearly separated by sand-bars from the ocean, deposits are often formed in which brackish and salt-water shells are included.

In Egypt, where the Nile is always adding to its delta by filling up part of the Mediterranean with mud, the newly deposited sediment is STRATIFIED, the thin layer thrown down in one season differing slightly in colour from that of a previous year, and being separable from it, as has been observed in excavations at Cairo and other places. (See "Principles of Geology" by the Author Index "Nile" "Rivers" etc.)

When beds of sand, clay, and marl, containing shells and vegetable matter, are found arranged in a similar manner in the interior of the earth, we ascribe to them a similar origin; and the more we examine their characters in minute detail, the more exact do we find the resemblance. Thus, for example, at various heights and depths in the earth, and often far from seas, lakes, and rivers, we meet with layers of rounded pebbles composed of flint, limestone, granite, or other rocks, resembling the shingles of a sea-beach or the gravel in a torrent's bed. Such layers of pebbles frequently alternate with others formed of sand or fine sediment, just as we may see in the channel of a river descending from hills bordering a coast, where the current sweeps down at one season coarse sand and gravel, while at another, when the waters are low and less rapid, fine mud and sand alone are carried seaward. (See Figure 7 Chapter 2.)

If a stratified arrangement, and the rounded form of pebbles, are alone sufficient to lead us to the conclusion that certain rocks originated under water, this opinion is farther confirmed by the distinct and independent evidence of FOSSILS, so abundantly included in the earth's crust. By a FOSSIL is meant any body, or the traces of the existence of any body, whether animal or vegetable, which has been buried in the earth by natural causes. Now the remains of animals, especially of aquatic species, are found almost everywhere imbedded in stratified rocks, and sometimes, in the case of limestone, they are in such abundance as to constitute the entire mass of the rock itself. Shells and corals are the most frequent, and with them are often associated the bones and teeth of fishes, fragments of wood, impressions of leaves, and other organic substances. Fossil shells, of forms such as now abound in the sea, are met with far inland, both near the surface, and at great depths below it. They occur at all heights above the level of the ocean, having been observed at elevations of more than 8000 feet in the Pyrenees, 10,000 in the Alps, 13,000 in the Andes, and above 18,000 feet in the Himalaya. (Colonel R.J. Strachey found oolitic fossils 18,400 feet high in the Himalaya.)

These shells belong mostly to marine testacea, but in some places exclusively to forms characteristic of lakes and rivers. Hence it is concluded that some ancient strata were deposited at the bottom of the sea, and others in lakes and estuaries.

We have now pointed out one great class of rocks, which, however they may vary in mineral composition, colour, grain, or other characters, external and internal, may nevertheless be grouped together as having a common origin. They have all been formed under water, in the same manner as modern accumulations of sand, mud, shingle, banks of shells, reefs of coral, and the like, and are all characterised by stratification or fossils, or by both.

VOLCANIC ROCKS.

The division of rocks which we may next consider are the volcanic, or those which have been produced at or near the surface whether in ancient or modern times, not by water, but by the action of fire or subterranean heat. These rocks are for the most part unstratified, and are devoid of fossils. They are more partially distributed than aqueous formations, at least in respect to horizontal extension. Among those parts of Europe where they exhibit characters not to be mistaken, I may mention not only Sicily and the country round Naples, but Auvergne, Velay, and Vivarais, now the departments of Puy de Dome, Haute Loire, and Ardeche, towards the centre and south of France, in which are several hundred conical hills having the forms of modern volcanoes, with craters more or less perfect on many of their summits. These cones are composed moreover of lava, sand, and ashes, similar to those of active volcanoes. Streams of lava may sometimes be traced from the cones into the adjoining valleys, where they have choked up the ancient channels of rivers with solid rock, in the same manner as some modern flows of lava in Iceland have been known to do, the rivers either flowing beneath or cutting out a narrow passage on one side of the lava. Although none of these French volcanoes have been in activity within the period of history or tradition, their forms are often very perfect. Some, however, have been compared to the mere skeletons of volcanoes, the rains and torrents having washed their sides, and removed all the loose sand and scoriae, leaving only the harder and more solid materials. By this erosion, and by earthquakes, their internal structure has occasionally been laid open to view, in fissures and ravines; and we then behold not only many successive beds and masses of porous lava, sand, and scoriae, but also perpendicular walls, or DIKES, as they are called, of volcanic rock, which have burst through the other materials. Such dikes are also observed in the structure of Vesuvius, Etna, and other active volcanoes. They have been formed by the pouring of melted matter, whether from above or below, into open fissures, and they commonly traverse deposits of VOLCANIC TUFF, a substance produced by the showering down from the air, or incumbent waters, of sand and cinders, first shot up from the interior of the earth by the explosions of volcanic gases.

Besides the parts of France above alluded to, there are other countries, as the north of Spain, the south of Sicily, the Tuscan territory of Italy, the lower Rhenish provinces, and Hungary, where spent volcanoes may be seen, still preserving in many cases a conical form, and having craters and often lava- streams connected with them.

There are also other rocks in England, Scotland, Ireland, and almost every country in Europe, which we infer to be of igneous origin, although they do not form hills with cones and craters. Thus, for example, we feel assured that the rock of Staffa, and that of the Giant's Causeway, called basalt, is volcanic, because it agrees in its columnar structure and mineral composition with streams of lava which we know to have flowed from the craters of volcanoes. We find also similar basaltic and other igneous rocks associated with beds of TUFF in various parts of the British Isles, and forming DIKES, such as have been spoken of; and some of the strata through which these dikes cut are occasionally altered at the point of contact, as if they had been exposed to the intense heat of melted matter.

The absence of cones and craters, and long narrow streams of superficial lava, in England and many other countries, is principally to be attributed to the eruptions having been submarine, just as a considerable proportion of volcanoes in our own times burst out beneath the sea. But this question must be enlarged upon more fully in the chapters on Igneous Rocks, in which it will also be shown, that as different sedimentary formations, containing each their characteristic fossils, have been deposited at successive periods, so also volcanic sand and scoriae have been thrown out, and lavas have flowed over the land or bed of the sea, at many different epochs, or have been injected into fissures; so that the igneous as well as the aqueous rocks may be classed as a chronological series of monuments, throwing light on a succession of events in the history of the earth.

PLUTONIC ROCKS (GRANITE ETC).

We have now pointed out the existence of two distinct orders of mineral masses, the aqueous and the volcanic: but if we examine a large portion of a continent, especially if it contain within it a lofty mountain range, we rarely fail to discover two other classes of rocks, very distinct from either of those above alluded to, and which we can neither assimilate to deposits such as are now accumulated in lakes or seas, nor to those generated by ordinary volcanic action. The members of both these divisions of rocks agree in being highly crystalline and destitute of organic remains. The rocks of one division have been called Plutonic, comprehending all the granites and certain porphyries, which are nearly allied in some of their characters to volcanic formations. The members of the other class are stratified and often slaty, and have been called by some the CRYSTALLINE SCHISTS, in which group are included gneiss, micaceous- schist (or mica-slate), hornblende-schist, statuary marble, the finer kinds of roofing slate, and other rocks afterwards to be described.

As it is admitted that nothing strictly analogous to these crystalline productions can now be seen in the progress of formation on the earth's surface, it will naturally be asked, on what data we can find a place for them in a system of classification founded on the origin of rocks. I can not, in reply to this question, pretend to give the student, in a few words, an intelligible account of the long chain of facts and reasonings from which geologists have been led to infer the nature of the rocks in question. The result, however, may be briefly stated. All the various kinds of granites which constitute the Plutonic family are supposed to be of igneous or aqueo-igneous origin, and to have been formed under great pressure, at a considerable depth in the earth, or sometimes, perhaps, under a certain weight of incumbent ocean. Like the lava of volcanoes, they have been melted, and afterwards cooled and crystallised, but with extreme slowness, and under conditions very different from those of bodies cooling in the open air. Hence they differ from the volcanic rocks, not only by their more crystalline texture, but also by the absence of tuffs and breccias, which are the products of eruptions at the earth's surface, or beneath seas of inconsiderable depth. They differ also by the absence of pores or cellular cavities, to which the expansion of the entangled gases gives rise in ordinary lava.

METAMORPHIC, OR STRATIFIED CRYSTALLINE ROCKS.

The fourth and last great division of rocks are the crystalline strata and slates, or schists, called gneiss, mica-schist, clay-slate, chlorite-schist, marble, and the like, the origin of which is more doubtful than that of the other three classes. They contain no pebbles, or sand, or scoriae, or angular pieces of imbedded stone, and no traces of organic bodies, and they are often as crystalline as granite, yet are divided into beds, corresponding in form and arrangement to those of sedimentary formations, and are therefore said to be stratified. The beds sometimes consist of an alternation of substances varying in colour, composition, and thickness, precisely as we see in stratified fossiliferous deposits. According to the Huttonian theory, which I adopt as the most probable, and which will be afterwards more fully explained, the materials of these strata were originally deposited from water in the usual form of sediment, but they were subsequently so altered by subterranean heat, as to assume a new texture. It is demonstrable, in some cases at least, that such a complete conversion has actually taken place, fossiliferous strata having exchanged an earthy for a highly crystalline texture for a distance of a quarter of a mile from their contact with granite. In some cases, dark limestones, replete with shells and corals, have been turned into white statuary marble; and hard clays, containing vegetable or other remains, into slates called mica- schist or hornblende-schist, every vestige of the organic bodies having been obliterated.

Although we are in a great degree ignorant of the precise nature of the influence exerted in these cases, yet it evidently bears some analogy to that which volcanic heat and gases are known to produce; and the action may be conveniently called Plutonic, because it appears to have been developed in those regions where Plutonic rocks are generated, and under similar circumstances of pressure and depth in the earth. Intensely heated water or steam permeating stratified masses under great pressure have no doubt played their part in producing the crystalline texture and other changes, and it is clear that the transforming influence has often pervaded entire mountain masses of strata.

In accordance with the hypothesis above alluded to, I proposed in the first edition of the Principles of Geology (1833) the term "Metamorphic" for the altered strata, a term derived from meta, trans, and morphe, forma.

Hence there are four great classes of rocks considered in reference to their origin— the aqueous, the volcanic, the Plutonic, and the metamorphic. In the course of this work it will be shown that portions of each of these four distinct classes have originated at many successive periods. They have all been produced contemporaneously, and may even now be in the progress of formation on a large scale. It is not true, as was formerly supposed, that all granites, together with the crystalline or metamorphic strata, were first formed, and therefore entitled to be called "primitive," and that the aqueous and volcanic rocks were afterwards superimposed, and should, therefore, rank as secondary in the order of time. This idea was adopted in the infancy of the science, when all formations, whether stratified or unstratified, earthy or crystalline, with or without fossils, were alike regarded as of aqueous origin. At that period it was naturally argued that the foundation must be older than the superstructure; but it was afterwards discovered that this opinion was by no means in every instance a legitimate deduction from facts; for the inferior parts of the earth's crust have often been modified, and even entirely changed, by the influence of volcanic and other subterranean causes, while superimposed formations have not been in the slightest degree altered. In other words, the destroying and renovating processes have given birth to new rocks below, while those above, whether crystalline or fossiliferous, have remained in their ancient condition. Even in cities, such as Venice and Amsterdam, it cannot be laid down as universally true that the upper parts of each edifice, whether of brick or marble, are more modern than the foundations on which they rest, for these often consist of wooden piles, which may have rotted and been replaced one after the other, without the least injury to the buildings above; meanwhile, these may have required scarcely any repair, and may have been constantly inhabited. So it is with the habitable surface of our globe, in its relation to large masses of rock immediately below; it may continue the same for ages, while subjacent materials, at a great depth, are passing from a solid to a fluid state, and then reconsolidating, so as to acquire a new texture.

As all the crystalline rocks may, in some respects, be viewed as belonging to one great family, whether they be stratified or unstratified, metamorphic or Plutonic, it will often be convenient to speak of them by one common name. It being now ascertained, as above stated, that they are of very different ages, sometimes newer than the strata called secondary, the terms primitive and primary which were formerly used for the whole must be abandoned, as they would imply a manifest contradiction. It is indispensable, therefore, to find a new name, one which must not be of chronological import, and must express, on the one hand, some peculiarity equally attributable to granite and gneiss (to the Plutonic as well as the ALTERED rocks), and, on the other, must have reference to characters in which those rocks differ, both from the volcanic and from the UNALTERED sedimentary strata. I proposed in the Principles of Geology (first edition volume 3) the term "hypogene" for this purpose, derived from upo, under, and ginomai, to be, or to be born; a word implying the theory that granite, gneiss, and the other crystalline formations are alike NETHERFORMED rocks, or rocks which have not assumed their present form and structure at the surface. They occupy the lowest place in the order of superposition. Even in regions such as the Alps, where some masses of granite and gneiss can be shown to be of comparatively modern date, belonging, for example, to the period hereafter to be described as tertiary, they are still UNDERLYING rocks. They never repose on the volcanic or trappean formations, nor on strata containing organic remains. They are HYPOGENE, as "being under" all the rest.

From what has now been said, the reader will understand that each of the four great classes of rocks may be studied under two distinct points of view; first, they may be studied simply as mineral masses deriving their origin from particular causes, and having a certain composition, form, and position in the earth's crust, or other characters both positive and negative, such as the presence or absence of organic remains. In the second place, the rocks of each class may be viewed as a grand chronological series of monuments, attesting a succession of events in the former history of the globe and its living inhabitants.

I shall accordingly proceed to treat of each family of rocks; first, in reference to those characters which are not chronological, and then in particular relation to the several periods when they were formed.

CHAPTER II.
AQUEOUS ROCKS.— THEIR COMPOSITION AND FORMS OF STRATIFICATION.

Mineral Composition of Strata.
Siliceous Rocks.
Argillaceous.
Calcareous.
Gypsum.
Forms of Stratification.
Original Horizontality.
Thinning out.
Diagonal Arrangement.
Ripple-mark.

In pursuance of the arrangement explained in the last chapter, we shall begin by examining the aqueous or sedimentary rocks, which are for the most part distinctly stratified, and contain fossils. We may first study them with reference to their mineral composition, external appearance, position, mode of origin, organic contents, and other characters which belong to them as aqueous formations, independently of their age, and we may afterwards consider them chronologically or with reference to the successive geological periods when they originated.

I have already given an outline of the data which led to the belief that the stratified and fossiliferous rocks were originally deposited under water; but, before entering into a more detailed investigation, it will be desirable to say something of the ordinary materials of which such strata are composed. These may be said to belong principally to three divisions, the siliceous, the argillaceous, and the calcareous, which are formed respectively of flint, clay, and carbonate of lime. Of these, the siliceous are chiefly made up of sand or flinty grains; the argillaceous, or clayey, of a mixture of siliceous matter with a certain proportion, about a fourth in weight, of aluminous earth; and, lastly, the calcareous rocks, or limestones, of carbonic acid and lime.

SILICEOUS AND ARENACEOUS ROCKS.

To speak first of the sandy division: beds of loose sand are frequently met with, of which the grains consist entirely of silex, which term comprehends all purely siliceous minerals, as quartz and common flint. Quartz is silex in its purest form. Flint usually contains some admixture of alumina and oxide of iron. The siliceous grains in sand are usually rounded, as if by the action of running water. Sandstone is an aggregate of such grains, which often cohere together without any visible cement, but more commonly are bound together by a slight quantity of siliceous or calcareous matter, or by oxide of iron or clay.

Pure siliceous rocks may be known by not effervescing when a drop of nitric, sulphuric or other acid is applied to them, or by the grains not being readily scratched or broken by ordinary pressure. In nature there is every intermediate gradation, from perfectly loose sand to the hardest sandstone. In MICACEOUS SANDSTONES mica is very abundant; and the thin silvery plates into which that mineral divides are often arranged in layers parallel to the planes of stratification, giving a slaty or laminated texture to the rock.

When sandstone is coarse-grained, it is usually called GRIT. If the grains are rounded, and large enough to be called pebbles, it becomes a CONGLOMERATE or PUDDING-STONE, which may consist of pieces of one or of many different kinds of rock. A conglomerate, therefore, is simply gravel bound together by cement.

ARGILLACEOUS ROCKS.

Clay, strictly speaking, is a mixture of silex or flint with a large proportion, usually about one fourth, of alumina, or argil; but in common language, any earth which possesses sufficient ductility, when kneaded up with water, to be fashioned like paste by the hand, or by the potter's lathe, is called a CLAY; and such clays vary greatly in their composition, and are, in general, nothing more than mud derived from the decomposition or wearing down of rocks. The purest clay found in nature is porcelain clay, or kaolin, which results from the decomposition of a rock composed of feldspar and quartz, and it is almost always mixed with quartz. The kaolin of China consists of 71.15 parts of silex, 15.86 of alumine, 1.92 of lime, and 6.73 of water (W. Phillips Mineralogy page 33.); but other porcelain clays differ materially, that of Cornwall being composed, according to Boase, of nearly equal parts of silica and alumine, with 1 per cent of magnesia. (Phil. Mag. volume 10 1837.) SHALE has also the property, like clay, of becoming plastic in water: it is a more solid form of clay, or argillaceous matter, condensed by pressure. It always divides into laminae more or less regular.

One general character of all argillaceous rocks is to give out a peculiar, earthy odour when breathed upon, which is a test of the presence of alumine, although it does not belong to pure alumine, but, apparently, to the combination of that substance with oxide of iron. (See W. Phillips Mineralogy "Alumine.")

CALCAREOUS ROCKS.

This division comprehends those rocks which, like chalk, are composed chiefly of lime and carbonic acid. Shells and corals are also formed of the same elements, with the addition of animal matter. To obtain pure lime it is necessary to calcine these calcareous substances, that is to say, to expose them to heat of sufficient intensity to drive off the carbonic acid, and other volatile matter. White chalk is sometimes pure carbonate of lime; and this rock, although usually in a soft and earthy state, is occasionally sufficiently solid to be used for building, and even passes into a COMPACT stone, or a stone of which the separate parts are so minute as not to be distinguishable from each other by the naked eye.

Many limestones are made up entirely of minute fragments of shells and coral, or of calcareous sand cemented together. These last might be called "calcareous sandstones;" but that term is more properly applied to a rock in which the grains are partly calcareous and partly siliceous, or to quartzose sandstones, having a cement of carbonate of lime.

The variety of limestone called OOLITE is composed of numerous small egg-like grains, resembling the roe of a fish, each of which has usually a small fragment of sand as a nucleus, around which concentric layers of calcareous matter have accumulated.

Any limestone which is sufficiently hard to take a fine polish is called MARBLE. Many of these are fossiliferous; but statuary marble, which is also called saccharoid limestone, as having a texture resembling that of loaf-sugar, is devoid of fossils, and is in many cases a member of the metamorphic series.

SILICEOUS LIMESTONE is an intimate mixture of carbonate of lime and flint, and is harder in proportion as the flinty matter predominates.

The presence of carbonate of lime in a rock may be ascertained by applying to the surface a small drop of diluted sulphuric, nitric, or muriatic acid, or strong vinegar; for the lime, having a greater chemical affinity for any one of these acids than for the carbonic, unites immediately with them to form new compounds, thereby becoming a sulphate, nitrate or muriate of lime. The carbonic acid, when thus liberated from its union with the lime, escapes in a gaseous form, and froths up or effervesces as it makes its way in small bubbles through the drop of liquid. This effervescence is brisk or feeble in proportion as the limestone is pure or impure, or, in other words, according to the quantity of foreign matter mixed with the carbonate of lime. Without the aid of this test, the most experienced eye can not always detect the presence of carbonate of lime in rocks.

The above-mentioned three classes of rocks, the siliceous, argillaceous, and calcareous, pass continually into each other, and rarely occur in a perfectly separate and pure form. Thus it is an exception to the general rule to meet with a limestone as pure as ordinary white chalk, or with clay as aluminous as that used in Cornwall for porcelain, or with sand so entirely composed of siliceous grains as the white sand of Alum Bay, in the Isle of Wight, employed in the manufacture of glass, or sandstone so pure as the grit of Fontainebleau, used for pavement in France. More commonly we find sand and clay, or clay and marl, intermixed in the same mass. When the sand and clay are each in considerable quantity, the mixture is called LOAM. If there is much calcareous matter in clay it is called MARL; but this term has unfortunately been used so vaguely, as often to be very ambiguous. It has been applied to substances in which there is no lime; as, to that red loam usually called red marl in certain parts of England. Agriculturists were in the habit of calling any soil a marl which, like true marl, fell to pieces readily on exposure to the air. Hence arose the confusion of using this name for soils which, consisting of loam, were easily worked by the plough, though devoid of lime.

MARL SLATE bears the same relation to marl which shale bears to clay, being a calcareous shale. It is very abundant in some countries, as in the Swiss Alps. Argillaceous or marly limestone is also of common occurrence.

There are few other kinds of rock which enter so largely into the composition of sedimentary strata as to make it necessary to dwell here on their characters. I may, however, mention two others— magnesian limestone or dolomite, and gypsum. MAGNESIAN LIMESTONE is composed of carbonate of lime and carbonate of magnesia; the proportion of the latter amounting in some cases to nearly one half. It effervesces much more slowly and feebly with acids than common limestone. In England this rock is generally of a yellowish colour; but it varies greatly in mineralogical character, passing from an earthy state to a white compact stone of great hardness. DOLOMITE, so common in many parts of Germany and France, is also a variety of magnesian limestone, usually of a granular texture.

Gypsum is a rock composed of sulphuric acid, lime, and water. It is usually a soft whitish-yellow rock, with a texture resembling that of loaf-sugar, but sometimes it is entirely composed of lenticular crystals. It is insoluble in acids, and does not effervesce like chalk and dolomite, because it does not contain carbonic acid gas, or fixed air, the lime being already combined with sulphuric acid, for which it has a stronger affinity than for any other. Anhydrous gypsum is a rare variety, into which water does not enter as a component part. GYPSEOUS MARL is a mixture of gypsum and marl. ALABASTER is a granular and compact variety of gypsum found in masses large enough to be used in sculpture and architecture. It is sometimes a pure snow-white substance, as that of Volterra in Tuscany, well known as being carved for works of art in Florence and Leghorn. It is a softer stone than marble, and more easily wrought.

FORMS OF STRATIFICATION.

A series of strata sometimes consists of one of the above rocks, sometimes of two or more in alternating beds.

Thus, in the coal districts of England, for example, we often pass through several beds of sandstone, some of finer, others of coarser grain, some white, others of a dark colour, and below these, layers of shale and sandstone or beds of shale, divisible into leaf-like laminae, and containing beautiful impressions of plants. Then again we meet with beds of pure and impure coal, alternating with shales and sandstones, and underneath the whole, perhaps, are calcareous strata, or beds of limestone, filled with corals and marine shells, each bed distinguishable from another by certain fossils, or by the abundance of particular species of shells or zoophytes.

This alternation of different kinds of rock produces the most distinct stratification; and we often find beds of limestone and marl, conglomerate and sandstone, sand and clay, recurring again and again, in nearly regular order, throughout a series of many hundred strata. The causes which may produce these phenomena are various, and have been fully discussed in my treatise on the modern changes of the earth's surface. (Consult Index to Principles of Geology, "Stratification" "Currents" "Deltas" "Water" etc.) It is there seen that rivers flowing into lakes and seas are charged with sediment, varying in quantity, composition, colour, and grain according to the seasons; the waters are sometimes flooded and rapid, at other periods low and feeble; different tributaries, also, draining peculiar countries and soils, and therefore charged with peculiar sediment, are swollen at distinct periods. It was also shown that the waves of the sea and currents undermine the cliffs during wintry storms, and sweep away the materials into the deep, after which a season of tranquillity succeeds, when nothing but the finest mud is spread by the movements of the ocean over the same submarine area.

It is not the object of the present work to give a description of these operations, repeated as they are, year after year, and century after century; but I may suggest an explanation of the manner in which some micaceous sandstones have originated, namely, those in which we see innumerable thin layers of mica dividing layers of fine quartzose sand. I observed the same arrangement of materials in recent mud deposited in the estuary of Laroche St. Bernard in Brittany, at the mouth of the Loire. The surrounding rocks are of gneiss, which, by its waste, supplies the mud: when this dries at low water, it is found to consist of brown laminated clay, divided by thin seams of mica. The separation of the mica in this case, or in that of micaceous sandstones, may be thus understood. If we take a handful of quartzose sand, mixed with mica, and throw it into a clear running stream, we see the materials immediately sorted by the water, the grains of quartz falling almost directly to the bottom, while the plates of mica take a much longer time to reach the bottom, and are carried farther down the stream. At the first instant the water is turbid, but immediately after the flat surfaces of the plates of mica are seen all alone, reflecting a silvery light, as they descend slowly, to form a distinct micaceous lamina. The mica is the heavier mineral of the two; but it remains a longer time suspended in the fluid, owing to its greater extent of surface. It is easy, therefore, to perceive that where such mud is acted upon by a river or tidal current, the thin plates of mica will be carried farther, and not deposited in the same places as the grains of quartz; and since the force and velocity of the stream varies from time to time, layers of mica or of sand will be thrown down successively on the same area.

ORIGINAL HORIZONTALITY.

It is said generally that the upper and under surfaces of strata, or the "planes of stratification," are parallel. Although this is not strictly true, they make an approach to parallelism, for the same reason that sediment is usually deposited at first in nearly horizontal layers. Such an arrangement can by no means be attributed to an original evenness or horizontality in the bed of the sea: for it is ascertained that in those places where no matter has been recently deposited, the bottom of the ocean is often as uneven as that of the dry land, having in like manner its hills, valleys, and ravines. Yet if the sea should go down, or be removed from near the mouth of a large river where a delta has been forming, we should see extensive plains of mud and sand laid dry, which, to the eye, would appear perfectly level, although, in reality, they would slope gently from the land towards the sea.

This tendency in newly-formed strata to assume a horizontal position arises principally from the motion of the water, which forces along particles of sand or mud at the bottom, and causes them to settle in hollows or depressions where they are less exposed to the force of a current than when they are resting on elevated points. The velocity of the current and the motion of the superficial waves diminish from the surface downward, and are least in those depressions where the water is deepest.

(FIGURE 1. Layers of sand and ashes on uneven ground.)

A good illustration of the principle here alluded to may be sometimes seen in the neighbourhood of a volcano, when a section, whether natural or artificial, has laid open to view a succession of various-coloured layers of sand and ashes, which have fallen in showers upon uneven ground. Thus let A B (Figure 1) be two ridges, with an intervening valley. These original inequalities of the surface have been gradually effaced by beds of sand and ashes c, d, e, the surface at e being quite level. It will be seen that, although the materials of the first layers have accommodated themselves in a great degree to the shape of the ground A B, yet each bed is thickest at the bottom. At first a great many particles would be carried by their own gravity down the steep sides of A and B, and others would afterwards be blown by the wind as they fell off the ridges, and would settle in the hollow, which would thus become more and more effaced as the strata accumulated from c to e. Now, water in motion can exert this levelling power on similar materials more easily than air, for almost all stones lose in water more than a third of the weight which they have in air, the specific gravity of rocks being in general as 2 1/2 when compared to that of water, which is estimated at 1. But the buoyancy of sand or mud would be still greater in the sea, as the density of salt-water exceeds that of fresh.

(FIGURE 2. Section of strata of sandstone, grit, and conglomerate.)

Yet, however uniform and horizontal may be the surface of new deposits in general, there are still many disturbing causes, such as eddies in the water, and currents moving first in one and then in another direction, which frequently cause irregularities. We may sometimes follow a bed of limestone, shale, or sandstone, for a distance of many hundred yards continuously; but we generally find at length that each individual stratum thins out, and allows the beds which were previously above and below it to meet. If the materials are coarse, as in grits and conglomerates, the same beds can rarely be traced many yards without varying in size, and often coming to an end abruptly. (See Figure 2.)

DIAGONAL OR CROSS STRATIFICATION.

(FIGURE 3. Section of sand at Sandy Hill, near Biggleswade, Bedfordshire. Height 20 feet. (Green-sand formation.))

(FIGURE 4. Layers of sediment on a bank.)

(FIGURE 5. Nearly horizontal layers of sediment over sloping strata.)

(FIGURE 6. Cliff between mismer and Dunwich.)

There is also another phenomenon of frequent occurrence. We find a series of larger strata, each of which is composed of a number of minor layers placed obliquely to the general planes of stratification. To this diagonal arrangement the name of "false or cross bedding" has been given. Thus in the section (Figure 3) we see seven or eight large beds of loose sand, yellow and brown, and the lines a, b, c mark some of the principal planes of stratification, which are nearly horizontal. But the greater part of the subordinate laminae do not conform to these planes, but have often a steep slope, the inclination being sometimes towards opposite points of the compass. When the sand is loose and incoherent, as in the case here represented, the deviation from parallelism of the slanting laminae can not possibly be accounted for by any rearrangement of the particles acquired during the consolidation of the rock. In what manner, then, can such irregularities be due to original deposition? We must suppose that at the bottom of the sea, as well as in the beds of rivers, the motions of waves, currents, and eddies often cause mud, sand, and gravel to be thrown down in heaps on particular spots, instead of being spread out uniformly over a wide area. Sometimes, when banks are thus formed, currents may cut passages through them, just as a river forms its bed. Suppose the bank A (Figure 4) to be thus formed with a steep sloping side, and, the water being in a tranquil state, the layer of sediment No. 1 is thrown down upon it, conforming nearly to its surface. Afterwards the other layers, 2, 3, 4, may be deposited in succession, so that the bank B C D is formed. If the current then increases in velocity, it may cut away the upper portion of this mass down to the dotted line e, and deposit the materials thus removed farther on, so as to form the layers 5, 6, 7, 8. We have now the bank B, C, D, E (Figure 5), of which the surface is almost level, and on which the nearly horizontal layers, 9, 10, 11, may then accumulate. It was shown in Figure 3 that the diagonal layers of successive strata may sometimes have an opposite slope. This is well seen in some cliffs of loose sand on the Suffolk coast. A portion of one of these is represented in Figure 6, where the layers, of which there are about six in the thickness of an inch, are composed of quartzose grains. This arrangement may have been due to the altered direction of the tides and currents in the same place.

(FIGURE 7. Section from Monte Calvo to the sea by the valley of the Magnan, near Nice. A. Dolomite and sandstone. (Green-sand formation?) a, b, d. Beds of gravel and sand. c. Fine marl and sand of Ste. Madeleine, with marine (Pliocene) shells.)

The description above given of the slanting position of the minor layers constituting a single stratum is in certain cases applicable on a much grander scale to masses several hundred feet thick, and many miles in extent. A fine example may be seen at the base of the Maritime Alps near Nice. The mountains here terminate abruptly in the sea, so that a depth of one hundred fathoms is often found within a stone's throw of the beach, and sometimes a depth of 3000 feet within half a mile. But at certain points, strata of sand, marl, or conglomerate intervene between the shore and the mountains, as in the section (Figure 7), where a vast succession of slanting beds of gravel and sand may be traced from the sea to Monte Calvo, a distance of no less than nine miles in a straight line. The dip of these beds is remarkably uniform, being always southward or towards the Mediterranean, at an angle of about 25 degrees. They are exposed to view in nearly vertical precipices, varying from 200 to

600 feet in height, which bound the valley through which the river Magnan flows. Although, in a general view, the strata appear to be parallel and uniform, they are nevertheless found, when examined closely, to be wedge-shaped, and to thin out when followed for a few hundred feet or yards, so that we may suppose them to have been thrown down originally upon the side of a steep bank where a river or Alpine torrent discharged itself into a deep and tranquil sea, and formed a delta, which advanced gradually from the base of Monte Calvo to a distance of nine miles from the original shore. If subsequently this part of the Alps and bed of the sea were raised 700 feet, the delta may have emerged, a deep channel may then have been cut through it by the river, and the coast may at the same time have acquired its present configuration.

(FIGURE 8. Slab of ripple-marked (New Red) sandstone from Cheshire.)

It is well known that the torrents and streams which now descend from the Alpine declivities to the shore, bring down annually, when the snow melts, vast quantities of shingle and sand, and then, as they subside, fine mud, while in summer they are nearly or entirely dry; so that it may be safely assumed that deposits like those of the valley of the Magnan, consisting of coarse gravel alternating with fine sediment, are still in progress at many points, as, for instance, at the mouth of the Var. They must advance upon the Mediterranean in the form of great shoals terminating in a steep talus; such being the original mode of accumulation of all coarse materials conveyed into deep water, especially where they are composed in great part of pebbles, which can not be transported to indefinite distances by currents of moderate velocity. By inattention to facts and inferences of this kind, a very exaggerated estimate has sometimes been made of the supposed depth of the ancient ocean. There can be no doubt, for example, that the strata a, Figure 7, or those nearest to Monte Calvo, are older than those indicated by b, and these again were formed before c; but the vertical depth of gravel and sand in any one place can not be proved to amount even to 1000 feet, although it may perhaps be much greater, yet probably never exceeding at any point 3000 or 4000 feet. But were we to assume that all the strata were once horizontal, and that their present dip or inclination was due to subsequent movements, we should then be forced to conclude that a sea several miles deep had been filled up with alternate layers of mud and pebbles thrown down one upon another.

In the locality now under consideration, situated a few miles to the west of Nice, there are many geological data, the details of which can not be given in this place, all leading to the opinion that, when the deposit of the Magnan was formed, the shape and outline of the Alpine declivities and the shore greatly resembled what we now behold at many points in the neighbourhood. That the beds a, b, c, d are of comparatively modern date is proved by this fact, that in seams of loamy marl intervening between the pebbly beds are fossil shells, half of which belong to species now living in the Mediterranean.

RIPPLE-MARK.

The ripple-mark, so common on the surface of sandstones of all ages (see Figure 8), and which is so often seen on the sea-shore at low tide, seems to originate in the drifting of materials along the bottom of the water, in a manner very similar to that which may explain the inclined layers above described. This ripple is not entirely confined to the beach between high and low water mark, but is also produced on sands which are constantly covered by water. Similar undulating ridges and furrows may also be sometimes seen on the surface of drift snow and blown sand.

The ripple-mark is usually an indication of a sea-beach, or of water from six to ten feet deep, for the agitation caused by waves even during storms extends to a very slight depth. To this rule, however, there are some exceptions, and recent ripple-marks have been observed at the depth of 60 or 70 feet. It has also been ascertained that currents or large bodies of water in motion may disturb mud and sand at the depth of 300 or even 450 feet. (Darwin Volcanic Islands page 134.) Beach ripple, however, may usually be distinguished from current ripple by frequent changes in its direction. In a slab of sandstone, not more than an inch thick, the furrows or ridges of an ancient ripple may often be seen in several successive laminae to run towards different points of the compass.

CHAPTER III.

ARRANGEMENT OF FOSSILS IN STRATA.— FRESH-WATER AND MARINE FOSSILS.

Successive Deposition indicated by Fossils.
Limestones formed of Corals and Shells.
Proofs of gradual Increase of Strata derived from Fossils.
Serpula attached to Spatangus.
Wood bored by Teredina.
Tripoli formed of Infusoria.
Chalk derived principally from Organic Bodies.
Distinction of Fresh-water from Marine Formations.
Genera of Fresh-water and Land Shells.
Rules for recognising Marine Testacea.
Gyrogonite and Chara.
Fresh-water Fishes.
Alternation of Marine and Fresh-water Deposits.
Lym-Fiord.

Having in the last chapter considered the forms of stratification so far as they are determined by the arrangement of inorganic matter, we may now turn our attention to the manner in which organic remains are distributed through stratified deposits. We should often be unable to detect any signs of stratification or of successive deposition, if particular kinds of fossils did not occur here and there at certain depths in the mass. At one level, for example, univalve shells of some one or more species predominate; at another, bivalve shells; and at a third, corals; while in some formations we find layers of vegetable matter, commonly derived from land plants, separating strata.

It may appear inconceivable to a beginner how mountains, several thousand feet thick, can have become full of fossils from top to bottom; but the difficulty is removed, when he reflects on the origin of stratification, as explained in the last chapter, and allows sufficient time for the accumulation of sediment. He must never lose sight of the fact that, during the process of deposition, each separate layer was once the uppermost, and immediately in contact with the water in which aquatic animals lived. Each stratum, in fact, however far it may now lie beneath the surface, was once in the state of shingle, or loose sand or soft mud at the bottom of the sea, in which shells and other bodies easily became enveloped.

RATE OF DEPOSITION INDICATED BY FOSSILS.

By attending to the nature of these remains, we are often enabled to determine whether the deposition was slow or rapid, whether it took place in a deep or shallow sea, near the shore or far from land, and whether the water was salt, brackish, or fresh. Some limestones consist almost exclusively of corals, and in many cases it is evident that the present position of each fossil zoophyte has been determined by the manner in which it grew originally. The axis of the coral, for example, if its natural growth is erect, still remains at right angles to the plane of stratification. If the stratum be now horizontal, the round spherical heads of certain species continue uppermost, and their points of attachment are directed downward. This arrangement is sometimes repeated throughout a great succession of strata. From what we know of the growth of similar zoophytes in modern reefs, we infer that the rate of increase was extremely slow, and some of the fossils must have flourished for ages like forest-trees, before they attained so large a size. During these ages, the water must have been clear and transparent, for such corals can not live in turbid water.

(FIGURE 9. Fossil Gryphaea, covered both on the outside and inside with fossil Serpulae.)

In like manner, when we see thousands of full-grown shells dispersed everywhere throughout a long series of strata, we can not doubt that time was required for the multiplication of successive generations; and the evidence of slow accumulation is rendered more striking from the proofs, so often discovered, of fossil bodies having lain for a time on the floor of the ocean after death before they were imbedded in sediment. Nothing, for example, is more common than to see fossil oysters in clay, with Serpulae, or barnacles (acorn-shells), or corals, and other creatures, attached to the inside of the valves, so that the mollusk was certainly not buried in argillaceous mud the moment it died. There must have been an interval during which it was still surrounded with clear water, when the creatures whose remains now adhere to it grew from an embryonic to a mature state. Attached shells which are merely external, like some of the Serpulae (a) in Figure 9, may often have grown upon an oyster or other shell while the animal within was still living; but if they are found on the inside, it could only happen after the death of the inhabitant of the shell which affords the support. Thus, in Figure 9, it will be seen that two Serpulae have grown on the interior, one of them exactly on the place where the adductor muscle of the Gryphaea (a kind of oyster) was fixed.

(FIGURE 10. Serpula attached to a fossil Micraster from the Chalk.)

(FIGURE 11. Recent Spatangus with the spines removed from one side. b. Spine and tubercles, natural size. a. The same magnified.)

Some fossil shells, even if simply attached to the OUTSIDE of others, bear full testimony to the conclusion above alluded to, namely, that an interval elapsed between the death of the creature to whose shell they adhere, and the burial of the same in mud or sand. The sea-urchins, or Echini, so abundant in white chalk, afford a good illustration. It is well known that these animals, when living, are invariably covered with spines supported by rows of tubercles. These last are only seen after the death of the sea-urchin, when the spines have dropped off. In Figure 11 a living species of Spatangus, common on our coast, is represented with one half of its shell stripped of the spines. In Figure 10 a fossil of a similar and allied genus from the white chalk of England shows the naked surface which the individuals of this family exhibit when denuded of their bristles. The full-grown Serpula, therefore, which now adheres externally, could not have begun to grow till the Micraster had died, and the spines became detached.

(FIGURE 12. a. Ananchytes from the chalk with lower valve of Crania attached. b. Upper valve of Crania detached.)

Now the series of events here attested by a single fossil may be carried a step farther. Thus, for example, we often meet with a sea-urchin (Ananchytes) in the chalk (see Figure 12) which has fixed to it the lower valve of a Crania, a genus of bivalve mollusca. The upper valve (b, Figure 12) is almost invariably wanting, though occasionally found in a perfect state of preservation in white chalk at some distance. In this case, we see clearly that the sea-urchin first lived from youth to age, then died and lost its spines, which were carried away. Then the young Crania adhered to the bared shell, grew and perished in its turn; after which the upper valve was separated from the lower before the Ananchytes became enveloped in chalky mud.

(FIGURES 13 AND 14. Fossil and recent wood drilled by perforating Mollusca.

(FIGURE 13. a. Fossil wood from London Clay, bored by Teredina. b. Shell and tube of Teredina personata, the right-hand figure the ventral, the left the dorsal view.)

(FIGURE 14. e. Recent wood bored by Toredo. d. Shell and tube of Teredo navalis, from the same. c. Anterior and posterior view of the valves of same detached from the tube.))

It may be well to mention one more illustration of the manner in which single fossils may sometimes throw light on a former state of things, both in the bed of the ocean and on some adjoining land. We meet with many fragments of wood bored by ship-worms at various depths in the clay on which London is built. Entire branches and stems of trees, several feet in length, are sometimes found drilled all over by the holes of these borers, the tubes and shells of the mollusk still remaining in the cylindrical hollows. In Figure 14, e, a representation is given of a piece of recent wood pierced by the Teredo navalis, or common ship-worm, which destroys wooden piles and ships. When the cylindrical tube d has been extracted from the wood, the valves are seen at the larger or anterior extremity, as shown at c. In like manner, a piece of fossil wood (a, Figure 13) has been perforated by a kindred but extinct genus, the Teredina of Lamarck. The calcareous tube of this mollusk was united and, as it were, soldered on to the valves of the shell (b), which therefore can not be detached from the tube, like the valves of the recent Teredo. The wood in this fossil specimen is now converted into a stony mass, a mixture of clay and lime; but it must once have been buoyant and floating in the sea, when the Teredinae lived upon, and perforated it. Again, before the infant colony settled upon the drift wood, part of a tree must have been floated down to the sea by a river, uprooted, perhaps, by a flood, or torn off and cast into the waves by the wind: and thus our thoughts are carried back to a prior period, when the tree grew for years on dry land, enjoying a fit soil and climate.

STRATA OF ORGANIC ORIGIN.
(FIGURE 15. Gaillonella ferruginea, Ehb.)
(FIGURE 16. Gaillonella distans, Ehb.)
(FIGURE 17. Bacillaria paradoxa. a. Front view. b. Side view.)

It has been already remarked that there are rocks in the interior of continents, at various depths in the earth, and at great heights above the sea, almost entirely made up of the remains of zoophytes and testacea. Such masses may be compared to modern oyster-beds and coral-reefs; and, like them, the rate of increase must have been extremely gradual. But there are a variety of stone deposits in the earth's crust, now proved to have been derived from plants and animals of which the organic origin was not suspected until of late years, even by naturalists. Great surprise was therefore created some years since by the discovery of Professor Ehrenberg, of Berlin, that a certain kind of siliceous stone, called tripoli, was entirely composed of millions of the remains of organic beings, which were formerly referred to microscopic Infusoria, but which are now admitted to be plants. They abound in rivulets, lakes, and ponds in England and other countries, and are termed Diatomaceae by those naturalists who believe in their vegetable origin. The subject alluded to has long been well- known in the arts, under the name of infusorial earth or mountain meal, and is used in the form of powder for polishing stones and metals. It has been procured, among other places, from the mud of a lake at Dolgelly, in North Wales, and from Bilin, in Bohemia, in which latter place a single stratum, extending over a wide area, is no less than fourteen feet thick. This stone, when examined with a powerful microscope, is found to consist of the siliceous plates or frustules of the above-figured Diatomaceae, united together without any visible cement. It is difficult to convey an idea of their extreme minuteness; but Ehrenberg estimates that in the Bilin tripoli there are 41,000 millions of individuals of the Gaillonella distans (see Figure 16) in every cubic inch (which weighs about 220 grains), or about 187 millions in a single grain. At every stroke, therefore, that we make with this polishing powder, several millions, perhaps tens of millions, of perfect fossils are crushed to atoms.

A well-known substance, called bog-iron ore, often met with in peat-mosses, has often been shown by Ehrenberg to consist of innumerable articulated threads, of a yellow ochre colour, composed of silica, argillaceous matter, and peroxide of iron. These threads are the cases of a minute microscopic body, called Gaillonella ferruginea (Figure 15), associated with the siliceous frustules of other fresh-water algae. Layers of this iron ore occurring in Scotch peat bogs are often called "the pan," and are sometimes of economical value.

It is clear much time must have been required for the accumulation of strata to which countless generations of Diatomaceae have contributed their remains; and these discoveries lead us naturally to suspect that other deposits, of which the materials have been supposed to be inorganic, may in reality be composed chiefly of microscopic organic bodies. That this is the case with the white chalk, has often been imagined, and is now proved to be the fact. It has, moreover, been lately discovered that the chambers into which these Foraminifera are divided are actually often filled with thousands of well-preserved organic bodies, which abound in every minute grain of chalk, and are especially apparent in the white coating of flints, often accompanied by innumerable needle-shaped spiculae of sponges (see Chapter 17.).

"The dust we tread upon was once alive!"— Byron.

How faint an idea does this exclamation of the poet convey of the real wonders of nature! for here we discover proofs that the calcareous and siliceous dust of which hills are composed has not only been once alive, but almost every particle, albeit invisible to the naked eye, still retains the organic structure which, at periods of time incalculably remote, was impressed upon it by the powers of life.

FRESH-WATER AND MARINE FOSSILS.

Strata, whether deposited in salt or fresh water, have the same forms; but the imbedded fossils are very different in the two cases, because the aquatic animals which frequent lakes and rivers are distinct from those inhabiting the sea. In the northern part of the Isle of Wight formations of marl and limestone, more than 50 feet thick occur, in which the shells are of extinct species. Yet we recognise their fresh-water origin, because they are of the same genera as those now abounding in ponds, lakes, and rivers, either in our own country or in warmer latitudes.

In many parts of France— in Auvergne, for example— strata occur of limestone, marl, and sandstone hundreds of feet thick, which contain exclusively fresh- water and land shells, together with the remains of terrestrial quadrupeds. The number of land-shells scattered through some of these fresh-water deposits is exceedingly great; and there are districts in Germany where the rocks scarcely contain any other fossils except snail-shells (helices); as, for instance, the limestone on the left bank of the Rhine, between Mayence and Worms, at Oppenheim, Findheim, Budenheim, and other places. In order to account for this phenomenon, the geologist has only to examine the small deltas of torrents which enter the Swiss lakes when the waters are low, such as the newly-formed plain where the Kander enters the Lake of Thun. He there sees sand and mud strewn over with innumerable dead land-shells, which have been brought down from the valleys in the Alps in the preceding spring, during the melting of the snows. Again, if we search the sands on the borders of the Rhine, in the lower part of its course, we find countless land-shells mixed with others of species belonging to lakes, stagnant pools, and marshes. These individuals have been washed away from the alluvial plains of the great river and its tributaries, some from mountainous regions, others from the low country.

Although fresh-water formations are often of great thickness, yet they are usually very limited in area when compared to marine deposits, just as lakes and estuaries are of small dimensions in comparison with seas.

The absence of many fossil forms usually met with in marine strata, affords a useful negative indication of the fresh-water origin of a formation. For example, there are no sea-urchins, no corals, no chambered shells, such as the nautilus, nor microscopic Foraminifera in lacustrine or fluviatile deposits. In distinguishing the latter from formations accumulated in the sea, we are chiefly guided by the forms of the mollusca. In a fresh-water deposit, the number of individual shells is often as great as in a marine stratum, if not greater; but there is a smaller variety of species and genera. This might be anticipated from the fact that the genera and species of recent fresh-water and land shells are few when contrasted with the marine. Thus, the genera of true mollusca according to Woodward's system, excluding those altogether extinct and those without shells, amount to 446 in number, of which the terrestrial and fresh-water genera scarcely form more than a fifth. (See Woodward's Manual of Mollusca 1856.)

(FIGURE 18. Cyrena obovata, Sowerby; fossil. Hants.)
(FIGURE 19. Cyrena (Corbicella) fluminalis, Moll.; fossil. Grays, Essex.)
(FIGURE 20. Anodonta Cordierii; D'Orbigny; fossil. Paris.)
(FIGURE 21. Anodonta latimarginata; recent. Bahia.)
(FIGURE 22. Unio littoralis. Lamarck; recent. Auvergne.)
(FIGURE 23. Gryphaea incurva, Sowerby; (G. arcuata, Lamarck) upper valve. Lias.)

Almost all bivalve shells, or those of acephalous mollusca, are marine, about sixteen only out of 140 genera being fresh-water. Among these last, the four most common forms, both recent and fossil, are Cyclas, Cyrena, Unio, and Anodonta (see Figures 18-22); the two first and two last of which are so nearly allied as to pass into each other.

Lamarck divided the bivalve mollusca into the Dimyary, or those having two large muscular impressions in each valve, as a, b in the Cyclas, Figure 18, and Unio, Figure 22, and the Monomyary, such as the oyster and scallop, in which there is only one of these impressions, as is seen in Figure 23. Now, as none of these last, or the unimuscular bivalves, are fresh-water, we may at once presume a deposit containing any of them to be marine. (The fresh-water Mulleria, when young, forms a single exception to the rule, as it then has two muscular impressions, but it has only one in the adult state.)

(FIGURE 24. Planorbis euomphalus, Sowerby; fossil. Isle of Wight.)

(FIGURE 25. Limnaea longiscala, Brongniart; fossil. Isle of Wight.)
(FIGURE 26. Paludina lenta, Brand.; fossil. Isle of Wight.)
(FIGURE 27. Succinea amphibia, Drap. (S. putris, L.); fossil. Loess, Rhine.)
(FIGURE 28. Ancylus velletia (A. elegans), Sowerby; fossil. Isle of Wight.)
(FIGURE 29. Valvata piscinalis, Mull.; fossil. Grays, Essex.)
(FIGURE 30. Physa hypnorum, Linne; recent. Isle of Wight.)
(FIGURE 31. Auricula; recent. Ava.)
(FIGURE 32. Melania inquinata, Def. Paris basin.)
(FIGURE 33. Physa columnaris, Desh. Paris basin.)
(FIGURE 34. Melanopsis buccinoidea, Ferr.; recent. Asia.)

The univalve shells most characteristic of fresh-water deposits are, Planorbis, Limnaea, and Paludina. (See Figures 24-26.) But to these are occasionally added Physa, Succinea, Ancylus, Valvata, Melanopsis, Melania, Potamides, and Neritina (see Figures 27-34), the four last being usually found in estuaries.

(FIGURE 35. Neritina globulus, Def. Paris basin.)
(FIGURE 36. Nerita granulosa, Desh. Paris basin.)

Some naturalists include Neritina (Figure 35) and the marine Nerita (Figure 36) in the same genus, it being scarcely possible to distinguish the two by good generic characters. But, as a general rule, the fluviatile species are smaller, smoother, and more globular than the marine; and they have never, like the Neritae, the inner margin of the outer lip toothed or crenulated. (See Figure 36.)

(FIGURE 37. Potamides cinctus, Sowerby. Paris basin.)

The Potamides inhabit the mouths of rivers in warm latitudes, and are distinguishable from the marine Cerithia by their orbicular and multispiral opercula. The genus Auricula (Figure 31) is amphibious, frequenting swamps and marshes within the influence of the tide.

(FIGURE 38. Helix Turonensis, Desh.; faluns, Touraine.)
(FIGURE 39. Cyclostoma elegans, Mull.; Loess.)
(FIGURE 40. Pupa tridens, Drap.; Loess.)
(FIGURE 41. Clausilia bidens, Drap.; Loess.)
(FIGURE 42. Bulimus lubricus, Mull.; Loess, Rhine.)

The terrestrial shells are all univalves. The most important genera among these, both in a recent and fossil state, are Helix (Figure 38), Cyclostoma (Figure 39), Pupa (Figure 40), Clausilia (Figure 41) Bulimus (Figure 42), Glandina and Achatina.

(FIGURE 43. Ampullaria glauca, from the Jumna.)

Ampullaria (Figure 43) is another genus of shells inhabiting rivers and ponds in hot countries. Many fossil species formerly referred to this genus, and which have been met with chiefly in marine formations, are now considered by conchologists to belong to Natica and other marine genera.

(FIGURE 44. Pleurotoma exorta, Brand. Upper and Middle Eocene. Barton and Bracklesham.)
(FIGURE 45. Ancillaria subulata, Sowerby. Barton clay. Eocene.)

All univalve shells of land and fresh-water species, with the exception of Melanopsis (Figure 34), and Achatina, which has a slight indentation, have entire mouths; and this circumstance may often serve as a convenient rule for distinguishing fresh-water from marine strata; since, if any univalves occur of which the mouths are not entire, we may presume that the formation is marine. The aperture is said to be entire in such shells as the fresh-water Ampullaria and the land-shells (Figures 38-42), when its outline is not interrupted by an indentation or notch, such as that seen at b in Ancillaria (Figure 45); or is not prolonged into a canal, as that seen at a in Pleurotoma (Figure 44).

The mouths of a large proportion of the marine univalves have these notches or canals, and almost all species are carnivorous; whereas nearly all testacea having entire mouths are plant-eaters, whether the species be marine, fresh- water, or terrestrial.

There is, however, one genus which affords an occasional exception to one of the above rules. The Potamides (Figure 37), a subgenus of Cerithium, although provided with a short canal, comprises some species which inhabit salt, others brackish, and others fresh-water, and they are said to be all plant-eaters.

Among the fossils very common in fresh-water deposits are the shells of Cypris, a minute bivalve crustaceous animal. (For figures of fossil species of Purbeck see below, Chapter 19.) Many minute living species of this genus swarm in lakes and stagnant pools in Great Britain; but their shells are not, if considered separately, conclusive as to the fresh-water origin of a deposit, because the majority of species in another kindred genus of the same order, the Cytherina of Lamarck, inhabit salt-water; and, although the animal differs slightly, the shell is scarcely distinguishable from that of the Cypris.

FRESH-WATER FOSSIL PLANTS.

(FIGURE 46. Chara medicaginula; fossil. Upper Eocene, Isle of Wight.)

The seed-vessels and stems of Chara, a genus of aquatic plants, are very frequent in fresh-water strata. These seed-vessels were called, before their true nature was known, gyrogonites, and were supposed to be foraminiferous shells. (See Figure 46, a.)

(FIGURE 47. Chara elastica; recent, Italy. a. Sessile seed-vessel between the divisions of the leaves of the female plant. b. Magnified transverse section of a branch, with five seed-vessels, seen from below upward.)

The Charae inhabit the bottom of lakes and ponds, and flourish mostly where the water is charged with carbonate of lime. Their seed-vessels are covered with a very tough integument, capable of resisting decomposition; to which circumstance we may attribute their abundance in a fossil state. Figure 47 represents a branch of one of many new species found by Professor Amici in the lakes of Northern Italy. The seed-vessel in this plant is more globular than in the British Charae, and therefore more nearly resembles in form the extinct fossil species found in England, France, and other countries. The stems, as well as the seed-vessels, of these plants occur both in modern shell-marl and in ancient fresh-water formations. They are generally composed of a large central tube surrounded by smaller ones; the whole stem being divided at certain intervals by transverse partitions or joints. (See b, Figure 46.)

It is not uncommon to meet with layers of vegetable matter, impressions of leaves, and branches of trees, in strata containing fresh-water shells; and we also find occasionally the teeth and bones of land quadrupeds, of species now unknown. The manner in which such remains are occasionally carried by rivers into lakes, especially during floods, has been fully treated of in the "Principles of Geology."

FRESH-WATER AND MARINE FISH.

The remains of fish are occasionally useful in determining the fresh-water origin of strata. Certain genera, such as carp, perch, pike, and loach (Cyprinus, Perca, Esox, and Cobitis), as also Lebias, being peculiar to fresh- water. Other genera contain some fresh-water and some marine species, as Cottus, Mugil, and Anguilla, or eel. The rest are either common to rivers and the sea, as the salmon; or are exclusively characteristic of salt-water. The above observations respecting fossil fishes are applicable only to the more modern or tertiary deposits; for in the more ancient rocks the forms depart so widely from those of existing fishes, that it is very difficult, at least in the present state of science, to derive any positive information from ichthyolites respecting the element in which strata were deposited.

The alternation of marine and fresh-water formations, both on a small and large scale, are facts well ascertained in geology. When it occurs on a small scale, it may have arisen from the alternate occupation of certain spaces by river- water and the sea; for in the flood season the river forces back the ocean and freshens it over a large area, depositing at the same time its sediment; after which the salt-water again returns, and, on resuming its former place, brings with it sand, mud, and marine shells.

There are also lagoons at the mouth of many rivers, as the Nile and Mississippi, which are divided off by bars of sand from the sea, and which are filled with salt and fresh water by turns. They often communicate exclusively with the river for months, years, or even centuries; and then a breach being made in the bar of sand, they are for long periods filled with salt-water.

LYM-FIORD.

The Lym-Fiord in Jutland offers an excellent illustration of analogous changes; for, in the course of the last thousand years, the western extremity of this long frith, which is 120 miles in length, including its windings, has been four times fresh and four times salt, a bar of sand between it and the ocean having been often formed and removed. The last irruption of salt water happened in 1824, when the North Sea entered, killing all the fresh-water shells, fish, and plants; and from that time to the present, the sea-weed Fucus vesiculosus, together with oysters and other marine mollusca, have succeeded the Cyclas, Lymnaea, Paludina, and Charae. (See Principles Index "Lym-Fiord.")

But changes like these in the Lym-Fiord, and those before mentioned as occurring at the mouths of great rivers, will only account for some cases of marine deposits of partial extent resting on fresh-water strata. When we find, as in the south-east of England (Chapter 18), a great series of fresh-water beds, 1000 feet in thickness, resting upon marine formations and again covered by other rocks, such as the Cretaceous, more than 1000 feet thick, and of deep-sea origin, we shall find it necessary to seek for a different explanation of the phenomena.

CHAPTER IV.
CONSOLIDATION OF STRATA AND PETRIFACTION OF FOSSILS.

Chemical and Mechanical Deposits.
Cementing together of Particles.
Hardening by Exposure to Air.
Concretionary Nodules.
Consolidating Effects of Pressure.
Mineralization of Organic Remains.
Impressions and Casts: how formed.
Fossil Wood.
Goppert's Experiments.
Precipitation of Stony Matter most rapid where Putrefaction is going on.
Sources of Lime and Silex in Solution.

Having spoken in the preceding chapters of the characters of sedimentary formations, both as dependent on the deposition of inorganic matter and the distribution of fossils, I may next treat of the consolidation of stratified rocks, and the petrifaction of imbedded organic remains.

CHEMICAL AND MECHANICAL DEPOSITS.

A distinction has been made by geologists between deposits of a mechanical, and those of a chemical, origin. By the name mechanical are designated beds of mud, sand, or pebbles produced by the action of running water, also accumulations of stones and scoriae thrown out by a volcano, which have fallen into their present place by the force of gravitation. But the matter which forms a chemical deposit has not been mechanically suspended in water, but in a state of solution until separated by chemical action. In this manner carbonate of lime is occasionally precipitated upon the bottom of lakes in a solid form, as may be well seen in many parts of Italy, where mineral springs abound, and where the calcareous stone, called travertin, is deposited. In these springs the lime is usually held in solution by an excess of carbonic acid, or by heat if it be a hot spring, until the water, on issuing from the earth, cools or loses part of its acid. The calcareous matter then falls down in a solid state, incrusting shells, fragments of wood and leaves, and binding them together.

That similar travertin is formed at some points in the bed of the sea where calcareous springs issue can not be doubted, but as a general rule the quantity of lime, according to Bischoff, spread through the waters of the ocean is very small, the free carbonic acid gas in the same waters being five times as much as is necessary to keep the lime in a fluid state. Carbonate of lime, therefore, can rarely be precipitated at the bottom of the sea by chemical action alone, but must be produced by vital agency as in the case of coral reefs.

In such reefs, large masses of limestone are formed by the stony skeletons of zoophytes; and these, together with shells, become cemented together by carbonate of lime, part of which is probably furnished to the sea-water by the decomposition of dead corals. Even shells, of which the animals are still living on these reefs, are very commonly found to be incrusted over with a hard coating of limestone.

If sand and pebbles are carried by a river into the sea, and these are bound together immediately by carbonate of lime, the deposit may be described as of a mixed origin, partly chemical, and partly mechanical.

Now, the remarks already made in Chapter 2 on the original horizontality of strata are strictly applicable to mechanical deposits, and only partially to those of a mixed nature. Such as are purely chemical may be formed on a very steep slope, or may even incrust the vertical walls of a fissure, and be of equal thickness throughout; but such deposits are of small extent, and for the most part confined to vein-stones.

CONSOLIDATION OF STRATA.

It is chiefly in the case of calcareous rocks that solidification takes place at the time of deposition. But there are many deposits in which a cementing process comes into operation long afterwards. We may sometimes observe, where the water of ferruginous or calcareous springs has flowed through a bed of sand or gravel, that iron or carbonate of lime has been deposited in the interstices between the grains or pebbles, so that in certain places the whole has been bound together into a stone, the same set of strata remaining in other parts loose and incoherent.

Proofs of a similar cementing action are seen in a rock at Kelloway, in Wiltshire. A peculiar band of sandy strata belonging to the group called Oolite by geologists may be traced through several counties, the sand being for the most part loose and unconsolidated, but becoming stony near Kelloway. In this district there are numerous fossil shells which have decomposed, having for the most part left only their casts. The calcareous matter hence derived has evidently served, at some former period, as a cement to the siliceous grains of sand, and thus a solid sandstone has been produced. If we take fragments of many other argillaceous grits, retaining the casts of shells, and plunge them into dilute muriatic or other acid, we see them immediately changed into common sand and mud; the cement of lime, derived from the shells, having been dissolved by the acid.

Traces of impressions and casts are often extremely faint. In some loose sands of recent date we meet with shells in so advanced a stage of decomposition as to crumble into powder when touched. It is clear that water percolating such strata may soon remove the calcareous matter of the shell; and unless circumstances cause the carbonate of lime to be again deposited, the grains of sand will not be cemented together; in which case no memorial of the fossil will remain.

In what manner silex and carbonate of lime may become widely diffused in small quantities through the waters which permeate the earth's crust will be spoken of presently, when the petrifaction of fossil bodies is considered; but I may remark here that such waters are always passing in the case of thermal springs from hotter to colder parts of the interior of the earth; and, as often as the temperature of the solvent is lowered, mineral matter has a tendency to separate from it and solidify. Thus a stony cement is often supplied to sand, pebbles, or any fragmentary mixture. In some conglomerates, like the pudding-stone of Hertfordshire (a Lower Eocene deposit), pebbles of flint and grains of sand are united by a siliceous cement so firmly, that if a block be fractured, the rent passes as readily through the pebbles as through the cement.

It is probable that many strata became solid at the time when they emerged from the waters in which they were deposited, and when they first formed a part of the dry land. A well-known fact seems to confirm this idea: by far the greater number of the stones used for building and road-making are much softer when first taken from the quarry than after they have been long exposed to the air; and these, when once dried, may afterwards be immersed for any length of time in water without becoming soft again. Hence it is found desirable to shape the stones which are to be used in architecture while they are yet soft and wet, and while they contain their "quarry-water," as it is called; also to break up stone intended for roads when soft, and then leave it to dry in the air for months that it may harden. Such induration may perhaps be accounted for by supposing the water, which penetrates the minutest pores of rocks, to deposit, on evaporation, carbonate of lime, iron, silex, and other minerals previously held in solution, and thereby to fill up the pores partially. These particles, on crystallising, would not only be themselves deprived of freedom of motion, but would also bind together other portions of the rock which before were loosely aggregated. On the same principle wet sand and mud become as hard as stone when frozen; because one ingredient of the mass, namely, the water, has crystallised, so as to hold firmly together all the separate particles of which the loose mud and sand were composed.

Dr. MacCulloch mentions a sandstone in Skye, which may be moulded like dough when first found; and some simple minerals, which are rigid and as hard as glass in our cabinets, are often flexible and soft in their native beds: this is the case with asbestos, sahlite, tremolite, and chalcedony, and it is reported also to happen in the case of the beryl. (Dr. MacCulloch System of Geology volume 1 page 123.)

The marl recently deposited at the bottom of Lake Superior, in North America, is soft, and often filled with fresh-water shells; but if a piece be taken up and dried, it becomes so hard that it can only be broken by a smart blow of the hammer. If the lake, therefore, was drained, such a deposit would be found to consist of strata of marlstone, like that observed in many ancient European formations, and, like them, containing fresh-water shells.

CONCRETIONARY STRUCTURE.

(FIGURE 48. Calcareous nodules in Lias.)

It is probable that some of the heterogeneous materials which rivers transport to the sea may at once set under water, like the artificial mixture called pozzolana, which consists of fine volcanic sand charged with about twenty per cent of oxide of iron, and the addition of a small quantity of lime. This substance hardens, and becomes a solid stone in water, and was used by the Romans in constructing the foundations of buildings in the sea. Consolidation in such cases is brought about by the action of chemical affinity on finely comminuted matter previously suspended in water. After deposition similar particles seem often to exert a mutual attraction on each other, and congregate together in particular spots, forming lumps, nodules, and concretions. Thus in many argillaceous deposits there are calcareous balls, or spherical concretions, ranged in layers parallel to the general stratification; an arrangement which took place after the shale or marl had been thrown down in successive laminae; for these laminae are often traceable through the concretions, remaining parallel to those of the surrounding unconsolidated rock. (See Figure 48.) Such nodules of limestone have often a shell or other foreign body in the centre.

(FIGURE 49. Spheroidal concretions in magnesian limestone.)

Among the most remarkable examples of concretionary structure are those described by Professor Sedgwick as abounding in the magnesian limestone of the north of England. The spherical balls are of various sizes, from that of a pea to a diameter of several feet, and they have both a concentric and radiated structure, while at the same time the laminae of original deposition pass uninterruptedly through them. In some cliffs this limestone resembles a great irregular pile of cannon-balls. Some of the globular masses have their centre in one stratum, while a portion of their exterior passes through to the stratum above or below. Thus the larger spheroid in the section (Figure 49) passes from the stratum b upward into a. In this instance we must suppose the deposition of a series of minor layers, first forming the stratum b, and afterwards the incumbent stratum a; then a movement of the particles took place, and the carbonates of lime and magnesia separated from the more impure and mixed matter forming the still unconsolidated parts of the stratum. Crystallisation, beginning at the centre, must have gone on forming concentric coats around the original nucleus without interfering with the laminated structure of the rock.

(FIGURE 50. Section through strata of grit.)

When the particles of rocks have been thus rearranged by chemical forces, it is sometimes difficult or impossible to ascertain whether certain lines of division are due to original deposition or to the subsequent aggregation of several particles. Thus suppose three strata of grit, A, B, C, are charged unequally with calcareous matter, and that B is the most calcareous. If consolidation takes place in B, the concretionary action may spread upward into a part of A, where the carbonate of lime is more abundant than in the rest; so that a mass, d e f, forming a portion of the superior stratum, becomes united with B into one solid mass of stone. The original line of division, d e, being thus effaced, the line d f would generally be considered as the surface of the bed B, though not strictly a true plane of stratification. (Figure 50.)

PRESSURE AND HEAT.

When sand and mud sink to the bottom of a deep sea, the particles are not pressed down by the enormous weight of the incumbent ocean; for the water, which becomes mingled with the sand and mud, resists pressure with a force equal to that of the column of fluid above. The same happens in regard to organic remains which are filled with water under great pressure as they sink, otherwise they would be immediately crushed to pieces and flattened. Nevertheless, if the materials of a stratum remain in a yielding state, and do not set or solidify, they will be gradually squeezed down by the weight of other materials successively heaped upon them, just as soft clay or loose sand on which a house is built may give way. By such downward pressure particles of clay, sand, and marl may become packed into a smaller space, and be made to cohere together permanently.

Analogous effects of condensation may arise when the solid parts of the earth's crust are forced in various directions by those mechanical movements hereafter to be described, by which strata have been bent, broken, and raised above the level of the sea. Rocks of more yielding materials must often have been forced against others previously consolidated, and may thus by compression have acquired a new structure. A recent discovery may help us to comprehend how fine sediment derived from the detritus of rocks may be solidified by mere pressure. The graphite or "black lead" of commerce having become very scarce, Mr. Brockedon contrived a method by which the dust of the purer portions of the mineral found in Borrowdale might be recomposed into a mass as dense and compact as native graphite. The powder of graphite is first carefully prepared and freed from air, and placed under a powerful press on a strong steel die, with air- tight fittings. It is then struck several blows, each of a power of 1000 tons; after which operation the powder is so perfectly solidified that it can be cut for pencils, and exhibits when broken the same texture as native graphite.

But the action of heat at various depths in the earth is probably the most powerful of all causes in hardening sedimentary strata. To this subject I shall refer again when treating of the metamorphic rocks, and of the slaty and jointed structure.

MINERALISATION OF ORGANIC REMAINS.

(FIGURE 51. Phasianella Heddingtonensis, and cast of the same. Coral Rag.)

(FIGURE 52. Pleurotomaria Anglica, and cast. Lias.)

The changes which fossil organic bodies have undergone since they were first imbedded in rocks, throw much light on the consolidation of strata. Fossil shells in some modern deposits have been scarcely altered in the course of centuries, having simply lost a part of their animal matter. But in other cases the shell has disappeared, and left an impression only of its exterior, or, secondly, a cast of its interior form, or, thirdly, a cast of the shell itself, the original matter of which has been removed. These different forms of fossilisation may easily be understood if we examine the mud recently thrown out from a pond or canal in which there are shells. If the mud be argillaceous, it acquires consistency on drying, and on breaking open a portion of it we find that each shell has left impressions of its external form. If we then remove the shell itself, we find within a solid nucleus of clay, having the form of the interior of the shell. This form is often very different from that of the outer shell. Thus a cast such as a, Figure 51, commonly called a fossil screw, would never be suspected by an inexperienced conchologist to be the internal shape of the fossil univalve, b, Figure 51. Nor should we have imagined at first sight that the shell a and the cast b, Figure 52, belong to one and the same fossil. The reader will observe, in the last-mentioned figure (b, Figure 52), that an empty space shaded dark, which the SHELL ITSELF once occupied, now intervenes between the enveloping stone and the cast of the smooth interior of the whorls. In such cases the shell has been dissolved and the component particles removed by water percolating the rock. If the nucleus were taken out, a hollow mould would remain, on which the external form of the shell with its tubercles and striae, as seen in a, Figure 52, would be seen embossed. Now if the space alluded to between the nucleus and the impression, instead of being left empty, has been filled up with calcareous spar, flint, pyrites, or other mineral, we then obtain from the mould an exact cast both of the external and internal form of the original shell. In this manner silicified casts of shells have been formed; and if the mud or sand of the nucleus happen to be incoherent, or soluble in acid, we can then procure in flint an empty shell, which in shape is the exact counterpart of the original. This cast may be compared to a bronze statue, representing merely the superficial form, and not the internal organisation; but there is another description of petrifaction by no means uncommon, and of a much more wonderful kind, which may be compared to certain anatomical models in wax, where not only the outward forms and features, but the nerves, blood-vessels, and other internal organs are also shown. Thus we find corals, originally calcareous, in which not only the general shape, but also the minute and complicated internal organisation is retained in flint.

(FIGURE 53. Section of a tree from the coal-measures, magnified (Witham), showing texture of wood.)

Such a process of petrifaction is still more remarkably exhibited in fossil wood, in which we often perceive not only the rings of annual growth, but all the minute vessels and medullary rays. Many of the minute cells and fibres of plants, and even those spiral vessels which in the living vegetable can only be discovered by the microscope, are preserved. Among many instances, I may mention a fossil tree, seventy-two feet in length, found at Gosforth, near Newcastle, in sandstone strata associated with coal. By cutting a transverse slice so thin as to transmit light, and magnifying it about fifty-five times, the texture, as seen in Figure 53, is exhibited. A texture equally minute and complicated has been observed in the wood of large trunks of fossil trees found in the Craigleith quarry near Edinburgh, where the stone was not in the slightest degree siliceous, but consisted chiefly of carbonate of lime, with oxide of iron, alumina, and carbon. The parallel rows of vessels here seen are the rings of annual growth, but in one part they are imperfectly preserved, the wood having probably decayed before the mineralising matter had penetrated to that portion of the tree.

In attempting to explain the process of petrifaction in such cases, we may first assume that strata are very generally permeated by water charged with minute portions of calcareous, siliceous, and other earths in solution. In what manner they become so impregnated will be afterwards considered. If an organic substance is exposed in the open air to the action of the sun and rain, it will in time putrefy, or be dissolved into its component elements, consisting usually of oxygen, hydrogen, nitrogen, and carbon. These will readily be absorbed by the atmosphere or be washed away by rain, so that all vestiges of the dead animal or plant disappear. But if the same substances be submerged in water, they decompose more gradually; and if buried in earth, still more slowly; as in the familiar example of wooden piles or other buried timber. Now, if as fast as each particle is set free by putrefaction in a fluid or gaseous state, a particle equally minute of carbonate of lime, flint, or other mineral, is at hand ready to be precipitated, we may imagine this inorganic matter to take the place just before left unoccupied by the organic molecule. In this manner a cast of the interior of certain vessels may first be taken, and afterwards the more solid walls of the same may decay and suffer a like transmutation. Yet when the whole is lapidified, it may not form one homogeneous mass of stone or metal. Some of the original ligneous, osseous, or other organic elements may remain mingled in certain parts, or the lapidifying substance itself may be differently coloured at different times, or so crystallised as to reflect light differently, and thus the texture of the original body may be faithfully exhibited.

The student may perhaps ask whether, on chemical principles, we have any ground to expect that mineral matter will be thrown down precisely in those spots where organic decomposition is in progress? The following curious experiments may serve to illustrate this point: Professor Goppert of Breslau, with a view of imitating the natural process of petrifaction, steeped a variety of animal and vegetable substances in waters, some holding siliceous, others calcareous, others metallic matter in solution. He found that in the period of a few weeks, or sometimes even days, the organic bodies thus immersed were mineralised to a certain extent. Thus, for example, thin vertical slices of deal, taken from the Scotch fir (Pinus sylvestris), were immersed in a moderately strong solution of sulphate of iron. When they had been thoroughly soaked in the liquid for several days they were dried and exposed to a red-heat until the vegetable matter was burnt up and nothing remained but an oxide of iron, which was found to have taken the form of the deal so exactly that casts even of the dotted vessels peculiar to this family of plants were distinctly visible under the microscope.

The late Dr. Turner observes, that when mineral matter is in a "nascent state," that is to say, just liberated from a previous state of chemical combination, it is most ready to unite with other matter, and form a new chemical compound. Probably the particles or atoms just set free are of extreme minuteness, and therefore move more freely, and are more ready to obey any impulse of chemical affinity. Whatever be the cause, it clearly follows, as before stated, that where organic matter newly imbedded in sediment is decomposing, there will chemical changes take place most actively.

An analysis was lately made of the water which was flowing off from the rich mud deposited by the Hooghly River in the Delta of the Ganges after the annual inundation. This water was found to be highly charged with carbonic acid holding lime in solution. (Piddington Asiatic Researches volume 18 page 226.) Now if newly-deposited mud is thus proved to be permeated by mineral matter in a state of solution, it is not difficult to perceive that decomposing organic bodies, naturally imbedded in sediment, may as readily become petrified as the substances artificially immersed by Professor Goppert in various fluid mixtures.

It is well known that the waters of all springs are more or less charged with earthy, alkaline, or metallic ingredients derived from the rocks and mineral veins through which they percolate. Silex is especially abundant in hot springs, and carbonate of lime is almost always present in greater or less quantity. The materials for the petrifaction of organic remains are, therefore, usually at hand in a state of chemical solution wherever organic remains are imbedded in new strata.

CHAPTER V.
ELEVATION OF STRATA ABOVE THE SEA.— HORIZONTAL AND INCLINED STRATIFICATION.

Why the Position of Marine Strata, above the Level of the Sea, should be referred to the rising up of the Land, not to the going down of the Sea.
Strata of Deep-sea and Shallow-water Origin alternate.
Also Marine and Fresh-water Beds and old Land Surfaces.
Vertical, inclined, and folded Strata.
Anticlinal and Synclinal Curves.
Theories to explain Lateral Movements.
Creeps in Coal-mines.
Dip and Strike.
Structure of the Jura.
Various Forms of Outcrop.
Synclinal Strata forming Ridges.
Connection of Fracture and Flexure of Rocks.
Inverted Strata.
Faults described.
Superficial Signs of the same obliterated by Denudation.
Great Faults the Result of repeated Movements.
Arrangement and Direction of parallel Folds of Strata.
Unconformability.
Overlapping Strata.

LAND HAS BEEN RAISED, NOT THE SEA LOWERED.

It has been already stated that the aqueous rocks containing marine fossils extend over wide continental tracts, and are seen in mountain chains rising to great heights above the level of the sea (Chapter 1). Hence it follows, that what is now dry land was once under water. But if we admit this conclusion, we must imagine, either that there has been a general lowering of the waters of the ocean, or that the solid rocks, once covered by water, have been raised up bodily out of the sea, and have thus become dry land. The earlier geologists, finding themselves reduced to this alternative, embraced the former opinion, assuming that the ocean was originally universal, and had gradually sunk down to its actual level, so that the present islands and continents were left dry. It seemed to them far easier to conceive that the water had gone down, than that solid land had risen upward into its present position. It was, however, impossible to invent any satisfactory hypothesis to explain the disappearance of so enormous a body of water throughout the globe, it being necessary to infer that the ocean had once stood at whatever height marine shells might be detected. It moreover appeared clear, as the science of geology advanced, that certain spaces on the globe had been alternately sea, then land, then estuary, then sea again, and, lastly, once more habitable land, having remained in each of these states for considerable periods. In order to account for such phenomena without admitting any movement of the land itself, we are required to imagine several retreats and returns of the ocean; and even then our theory applies merely to cases where the marine strata composing the dry land are horizontal, leaving unexplained those more common instances where strata are inclined, curved, or placed on their edges, and evidently not in the position in which they were first deposited.

Geologists, therefore, were at last compelled to have recourse to the doctrine that the solid land has been repeatedly moved upward or downward, so as permanently to change its position relatively to the sea. There are several distinct grounds for preferring this conclusion. First, it will account equally for the position of those elevated masses of marine origin in which the stratification remains horizontal, and for those in which the strata are disturbed, broken, inclined, or vertical. Secondly, it is consistent with human experience that land should rise gradually in some places and be depressed in others. Such changes have actually occurred in our own days, and are now in progress, having been accompanied in some cases by violent convulsions, while in others they have proceeded so insensibly as to have been ascertainable only by the most careful scientific observations, made at considerable intervals of time. On the other hand, there is no evidence from human experience of a rising or lowering of the sea's level in any region, and the ocean can not be raised or depressed in one place without its level being changed all over the globe.

These preliminary remarks will prepare the reader to understand the great theoretical interest attached to all facts connected with the position of strata, whether horizontal or inclined, curved or vertical.

Now the first and most simple appearance is where strata of marine origin occur above the level of the sea in horizontal position. Such are the strata which we meet with in the south of Sicily, filled with shells for the most part of the same species as those now living in the Mediterranean. Some of these rocks rise to the height of more than 2000 feet above the sea. Other mountain masses might be mentioned, composed of horizontal strata of high antiquity, which contain fossil remains of animals wholly dissimilar from any now known to exist. In the south of Sweden, for example, near Lake Wener, the beds of some of the oldest fossiliferous deposits, called Silurian and Cambrian by geologists, occur in as level a position as if they had recently formed part of the delta of a great river, and been left dry on the retiring of the annual floods. Aqueous rocks of equal antiquity extend for hundreds of miles over the lake-district of North America, and exhibit in like manner a stratification nearly undisturbed. The Table Mountain at the Cape of Good Hope is another example of highly elevated yet perfectly horizontal strata, no less than 3500 feet in thickness, and consisting of sandstone of very ancient date.

Instead of imagining that such fossiliferous rocks were always at their present level, and that the sea was once high enough to cover them, we suppose them to have constituted the ancient bed of the ocean, and to have been afterwards uplifted to their present height. This idea, however startling it may at first appear, is quite in accordance, as before stated, with the analogy of changes now going on in certain regions of the globe. Thus, in parts of Sweden, and the shores and islands of the Gulf of Bothnia, proofs have been obtained that the land is experiencing, and has experienced for centuries, a slow upheaving movement. (See "Principles of Geology" 1867 page 314.)

It appears from the observations of Mr. Darwin and others, that very extensive regions of the continent of South America have been undergoing slow and gradual upheaval, by which the level plains of Patagonia, covered with recent marine shells, and the Pampas of Buenos Ayres, have been raised above the level of the sea. On the other hand, the gradual sinking of the west coast of Greenland, for the space of more than 600 miles from north to south, during the last four centuries, has been established by the observations of a Danish naturalist, Dr. Pingel. And while these proofs of continental elevation and subsidence, by slow and insensible movements, have been recently brought to light, the evidence has been daily strengthened of continued changes of level effected by violent convulsions in countries where earthquakes are frequent. There the rocks are rent from time to time, and heaved up or thrown down several feet at once, and disturbed in such a manner as to show how entirely the original position of strata may be modified in the course of centuries.

Mr. Darwin has also inferred that, in those seas where circular coral islands and barrier reefs abound, there is a slow and continued sinking of the submarine mountains on which the masses of coral are based; while there are other areas of the South Sea where the land is on the rise, and where coral has been upheaved far above the sea-level.

ALTERNATIONS OF MARINE AND FRESH-WATER STRATA.

It has been shown in the third chapter that there is such a difference between land, fresh-water, and marine fossils as to enable the geologist to determine whether particular groups of strata were formed at the bottom of the ocean or in estuaries, rivers, or lakes. If surprise was at first created by the discovery of marine corals and shells at the height of several miles above the sea-level, the imagination was afterwards not less startled by observing that in the successive strata composing the earth's crust, especially if their total thickness amounted to thousands of feet, they comprised in some parts formations of shallow-sea as well as of deep-sea origin; also beds of brackish or even of purely fresh-water formation, as well as vegetable matter or coal accumulated on ancient land. In these cases we as frequently find fresh-water beds below a marine set or shallow-water under those of deep-sea origin as the reverse. Thus, if we bore an artesian well below London, we pass through a marine clay, and there reach, at the depth of several hundred feet, a shallow-water and fluviatile sand, beneath which comes the white chalk originally formed in a deep sea. Or if we bore vertically through the chalk of the North Downs, we come, after traversing marine chalky strata, upon a fresh-water formation many hundreds of feet thick, called the Wealden, such as is seen in Kent and Surrey, which is known in its turn to rest on purely marine beds. In like manner, in various parts of Great Britain we sink vertical shafts through marine deposits of great thickness, and come upon coal which was formed by the growth of plants on an ancient land-surface sometimes hundreds of square miles in extent.

VERTICAL, INCLINED, AND CURVED STRATA.

(FIGURE 54. Vertical conglomerate and sandstone.)

It has been stated that marine strata of different ages are sometimes found at a considerable height above the sea, yet retaining their original horizontality; but this state of things is quite exceptional. As a general rule, strata are inclined or bent in such a manner as to imply that their original position has been altered.

(FIGURE 55. Section of Forfarshire, from N.W. to S.E., from the foot of the Grampians to the sea at Arbroath (volcanic or trap rocks omitted). Length of section twenty miles.
From S.E. (left) Sea: Whiteness, Arbroath: Strata a, 2, 3: Leys Mill: Strata 4: Sidlaw Hills. Viney R.: Strata B: Pitmuies: Strata 4: Position and nature of the rocks below No. 4 unknown: Turin: Findhaven: Strata 3, 2, A: Valley of Strathmore: Strata 1, 2, 3: W. Ogle: Strata 4 and Clay-Slate: to N.W. (right).)

The most unequivocal evidence of such a change is afforded by their standing up vertically, showing their edges, which is by no means a rare phenomenon, especially in mountainous countries. Thus we find in Scotland, on the southern skirts of the Grampians, beds of pudding-stone alternating with thin layers of fine sand, all placed vertically to the horizon. When Saussure first observed certain conglomerates in a similar position in the Swiss Alps, he remarked that the pebbles, being for the most part of an oval shape, had their longer axes parallel to the planes of stratification (see Figure 54). From this he inferred that such strata must, at first, have been horizontal, each oval pebble having settled at the bottom of the water, with its flatter side parallel to the horizon, for the same reason that an egg will not stand on either end if unsupported. Some few, indeed, of the rounded stones in a conglomerate occasionally afford an exception to the above rule, for the same reason that in a river's bed, or on a shingle beach, some pebbles rest on their ends or edges; these having been shoved against or between other stones by a wave or current, so as to assume this position.

ANTICLINAL AND SYNCLINAL CURVES.

Vertical strata, when they can be traced continuously upward or downward for some depth, are almost invariably seen to be parts of great curves, which may have a diameter of a few yards, or of several miles. I shall first describe two curves of considerable regularity, which occur in Forfarshire, extending over a country twenty miles in breadth, from the foot of the Grampians to the sea near Arbroath.

The mass of strata here shown may be 2000 feet in thickness, consisting of red and white sandstone, and various coloured shales, the beds being distinguishable into four principal groups, namely, No. 1, red marl or shale; No. 2, red sandstone, used for building; No. 3, conglomerate; and No. 4, grey paving-stone, and tile-stone, with green and reddish shale, containing peculiar organic remains. A glance at the section (Figure 55.) will show that each of the formations 2, 3, 4 are repeated thrice at the surface, twice with a southerly, and once with a northerly inclination or DIP, and the beds in No. 1, which are nearly horizontal, are still brought up twice by a slight curvature to the surface, once on each side of A. Beginning at the north-west extremity, the tile-stones and conglomerates, No. 4 and No. 3, are vertical, and they generally form a ridge parallel to the southern skirts of the Grampians. The superior strata, Nos. 2 and 1, become less and less inclined on descending to the valley of Strathmore, where the strata, having a concave bend, are said by geologists to lie in a "trough" or "basin." Through the centre of this valley runs an imaginary line A, called technically a "synclinal line," where the beds, which are tilted in opposite directions, may be supposed to meet. It is most important for the observer to mark such lines, for he will perceive by the diagram that, in travelling from the north to the centre of the basin, he is always passing from older to newer beds; whereas, after crossing the line A, and pursuing his course in the same southerly direction, he is continually leaving the newer, and advancing upon older strata. All the deposits which he had before examined begin then to recur in reversed order, until he arrives at the central axis of the Sidlaw hills, where the strata are seen to form an arch, or SADDLE, having an ANTICLINAL line, B, in the centre. On passing this line, and continuing towards the S.E., the formations 4, 3, and 2, are again repeated, in the same relative order of superposition, but with a southerly dip. At Whiteness (see Figure 55) it will be seen that the inclined strata are covered by a newer deposit, a, in horizontal beds. These are composed of red conglomerate and sand, and are newer than any of the groups, 1, 2, 3, 4, before described, and rest UNCONFORMABLY upon strata of the sandstone group, No. 2.

An example of curved strata, in which the bends or convolutions of the rock are sharper and far more numerous within an equal space, has been well described by Sir James Hall. (Edinburgh Transactions volume 7 plate 3.) It occurs near St. Abb's Head, on the east coast of Scotland, where the rocks consist principally of a bluish slate, having frequently a ripple-marked surface. The undulations of the beds reach from the top to the bottom of cliffs from 200 to 300 feet in height, and there are sixteen distinct bendings in the course of about six miles, the curvatures being alternately concave and convex upward.

FOLDING BY LATERAL MOVEMENT.

(FIGURE 56. Curved strata of slate near St. Abb's Head, Berwickshire. (Sir J. Hall.)

(FIGURE 57. Curved strata in line of cliff.)

(FIGURE 58. Folded cloths imitating bent strata.)

An experiment was made by Sir James Hall, with a view of illustrating the manner in which such strata, assuming them to have been originally horizontal, may have been forced into their present position. A set of layers of clay were placed under a weight, and their opposite ends pressed towards each other with such force as to cause them to approach more nearly together. On the removal of the weight, the layers of clay were found to be curved and folded, so as to bear a miniature resemblance to the strata in the cliffs. We must, however, bear in mind that in the natural section or sea-cliff we only see the foldings imperfectly, one part being invisible beneath the sea, and the other, or upper portion, being supposed to have been carried away by DENUDATION, or that action of water which will be explained in the next chapter. The dark lines in the plan (Figure 57) represent what is actually seen of the strata in the line of cliff alluded to; the fainter lines, that portion which is concealed beneath the sea- level, as also that which is supposed to have once existed above the present surface.

We may still more easily illustrate the effects which a lateral thrust might produce on flexible strata, by placing several pieces of differently coloured cloths upon a table, and when they are spread out horizontally, cover them with a book. Then apply other books to each end, and force them towards each other. The folding of the cloths (see Figure 58) will imitate those of the bent strata; the incumbent book being slightly lifted up, and no longer touching the two volumes on which it rested before, because it is supported by the tops of the anticlinal ridges formed by the curved cloths. In like manner there can be no doubt that the squeezed strata, although laterally condensed and more closely packed, are yet elongated and made to rise upward, in a direction perpendicular to the pressure.

Whether the analogous flexures in stratified rocks have really been due to similar sideway movements is a question which we can not decide by reference to our own observation. Our inability to explain the nature of the process is, perhaps, not simply owing to the inaccessibility of the subterranean regions where the mechanical force is exerted, but to the extreme slowness of the movement. The changes may sometimes be due to variation in the temperature of mountain masses of rock causing them, while still solid, to expand or contract; or melting them, and then again cooling them and allowing them to crystallise. If such be the case, we have scarcely more reason to expect to witness the operation of the process within the limited periods of our scientific observation than to see the swelling of the roots of a tree, by which, in the course of years, a wall of solid masonry may be lifted up, rent or thrown down. In both instances the force may be irresistible, but though adequate, it need not be visible by us, provided the time required for its development be very great. The lateral pressure arising from the unequal expansion of rocks by heat may cause one mass lying in the same horizontal plane gradually to occupy a larger space, so as to press upon another rock, which, if flexible, may be squeezed into a bent and folded form. It will also appear, when the volcanic and granitic rocks are described, that some of them have, when melted in the interior of the earth's crust, been injected forcibly into fissures, and after the solidification of such intruded matter, other sets of rents, crossing the first, have been formed and in their turn filled by melted rock. Such repeated injections imply a stretching, and often upheaval, of the whole mass.

We also know, especially by the study of regions liable to earthquakes, that there are causes at work in the interior of the earth capable of producing a sinking in of the ground, sometimes very local, but often extending over a wide area. The continuance of such a downward movement, especially if partial and confined to linear areas, may produce regular folds in the strata.

CREEPS IN COAL-MINES.

The "creeps," as they are called in coal-mines, afford an excellent illustration of this fact.— First, it may be stated generally, that the excavation of coal at a considerable depth causes the mass of overlying strata to sink down bodily, even when props are left to support the roof of the mine. "In Yorkshire," says Mr. Buddle, "three distinct subsidences were perceptible at the surface, after the clearing out of three seams of coal below, and innumerable vertical cracks were caused in the incumbent mass of sandstone and shale which thus settled down." (Proceedings of Geological Society volume 3 page 148.) The exact amount of depression in these cases can only be accurately measured where water accumulates on the surface, or a railway traverses a coal-field.

(FIGURE 59. Section of carboniferous strata at Wallsend, Newcastle, showing "creeps." (J. Buddle, Esq.) Horizontal length of section 174 feet. The upper seam, or main coal, here worked out, was 630 feet below the surface. Section through, from top to bottom: Siliceous sandstone. Shale. 1. Main coal, 6 feet 6 inches, with creeps a, b, c, d. Shale eighteen yards thick. 2. Metal coal, 3 feet, with fractures e, f, g, h.)

When a bed of coal is worked out, pillars or rectangular masses of coal are left at intervals as props to support the roof, and protect the colliers. Thus in Figure 59, representing a section at Wallsend, Newcastle, the galleries which have been excavated are represented by the white spaces a, b, while the adjoining dark portions are parts of the original coal seam left as props, beds of sandy clay or shale constituting the floor of the mine. When the props have been reduced in size, they are pressed down by the weight of overlying rocks (no less than 630 feet thick) upon the shale below, which is thereby squeezed and forced up into the open spaces.

Now it might have been expected that, instead of the floor rising up, the ceiling would sink down, and this effect, called a "thrust," does, in fact, take place where the pavement is more solid than the roof. But it usually happens, in coalmines, that the roof is composed of hard shale, or occasionally of sandstone, more unyielding than the foundation, which often consists of clay. Even where the argillaceous substrata are hard at first, they soon become softened and reduced to a plastic state when exposed to the contact of air and water in the floor of a mine.

The first symptom of a "creep," says Mr. Buddle, is a slight curvature at the bottom of each gallery, as at a, Figure 59: then the pavement, continuing to rise, begins to open with a longitudinal crack, as at b; then the points of the fractured ridge reach the roof, as at c; and, lastly, the upraised beds close up the whole gallery, and the broken portions of the ridge are reunited and flattened at the top, exhibiting the flexure seen at d. Meanwhile the coal in the props has become crushed and cracked by pressure. It is also found that below the creeps a, b, c, d, an inferior stratum, called the "metal coal," which is 3 feet thick, has been fractured at the points e, f, g, h, and has risen, so as to prove that the upward movement, caused by the working out of the "main coal," has been propagated through a thickness of 54 feet of argillaceous beds, which intervene between the two coal-seams. This same displacement has also been traced downward more than 150 feet below the metal coal, but it grows continually less and less until it becomes imperceptible.

No part of the process above described is more deserving of our notice than the slowness with which the change in the arrangement of the beds is brought about. Days, months, or even years, will sometimes elapse between the first bending of the pavement and the time of its reaching the roof. Where the movement has been most rapid, the curvature of the beds is most regular, and the reunion of the fractured ends most complete; whereas the signs of displacement or violence are greatest in those creeps which have required months or years for their entire accomplishment. Hence we may conclude that similar changes may have been wrought on a larger scale in the earth's crust by partial and gradual subsidences, especially where the ground has been undermined throughout long periods of time; and we must be on our guard against inferring sudden violence, simply because the distortion of the beds is excessive.

Engineers are familiar with the fact that when they raise the level of a railway by heaping stone or gravel on a foundation of marsh, quicksand, or other yielding formation, the new mound often sinks for a time as fast as they attempt to elevate it; when they have persevered so as to overcome this difficulty, they frequently find that some of the adjoining flexible ground has risen up in one or more parallel arches or folds, showing that the vertical pressure of the sinking materials has given rise to a lateral folding movement.

In like manner, in the interior of the earth, the solid parts of the earth's crust may sometimes, as before mentioned, be made to expand by heat, or may be pressed by the force of steam against flexible strata loaded with a great weight of incumbent rocks. In this case the yielding mass, squeezed, but unable to overcome the resistance which it meets with in a vertical direction, may be gradually relieved by lateral folding.

DIP AND STRIKE.

(FIGURE 60. Series of inclined strata dipping to the north at an angle of 45 degrees.)

In describing the manner in which strata depart from their original horizontality, some technical terms, such as "dip" and "strike," "anticlinal" and "synclinal" line or axis, are used by geologists. I shall now proceed to explain some of these to the student. If a stratum or bed of rock, instead of being quite level, be inclined to one side, it is said to DIP; the point of the compass to which it is inclined is called the POINT OF DIP, and the degree of deviation from a level or horizontal line is called THE AMOUNT OF DIP, or THE ANGLE OF DIP. Thus, in the diagram (Figure 60), a series of strata are inclined, and they dip to the north at an angle of forty-five degrees. The STRIKE, or LINE OF BEARING, is the prolongation or extension of the strata in a direction AT RIGHT ANGLES to the dip; and hence it is sometimes called the DIRECTION of the strata. Thus, in the above instance of strata dipping to the north, their strike must necessarily be east and west. We have borrowed the word from the German geologists, streichen signifying to extend, to have a certain direction. Dip and strike may be aptly illustrated by a row of houses running east and west, the long ridge of the roof representing the strike of the stratum of slates, which dip on one side to the north, and on the other to the south.

A stratum which is horizontal, or quite level in all directions, has neither dip nor strike.

It is always important for the geologist, who is endeavouring to comprehend the structure of a country, to learn how the beds dip in every part of the district; but it requires some practice to avoid being occasionally deceived, both as to the point of dip and the amount of it.

(FIGURE 61. Apparent horizontality of inclined strata.)

If the upper surface of a hard stony stratum be uncovered, whether artificially in a quarry, or by waves at the foot of a cliff, it is easy to determine towards what point of the compass the slope is steepest, or in what direction water would flow if poured upon it. This is the true dip. But the edges of highly inclined strata may give rise to perfectly horizontal lines in the face of a vertical cliff, if the observer see the strata in the line of the strike, the dip being inward from the face of the cliff. If, however, we come to a break in the cliff, which exhibits a section exactly at right angles to the line of the strike, we are then able to ascertain the true dip. In the drawing (Figure 61), we may suppose a headland, one side of which faces to the north, where the beds would appear perfectly horizontal to a person in the boat; while in the other side facing the west, the true dip would be seen by the person on shore to be at an angle of 40 degrees. If, therefore, our observations are confined to a vertical precipice facing in one direction, we must endeavour to find a ledge or portion of the plane of one of the beds projecting beyond the others, in order to ascertain the true dip.

(FIGURE 62. Two hands used to determine the inclination of strata.)

If not provided with a clinometer, a most useful instrument, when it is of consequence to determine with precision the inclination of the strata, the observer may measure the angle within a few degrees by standing exactly opposite to a cliff where the true dip is exhibited, holding the hands immediately before the eyes, and placing the fingers of one in a perpendicular, and of the other in a horizontal position, as in Figure 62. It is thus easy to discover whether the lines of the inclined beds bisect the angle of 90 degrees, formed by the meeting of the hands, so as to give an angle of 45 degrees, or whether it would divide the space into two equal or unequal portions. You have only to change hands to get the line of dip on the upper side of the horizontal hand.

(FIGURE 63. Section illustrating the structure of the Swiss Jura.)

It has been already seen, in describing the curved strata on the east coast of Scotland, in Forfarshire and Berwickshire, that a series of concave and convex bendings are occasionally repeated several times. These usually form part of a series of parallel waves of strata, which are prolonged in the same direction, throughout a considerable extent of country. Thus, for example, in the Swiss Jura, that lofty chain of mountains has been proved to consist of many parallel ridges, with intervening longitudinal valleys, as in Figure 63, the ridges being formed by curved fossiliferous strata, of which the nature and dip are occasionally displayed in deep transverse gorges, called "cluses," caused by fractures at right angles to the direction of the chain. (Thurmann "Essai sur les Soulevemens Jurassiques de Porrentruy" Paris 1832.) Now let us suppose these ridges and parallel valleys to run north and south, we should then say that the STRIKE of the beds is north and south, and the DIP east and west. Lines drawn along the summits of the ridges, A, B, would be anticlinal lines, and one following the bottom of the adjoining valleys a synclinal line.

OUTCROP OF STRATA.

(FIGURE 64. Ground-plan of the denuded ridge C, Figure 63.)

(FIGURE 65. Transverse section of the denuded ridge C, Figure 63..)

It will be observed that some of these ridges, A, B, are unbroken on the summit, whereas one of them, C, has been fractured along the line of strike, and a portion of it carried away by denudation, so that the ridges of the beds in the formations a, b, c come out to the day, or, as the miners say, CROP OUT, on the sides of a valley. The ground-plan of such a denuded ridge as C, as given in a geological map, may be expressed by the diagram, Figure 64, and the cross-section of the same by Figure 65. The line D E, Figure 64, is the anticlinal line, on each side of which the dip is in opposite directions, as expressed by the arrows. The emergence of strata at the surface is called by miners their OUTCROP, or BASSET.

If, instead of being folded into parallel ridges, the beds form a boss or dome- shaped protuberance, and if we suppose the summit of the dome carried off, the ground-plan would exhibit the edges of the strata forming a succession of circles, or ellipses, round a common centre. These circles are the lines of strike, and the dip being always at right angles is inclined in the course of the circuit to every point of the compass, constituting what is termed a quaquaversal dip— that is, turning every way.

There are endless variations in the figures described by the basset-edges of the strata, according to the different inclination of the beds, and the mode in which they happen to have been denuded. One of the simplest rules, with which every geologist should be acquainted, relates to the V-like form of the beds as they crop out in an ordinary valley. First, if the strata be horizontal, the V- like form will be also on a level, and the newest strata will appear at the greatest heights.

(FIGURE 66. Slope of valley 40 degrees, dip of strata 20 degrees.)

Secondly, if the beds be inclined and intersected by a valley sloping in the same direction, and the dip of the beds be less steep than the slope of the valley, then the V's, as they are often termed by miners, will point upward (see Figure 66), those formed by the newer beds appearing in a superior position, and extending highest up the valley, as A is seen above B.

(FIGURE 67. Slope of valley 20 degrees, dip of strata 50 degrees.)

Thirdly, if the dip of the beds be steeper than the slope of the valley, then the V's will point downward (see Figure 67), and those formed of the older beds will now appear uppermost, as B appears above A.

(FIGURE 68. Slope of valley 20 degrees, dip of strata 20 degrees, in opposite directions.)

Fourthly, in every case where the strata dip in a contrary direction to the slope of the valley, whatever be the angle of inclination, the newer beds will appear the highest, as in the first and second cases. This is shown by the drawing (Figure 68), which exhibits strata rising at an angle of 20 degrees, and crossed by a valley, which declines in an opposite direction at 20 degrees.

These rules may often be of great practical utility; for the different degrees of dip occurring in the two cases represented in Figures 66 and 67 may occasionally be encountered in following the same line of flexure at points a few miles distant from each other. A miner unacquainted with the rule, who had first explored the valley Figure 66, may have sunk a vertical shaft below the coal-seam A, until he reached the inferior bed, B. He might then pass to the valley, Figure 67, and discovering there also the outcrop of two coal-seams, might begin his workings in the uppermost in the expectation of coming down to the other bed A, which would be observed cropping out lower down the valley. But a glance at the section will demonstrate the futility of such hopes. (I am indebted to the kindness of T. Sopwith, Esq., for three models which I have copied in the above diagrams; but the beginner may find it by no means easy to understand such copies, although, if he were to examine and handle the originals, turning them about in different ways, he would at once comprehend their meaning, as well as the import of others far more complicated, which the same engineer has constructed to illustrate FAULTS.)

SYNCLINAL STRATA FORMING RIDGES.

(FIGURE 69. Section of carboniferous rocks of Lancashire. (E. Hull. (Edward Hull, Quarterly Geological Journal volume 24 page 324. 1868.)) a. Synclinal. Grits and shales. c. Anticlinal. Mountain limestone. b. Synclinal. Grits and shales.)

Although in many cases an anticlinal axis forms a ridge, and a synclinal axis a valley, as in A B, Figure 63, yet this can by no means be laid down as a general rule, as the beds very often slope inward from either side of a mountain, as at a, b, Figure 69, while in the intervening valley, c, they slope upward, forming an arch.

It would be natural to expect the fracture of solid rocks to take place chiefly where the bending of the strata has been sharpest, and such rending may produce ravines giving access to running water and exposing the surface to atmospheric waste. The entire absence, however, of such cracks at points where the strain must have been greatest, as at a, Figure 63, is often very remarkable, and not always easy of explanation. We must imagine that many strata of limestone, chert, and other rocks which are now brittle, were pliant when bent into their present position. They may have owed their flexibility in part to the fluid matter which they contained in their minute pores, as before described, and in part to the permeation of sea-water while they were yet submerged.

(FIGURE 70. Strata of chert, grit, and marl, near St. Jean de Luz.)

At the western extremity of the Pyrenees, great curvatures of the strata are seen in the sea-cliffs, where the rocks consist of marl, grit, and chert. At certain points, as at a, Figure 70, some of the bendings of the flinty chert are so sharp that specimens might be broken off well fitted to serve as ridge-tiles on the roof of a house. Although this chert could not have been brittle as now, when first folded into this shape, it presents, nevertheless, here and there, at the points of greatest flexure, small cracks, which show that it was solid, and not wholly incapable of breaking at the period of its displacement. The numerous rents alluded to are not empty, but filled with chalcedony and quartz.

(FIGURE 71. Bent and undulating gypseous marl. g. Gypsum. m. Marl.)

Between San Caterina and Castrogiovanni, in Sicily, bent and undulating gypseous marls occur, with here and there thin beds of solid gypsum interstratified. Sometimes these solid layers have been broken into detached fragments, still preserving their sharp edges (g, g, Figure 71), while the continuity of the more pliable and ductile marls, m, m, has not been interrupted.

(FIGURE 72. Folded strata.)
(FIGURE 73. Folded strata.)

We have already explained, Figure 69, that stratified rocks have usually their strata bent into parallel folds forming anticlinal and synclinal axes, a group of several of these folds having often been subjected to a common movement, and having acquired a uniform strike or direction. In some disturbed regions these folds have been doubled back upon themselves in such a manner that it is often difficult for an experienced geologist to determine correctly the relative age of the beds by superposition. Thus, if we meet with the strata seen in the section, Figure 72, we should naturally suppose that there were twelve distinct beds, or sets of beds, No. 1 being the newest, and No. 12 the oldest of the series. But this section may perhaps exhibit merely six beds, which have been folded in the manner seen in Figure 73, so that each of them is twice repeated, the position of one half being reversed, and part of No. 1, originally the uppermost, having now become the lowest of the series.

These phenomena are observable on a magnificent scale in certain regions in Switzerland, in precipices often more than 2000 feet in perpendicular height, and there are flexures not inferior in dimensions in the Pyrenees. The upper part of the curves seen in this diagram, Figure 73, and expressed in fainter lines, has been removed by what is called denudation, to be afterwards explained.

FRACTURES OF THE STRATA AND FAULTS.

Numerous rents may often be seen in rocks which appear to have been simply broken, the fractured parts still remaining in contact; but we often find a fissure, several inches or yards wide, intervening between the disunited portions. These fissures are usually filled with fine earth and sand, or with angular fragments of stone, evidently derived from the fracture of the contiguous rocks.

The face of each wall of the fissure is often beautifully polished, as if glazed, striated, or scored with parallel furrows and ridges, such as would be produced by the continued rubbing together of surfaces of unequal hardness. These polished surfaces are called by miners "slickensides." It is supposed that the lines of the striae indicate the direction in which the rocks were moved. During one of the minor earthquakes in Chili, in 1840, the brick walls of a building were rent vertically in several places, and made to vibrate for several minutes during each shock, after which they remained uninjured, and without any opening, although the line of each crack was still visible. When all movement had ceased, there were seen on the floor of the house, at the bottom of each rent, small heaps of fine brick-dust, evidently produced by trituration.

(FIGURE 74. Faults. A B perpendicular, C D oblique to the horizon.)

(FIGURE 75. E F, fault or fissure filled with rubbish, on each side of which the shifted strata are not parallel.)

It is not uncommon to find the mass of rock on one side of a fissure thrown up above or down below the mass with which it was once in contact on the other side. "This mode of displacement is called a fault, shift, slip, or throw." "The miner," says Playfair, describing a fault, "is often perplexed, in his subterranean journey, by a derangement in the strata, which changes at once all those lines and bearings which had hitherto directed his course. When his mine reaches a certain plane, which is sometimes perpendicular, as in A B, Figure 74, sometimes oblique to the horizon (as in C D, ibid.), he finds the beds of rock broken asunder, those on the one side of the plane having changed their place, by sliding in a particular direction along the face of the others. In this motion they have sometimes preserved their parallelism, as in Figure 74, so that the strata on each side of faults A B, C D, continue parallel to one another; in other cases, the strata on each side are inclined, as in a, b, c, d (Figure 75), though their identity is still to be recognised by their possessing the same thickness and the same internal characters." (Playfair, Illustration of Hutt. Theory paragraph 42.)

In Coalbrook Dale, says Mr. Prestwich (Geological Transactions second series volume 5 page 452.), deposits of sandstone, shale, and coal, several thousand feet thick, and occupying an area of many miles, have been shivered into fragments, and the broken remnants have been placed in very discordant positions, often at levels differing several hundred feet from each other. The sides of the faults, when perpendicular, are commonly several yards apart, and are sometimes as much as 50 yards asunder, the interval being filled with broken debris of the strata. In following the course of the same fault it is sometimes found to produce in different places very unequal changes of level, the amount of shift being in one place 300, and in another 700 feet, which arises from the union of two or more faults. In other words, the disjointed strata have in certain districts been subjected to renewed movements, which they have not suffered elsewhere.

We may occasionally see exact counterparts of these slips, on a small scale, in pits of loose sand and gravel, many of which have doubtless been caused by the drying and shrinking of argillaceous and other beds, slight subsidences having taken place from failure of support. Sometimes, however, even these small slips may have been produced during earthquakes; for land has been moved, and its level, relatively to the sea, considerably altered, within the period when much of the alluvial sand and gravel now covering the surface of continents was deposited.

I have already stated that a geologist must be on his guard, in a region of disturbed strata, against inferring repeated alternations of rocks, when, in fact, the same strata, once continuous, have been bent round so as to recur in the same section, and with the same dip. A similar mistake has often been occasioned by a series of faults.

(FIGURE 76. Apparent alternations of strata caused by vertical faults.)

If, for example, the dark line A H (Figure 76) represent the surface of a country on which the strata a, b, c frequently crop out, an observer who is proceeding from H to A might at first imagine that at every step he was approaching new strata, whereas the repetition of the same beds has been caused by vertical faults, or downthrows. Thus, suppose the original mass, A, B, C, D, to have been a set of uniformly inclined strata, and that the different masses under E F, F G, and G D sank down successively, so as to leave vacant the spaces marked in the diagram by dotted lines, and to occupy those marked by the continuous lines, then let denudation take place along the line A H, so that the protruding masses indicated by the fainter lines are swept away— a miner, who has not discovered the faults, finding the mass a, which we will suppose to be a bed of coal four times repeated, might hope to find four beds, workable to an indefinite depth, but first, on arriving at the fault G, he is stopped suddenly in his workings, for he comes partly upon the shale b, and partly on the sandstone c; the same result awaits him at the fault F, and on reaching E he is again stopped by a wall composed of the rock d.

The very different levels at which the separated parts of the same strata are found on the different sides of the fissure, in some faults, is truly astonishing. One of the most celebrated in England is that called the "ninety- fathom dike," in the coal-field of Newcastle.

This name has been given to it, because the same beds are ninety fathoms (540 feet) lower on the northern than they are on the southern side. The fissure has been filled by a body of sand, which is now in the state of sandstone, and is called the dike, which is sometimes very narrow, but in other places more than twenty yards wide. (Conybeare and Phillips Outlines, etc. page 376.) The walls of the fissure are scored by grooves, such as would have been produced if the broken ends of the rock had been rubbed along the plane of the fault. (Phillips Geology Lardner's Cyclop. page 41.) In the Tynedale and Craven faults, in the north of England, the vertical displacement is still greater, and the fracture has extended in a horizontal direction for a distance of thirty miles or more.

GREAT FAULTS THE RESULT OF REPEATED MOVEMENTS.

It must not, however, be supposed that faults generally consist of single linear rents; there are usually a number of faults springing off from the main one, and sometimes a long strip of country seems broken up into fragments by sets of parallel and connecting transverse faults. Oftentimes a great line of fault has been repeated, or the movements have been continued through successive periods, so that, newer deposits having covered the old line of displacement, the strata both newer and older have given way along the old line of fracture. Some geologists have considered it necessary to imagine that the upward or downward movement in these cases was accomplished at a single stroke, and not by a series of sudden but interrupted movements. They appear to have derived this idea from a notion that the grooved walls have merely been rubbed in one direction, which is far from being a constant phenomenon. Not only are some sets of striae not parallel to others, but the clay and rubbish between the walls, when squeezed or rubbed, have been streaked in different directions, the grooves which the harder minerals have impressed on the softer being frequently curved and irregular.

(FIGURE 77. Faults and denuded coal-strata, Ashby de la Zouch. (Mammatt.))

The usual absence of protruding masses of rock forming precipices or ridges along the lines of great faults has already been alluded to in explaining Figure 76, and the same remarkable fact is well exemplified in every coal-field which has been extensively worked. It is in such districts that the former relation of the beds which have been shifted is determinable with great accuracy. Thus in the coal-field of Ashby de la Zouch, in Leicestershire (see Figure 77), a fault occurs, on one side of which the coal-beds a, b, c, d must once have risen to the height of 500 feet above the corresponding beds on the other side. But the uplifted strata do not stand up 500 feet above the general surface; on the contrary, the outline of the country, as expressed by the line z z, is uniformly undulating, without any break, and the mass indicated by the dotted outline must have been washed away. (See Mammatt's Geological Facts etc. page 90 and plate.)

The student may refer to Mr. Hull's measurement of faults, observed in the Lancashire coal-field, where the vertical displacement has amounted to thousands of feet, and yet where all the superficial inequalities which must have resulted from such movements have been obliterated by subsequent denudation. In the same memoir proofs are afforded of there having been two periods of vertical movement in the same fault— one, for example, before, and another after, the Triassic epoch. (Hull Quarterly Geological Journal volume 24 page 318. 1868.)

The shifting of the beds by faults is often intimately connected with those same foldings which constitute the anticlinal and synclinal axes before alluded to, and there is no doubt that the subterranean causes of both forms of disturbance are to a great extent the same. A fault in Virginia, believed to imply a displacement of several thousand feet, has been traced for more than eighty miles in the same direction as the foldings of the Appalachian chain. (H.D. Rogers Geology of Pennsylvania page 897.) An hypothesis which attributes such a change of position to a succession of movements, is far preferable to any theory which assumes each fault to have been accomplished by a single upcast or downthrow of several thousand feet. For we know that there are operations now in progress, at great depths in the interior of the earth, by which both large and small tracts of ground are made to rise above and sink below their former level, some slowly and insensibly, others suddenly and by starts, a few feet or yards at a time; whereas there are no grounds for believing that, during the last 3000 years at least, any regions have been either upheaved or depressed, at a single stroke, to the amount of several hundred, much less several thousand feet.

It is certainly not easy to understand how in the subterranean regions one mass of solid rock should have been folded up by a continued series of movements, while another mass in contact, or only separated by a line of fissure, has remained stationary or has perhaps subsided. But every volcano, by the intermittent action of the steam, gases, and lava evolved during an eruption, helps us to form some idea of the manner in which such operations take place. For eruptions are repeated at uncertain intervals throughout the whole or a large part of a geological period, some of the surrounding and contiguous districts remaining quite undisturbed. And in most of the instances with which we are best acquainted the emission of lava, scoria, and steam is accompanied by the uplifting of the solid crust. Thus in Vesuvius, Etna, the Madeiras, the Canary Islands, and the Azores there is evidence of marine deposits of recent and tertiary date having been elevated to the height of a thousand feet, and sometimes more, since the commencement of the volcanic explosions. There is, moreover, a general tendency in contemporaneous volcanic vents to affect a linear arrangement, extending in some instances, as in the Andes or the Indian Archipelago, to distances equalling half the circumference of the globe. Where volcanic heat, therefore, operates at such a depth as not to obtain vent at the surface, in the form of an eruption, it may nevertheless be conceived to give rise to upheavals, foldings, and faults in certain linear tracts. And marine denudation, to be treated of in the next chapter, will help us to understand why that which should be the protruding portion of the faulted rocks is missing at the surface.

ARRANGEMENT AND DIRECTION OF PARALLEL FOLDS OF STRATA.

The possible causes of the folding of strata by lateral movements have been considered in a former part of this chapter. No European chain of mountains affords so remarkable an illustration of the persistency of such flexures for a great distance as the Appalachians before alluded to, and none has been studied and described by many good observers with more accuracy. The chain extends from north to south, or rather N.N.E. to S.S.W., for nearly 1500 miles, with a breadth of 50 miles, throughout which the Palaeozoic strata have been so bent as to form a series of parallel anticlinal and synclinal ridges and troughs, comprising usually three or four principal and many smaller plications, some of them forming broad and gentle arches, others narrower and steeper ones, while some, where the bending has been greatest, have the position of their beds inverted, as before shown in Figure 73.

The strike of the parallel ridges, after continuing in a straight line for many hundred miles, is then found to vary for a more limited distance as much as 30 degrees, the folds wheeling round together in the new direction and continuing to be parallel, as if they had all obeyed the same movement. The date of the movements by which the great flexures were brought about must, of course, be subsequent to the formation of the uppermost part of the coal or the newest of the bent rocks, but the disturbance must have ceased before the Triassic strata were deposited on the denuded edges of the folded beds.

The manner in which the numerous parallel folds, all simultaneously formed, assume a new direction common to the whole of them, and sometimes varying at an angle of 30 degrees from the normal strike of the chain, shows what deviation from an otherwise uniform strike of the beds may be experienced when the geographical area through which they are traced is on so vast a scale.

The disturbances in the case here adverted to occurred between the Carboniferous period and that of the Trias, and this interval is so vast that they may have occupied a great lapse of time, during which their parallelism was always preserved. But, as a rule, wherever after a long geological interval the recurrence of lateral movements gives rise to a new set of folds, the strike of these last is different. Thus, for example, Mr. Hull has pointed out that three principal lines of disturbance, all later than the Carboniferous period, have affected the stratified rocks of Lancashire. The first of these, having an E.N.E. direction, took place at the close of the Carboniferous period. The next, running north and south, at the close of the

Permian, and the third, having a N.N.W. direction, at the close of the Jurassic period. (Edward Hull Quarterly Geological Journal volume 24 page 323.)

UNCONFORMABILITY OF STRATA.

(FIGURE 78. Unconformable junction of old red sandstone and Silurian schist at the Siccar Point, near St. Abb's Head, Berwickshire.)

Strata are said to be unconformable when one series is so placed over another that the planes of the superior repose on the edges of the inferior (see Figure 78.) In this case it is evident that a period had elapsed between the production of the two sets of strata, and that, during this interval, the older series had been tilted and disturbed. Afterwards the upper series was thrown down in horizontal strata upon it. If these superior beds, d d Figure 78, are also inclined, it is plain that the lower strata a a, have been twice displaced; first, before the deposition of the newer beds, d d, and a second time when these same strata were upraised out of the sea, and thrown slightly out of the horizontal position.

(FIGURE 79. Junction of unconformable strata near Mons, in Belgium.)

It often happens that in the interval between the deposition of two sets of unconformable strata, the inferior rock has not only been denuded, but drilled by perforating shells. Thus, for example, at Autreppe and Gusigny, near Mons, beds of an ancient (primary or palaeozoic) limestone, highly inclined, and often bent, are covered with horizontal strata of greenish and whitish marls of the Cretaceous formation. The lowest, and therefore the oldest, bed of the horizontal series is usually the sand and conglomerate, a, in which are rounded fragments of stone, from an inch to two feet in diameter. These fragments have often adhering shells attached to them, and have been bored by perforating mollusca. The solid surface of the inferior limestone has also been bored, so as to exhibit cylindrical and pear-shaped cavities, as at c, the work of saxicavous mollusca; and many rents, as at b, which descend several feet or yards into the limestone, have been filled with sand and shells, similar to those in the stratum a.

OVERLAPPING STRATA.

Strata are said to overlap when an upper bed extends beyond the limits of a lower one. This may be produced in various ways; as, for example, when alterations of physical geography cause the arms of a river or channels of discharge to vary, so that sediment brought down is deposited over a wider area than before, or when the sea-bottom has been raised up and again depressed without disturbing the horizontal position of the strata. In this case the newer strata may rest for the most part conformably on the older, but, extending farther, pass over their edges. Every intermediate state between unconformable and overlapping beds may occur, because there may be every gradation between a slight derangement of position, and a considerable disturbance and denudation of the older formation before the newer beds come on.

CHAPTER VI.
DENUDATION.

Denudation defined.
Its Amount more than equal to the entire Mass of Stratified Deposits in the
Earth's Crust.
Subaerial Denudation.
Action of the Wind.
Action of Running Water.
Alluvium defined.
Different Ages of Alluvium.
Denuding Power of Rivers affected by Rise or Fall of Land.
Littoral Denudation.
Inland Sea-Cliffs.
Escarpments.
Submarine Denudation.
Dogger-bank.
Newfoundland Bank.
Denuding Power of the Ocean during Emergence of Land.

Denudation, which has been occasionally spoken of in the preceding chapters, is the removal of solid matter by water in motion, whether of rivers or of the waves and currents of the sea, and the consequent laying bare of some inferior rock. This operation has exerted an influence on the structure of the earth's crust as universal and important as sedimentary deposition itself; for denudation is the necessary antecedent of the production of all new strata of mechanical origin. The formation of every new deposit by the transport of sediment and pebbles necessarily implies that there has been, somewhere else, a grinding down of rock into rounded fragments, sand, or mud, equal in quantity to the new strata. All deposition, therefore, except in the case of a shower of volcanic ashes, and the outflow of lava, and the growth of certain organic formations, is the sign of superficial waste going on contemporaneously, and to an equal amount, elsewhere. The gain at one point is no more than sufficient to balance the loss at some other. Here a lake has grown shallower, there a ravine has been deepened. Here the depth of the sea has been augmented by the removal of a sandbank during a storm, there its bottom has been raised and shallowed by the accumulation in its bed of the same sand transported from the bank.

When we see a stone building, we know that somewhere, far or near, a quarry has been opened. The courses of stone in the building may be compared to successive strata, the quarry to a ravine or valley which has suffered denudation. As the strata, like the courses of hewn stone, have been laid one upon another gradually, so the excavation both of the valley and quarry have been gradual. To pursue the comparison still farther, the superficial heaps of mud, sand, and gravel, usually called alluvium, may be likened to the rubbish of a quarry which has been rejected as useless by the workmen, or has fallen upon the road between the quarry and the building, so as to lie scattered at random over the ground.

But we occasionally find in a conglomerate large rounded pebbles of an older conglomerate, which had previously been derived from a variety of different rocks. In such cases we are reminded that, the same materials having been used over and over again, it is not enough to affirm that the entire mass of stratified deposits in the earth's crust affords a monument and measure of the denudation which has taken place, for in truth the quantity of matter now extant in the form of stratified rock represents but a fraction of the material removed by water and redeposited in past ages.

SUBAERIAL DENUDATION.

Denudation may be divided into subaerial, or the action of wind, rain, and rivers; and submarine, or that effected by the waves of the sea, and its tides and currents. With the operation of the first of these we are best acquainted, and it may be well to give it our first attention.

ACTION OF THE WIND.

In desert regions where no rain falls, or where, as in parts of the Sahara, the soil is so salt as to be without any covering of vegetation, clouds of dust and sand attest the power of the wind to cause the shifting of the unconsolidated or disintegrated rock.

In examining volcanic countries I have been much struck with the great superficial changes brought about by this power in the course of centuries. The highest peak of Madeira is about 6050 feet above the sea, and consists of the skeleton of a volcanic cone now 250 feet high, the beds of which once dipped from a centre in all directions at an angle of more than 30 degrees. The summit is formed of a dike of basalt with much olivine, fifteen feet wide, apparently the remains of a column of lava which once rose to the crater. Nearly all the scoriae of the upper part of the cone have been swept away, those portions only remaining which were hardened by the contact or proximity of the dike. While I was myself on this peak on January 25, 1854, I saw the wind, though it was not stormy weather, removing sand and dust derived from the decomposing scoriae. There had been frost in the night, and some ice was still seen in the crevices of the rock.

On the highest platform of the Grand Canary, at an elevation of 6000 feet, there is a cylindrical column of hard lava, from which the softer matter has been carried away; and other similar remnants of the dikes of cones of eruption attest the denuding power of the wind at points where running water could never have exerted any influence. The waste effected by wind aided by frost and snow, may not be trifling, even in a single winter, and when multiplied by centuries may become indefinitely great.

ACTION OF RUNNING WATER.

(FIGURE 80. Section through several eroded formations. a. Older alluvium or drift. b. Modern alluvium.)

There are different classes of phenomena which attest in a most striking manner the vast spaces left vacant by the erosive power of water. I may allude, first, to those valleys on both sides of which the same strata are seen following each other in the same order, and having the same mineral composition and fossil contents. We may observe, for example, several formations, as Nos. 1, 2, 3, 4, in the diagram (Figure 80): No. 1, conglomerate, No. 2, clay, No. 3, grit, and No. 4, limestone, each repeated in a series of hills separated by valleys varying in depth. When we examine the subordinate parts of these four formations, we find, in like manner, distinct beds in each, corresponding, on the opposite sides of the valleys, both in composition and order of position. No one can doubt that the strata were originally continuous, and that some cause has swept away the portions which once connected the whole series. A torrent on the side of a mountain produces similar interruptions; and when we make artificial cuts in lowering roads, we expose, in like manner, corresponding beds on either side. But in nature, these appearances occur in mountains several thousand feet high, and separated by intervals of many miles or leagues in extent.

In the "Memoirs of the Geological Survey of Great Britain" (volume 1), Professor Ramsay has shown that the missing beds, removed from the summit of the Mendips, must have been nearly a mile in thickness; and he has pointed out considerable areas in South Wales and some of the adjacent counties of England, where a series of primary (or palaeozoic) strata, no less than 11,000 feet in thickness, have been stripped off. All these materials have of course been transported to new regions, and have entered into the composition of more modern formations. On the other hand, it is shown by observations in the same "Survey," that the Palaeozoic strata are from 20,000 to 30,000 feet thick. It is clear that such rocks, formed of mud and sand, now for the most part consolidated, are the monuments of denuding operations, which took place on a grand scale at a very remote period in the earth's history. For, whatever has been given to one area must always have been borrowed from another; a truth which, obvious as it may seem when thus stated, must be repeatedly impressed on the student's mind, because in many geological speculations it is taken for granted that the external crust of the earth has been always growing thicker in consequence of the accumulation, period after period, of sedimentary matter, as if the new strata were not always produced at the expense of pre-existing rocks, stratified or unstratified. By duly reflecting on the fact that all deposits of mechanical origin imply the transportation from some other region, whether contiguous or remote, of an equal amount of solid matter, we perceive that the stony exterior of the planet must always have grown thinner in one place, whenever, by accessions of new strata, it was acquiring thickness in another.

It is well known that generally at the mouths of large rivers, deltas are forming and the land is encroaching upon the sea; these deltas are monuments of recent denudation and deposition; and it is obvious that if the mud, sand, and gravel were taken from them and restored to the continents they would fill up a large part of the gullies and valleys which are due to the excavating and transporting power of torrents and rivers.

ALLUVIUM.

Between the superficial covering of vegetable mould and the subjacent rock there usually intervenes in every district a deposit of loose gravel, sand, and mud, to which when it occurs in valleys the name of alluvium has been popularly applied. The term is derived from alluvio, an inundation, or alluo, to wash, because the pebbles and sand commonly resemble those of a river's bed or the mud and gravel washed over low lands by a flood.

In the course of those changes in physical geography which may take place during the gradual emergence of the bottom of the sea and its conversion into dry land, any spot may either have been a sunken reef, or a bay, or estuary, or sea-shore, or the bed of a river. The drainage, moreover, may have been deranged again and again by earthquakes, during which temporary lakes are caused by landslips, and partial deluges occasioned by the bursting of the barriers of such lakes. For this reason it would be unreasonable to hope that we should ever be able to account for all the alluvial phenomena of each particular country, seeing that the causes of their origin are so various. Besides, the last operations of water have a tendency to disturb and confound together all pre-existing alluviums. Hence we are always in danger of regarding as the work of a single era, and the effect of one cause, what has in reality been the result of a variety of distinct agents, during a long succession of geological epochs. Much useful instruction may therefore be gained from the exploration of a country like Auvergne, where the superficial gravel of very different eras happens to have been preserved and kept separate by sheets of lava, which were poured out one after the other at periods when the denudation, and probably the upheaval, of rocks were in progress. That region had already acquired in some degree its present configuration before any volcanoes were in activity, and before any igneous matter was superimposed upon the granitic and fossiliferous formations. The pebbles therefore in the older gravels are exclusively constituted of granite and other aboriginal rocks; and afterwards, when volcanic vents burst forth into eruption, those earlier alluviums were covered by streams of lava, which protected them from intermixture with gravel of subsequent date. In the course of ages, a new system of valleys was excavated, so that the rivers ran at lower levels than those at which the first alluviums and sheets of lava were formed. When, therefore, fresh eruptions gave rise to new lava, the melted matter was poured out over lower grounds; and the gravel of these plains differed from the first or upland alluvium, by containing in it rounded fragments of various volcanic rocks, and often fossil bones belonging to species of land animals different from those which had previously flourished in the same country and been buried in older gravels.

(FIGURE 81. Lavas of Auvergne resting on alluviums of different ages.)

Figure 81 will explain the different heights at which beds of lava and gravel, each distinct from the other in composition and age, are observed, some on the flat tops of hills, 700 or 800 feet high, others on the slope of the same hills, and the newest of all in the channel of the existing river where there is usually gravel alone, although in some cases a narrow strip of solid lava shares the bottom of the valley with the river.

The proportion of extinct species of quadrupeds is more numerous in the fossil remains of the gravel No. 1 than in that indicated as No. 2; and in No. 3 they agree more closely, sometimes entirely, with those of the existing fauna. The usual absence or rarity of organic remains in beds of loose gravel and sand is partly owing to the friction which originally ground down the rocks into small fragments, and partly to the porous nature of alluvium, which allows the free percolation through it of rain-water, and promotes the decomposition and removal of fossil remains.

The loose transported matter on the surface of a large part of the land now existing in the temperate and arctic regions of the northern hemisphere, must be regarded as being in a somewhat exceptional state, in consequence of the important part which ice has played in comparatively modern geological times. This subject will be more specially alluded to when we describe, in the eleventh chapter, the deposits called "glacial."

DENUDING POWER OF RIVERS AFFECTED BY RISE OR FALL OF LAND.

It has long been a matter of common observation that most rivers are now cutting their channels through alluvial deposits of greater depth and extent than could ever have been formed by the present streams. From this fact it has been inferred that rivers in general have grown smaller, or become less liable to be flooded than formerly. It may be true that in the history of almost every country the rivers have been both larger and smaller than they are at the present moment. For the rainfall in particular regions varies according to climate and physical geography, and is especially governed by the elevation of the land above the sea, or its distance from it and other conditions equally fluctuating in the course of time. But the phenomenon alluded to may sometimes be accounted for by oscillations in the level of the land, experienced since the existing valleys originated, even where no marked diminution in the quantity of rain and in the size of the rivers has occurred.

We know that many large areas of land are rising and others sinking, and unless it could be assumed that both the upward and downward movements are everywhere uniform, many of the existing hydrographical basins ought to have the appearance of having been temporary lakes first filled with fluviatile strata and then partially re-excavated.

Suppose, for example, part of a continent, comprising within it a large hydrographical basin like that of the Mississippi, to subside several inches or feet in a century, as the west coast of Greenland, extending 600 miles north and south, has been sinking for three or four centuries, between the latitudes 60 and 69 degrees N. (Principles of Geology 7th edition page 506; 10th edition volume 2 page 196.) It will rarely happen that the rate of subsidence will be everywhere equal, and in many cases the amount of depression in the interior will regularly exceed that of the region nearer the sea. Whenever this happens, the fall of the waters flowing from the upland country will be diminished, and each tributary stream will have less power to carry its sand and sediment into the main river, and the main river less power to convey its annual burden of transported matter to the sea. All the rivers, therefore, will proceed to fill up partially their ancient channels, and, during frequent inundations, will raise their alluvial plains by new deposits. If then the same area of land be again upheaved to its former height, the fall, and consequently the velocity, of every river will begin to augment. Each of them will be less given to overflow its alluvial plain; and their power of carrying earthy matter seaward, and of scouring out and deepening their channels, will be sustained till, after a lapse of many thousand years, each of them has eroded a new channel or valley through a fluviatile formation of comparatively modern date. The surface of what was once the river-plain at the period of greatest depression, will then remain fringing the valley-sides in the form of a terrace apparently flat, but in reality sloping down with the general inclination of the river. Everywhere this terrace will present cliffs of gravel and sand, facing the river. That such a series of movements has actually taken place in the main valley of the Mississippi and in its tributary valleys during oscillations of level, I have endeavoured to show in my description of that country (Second Visit to the United States volume 1 chapter 34.); and the fresh-water shells of existing species and bones of land quadrupeds, partly of extinct races, preserved in the terraces of fluviatile origin, attest the exclusion of the sea during the whole process of filling up and partial re-excavation.

LITTORAL DENUDATION.

Part of the action of the waves between high and low watermark must be included in subaerial denudation, more especially as the undermining of cliffs by the waves is facilitated by land-springs, and these often lead to the sliding down of great masses of land into the sea. Along our coasts we find numerous submerged forests, only visible at low water, having the trunks of the trees erect and their roots attached to them and still spreading through the ancient soil as when they were living. They occur in too many places, and sometimes at too great a depth, to be explained by a mere change in the level of the tides, although as the coasts waste away and alter in shape, the height to which the tides rise and fall is always varying, and the level of high tide at any given point may, in the course of many ages, differ by several feet or even fathoms. It is this fluctuation in the height of the tides, and the erosion and destruction of the sea-coast by the waves, that makes it exceedingly difficult for us in a few centuries, or even perhaps in a few thousand years, to determine whether there is a change by subterranean movement in the relative level of sea and land.

We often behold, as on the coasts of Devonshire and Pembrokeshire, facts which appear to lead to opposite conclusions. In one place a raised beach with marine littoral shells, and in another immediately adjoining a submerged forest. These phenomena indicate oscillations of level, and as the movements are very gradual, they must give repeated opportunities to the breakers to denude the land which is thus again and again exposed to their fury, although it is evident that the submergence is sometimes effected in such a manner as to allow the trees which border the coast not to be carried away.

INLAND SEA-CLIFFS.

In countries where hard limestone rocks abound, inland cliffs have often retained faithfully for ages the characters which they acquired when they constituted the boundary of land and sea. Thus, in the Morea, no less than three or even four ranges of cliffs are well-preserved, rising one above the other at different distances from the actual shore, the summit of the highest and oldest occasionally attaining 1000 feet in elevation. A consolidated beach with marine shells is usually found at the base of each cliff, and a line of littoral caverns. These ranges of cliff probably imply pauses in the process of upheaval when the waves and currents had time to undermine and clear away considerable masses of rock.

But the beginner should be warned not to expect to find evidence of the former sojourn of the sea on all those lands which we are nevertheless sure have been submerged at periods comparatively modern; for notwithstanding the enduring nature of the marks left by littoral action on some rocks, especially limestones, we can by no means detect sea-beaches and inland cliffs everywhere. On the contrary, they are, upon the whole, extremely partial, and are often entirely wanting in districts composed of argillaceous and sandy formations, which must, nevertheless, have been upheaved at the same time, and by the same intermittent movements, as the adjoining harder rocks.

ESCARPMENTS.

Besides the inland cliffs above alluded to which mark the ancient limits of the sea, there are other abrupt terminations of rocks of various kinds which resemble sea-cliffs, but which have in reality been due to subaerial denudation. These have been called "escarpments," a term which it is useful to confine to the outcrop of particular formations having a scarped outline, as distinct from cliffs due to marine action.

I formerly supposed that the steep line of cliff-like slopes seen along the outcrop of the chalk, when we follow the edge of the North or South Downs, was due to marine action; but Professor Ramsay has shown (Physical Geography and Geology of Great Britain page 78 1864.) that the present outline of the physical geography is more in favour of the idea of the escarpments having been due to gradual waste since the rocks were exposed in the atmosphere to the action of rain and rivers.

Mr. Whittaker has given a good summary of the grounds for ascribing these apparent sea-cliffs to waste in the open air. 1. There is an absence of all signs of ancient sea-beaches or littoral deposits at the base of the escarpment. 2. Great inequality is observed in the level of the base line. 3. The escarpments do not intersect, like sea-cliffs, a series of distinct rocks, but are always confined to the boundary-line of the same formation. 4. There are sometimes different contiguous and parallel escarpments— those, for example, of the greensand and chalk— which are so near each other, and occasionally so similar in altitude, that we can not imagine any existing archipelago if converted into dry land to present a like outline.

The above theory is by no means inconsistent with the opinion that the limits of the outcrop of the chalk and greensand which the escarpments now follow, were originally determined by marine denudation. When the south-east of England last emerged from beneath the level of the sea, it was acted upon, no doubt, by the tide, waves, and currents, and the chalk would form from the first a mass projecting above the more destructible clay called Gault. Still the present escarpments so much resembling sea-cliffs have no doubt, for reasons above stated, derived their most characteristic features subsequently to emergence from subaerial waste by rain and rivers.

SUBMARINE DENUDATION.

When we attempt to estimate the amount of submarine denudation, we become sensible of the disadvantage under which we labour from our habitual incapacity of observing the action of marine currents on the bed of the sea. We know that the agitation of the waves, even during storms, diminishes at a rapid rate, so as to become very insignificant at the depth of a few fathoms, and is quite imperceptible at the depth of about sixteen fathoms; but when large bodies of water are transferred by a current from one part of the ocean to another, they are known to maintain at great depths such a velocity as must enable them to remove the finer, and sometimes even the coarser, materials of the rocks over which they flow. As the Mississippi when more than 150 feet deep can keep open its channel and even carry down gravel and sand to its delta, the surface velocity being not more than two or three miles an hour, so a gigantic current, like the Gulf Stream, equal in volume to many hundred Mississippis, and having in parts a surface velocity of more than three miles, may act as a propelling and abrading power at still greater depths. But the efficacy of the sea as a denuding agent, geologically considered, is not dependent on the power of currents to preserve at great depths a velocity sufficient to remove sand and mud, because, even where the deposition or removal of sediment is not in progress, the depth of water does not remain constant throughout geological time. Every page of the geological record proves to us that the relative levels of land and sea, and the position of the ocean and of continents and islands, has been always varying, and we may feel sure that some portions of the submarine area are now rising and others sinking. The force of tidal and other currents and of the waves during storms is sufficient to prevent the emergence of many lands, even though they may be undergoing continual upheaval. It is not an uncommon error to imagine that the waste of sea-cliffs affords the measure of the amount of marine denudation of which it probably constitutes an insignificant portion.

DOGGER-BANK.

That great shoal called the Dogger-bank, about sixty miles east of the coast of Northumberland, and occupying an area about as large as Wales, has nowhere a depth of more than ninety feet, and in its shallower parts is less than forty feet under water. It might contribute towards the safety of the navigation of our seas to form an artificial island, and to erect a light-house on this bank; but no engineer would be rash enough to attempt it, as he would feel sure that the ocean in the first heavy gale would sweep it away as readily as it does every temporary shoal that accumulates from time to time around a sunk vessel on the same bank. (Principles 10th edition volume 1 page 569.)

No observed geographical changes in historical times entitle us to assume that where upheaval may be in progress it proceeds at a rapid rate. Three or four feet rather than as many yards in a century may probably be as much as we can reckon upon in our speculations; and if such be the case, the continuance of the upward movement might easily be counteracted by the denuding force of such currents aided by such waves as, during a gale, are known to prevail in the German Ocean. What parts of the bed of the ocean are stationary at present, and what areas may be rising or sinking, is a matter of which we are very ignorant, as the taking of accurate soundings is but of recent date.

NEWFOUNDLAND BANK.

The great bank of Newfoundland may be compared in size to the whole of England. This part of the bottom of the Atlantic is surrounded on three sides by a rapidly deepening ocean, the bank itself being from twenty to fifty fathoms (or from 120 to 300 feet) under water. We are unable to determine by the comparison of different charts made at distant periods, whether it is undergoing any change of level, but if it be gradually rising we can not anticipate on that account that it will become land, because the breakers in an open sea would exercise a prodigious force even on solid rock brought up to within a few yards of the surface. We know, for example, that when a new volcanic island rose in the Mediterranean in 1831, the waves were capable in a few years of reducing it to a sunken rock.

In the same way currents which flow over the Newfoundland bank a great part of the year at the rate of two miles an hour, and are known to retain a considerable velocity to near the bottom, may carry away all loose sand and mud, and make the emergence of the shoal impossible, in spite of the accessions of mud, sand, and boulders derived occasionally from melting icebergs which, coming from the northern glaciers, are frequently stranded on various parts of the bank. They must often leave at the bottom large erratic blocks which the marine currents may be incapable of moving, but the same rocky fragments may be made to sink by the undermining of beds consisting of finer matter on which the blocks and gravel repose. In this way gravel and boulders may continue to overspread a submarine bottom after the latter has been lowered for hundreds of feet, the surface never having been able to emerge and become land. It is by no means improbable that the annual removal of an average thickness of half an inch of rock might counteract the ordinary upheaval which large submarine areas are undergoing; and the real enigma which the geologist has to solve is not the extensive denudation of the white chalk or of our tertiary sands and clays, but the fact that such incoherent materials have ever succeeded in lifting up their heads above water in an open sea. Why were they not swept away during storms into some adjoining abysses, the highest parts of each shoal being always planed off down to the depth of a few fathoms? The hardness and toughness of some rocks already exposed to windward and acting as breakwaters may perhaps have assisted; nor must we forget the protection afforded by a dense and unbroken covering of barnacles, limpets, and other creatures which flourish most between high and low water and shelter some newly risen coasts from the waves.

CHAPTER VII.

JOINT ACTION OF DENUDATION, UPHEAVAL, AND SUBSIDENCE IN REMODELLING THE EARTH'S CRUST.

How we obtain an Insight at the Surface, of the Arrangement of Rocks at great Depths.
Why the Height of the successive Strata in a given Region is so disproportionate to their Thickness.
Computation of the average annual Amount of subaerial Denudation.
Antagonism of Volcanic Force to the Levelling Power of running Water.
How far the Transfer of Sediment from the Land to a neighbouring Sea-bottom may affect Subterranean Movements.
Permanence of Continental and Oceanic Areas.

HOW WE OBTAIN AN INSIGHT AT THE SURFACE, OF THE ARRANGEMENT OF ROCKS AT GREAT DEPTHS.

The reader has been already informed that, in the structure of the earth's crust, we often find proofs of the direct superposition of marine to fresh-water strata, and also evidence of the alternation of deep-sea and shallow-water formations. In order to explain how such a series of rocks could be made to form our present continents and islands, we have not only to assume that there have been alternate upward and downward movements of great vertical extent, but that the upheaval in the areas which we at present inhabit has, in later geological times, sufficiently predominated over subsidence to cause these portions of the earth's crust to be land instead of sea. The sinking down of a delta beneath the sea-level may cause strata of fluviatile or even terrestrial origin, such as peat with trees proper to marshes, to be covered by deposits of deep-sea origin. There is also no end to the thickness of mud and sand which may accumulate in shallow water, provided that fresh sediment is brought down from the wasting land at a rate corresponding to that of the sinking of the bed of the sea. The latter, again, may sometimes sink so fast that the earthy matter, being intercepted in some new landward depression, may never reach its former resting-place, where, the water becoming clear may favour the growth of shells and corals, and calcareous rocks of organic origin may thus be superimposed on mechanical deposits.

The succession of strata here alluded to would be consistent with the occurrence of gradual downward and upward movements of the land and bed of the sea without any disturbance of the horizontality of the several formations. But the arrangement of rocks composing the earth's crust differs materially from that which would result from a mere series of vertical movements. Had the volcanic forces been confined to such movements, and had the stratified rocks been first formed beneath the sea and then raised above it, without any lateral compression, the geologist would never have obtained an insight into the monuments of various ages, some of extremely remote antiquity.

What we have said in Chapter 5 of dip and strike, of the folding and inversion of strata, of anticlinal and synclinal flexures, and in Chapter 6 of denudation at different periods, whether subaerial or submarine, must be understood before the student can comprehend what may at first seem to him an anomaly, but which it is his business particularly to understand. I allude to the small height above the level of the sea attained by strata often many miles in thickness, and about the chronological succession of which, in one and the same region, there is no doubt whatever. Had stratified rocks in general remained horizontal, the waves of the sea would have been enabled during oscillations of level to plane off entirely the uppermost beds as they rose or sank during the emergence or submergence of the land. But the occurrence of a series of formations of widely different ages, all remaining horizontal and in conformable stratification, is exceptional, and for this reason the total annihilation of the uppermost strata has rarely taken place. We owe, indeed, to the side way movements of LATERAL COMPRESSION those anticlinal and synclinal curves of the beds already described (Figure 55 Chapter 4), which, together with denudation, subaerial and submarine, enable us to investigate the structure of the earth's crust many miles below those points which the miner can reach. I have already shown in Figure 56 Chapter 4, how, at St. Abb's Head, a series of strata of indefinite thickness may become vertical, and then denuded, so that the edges of the beds alone shall be exposed to view, the altitude of the upheaved ridges being reduced to a moderate height above the sea-level; and it may be observed that although the incumbent strata of Old Red Sandstone are in that place nearly horizontal, yet these same newer beds will in other places be found so folded as to present vertical strata, the edges of which are abruptly cut off, as in 2, 3, 4 on the right-hand side of the diagram, Figure 55 Chapter 4.

WHY THE HEIGHT OF THE SUCCESSIVE STRATA IN A GIVEN REGION IS SO DISPROPORTIONATE TO THEIR THICKNESS.

We can not too distinctly bear in mind how dependent we are on the joint action of the volcanic and aqueous forces, the one in disturbing the original position of rocks, and the other in destroying large portions of them, for our power of consulting the different pages and volumes of those stony records of which the crust of the globe is composed. Why, it may be asked, if the ancient bed of the sea has been in many regions uplifted to the height of two or three miles, and sometimes twice that altitude, and if it can be proved that some single formations are of themselves two or three miles thick, do we so often find several important groups resting one upon the other, yet attaining only the height of a few hundred feet above the level of the sea?

The American geologists, after carefully studying the Allegheny or Appalachian mountains, have ascertained that the older fossiliferous rocks of that chain (from the Silurian to the Carboniferous inclusive) are not less than 42,000 feet thick, and if they were now superimposed on each other in the order in which they were thrown down, they ought to equal in height the Himalayas with the Alps piled upon them. Yet they rarely reach an altitude of 5000 feet, and their loftiest peaks are no more than 7000 feet high. The Carboniferous strata forming the highest member of the series, and containing beds of coal, can be shown to be of shallow-water origin, or even sometimes to have originated in swamps in the open air. But what is more surprising, the lowest part of this great Palaeozoic series, instead of having been thrown down at the bottom of an abyss more than 40,000 feet deep, consists of sediment (the Potsdam sandstone), evidently spread out on the bottom of a shallow sea, on which ripple-marked sands were occasionally formed. This vast thickness of 40,000 feet is not obtained by adding together the maximum density attained by each formation in distant parts of the chain, but by measuring the successive groups as they are exposed in a very limited area, and where the denuded edges of the vertical strata forming the parallel folds alluded to in Chapter 5 "crop out" at the surface. Our attention has been called by Mr. James Hall, Palaeontologist of New York, to the fact that these Palaeozoic rocks of the Appalachian chain, which are of such enormous density, where they are almost entirely of mechanical origin, thin out gradually as they are traced to the westward, where evidently the contemporaneous seas allowed organic rocks to be formed by corals, echinoderms, and encrinites in clearer water, and where, although the same successive periods are represented, the total mass of strata from the Silurian to the Carboniferous, instead of being 40,000 is only 4000 feet thick.

A like phenomenon is exhibited in every mountainous country, as, for example, in the European Alps; but we need not go farther than the north of England for its illustration. Thus in Lancashire and central England the thickness of the Carboniferous formation, including the Millstone Grit and Yoredale beds, is computed to be more than 18,000 feet; to this we may add the Mountain Limestone, at least 2000 feet in thickness, and the overlying Permian and Triassic formations, 3000 or 4000 feet thick. How then does it happen that the loftiest hills of Yorkshire and Lancashire, instead of being 24,000 feet high, never rise above 3000 feet? For here, as before pointed out in the Alleghenies, all the great thicknesses are sometimes found in close approximation and in a region only a few miles in diameter. It is true that these same sets of strata do not preserve their full force when followed for indefinite distances. Thus the 18,000 feet of Carboniferous grits and shales in Lancashire, before alluded to, gradually thin out, as Mr. Hull has shown, as they extend southward, by attenuation or original deficiency of sediment, and not in consequence of subsequent denudation, so that when we have followed them for about 100 miles into Leicestershire, they have dwindled away to a thickness of only 3000 feet. In the same region the Carboniferous limestone attains so unusual a thickness— namely, more than 4000 feet— as to appear to compensate in some measure for the deficiency of contemporaneous sedimentary rock. (Hull Quarterly Geological Journal volume 24 page 322 1868.)

(FIGURE 82. Unconformable Palaeozoic strata, Sutherlandshire (Murchison). Queenaig (2673 feet). 1. Laurentian gneiss. 2. Cambrian conglomerate and sandstone. 3, 3'. Quartzose Lower Silurian, with annelid burrows.)

It is admitted that when two formations are unconformable their fossil remains almost always differ considerably. The break in the continuity of the organic forms seems connected with a great lapse of time, and the same interval has allowed extensive disturbance of the strata, and removal of parts of them by denudation, to take place. The more we extend our investigations the more numerous do the proofs of these breaks become, and they extend to the most ancient rocks yet discovered. The oldest examples yet brought to light in the British Isles are on the borders of Rosshire and Sutherlandshire, and have been well described by Sir Roderick Murchison, by whom their chronological relations were admirably worked out, and proved to be very different from those which previous observers had imagined them to be. I had an opportunity in the autumn of 1869 of verifying the splendid section given in Figure 82 by climbing in a few hours from the banks of Loch Assynt to the summit of the mountain called Queenaig, 2673 feet high.

The formations 1, 2, 3, the Laurentian, Cambrian, and Silurian, to be explained in Chapters 25 and 26, not only occur in succession in this one mountain, but their unconformable junctions are distinctly exposed to view.

To begin with the oldest set of rocks, No. 1; they consist chiefly of hornblendic gneiss, and in the neighbouring Hebrides form whole islands, attaining a thickness of thousands of feet, although they have suffered such contortions and denudation that they seldom rise more than a few hundred feet above the sea-level. In discordant stratification upon the edges of this gneiss reposes No. 2, a group of conglomerate and purple sandstone referable to the Cambrian (or Longmynd) formation, which can elsewhere be shown to be characterised by its peculiar organic remains. On this again rests No. 3, a lower member of the important group called Silurian, an outlier of which, 3', caps the summit of Queenaig, attesting the removal by denudation of rocks of the same age, which once extended from the great mass 3 to 3'. Although this rock now consists of solid quartz, it is clear that in its original state it was formed of fine sand, perforated by numerous lob-worms or annelids, which left their burrows in the shape of tubular hollows (Chapter 26, Figure 563 of Arenicolites), hundreds, nay thousands, of which I saw as I ascended the mountain.

(FIGURE 83. Diagrammatic section of the same groups near Queenaig (Murchison) through west (left), Suilvein, Assynt and Ben More, east (right). 1. Laurentian gneiss. 2. Cambrian conglomerate and sandstone. 3, 3'. Quartzose Lower Silurian, with annelid burrows. 3a. Fossiliferous Silurian limestone. 3b. Quartzose, micaceous and gneissose rocks (altered Silurian).)

In Queenaig we only behold this single quartzose member of the Silurian series, but in the neighbouring country (see Figure 83) it is seen to the eastward to be followed by limestones, 3a, and schists, 3b, presenting numerous folds, and becoming more and more metamorphic and crystalline, until at length, although very different in age and strike, they much resemble in appearance the group No. 1. It is very seldom that in the same country one continuous formation, such as the Silurian, is, as in this case, more fossiliferous and less altered by volcanic heat in its older than in its newer strata, and still more rare to find an underlying and unconformable group like the Cambrian retaining its original condition of a conglomerate and sandstone more perfectly than the overlying formation. Here also we may remark in regard to the origin of these Cambrian rocks that they were evidently produced at the expense of the underlying Laurentian, for the rounded pebbles occurring in them are identical in composition and texture with that crystalline gneiss which constitutes the contorted beds of the inferior formation No. 1. When the reader has studied the chapter on metamorphism, and has become aware how much modification by heat, pressure, and chemical action is required before the conversion of sedimentary into crystalline strata can be brought about, he will appreciate the insight which we thus gain into the date of the changes which had already been effected in the Laurentian rocks long before the Cambrian pebbles of quartz and gneiss were derived from them. The Laurentian is estimated by Sir William Logan to amount in Canada to 30,000 feet in thickness. As to the Cambrian, it is supposed by Sir Roderick Murchison that the fragment left in Sutherlandshire is about 3500 feet thick, and in Wales and the borders of Shropshire this formation may equal 10,000 feet, while the Silurian strata No. 3, difficult as it may be to measure them in their various foldings to the eastward, where they have been invaded by intrusive masses of granite, are supposed many times to surpass the Cambrian in volume and density.

But although we are dealing here with stratified rocks, each of which would be several miles in thickness, if they were fully represented, the whole of them do not attain the elevation of a single mile above the level of the sea.

COMPUTATION OF THE AVERAGE ANNUAL AMOUNT OF SUBAERIAL DENUDATION.

The geology of the district above alluded to may assist our imagination in conceiving the extent to which groups of ancient rocks, each of which may in their turn have formed continents and oceanic basins, have been disturbed, folded, and denuded even in the course of a few out of many of those geological periods to which our imperfect records relate. It is not easy for us to overestimate the effects which causes in every day action must produce when the multiplying power of time is taken into account.

Attempts were made by Manfredi in 1736, and afterwards by Playfair in 1802, to calculate the time which it would require to enable the rivers to deliver over the whole of the land into the basin of the ocean. The data were at first too imperfect and vague to allow them even to approximate to safe conclusions. But in our own time similar investigations have been renewed with more prospect of success, the amount brought down by many large rivers to the sea having been more accurately ascertained. Mr. Alfred Tylor, in 1850, inferred that the quantity of detritus now being distributed over the sea-bottom would, at the end of 10,000 years, cause an elevation of the sea-level to the extent of at least three inches. (Tylor Philosophical Magazine 4th series page 268 1850.) Subsequently Mr. Croll, in 1867, and again, with more exactness, in 1868, deduced from the latest measurement of the sediment transported by European and American rivers the rate of subaerial denudation to which the surface of large continents is exposed, taking especially the hydrographical basin of the Mississippi as affording the best available measure of the average waste of the land. The conclusion arrived at in his able memoir was that the whole terrestrial surface is denuded at the rate of one foot in 6000 years (Croll Philosophical Magazine 1868 page 381.), and this opinion was simultaneously enforced by his fellow-labourer, Mr. Geikie, who, being jointly engaged in the same line of inquiry, published a luminous essay on the subject in 1868.

The student, by referring to my "Principles of Geology" (Volume 1 page 442 1867.) may see that Messrs. Humphrey and Abbot, during their survey of the Mississippi, attempted to make accurate measurements of the proportion of sediment carried down annually to the sea by that river, including not only the mud held in suspension, but also the sand and gravel forced along the bottom.

It is evident that when we know the dimensions of the area which is drained, and the annual quantity of earthy matter taken from it and borne into the sea, we can affirm how much on an average has been removed from the general surface in one year, and there seems no danger of our overrating the mean rate of waste by selecting the Mississippi as our example, for that river drains a country equal to more than half the continent of Europe, extends through twenty degrees of latitude, and therefore through regions enjoying a great variety of climate, and some of its tributaries descend from mountains of great height. The Mississippi is also more likely to afford us a fair test of ordinary denudation, because, unlike the St. Lawrence and its tributaries, there are no great lakes in which the fluviatile sediment is thrown down and arrested in its way to the sea. In striking a general average we have to remember that there are large deserts in which there is scarcely any rainfall, and tracts which are as rainless as parts of Peru, and these must not be neglected as counterbalancing others, in the tropics, where the quantity of rain is in excess. If then, argues Mr. Geikie, we assume that the Mississippi is lowering the surface of the great basin which it drains at the rate of one foot in 6000 years, 10 feet in 60,000 years, 100 feet in 600,000 years, and 1000 feet in 6,000,000 years, it would not require more than about 4,500,000 years to wear away the whole of the North American continent if its mean height is correctly estimated by Humboldt at 748 feet. And if the mean height of all the land now above the sea throughout the globe is 1000 feet, as some geographers believe, it would only require six million years to subject a mass of rock equal in volume to the whole of the land to the action of subaerial denudation. It may be objected that the annual waste is partial, and not equally derived from the general surface of the country, inasmuch as plains, water-sheds, and level ground at all heights remain comparatively unaltered; but this, as Mr. Geikie has well pointed out, does not affect our estimate of the sum total of denudation. The amount remains the same, and if we allow too little for the loss from the surface of table-lands we only increase the proportion of the loss sustained by the sides and bottoms of the valleys, and vice versa. (Transactions of the Geological Society Glasgow volume 3 page 169.)

ANTAGONISM OF VOLCANIC FORCE TO THE LEVELLING POWER OF RUNNING WATER.

In all these estimates it is assumed that the entire quantity of land above the sea-level remains on an average undiminished in spite of annual waste. Were it otherwise the subaerial denudation would be continually lessened by the diminution of the height and dimensions of the land exposed to waste. Unfortunately we have as yet no accurate data enabling us to measure the action of that force by which the inequalities of the surface of the earth's crust may be restored, and the height of the continents and depth of the seas made to continue unimpaired. I stated in 1830 in the "Principles of Geology" (1st edition chapter 10 page 167 1830; see also 10th edition volume 1 chapter 15 page 327 1867.), that running water and volcanic action are two antagonistic forces; the one labouring continually to reduce the whole of the land to the level of the sea, the other to restore and maintain the inequalities of the crust on which the very existence of islands and continents depends. I stated, however, that when we endeavour to form some idea of the relation of these destroying and renovating forces, we must always bear in mind that it is not simply by upheaval that subterranean movements can counteract the levelling force of running water. For whereas the transportation of sediment from the land to the ocean would raise the general sea-level, the subsidence of the sea-bottom, by increasing its capacity, would check this rise and prevent the submergence of the land. I have, indeed, endeavoured to show that unless we assume that there is, on the whole, more subsidence than upheaval, we must suppose the diameter of the planet to be always increasing, by that quantity of volcanic matter which is annually poured out in the shape of lava or ashes, whether on the land or in the bed of the sea, and which is derived from the interior of the earth. The abstraction of this matter causes, no doubt, subterranean vacuities and a corresponding giving way of the surface; if it were not so, the average density of parts of the interior would be always lessening and the size of the planet increasing. (Principles volume 2 page 237; also 1st edition page 447 1830.)

Our inability to estimate the amount or direction of the movements due to volcanic power by no means renders its efficacy as a land-preserving force in past times a mere matter of conjecture. The student will see in Chapter 24 that we have proofs of Carboniferous forests hundreds of miles in extent which grew on the lowlands or deltas near the sea, and which subsided and gave place to other forests, until in some regions fluviatile and shallow-water strata with occasional seams of coal were piled one over the other, till they attained a thickness of many thousand feet. Such accumulations, observed in Great Britain and America on opposite sides of the Atlantic, imply the long-continued existence of land vegetation, and of rivers draining a former continent placed where there is now deep sea.

It will be also seen in Chapter 25 that we have evidence of a rich terrestrial flora, the Devonian, even more ancient than the Carboniferous; while on the other hand, the later Triassic, Oolitic, Cretaceous, and successive Tertiary periods have all supplied us with fossil plants, insects, or terrestrial mammalia; showing that, in spite of great oscillations of level and continued changes in the position of land and sea, the volcanic forces have maintained a due proportion of dry land. We may appeal also to fresh-water formations, such as the Purbeck and Wealden, to prove that in the Oolitic and Neocomian eras there were rivers draining ancient lands in Europe in times when we know that other spaces, now above water, were submerged.

HOW FAR THE TRANSFER OF SEDIMENT FROM THE LAND TO A NEIGHBOURING SEA-BOTTOM MAY AFFECT SUBTERRANEAN MOVEMENTS.

Little as we understand at present the laws which govern the distribution of volcanic heat in the interior and crust of the globe, by which mountain chains, high table-lands, and the abysses of the ocean are formed, it seems clear that this heat is the prime mover on which all the grander features in the external configuration of the planet depend.

It has been suggested that the stripping off by denudation of dense masses from one part of a continent and the delivery of the same into the bed of the ocean must have a decided effect in causing changes of temperature in the earth's crust below, or, in other words, in causing the subterranean isothermals to shift their position. If this be so, one part of the crust may be made to rise, and another to sink, by the expansion and contraction of the rocks, of which the temperature is altered.

I can not, at present, discuss this subject, of which I have treated more fully elsewhere (Principles volume 2 page 229 1868.), but may state here that I believe this transfer of sediment to play a very subordinate part in modifying those movements on which the configuration of the earth's crust depends. In order that strata of shallow-water origin should be able to attain a thickness of several thousand feet, and so come to exert a considerable downward pressure, there must have been first some independent and antecedent causes at work which have given rise to the incipient shallow receptacle in which the sediment began to accumulate. The same causes there continuing to depress the sea-bottom, room would be made for fresh accessions of sediment, and it would only be by a long repetition of the depositing process that the new matter could acquire weight enough to affect the temperature of the rocks far below, so as to increase or diminish their volume.

PERMANENCE OF CONTINENTAL AND OCEANIC AREAS.

If the thickness of more than 40,000 feet of sedimentary strata before alluded to in the Appalachians proves a preponderance of downward movements in Palaeozoic times in a district now forming the eastern border of North America, it also proves, as before hinted, the continued existence and waste of some neighbouring continent, probably formed of Laurentian rocks, and situated where the Atlantic now prevails. Such an hypothesis would be in perfect harmony with the conclusions forced upon us by the study of the present configuration of our continents, and the relation of their height to the depth of the oceanic basins; also to the considerable elevation and extent sometimes reached by drift containing shells of recent species, and still more by the fact of sedimentary strata, several thousand feet thick, as those of central Sicily, or such as flank the Alps and Apennines, containing fossil Mollusca sometimes almost wholly identical with species still living.

I have remarked elsewhere (Principles volume 1 page 265 1867.) that upward and downward movements of 1000 feet or more would turn much land into sea and sea into land in the continental areas and their borders, whereas oscillations of equal magnitude would have no corresponding effect in the bed of the ocean generally, believed as it is to have a mean depth of 15,000 feet, and which, whether this estimate be correct or not, is certainly of great profundity. Subaerial denudation would not of itself lessen the area of the land, but would tend to fill up with sediment seas of moderate depth adjoining the coast. The coarser matter falls to the bottom near the shore in the first still water which it reaches, and whenever the sea-bottom on which this matter has been thrown is slightly elevated, it becomes land, and an upheaval of a thousand feet causes it to attain the mean elevation of continents in general.

Suppose, therefore, we had ascertained that the triturating power of subaerial denudation might in a given time— in three, or six, or a greater number of millions of years— pulverise a volume of rock equal in dimensions to all the present land, we might yet find, could we revisit the earth at the end of such a period, that the continents occupied very much the same position which they held before; we should find the rivers employed in carrying down to the sea the very same mud, sand, and pebbles with which they had been charged in our own time, the superficial alluvial matter as well as a great thickness of sedimentary strata would inclose shells, all or a great part of which we should recognise as specifically identical with those already known to us as living. Every geologist is aware that great as have been the geographical changes in the northern hemisphere since the commencement of the Glacial Period, there having been submergence and re-emergence of land to the extent of 1000 feet vertically, and in the temperate latitudes great vicissitudes of climate, the marine mollusca have not changed, and the same drift which had been carried down to the sea at the beginning of the period is now undergoing a second transportation in the same direction.

As when we have measured a fraction of time in an hour-glass we have only to reverse the position of our chronometer and we make the same sand measure over again the duration of a second equal period, so when the volcanic force has remoulded the form of a continent and the adjoining sea-bottom, the same materials are made to do duty a second time. It is true that at each oscillation of level the solid rocks composing the original continent suffer some fresh denudation, and do not remain unimpaired like the wooden and glass framework of the hour-glass, still the wear and tear suffered by the larger area exposed to subaerial denudation consists either of loose drift or of sedimentary strata, which were thrown down in seas near the land, and subsequently upraised, the same continents and oceanic basins remaining in existence all the while.

From all that we know of the extreme slowness of the upward and downward movements which bring about even slight geographical changes, we may infer that it would require a long succession of geological periods to cause the submarine and supramarine areas to change places, even if the ascending movements in the one region and the descending in the other were continuously in one direction. But we have only to appeal to the structure of the Alps, where there are so many shallow and deep water formations of various ages crowded into a limited area, to convince ourselves that mountain chains are the result of great oscillations of level. High land is not produced simply by uniform upheaval, but by a predominance of elevatory over subsiding movements. Where the ocean is extremely deep it is because the sinking of the bottom has been in excess, in spite of interruptions by upheaval.

Yet persistent as may be the leading features of land and sea on the globe, they are not immutable. Some of the finest mud is doubtless carried to indefinite distances from the coast by marine currents, and we are taught by deep-sea dredgings that in clear water at depths equalling the height of the Alps organic beings may flourish, and their spoils slowly accumulate on the bottom. We also occasionally obtain evidence that submarine volcanoes are pouring out ashes and streams of lava in mid-ocean as well as on land (see Principles volume 2 page 64), and that wherever mountains like Etna, Vesuvius, and the Canary Islands are now the site of eruptions, there are signs of accompanying upheaval, by which beds of ashes full of recent marine shells have been uplifted many hundred feet. We need not be surprised, therefore, if we learn from geology that the continents and oceans were not always placed where they now are, although the imagination may well be overpowered when it endeavours to contemplate the quantity of time required for such revolutions.

We shall have gained a great step if we can approximate to the number of millions of years in which the average aqueous denudation going on upon the land would convey seaward a quantity of matter equal to the average volume of our continents, and this might give us a gauge of the minimum of volcanic force necessary to counteract such levelling power of running water; but to discover a relation between these great agencies and the rate at which species of organic beings vary, is at present wholly beyond the reach of our computation, though perhaps it may not prove eventually to transcend the powers of man.

CHAPTER VIII.
CHRONOLOGICAL CLASSIFICATION OF ROCKS.

Aqueous, Plutonic, volcanic, and metamorphic Rocks considered chronologically.
Terms Primary, Secondary, and Tertiary; Palaeozoic, Mesozoic, and Cainozoic explained.
On the different Ages of the aqueous Rocks.
Three principal Tests of relative Age: Superposition, Mineral Character, and Fossils.
Change of Mineral Character and Fossils in the same continuous Formation.
Proofs that distinct Species of Animals and Plants have lived at successive Periods.
Distinct Provinces of indigenous Species.
Great Extent of single Provinces.

Similar Laws prevailed at successive Geological Periods.
Relative Importance of mineral and palaeontological Characters.
Test of Age by included Fragments.
Frequent Absence of Strata of intervening Periods.
Tabular Views of fossiliferous Strata.

CHRONOLOGY OF ROCKS.

In the first chapter it was stated that the four great classes of rocks, the aqueous, the volcanic, the Plutonic, and the metamorphic, would each be considered not only in reference to their mineral characters, and mode of origin, but also to their relative age. In regard to the aqueous rocks, we have already seen that they are stratified, that some are calcareous, others argillaceous or siliceous, some made up of sand, others of pebbles; that some contain freshwater, others marine fossils, and so forth; but the student has still to learn which rocks, exhibiting some or all of these characters, have originated at one period of the earth's history, and which at another.

To determine this point in reference to the fossiliferous formations is more easy than in any other class, and it is therefore the most convenient and natural method to begin by establishing a chronology for these strata, and then to refer as far as possible to the same divisions, the several groups of Plutonic, volcanic, and metamorphic rocks. Such a system of classification is not only recommended by its greater clearness and facility of application, but is also best fitted to strike the imagination by bringing into one view the contemporaneous revolutions of the inorganic and organic creations of former times. For the sedimentary formations are most readily distinguished by the different species of fossil animals and plants which they inclose, and of which one assemblage after another has flourished and then disappeared from the earth in succession.

In the present work, therefore, the four great classes of rocks, the aqueous, Plutonic, volcanic, and metamorphic, will form four parallel, or nearly parallel, columns in one chronological table. They will be considered as four sets of monuments relating to four contemporaneous, or nearly contemporaneous, series of events. I shall endeavour, in a subsequent chapter on the Plutonic rocks, to explain the manner in which certain masses belonging to each of the four classes of rocks may have originated simultaneously at every geological period, and how the earth's crust may have been continually remodelled, above and below, by aqueous and igneous causes, from times indefinitely remote. In the same manner as aqueous and fossiliferous strata are now formed in certain seas or lakes, while in other places volcanic rocks break out at the surface, and are connected with reservoirs of melted matter at vast depths in the bowels of the earth, so, at every era of the past, fossiliferous deposits and superficial igneous rocks were in progress contemporaneously with others of subterranean and Plutonic origin, and some sedimentary strata were exposed to heat, and made to assume a crystalline or metamorphic structure.

It can by no means be taken for granted, that during all these changes the solid crust of the earth has been increasing in thickness. It has been shown, that so far as aqueous action is concerned, the gain by fresh deposits, and the loss by denudation, must at each period have been equal (see Chapter 6); and in like manner, in the inferior portion of the earth's crust, the acquisition of new crystalline rocks, at each successive era, may merely have counterbalanced the loss sustained by the melting of materials previously consolidated. As to the relative antiquity of the crystalline foundations of the earth's crust, when compared to the fossiliferous and volcanic rocks which they support, I have already stated, in the first chapter, that to pronounce an opinion on this matter is as difficult as at once to decide which of the two, whether the foundations or superstructure of an ancient city built on wooden piles may be the oldest. We have seen that, to answer this question, we must first be prepared to say whether the work of decay and restoration had gone on most rapidly above or below; whether the average duration of the piles has exceeded that of the buildings, or the contrary. So also in regard to the relative age of the superior and inferior portions of the earth's crust; we can not hazard even a conjecture on this point, until we know whether, upon an average, the power of water above, or that of heat below, is most efficacious in giving new forms to solid matter.

The early geologists gave to all the crystalline and non-fossiliferous rocks the name of Primitive or Primary, under the idea that they were formed anterior to the appearance of life upon the earth, while the aqueous or fossiliferous strata were termed Secondary, and alluviums or other superficial deposits, Tertiary. The meaning of these terms, has, however, been gradually modified with advancing knowledge, and they are now used to designate three great chronological divisions under which all geological formations can be classed, each of them being characterised by the presence of distinctive groups of organic remains rather than by any mechanical peculiarities of the strata themselves. If, therefore, we retain the term "primary," it must not be held to designate a set of crystalline rocks some of which have been proved to be even of Tertiary age, but must be applied to all rocks older than the secondary formations. Some geologists, to avoid misapprehension, have introduced the term Palaeozoic for primary, from palaion, "ancient," and zoon, "an organic being," still retaining the terms secondary and tertiary; Mr. Phillips, for the sake of uniformity, has proposed Mesozoic, for secondary, from mesos, "middle," etc.; and Cainozoic, for tertiary, from kainos, "recent," etc.; but the terms primary, secondary, and tertiary have the claim of priority in their favour, and are of corresponding value.

It may perhaps be suggested that some metamorphic strata, and some granites, may be anterior in date to the oldest of the primary fossiliferous rocks. This opinion is doubtless true, and will be discussed in future chapters; but I may here observe, that when we arrange the four classes of rocks in four parallel columns in one table of chronology, it is by no means assumed that these columns are all of equal length; one may begin at an earlier period than the rest, and another may come down to a later point of time, and we may not be yet acquainted with the most ancient of the primary fossiliferous beds, or with the newest of the hypogene.

For reasons already stated, I proceed first to treat of the aqueous or fossiliferous formations considered in chronological order or in relation to the different periods at which they have been deposited.

There are three principal tests by which we determine the age of a given set of strata; first, superposition; secondly, mineral character; and, thirdly, organic remains. Some aid can occasionally be derived from a fourth kind of proof, namely, the fact of one deposit including in it fragments of a pre-existing rock, by which the relative ages of the two may, even in the absence of all other evidence, be determined.

SUPERPOSITION.

The first and principal test of the age of one aqueous deposit, as compared to another, is relative position. It has been already stated, that, where strata are horizontal, the bed which lies uppermost is the newest of the whole, and that which lies at the bottom the most ancient. So, of a series of sedimentary formations, they are like volumes of history, in which each writer has recorded the annals of his own times, and then laid down the book, with the last written page uppermost, upon the volume in which the events of the era immediately preceding were commemorated. In this manner a lofty pile of chronicles is at length accumulated; and they are so arranged as to indicate, by their position alone, the order in which the events recorded in them have occurred.

In regard to the crust of the earth, however, there are some regions where, as the student has already been informed, the beds have been disturbed, and sometimes extensively thrown over and turned upside down. (See Chapter 5.) But an experienced geologist can rarely be deceived by these exceptional cases. When he finds that the strata are fractured, curved, inclined, or vertical, he knows that the original order of superposition must be doubtful, and he then endeavours to find sections in some neighbouring district where the strata are horizontal, or only slightly inclined. Here, the true order of sequence of the entire series of deposits being ascertained, a key is furnished for settling the chronology of those strata where the displacement is extreme.

MINERAL CHARACTER.

The same rocks may often be observed to retain for miles, or even hundreds of miles, the same mineral peculiarities, if we follow the planes of stratification, or trace the beds, if they be undisturbed, in a horizontal direction. But if we pursue them vertically, or in any direction transverse to the planes of stratification, this uniformity ceases almost immediately.

In that case we can scarcely ever penetrate a stratified mass for a few hundred yards without beholding a succession of extremely dissimilar rocks, some of fine, others of coarse grain, some of mechanical, others of chemical origin; some calcareous, others argillaceous, and others siliceous. These phenomena lead to the conclusion that rivers and currents have dispersed the same sediment over wide areas at one period, but at successive periods have been charged, in the same region, with very different kinds of matter. The first observers were so astonished at the vast spaces over which they were able to follow the same homogeneous rocks in a horizontal direction, that they came hastily to the opinion, that the whole globe had been environed by a succession of distinct aqueous formations, disposed round the nucleus of the planet, like the concentric coats of an onion. But, although, in fact, some formations may be continuous over districts as large as half of Europe, or even more, yet most of them either terminate wholly within narrower limits, or soon change their lithological character. Sometimes they thin out gradually, as if the supply of sediment had failed in that direction, or they come abruptly to an end, as if we had arrived at the borders of the ancient sea or lake which served as their receptacle. It no less frequently happens that they vary in mineral aspect and composition, as we pursue them horizontally. For example, we trace a limestone for a hundred miles, until it becomes more arenaceous, and finally passes into sand, or sandstone. We may then follow this sandstone, already proved by its continuity to be of the same age, throughout another district a hundred miles or more in length.

ORGANIC REMAINS.

This character must be used as a criterion of the age of a formation, or of the contemporaneous origin of two deposits in distant places, under very much the same restrictions as the test of mineral composition.

First, the same fossils may be traced over wide regions, if we examine strata in the direction of their planes, although by no means for indefinite distances. Secondly, while the same fossils prevail in a particular set of strata for hundreds of miles in a horizontal direction, we seldom meet with the same remains for many fathoms, and very rarely for several hundred yards, in a vertical line, or a line transverse to the strata. This fact has now been verified in almost all parts of the globe, and has led to a conviction that at successive periods of the past, the same area of land and water has been inhabited by species of animals and plants even more distinct than those which now people the antipodes, or which now co-exist in the arctic, temperate, and tropical zones. It appears that from the remotest periods there has been ever a coming in of new organic forms, and an extinction of those which pre-existed on the earth; some species having endured for a longer, others for a shorter, time; while none have ever reappeared after once dying out. The law which has governed the succession of species, whether we adopt or reject the theory of transmutation, seems to be expressed in the verse of the poet:—

Natura il fece, e poi ruppe la stampa. Ariosto.
Nature made him, and then broke the die.

And this circumstance it is, which confers on fossils their highest value as chronological tests, giving to each of them, in the eyes of the geologist, that authority which belongs to contemporary medals in history.

The same can not be said of each peculiar variety of rock; for some of these, as red marl and red sandstone, for example, may occur at once at the top, bottom, and middle of the entire sedimentary series; exhibiting in each position so perfect an identity of mineral aspect as to be undistinguishable. Such exact repetitions, however, of the same mixtures of sediment have not often been produced, at distant periods, in precisely the same parts of the globe; and even where this has happened, we are seldom in any danger of confounding together the monuments of remote eras, when we have studied their imbedded fossils and their relative position.

ZOOLOGICAL PROVINCES.

It was remarked that the same species of organic remains can not be traced horizontally, or in the direction of the planes of stratifications for indefinite distances. This might have been expected from analogy; for when we inquire into the present distribution of living beings, we find that the habitable surface of the sea and land may be divided into a considerable number of distinct provinces, each peopled by a peculiar assemblage of animals and plants. In the "Principles of Geology," I have endeavoured to point out the extent and probable origin of these separate divisions; and it was shown that climate is only one of many causes on which they depend, and that difference of longitude as well as latitude is generally accompanied by a dissimilarity of indigenous species.

As different seas, therefore, and lakes are inhabited, at the same period, by different aquatic animals and plants, and as the lands adjoining these may be peopled by distinct terrestrial species, it follows that distinct fossils will be imbedded in contemporaneous deposits. If it were otherwise— if the same species abounded in every climate, or in every part of the globe where, so far as we can discover, a corresponding temperature and other conditions favourable to their existence are found— the identification of mineral masses of the same age, by means of their included organic contents, would be a matter of still greater certainty.

Nevertheless, the extent of some single zoological provinces, especially those of marine animals, is very great; and our geological researches have proved that the same laws prevailed at remote periods; for the fossils are often identical throughout wide spaces, and in detached deposits, consisting of rocks varying entirely in their mineral nature.

The doctrine here laid down will be more readily understood, if we reflect on what is now going on in the Mediterranean. That entire sea may be considered as one zoological province; for although certain species of testacea and zoophytes may be very local, and each region has probably some species peculiar to it, still a considerable number are common to the whole Mediterranean. If, therefore, at some future period, the bed of this inland sea should be converted into land, the geologist might be enabled, by reference to organic remains, to prove the contemporaneous origin of various mineral masses scattered over a space equal in area to half of Europe.

Deposits, for example, are well known to be now in progress in this sea in the deltas of the Po, Rhone, Nile, and other rivers, which differ as greatly from each other in the nature of their sediment as does the composition of the mountains which their drain. There are also other quarters of the Mediterranean, as off the coast of Campania, or near the base of Etna, in Sicily, or in the Grecian Archipelago, where another class of rocks is now forming; where showers of volcanic ashes occasionally fall into the sea, and streams of lava overflow its bottom; and where, in the intervals between volcanic eruptions, beds of sand and clay are frequently derived from the waste of cliffs, or the turbid waters of rivers. Limestones, moreover, such as the Italian travertins, are here and there precipitated from the waters of mineral springs, some of which rise up from the bottom of the sea. In all these detached formations, so diversified in their lithological characters, the remains of the same shells, corals, crustacea, and fish are becoming inclosed; or, at least, a sufficient number must be common to the different localities to enable the zoologist to refer them all to one contemporaneous assemblage of species.

There are, however, certain combinations of geographical circumstances which cause distinct provinces of animals and plants to be separated from each other by very narrow limits; and hence it must happen that strata will be sometimes formed in contiguous regions, differing widely both in mineral contents and organic remains. Thus, for example, the testacea, zoophytes, and fish of the Red Sea are, as a group, extremely distinct from those inhabiting the adjoining parts of the Mediterranean, although the two seas are separated only by the narrow isthmus of Suez. Calcareous formations have accumulated on a great scale in the Red Sea in modern times, and fossil shells of existing species are well preserved therein; and we know that at the mouth of the Nile large deposits of mud are amassed, including the remains of Mediterranean species. It follows, therefore, that if at some future period the bed of the Red Sea should be laid dry, the geologist might experience great difficulties in endeavouring to ascertain the relative age of these formations, which, although dissimilar both in organic and mineral characters, were of synchronous origin.

But, on the other hand, we must not forget that the north-western shores of the Arabian Gulf, the plains of Egypt, and the Isthmus of Suez, are all parts of one province of TERRESTRIAL species. Small streams, therefore, occasional land- floods, and those winds which drift clouds of sand along the deserts, might carry down into the Red Sea the same shells of fluviatile and land testacea which the Nile is sweeping into its delta, together with some remains of terrestrial plants and the bones of quadrupeds, whereby the groups of strata before alluded to might, notwithstanding the discrepancy of their mineral composition and MARINE organic fossils, be shown to have belonged to the same epoch.

Yet, while rivers may thus carry down the same fluviatile and terrestrial spoils into two or more seas inhabited by different marine species, it will much more frequently happen that the coexistence of terrestrial species of distinct zoological and botanical provinces will be proved by the identity of the marine beings which inhabited the intervening space. Thus, for example, the land quadrupeds and shells of the valley of the Mississippi, of central America, and of the West India islands differ very considerably, yet their remains are all washed down by rivers flowing from these three zoological provinces into the Gulf of Mexico.

In some parts of the globe, at the present period, the line of demarkation between distinct provinces of animals and plants is not very strongly marked, especially where the change is determined by temperature, as it is in seas extending from the temperate to the tropical zone, or from the temperate to the arctic regions. Here a gradual passage takes place from one set of species to another. In like manner the geologist, in studying particular formations of remote periods, has sometimes been able to trace the gradation from one ancient province to another, by observing carefully the fossils of all the intermediate places. His success in thus acquiring a knowledge of the zoological or botanical geography of very distant eras has been mainly owing to this circumstance, that the mineral character has no tendency to be affected by climate. A large river may convey yellow or red mud into some part of the ocean, where it may be dispersed by a current over an area several hundred leagues in length, so as to pass from the tropics into the temperate zone. If the bottom of the sea be afterwards upraised, the organic remains imbedded in such yellow or red strata may indicate the different animals or plants which once inhabited at the same time the temperate and equatorial regions.

It may be true, as a general rule, that groups of the same species of animals and plants may extend over wider areas than deposits of homogeneous composition; and if so, palaeontological characters will be of more importance in geological classification than the test of mineral composition; but it is idle to discuss the relative value of these tests, as the aid of both is indispensable, and it fortunately happens, that where the one criterion fails, we can often avail ourselves of the other.

TEST BY INCLUDED FRAGMENTS OF OLDER ROCKS.

It was stated, that proof may sometimes be obtained of the relative date of two formations by fragments of an older rock being included in a newer one. This evidence may sometimes be of great use, where a geologist is at a loss to determine the relative age of two formations from want of clear sections exhibiting their true order of position, or because the strata of each group are vertical. In such cases we sometimes discover that the more modern rock has been in part derived from the degradation of the older. Thus, for example, we may find chalk in one part of a country, and in another strata of clay, sand, and pebbles. If some of these pebbles consist of that peculiar flint, of which layers more or less continuous are characteristic of the chalk, and which include fossil shells, sponges, and foraminifera of cretaceous species, we may confidently infer that the chalk was the oldest of the two formations.

CHRONOLOGICAL GROUPS.

The number of groups into which the fossiliferous strata may be separated are more or less numerous, according to the views of classification which different geologists entertain; but when we have adopted a certain system of arrangement, we immediately find that a few only of the entire series of groups occur one upon the other in any single section or district.

(FIGURE 84. Seven fossiliferous groups.)

The thinning out of individual strata was before described (Chapter 2). But let the diagram (Figure 84) represent seven fossiliferous groups, instead of as many strata. It will then be seen that in the middle all the superimposed formations are present; but in consequence of some of them thinning out, No. 2 and No. 5 are absent at one extremity of the section, and No. 4 at the other.

(FIGURE 85. Section South of Bristol (A.C. Ramsay.) Dundry Hill. Length of section 4 miles. a-b. Level of the sea. 1. Inferior Oolite. 2. Lias. 3. New Red Sandstone. 4. Dolomitic or magnesian conglomerate. 5. Upper coal-measures (shales, etc.) 6. Pennant rock (sandstone.) 7. Lower coal-measures (shales, etc.) 8. Carboniferous or mountain limestone. 9. Old Red Sandstone.)

In another diagram (Figure 85), a real section of the geological formations in the neighbourhood of Bristol and the Mendip Hills is presented to the reader, as laid down on a true scale by Professor Ramsay, where the newer groups 1, 2, 3, 4 rest unconformably on the formations 5, 6, 7 and 8. At the southern end of the line of section we meet with the beds No. 3 (the New Red Sandstone) resting immediately on Nos. 7 and 8, while farther north as at Dundry Hill in Somersetshire, we behold eight groups superimposed one upon the other, comprising all the strata from the inferior Oolite, No. 1, to the coal and carboniferous limestone. The limited horizontal extension of the groups 1 and 2 is owing to denudation, as these formations end abruptly, and have left outlying patches to attest the fact of their having originally covered a much wider area.

In order, therefore, to establish a chronological succession of fossiliferous groups, a geologist must begin with a single section in which several sets of strata lie one upon the other. He must then trace these formations, by attention to their mineral character and fossils, continuously, as far as possible, from the starting-point. As often as he meets with new groups, he must ascertain by superposition their age relatively to those first examined, and thus learn how to intercalate them in a tabular arrangement of the whole.

By this means the German, French, and English geologists have determined the succession of strata throughout a great part of Europe, and have adopted pretty generally the following groups, almost all of which have their representatives in the British Islands.

ABRIDGED GENERAL TABLE OF FOSSILIFEROUS STRATA.
    1. RECENT.— POST-TERTIARY.— TERTIARY OR CAINOZOIC.— NEOZOIC.
    2. POST-PLIOCENE.— POST-TERTIARY.— TERTIARY OR CAINOZOIC.— NEOZOIC.
    3. NEWER-PLIOCENE.— PLIOCENE.— TERTIARY OR CAINOZOIC.— NEOZOIC.
    4. OLDER PLIOCENE.— PLIOCENE.— TERTIARY OR CAINOZOIC.— NEOZOIC.
    5. UPPER MIOCENE.— MIOCENE.— TERTIARY OR CAINOZOIC.— NEOZOIC.
    6. LOWER MIOCENE.— MIOCENE.— TERTIARY OR CAINOZOIC.— NEOZOIC.
    7. UPPER EOCENE.— EOCENE.— TERTIARY OR CAINOZOIC.— NEOZOIC.
    8. MIDDLE EOCENE.— EOCENE.— TERTIARY OR CAINOZOIC.— NEOZOIC.
    9. LOWER EOCENE.— EOCENE.— TERTIARY OR CAINOZOIC.— NEOZOIC.
    10. MAESTRICHT BEDS.— CRETACEOUS.— SECONDARY OR MESOZOIC.— NEOZOIC.
    11. WHITE CHALK.— CRETACEOUS.— SECONDARY OR MESOZOIC.— NEOZOIC.
    12. CHLORITIC SERIES.— CRETACEOUS.— SECONDARY OR MESOZOIC.— NEOZOIC.
    13. GAULT.— CRETACEOUS.— SECONDARY OR MESOZOIC.— NEOZOIC.
    14. NEOCOMIAN.— CRETACEOUS.— SECONDARY OR MESOZOIC.— NEOZOIC.

15. WEALDEN.— CRETACEOUS.— SECONDARY OR MESOZOIC.— NEOZOIC.
16. PURBECK BEDS.— JURASSIC.— SECONDARY OR MESOZOIC.— NEOZOIC.
17. PORTLAND STONE.— JURASSIC.— SECONDARY OR MESOZOIC.— NEOZOIC.
18. KIMMERIDGE CLAY.— JURASSIC.— SECONDARY OR MESOZOIC.— NEOZOIC.
19. CORAL RAG.— JURASSIC.— SECONDARY OR MESOZOIC.— NEOZOIC.
20. OXFORD CLAY.— JURASSIC.— SECONDARY OR MESOZOIC.— NEOZOIC.
21. GREAT or BATH OOLITE.— JURASSIC.— SECONDARY OR MESOZOIC.— NEOZOIC.
22. INFERIOR OOLITE.— JURASSIC.— SECONDARY OR MESOZOIC.— NEOZOIC.
23. LIAS.— JURASSIC.— SECONDARY OR MESOZOIC.— NEOZOIC.
24. UPPER TRIAS.— TRIASSIC.— SECONDARY OR MESOZOIC.— NEOZOIC.
25. MIDDLE TRIAS.— TRIASSIC.— SECONDARY OR MESOZOIC.— NEOZOIC.
26. LOWER TRIAS.— TRIASSIC.— SECONDARY OR MESOZOIC.— NEOZOIC.
27. PERMIAN.— PERMIAN.— PRIMARY OR PALAEOZOIC.— PALAEOZOIC.
28. COAL-MEASURES.— CARBONIFEROUS.— PRIMARY OR PALAEOZOIC.— PALAEOZOIC.
29. CARBONIFEROUS LIMESTONE.— CARBONIFEROUS.— — PRIMARY OR PALAEOZOIC.— PALAEOZOIC.
30. UPPER DEVONIAN.— DEVONIAN.— PRIMARY OR PALAEOZOIC.— PALAEOZOIC.
31. MIDDLE DEVONIAN.— DEVONIAN.— PRIMARY OR PALAEOZOIC.— PALAEOZOIC.
32. LOWER DEVONIAN.— DEVONIAN.— PRIMARY OR PALAEOZOIC.— PALAEOZOIC.
33. UPPER SILURIAN.— SILURIAN.— PRIMARY OR PALAEOZOIC.— PALAEOZOIC.
34. LOWER SILURIAN.— SILURIAN.— PRIMARY OR PALAEOZOIC.— PALAEOZOIC.
35. UPPER CAMBRIAN.— CAMBRIAN.— PRIMARY OR PALAEOZOIC.— PALAEOZOIC.
36. LOWER CAMBRIAN.— CAMBRIAN.— PRIMARY OR PALAEOZOIC.— PALAEOZOIC.
37. UPPER LAURENTIAN.— LAURENTIAN.— PRIMARY OR PALAEOZOIC.— PALAEOZOIC.
38. LOWER LAURENTIAN.— LAURENTIAN.— PRIMARY OR PALAEOZOIC.— PALAEOZOIC.

TABULAR VIEW OF THE FOSSILIFEROUS STRATA,
SHOWING THE ORDER OF SUPERPOSITION OR CHRONOLOGICAL SUCCESSION OF THE PRINCIPAL GROUPS DESCRIBED IN THIS WORK (CITING EXAMPLES).

POST-TERTIARY.
1. RECENT. Shells and mammalia, all of living species.
BRITISH.
Clyde marine strata, with canoes (Chapter 10.)

FOREIGN.
Danish kitchen middens (Chapter 10.)
Lacustrine mud, with remains of Swiss lake-dwellings (Chapter 10.)
Marine strata inclosing Temple of Serapis, at Puzzuoli (Chapter 10.)
    2. POST-PLIOCENE. Shells, recent mammalia in part extinct.
BRITISH.
Loam of Brixham cave, with flint implements and bones of extinct and living quadrupeds. (Chapter 10.)
Drift near Salisbury, with bones of mammoth, Spermophilus, and stone implements. (Chapter 10.)
Glacial drift of Scotland, with marine shells and remains of mammoth. (Chapter 11.)
Erratics of Pagham and Selsey Bill. (Chapter 11.)
Glacial drift of Wales, with marine fossil shells, about 1400 feet high, on Moel Tryfaen. (Chapter 11.)
    FOREIGN.
Dordogne caves of the reindeer period. (Chapter 10.)
Older valley-gravels of Amiens, with flint implements and bones of extinct mammalia. (Chapter 10.)
Loess of Rhine. (Chapter 10.)
Ancient Nile-mud forming river-terraces. (Chapter 10.)
Loam and breccia of Liege caverns, with human remains. (Chapter 10.)
Australian cave breccias, with bones of extinct marsupials. (Chapter 10.)
Glacial drift of Northern Europe. (Chapters 11 and 12.)

TERTIARY OR CAINOZOIC.
PLIOCENE.
    3. NEWER PLIOCENE. The shells almost all of living species.
BRITISH.
Bridlington beds, marine Arctic fauna. (Chapter 13.)
Glacial boulder formation of Norfolk cliffs. (Chapter 13.)
Forest-bed of Norfolk cliffs, with bones of Elephas meridionalis, etc. (Chapter 13.)
Chillesford and Aldeby beds, with marine shells, chiefly Arctic. (Chapter 13.)
Norwich Crag. (Chapter 13.)
    FOREIGN.
Eastern base of Mount Etna, with marine shells. (Chapter 13.)
Sicilian calcareous and tufaceous strata. (Chapter 13.)
Lacustrine strata of Upper Val d'Arno. (Chapter 13.)
Madeira leaf-bed and land-shells. (Chapter 29.)
    4. OLDER PLIOCENE. Extinct species of shells forming a large minority.
BRITISH.
Red crag of Suffolk, marine shells, some of northern forms. (Chapter 13.)
White or coralline crag of Suffolk. (Chapter 13.)
    FOREIGN.
Antwerp crag. (Chapter 13.)
Subapennine marls and sands. (Chapter 13.)

MIOCENE.
    5. UPPER MIOCENE. Majority of the shells extinct.
BRITISH.
Wanting.
    FOREIGN.
faluns of Touraine (Chapter 14.)
faluns, proper, of Bordeaux. (Chapter 14.)
Fresh-water strata of Gers. (Chapter 14.)
Swiss Oeningen beds, rich in plants and insects. (Chapter 14.)
Marine Molasse, Switzerland. (Chapter 14.)

Bolderberg beds of Belgium. (Chapter 14.)
Vienna basin. (Chapter 14.)
Beds of the Superga, near Turin. (Chapter 14.)
Deposit at Pikerme, near Athens. (Chapter 14.)
Strata of the Siwalik hills, India. (Chapter 14.)
Marine strata of the Atlantic border in the United States. (Chapter 14.)
Volcanic tuff and limestone of Madeira, the Canaries, and the Azores. (Chapter 30.)

      6. LOWER MIOCENE. Nearly all the shells extinct.
      BRITISH.
Hempstead beds, marine and fresh-water strata. (Chapter 15.)
Lignites and clays of Bovey Tracey. (Chapter 15.)
Isle of Mull leaf-bed, volcanic tuff. (Chapter 15.)
      FOREIGN.
Calcaire de la Beauce, etc. (Chapter 15.)
Gres de Fontainebleau. (Chapter 15.)
Lacustrine strata of the Limagne d'Auvergne, and the Cantal. (Chapter 15.)
Mayence basin. (Chapter 15.)
Radaboj beds of Croatia. (Chapter 15.)
Brown coal of Germany. (Chapter 15.)
Lower Molasse of Switzerland, fresh-water and brackish. (Chapter 15.)
Rupelmonde, Kleynspawen, and Tongrian beds of Belgium. (Chapter 15.)
Nebraska beds, United States. (Chapter 15.)
Lower Miocene beds of Italy. (Chapter 15.)
Miocene flora of North Greenland. (Chapter 15.)

      7. UPPER EOCENE.
      BRITISH.
Bembridge fluvio-marine strata. (Chapter 16.)
Osborne or St. Helen's series. (Chapter 16.)
Headon series, with marine and fresh-water shells. (Chapter 16.)
Barton sands and clays (Chapter 16.)
      FOREIGN.
Gypsum of Montmartre, fresh-water with Palaeotherium. (Chapter 16.)
Calcaire silicieux, or Travertin inferieur. (Chapter 16.)
Gres de Beauchamp, or Sables moyens. (Chapter 16.)

      8. MIDDLE EOCENE.
      BRITISH.
Bracklesham beds and Bagshot sands. (Chapter 16.)
White clays of Alum Bay and Bournemouth. (Chapter 16.)
      FOREIGN.
Calcaire grossier, miliolitic limestone. (Chapter 16.)
Soissonnais sands, or Lits coquilliers, with Nummulites planulata. (Chapter 16.)
Claiborne beds of the United States, with Orbitoides and Zeuglodon. (Chapter 16.)

      9. LOWER EOCENE.
      Nummulitic formation of Europe, Asia, etc. (Chapter 16.)
      BRITISH.
London Clay proper. (Chapter 16.)
Woolwich and Reading series, fluvio-marine. (Chapter 16.)
Thanet sands. (Chapter 16.)
      FOREIGN.
Argile de Londres, near Dunkirk. (Chapter 16.)
Argile plastique. (Chapter 16.)
Sables de Bracheux. (Chapter 16.)

      SECONDARY OR MESOZOIC.
      CRETACEOUS.

10. UPPER CRETACEOUS.
BRITISH.
Upper white chalk, with flints. (Chapter 17.)
Lower white chalk, without flints. (Chapter 17.)
Chalk marl. (Chapter 17.)
Chloritic series (or Upper Greensand), fire-stone of Surrey. (Chapter 17.)
Gault. (Chapter 17.)
Blackdown beds. (Chapter 17.)
FOREIGN.
Maestricht beds and Faxoe chalk. (Chapter 17.)
Pisolitic limestone of France. (Chapter 17.)
White chalk of France, Sweden, and Russia. (Chapter 17.)
Planer-kalk of Saxony. (Chapter 17.)
Sands and clays of Aix-la-Chapelle. (Chapter 17.)
Hippurite limestone of South of France. (Chapter 17.)
New Jersey, U.S., sands and marls. (Chapter 17.)

11. LOWER CRETACEOUS OR NEOCOMIAN.
BRITISH.
Sands of Folkestone, Sandgate, and Hythe. (Chapter 18.)
Atherfield clay, with Perna mulleti. (Chapter 18.)
Punfield marine beds, with Vicarya lujana. (Chapter 18.)
Speeton clay of Flamborough Head and Tealby. (Chapter 18.)
Weald clay of Surrey, Kent, and Sussex, fresh-water, with Cypris. (Chapter 18.)
Hastings sands.
FOREIGN.
Neocomian of Neufchatel, and Hils conglomerate of North Germany. (Chapter 18.)
Wealden beds of Hanover. (Chapter 18.)

OOLITE.
12. UPPER OOLITE.
BRITISH.
Upper Purbeck beds, fresh-water. (Chapter 19.)
Middle Purbeck, with numerous marsupial quadrupeds, etc. (Chapter 19.)
Lower Purbeck, fresh-water, with intercalated dirt-bed. (Chapter 19.)
Portland stone and sand. (Chapter 19.)
Kimmeridge clay. (Chapter 19.)
FOREIGN.
Marnes a gryphees virgules of Argonne. (Chapter 19.)
Lithographic-stone of Solenhofen, with Archaeopteryx. (Chapter 19.)

13. MIDDLE OOLITE.
BRITISH.
Coral rag of Berkshire, Wilts, and Yorkshire. (Chapter 19.)
Oxford clay, with belemnites and Ammonite. (Chapter 19.)
Kelloway rock of Wilts and Yorkshire. (Chapter 19.)
FOREIGN.
Nerinaean limestone of the Jura.

14. LOWER OOLITE.
BRITISH.
Cornbrash and forest marble. (Chapter 19.)
Great or Bath oolite of Bradford. (Chapter 19.)
Stonesfield slate, with marsupials and Araucaria. (Chapter 19.)
Fuller's earth of Bath. (Chapter 19.)
Inferior oolite. (Chapter 19.)

LIAS.
15. LIAS.
Upper Lias, argillaceous, with Ammonites striatulus. (Chapter 20.)
Shale and limestone, with Ammonites bifrons. (Chapter 20.)

Middle Lias or Marlstone series, with zones containing characteristic Ammonites. (Chapter 20.)
Lower Lias, also with zones characterised by peculiar Ammonites. (Chapter 20.)
TRIAS.
16. UPPER TRIAS.
BRITISH.
Rhaetic, Penarth or Avicula contorta beds (beds of passage). (Chapter 21.)
Keuper or Upper New Red sandstone, etc. (Chapter 21.)
Red shales of Cheshire and Lancashire, with rock-salt. (Chapter 21.)
Dolomite conglomerate of Bristol (Chapter 21.)
FOREIGN.
Keuper beds of Germany. (Chapter 21.)
St. Cassian or Hallstadt beds, with rich marine fauna. (Chapter 21.)
Coal-field of Richmond, Virginia. (Chapter 21.)
Chatham coal-field, North Carolina. (Chapter 21.)
17. MIDDLE TRIAS.
BRITISH.
Wanting.
FOREIGN.
Muschelkalk of Germany. (Chapter 21.)
18. LOWER TRIAS.
BRITISH.
Bunter or Lower New Red sandstone of Lancashire and Cheshire. (Chapter 21.)
FOREIGN.
Bunter-sandstein of Germany. (Chapter 21.)
Red sandstone of Connecticut Valley, with footprints of birds and reptiles. (Chapter 21.)
PRIMARY OR PALAEOZOIC.
PERMIAN.
19. PERMIAN.
BRITISH. Upper Permian of St. Bees' Head, Cumberland. (Chapter 22.) Middle Permian, magnesian limestone, and marl-slate of Durham and Yorkshire, with Protosaurus. (Chapter 22.) Lower Permian sandstones and breccias of Penrith and Dumfriesshire, intercalated. (Chapter 22.)
FOREIGN.
Dark-coloured shales of Thuringia. (Chapter 22.)
Zechstein or Dolomitic limestone. (Chapter 22.)
Mergel-schiefer or Kupfer-schiefer. (Chapter 22.)
Rothliegendes of Thuringia, with Psaronius. (Chapter 22.)
Magnesian limestones, etc., of Russia. (Chapter 22.)
CARBONIFEROUS.
20. UPPER CARBONIFEROUS.
BRITISH.
Coal-measures of South Wales, with underclays inclosing Stigmaria. (Chapter 23.)
Coal-measures of north and central England. (Chapter 23.)
Millstone grit. (Chapter 23.)
Yoredale series of Yorkshire. (Chapter 23.)
Coal-field of Kilkenny with Labyrinthodont. (Chapter 23.)
FOREIGN.
Coal-field of Saarbruck, with Archegosaurus. (Chapter 23.)
Carboniferous strata of South Joggins, Nova Scotia. (Chapter 23.)
Pennsylvania coal-field. (Chapter 23.)
21. LOWER CARBONIFEROUS.
BRITISH.
Mountain limestone of Wales and South of England. (Chapter 24.)
Same in Ireland. (Chapter 24.)

Carboniferous limestone of Scotland alternating with coal-bearing sandstones. (Chapter 23.)
Erect trees in volcanic ash in the Island of Arran. (Chapter 30.)
FOREIGN.
Mountain limestone of Belgium. (Chapter 24.)

DEVONIAN OR OLD RED SANDSTONE.
22. UPPER DEVONIAN.
BRITISH.

Yellow sandstone of Dura Den, with Holoptychius, etc. (Chapter 25.); and of Ireland with Anodon Jukesii. (Chapter 25.)
Sandstones of Forfarshire and Perthshire, with Holoptychius, etc. (Chapter 25.)
Pilton group of North Devon. (Chapter 25.)
Petherwyn group of Cornwall, with Clymenia and Cypridina. (Chapter 25.)
FOREIGN.
Clymenien-kalk and Cypridinen-schiefer of Germany. (Chapter 25.)

23. MIDDLE DEVONIAN.
BRITISH.

Bituminous schists of Gamrie, Caithness, etc., with numerous fish. (Chapter 25.)
Ilfracombe beds with peculiar trilobites and corals. (Chapter 25.)
Limestones of Torquay, with broad-winged Spirifers. (Chapter 25.)
FOREIGN. (Chapter 25.)
Eifel limestone, with underlying schists containing Calceola. (Chapter 25.)
Devonian strata of Russia. (Chapter 25.)

24. LOWER DEVONIAN.
BRITISH.

Arbroath paving-stones, with Cephalaspis and Pterygotus. (Chapter 25.)
Lower sandstones of Forfarshire, with Pterygotus. (Chapter 25.)
Sandstones and slates of the Foreland and Linton. (Chapter 25.)
FOREIGN.
Oriskany sandstone of Western Canada and New York. (Chapter 25.)
Sandstones of Gaspe, with Cephalaspis. (Chapter 25.)

SILURIAN.
25. UPPER SILURIAN.
BRITISH.

Upper Ludlow formation, Downton sandstone, with bone-bed. (Chapter 26.)
Lower Ludlow formation, with oldest known fish remains. (Chapter 26.)
Wenlock limestone and shale. (Chapter 26.)
Woolhope limestone and grit. (Chapter 26.)
Tarannon shales. (Chapter 26.)
Beds of passage between Upper and Lower Silurian:
Upper Llandovery, or May-hill sandstone, with Pentamerus oblongus, etc. (Chapter 26.)
Lower Llandovery slates. (Chapter 26.)
FOREIGN.
Niagara limestone, with Calymene, Homalonotus, etc. (Chapter 26.)
Clinton group of America, with Pentamerus oblongus, etc. (Chapter 26.)
Silurian strata of Russia, with Pentamerus. (Chapter 26.)

26. LOWER SILURIAN.
BRITISH.

Bala and Caradoc beds. (Chapter 26.)
Llandeilo flags. (Chapter 26.)
Arenig or Stiper-stones group (Lower Llandeilo of Murchison.) (Chapter 26.)
FOREIGN.
Ungulite or Obolus grit of Russia. (Chapter 26.)
Trenton limestone, and other Lower Silurian groups of North America. (Chapter

26.)
Lower Silurian of Sweden. (Chapter 26.)
CAMBRIAN.
27. UPPER CAMBRIAN.
BRITISH.
Tremadoc slates. (Chapter 27.)
Lingula flags, with Lingula Davisii. (Chapter 27.)
FOREIGN.
"Primordial" zone of Bohemia in part, with trilobites of the genera Paradoxides, etc. (Chapter 27.)
Alum schists of Sweden and Norway. (Chapter 27.)
Potsdam sandstone, with Dikelocephalus and Obolella. (Chapter 27.)
28. LOWER CAMBRIAN.
BRITISH.
Menevian beds of Wales, with Paradoxides Davidis, etc. (Chapter 27.)
Longmynd group, comprising the Harlech grits and Llanberis slates. (Chapter 27.)
FOREIGN.
Lower portion of Barrande's "Primordial" zone in Bohemia. (Chapter 27.)
Fucoid sandstones of Sweden. (Chapter 27.)
Huronian series of Canada? (Chapter 27.)
LAURENTIAN.
29. UPPER LAURENTIAN.
BRITISH.
Fundamental gneiss of the Hebrides? (Chapter 27.)
Hypersthene rocks of Skye? (Chapter 27.)
FOREIGN.
Labradorite series north of the river St. Lawrence in Canada. (Chapter 27.)
Adirondack mountains of New York. (Chapter 27.)
30. LOWER LAURENTIAN.
BRITISH.
Wanting?
FOREIGN. Beds of gneiss and quartzite, with interstratified limestones, in one of which, 1000 feet thick, occurs a foraminifer, Eozoon Canadense, the oldest known fossil. (Chapter 27.)
CHAPTER IX.
CLASSIFICATION OF TERTIARY FORMATIONS.
Order of Succession of Sedimentary Formations.
Frequent Unconformability of Strata.
Imperfection of the Record.
Defectiveness of the Monuments greater in Proportion to their Antiquity.
Reasons for studying the newer Groups first.
Nomenclature of Formations.
Detached Tertiary Formations scattered over Europe.
Value of the Shell-bearing Mollusca in Classification.
Classification of Tertiary Strata.
Eocene, Miocene, and Pliocene Terms explained.
By reference to the tables given at the end of the last chapter the reader will see that when the fossiliferous rocks are arranged chronologically, we have first to consider the Post-tertiary and then the Tertiary or Cainozoic formations, and afterwards to pass on to those of older date.
ORDER OF SUPERPOSITION.
(FIGURE 86. Section through Primary (left), Secondary, Tertiary and Post- tertiary (right) Strata. 1. Laurentian. 2. Cambrian. 3. Silurian. 4. Devonian. 5. Carboniferous. 6. Permian. 7. Triassic. 8. Jurassic. 9. Cretaceous. 10. Eocene. 11. Miocene. 12. Pliocene. 13. Post-pliocene. 14. Recent. Sea.)

The diagram (Figure 86.) will show the order of superposition of these deposits, assuming them all to be visible in one continuous section. In nature, as before hinted (Chapter 6), we have never an opportunity of seeing the whole of them so displayed in a single region; first, because sedimentary deposition is confined, during any one geological period, to limited areas; and secondly, because strata, after they have been formed, are liable to be utterly annihilated over wide areas by denudation. But wherever certain members of the series are present, they overlie one another in the order indicated in the diagram, though not always in the exact manner there represented, because some of them repose occasionally in unconformable stratification on others. This mode of superposition has been already explained (Chapters 5 and 7), where I pointed out that the discordance which implies a considerable lapse of time between two formations in juxtaposition is almost invariably accompanied by a great dissimilarity in the species of organic remains.

FREQUENT UNCONFORMABILITY OF STRATA.

Where the widest gaps appear in the sequence of the fossil forms, as between the Permian and Triassic rocks, or between the Cretaceous and Eocene, examples of such unconformability are very frequent. But they are also met with in some part or other of the world at the junction of almost all the other principal formations, and sometimes the subordinate divisions of any one of the leading groups may be found lying unconformably on another subordinate member of the same— the Upper, for example, on the Lower Silurian, or the superior division of the Old Red Sandstone on a lower member of the same, and so forth. Instances of such irregularities in the mode of succession of the strata are the more intelligible the more we extend our survey of the fossiliferous formations, for we are continually bringing to light deposits of intermediate date, which have to be intercalated between those previously known, and which reveal to us a long series of events, of which antecedently to such discoveries we had no knowledge.

But while unconformability invariably bears testimony to a lapse of unrepresented time, the conformability of two sets of strata in contact by no means implies that the newer formation immediately succeeded the older one. It simply implies that the ancient rocks were subjected to no movements of such a nature as to tilt, bend, or break them before the more modern formation was superimposed. It does not show that the earth's crust was motionless in the region in question, for there may have been a gradual sinking or rising, extending uniformly over a large surface, and yet, during such movement, the stratified rocks may have retained their original horizontality of position. There may have been a conversion of a wide area from sea into land and from land into sea, and during these changes of level some strata may have been slowly removed by aqueous action, and after this new strata may be superimposed, differing perhaps in date by thousands of years or centuries, and yet resting conformably on the older set. There may even be a blending of the materials constituting the older deposit with those of the newer, so as to give rise to a passage in the mineral character of the one rock into the other as if there had been no break or interruption in the depositing process.

IMPERFECTION OF THE RECORD.

Although by the frequent discovery of new sets of intermediate strata the transition from one type of organic remains to another is becoming less and less abrupt, yet the entire series of records appears to the geologists now living far more fragmentary and defective than it seemed to their predecessors half a century ago. The earlier inquirers, as often as they encountered a break in the regular sequence of formations, connected it theoretically with a sudden and violent catastrophe, which had put an end to the regular course of events that had been going on uninterruptedly for ages, annihilating at the same time all or nearly all the organic beings which had previously flourished, after which, order being re-established, a new series of events was initiated. In proportion as our faith in these views grows weaker, and the phenomena of the organic or inorganic world presented to us by geology seem explicable on the hypothesis of gradual and insensible changes, varied only by occasional convulsions, on a scale comparable to that witnessed in historical times; and in proportion as it is thought possible that former fluctuations in the organic world may be due to the indefinite modifiability of species without the necessity of assuming new and independent acts of creation, the number and magnitude of the gaps which still remain, or the extreme imperfection of the record, become more and more striking, and what we possess of the ancient annals of the earth's history appears as nothing when contrasted with that which has been lost.

When we examine a large area such as Europe, the average as well as the extreme height above the sea attained by the older formations is usually found to exceed that reached by the more modern ones, the primary or palaeozoic rising higher than the secondary, and these in their turn than the tertiary; while in reference to the three divisions of the tertiary, the lowest or Eocene group attains a higher summit-level than the Miocene, and these again a greater height than the Pliocene formations. Lastly, the post-tertiary deposits, such, at least, as are of marine origin, are most commonly restricted to much more moderate elevations above the sea-level than the tertiary strata.

It is also observed that strata, in proportion as they are of newer date, bear the nearest resemblance in mineral character to those which are now in the progress of formation in seas or lakes, the newest of all consisting principally of soft mud or loose sand, in some places full of shells, corals, or other organic bodies, animal or vegetable, in others wholly devoid of such remains. The farther we recede from the present time, and the higher the antiquity of the formations which we examine, the greater are the changes which the sedimentary deposits have undergone. Time, as I have explained in Chapters 5, 6, and 7, has multiplied the effects of condensation by pressure and cementation, and the modification produced by heat, fracture, contortion, upheaval, and denudation. The organic remains also have sometimes been obliterated entirely, or the mineral matter of which they were composed has been removed and replaced by other substances.

WHY NEWER GROUPS SHOULD BE STUDIED FIRST.

We likewise observe that the older the rocks the more widely do their organic remains depart from the types of the living creation. First, we find in the newer tertiary rocks a few species which no longer exist, mixed with many living ones, and then, as we go farther back, many genera and families at present unknown make their appearance, until we come to strata in which the fossil relics of existing species are nowhere to be detected, except a few of the lowest forms of invertebrate, while some orders of animals and plants wholly unrepresented in the living world begin to be conspicuous.

When we study, therefore, the geological records of the earth and its inhabitants, we find, as in human history, the defectiveness and obscurity of the monuments always increasing the remoter the era to which we refer, and the difficulty of determining the true chronological relations of rocks is more and more enhanced, especially when we are comparing those which were formed simultaneously in very distant regions of the globe. Hence we advance with securer steps when we begin with the study of the geological records of later times, proceeding from the newer to the older, or from the more to the less known.

In thus inverting what might at first seem to be the more natural order of historical research, we must bear in mind that each of the periods above enumerated, even the shortest, such as the Post-tertiary, or the Pliocene, Miocene, or Eocene, embrace a succession of events of vast extent, so that to give a satisfactory account of what we already know of any one of them would require many volumes. When, therefore, we approach one of the newer groups before endeavouring to decipher the monuments of an older one, it is like endeavouring to master the history of our own country and that of some contemporary nations, before we enter upon Roman History, or like investigating the annals of Ancient Italy and Greece before we approach those of Egypt and Assyria.

NOMENCLATURE.

The origin of the terms Primary and Secondary, and the synonymous terms Palaeozoic, and Mesozoic, were explained in Chapter 8.

The Tertiary or Cainozoic strata (see Chapter 8) were so called because they were all posterior in date to the Secondary series, of which last the Chalk of Cretaceous, No. 9, Figure 86, constitutes the newest group. The whole of them were at first confounded with the superficial alluviums of Europe; and it was long before their real extent and thickness, and the various ages to which they belong, were fully recognised. They were observed to occur in patches, some of fresh-water, others of marine origin, their geographical area being usually small as compared to the secondary formations, and their position often suggesting the idea of their having been deposited in different bays, lakes, estuaries, or inland seas, after a large portion of the space now occupied by Europe had already been converted into dry land.

The first deposits of this class, of which the characters were accurately determined, were those occurring in the neighbourhood of Paris, described in 1810 by MM. Cuvier and Brongniart. They were ascertained to consist of successive sets of strata, some of marine, others of fresh-water origin, lying one upon the other. The fossil shells and corals were perceived to be almost all of unknown species, and to have in general a near affinity to those now inhabiting warmer seas. The bones and skeletons of land animals, some of them of large size, and belonging to more than forty distinct species, were examined by Cuvier, and declared by him not to agree specifically, nor most of them even generically, with any hitherto observed in the living creation.

Strata were soon afterwards brought to light in the vicinity of London, and in Hampshire, which, although dissimilar in mineral composition, were justly inferred by Mr. T. Webster to be of the same age as those of Paris, because the greater number of the fossil shells were specifically identical. For the same reason, rocks found on the Gironde, in the South of France, and at certain points in the North of Italy, were suspected to be of contemporaneous origin.

Another important discovery was soon afterwards made by Brocchi in Italy, who investigated the argillaceous and sandy deposits, replete with shells, which form a low range of hills, flanking the Apennines on both sides, from the plains of the Po to Calabria. These lower hills were called by him the Subapennines, and were formed of strata chiefly marine, and newer than those of Paris and London.

Another tertiary group occurring in the neighbourhood of Bordeaux and Dax, in the South of France, was examined by M. de Basterot in 1825, who described and figured several hundred species of shells, which differed for the most part both from the Parisian series and those of the Subapennine hills. It was soon, therefore, suspected that this fauna might belong to a period intermediate between that of the Parisian and Subapennine strata, and it was not long before the evidence of superposition was brought to bear in support of this opinion; for other strata, contemporaneous with those of Bordeaux, were observed in one district (the Valley of the Loire), to overlie the Parisian formation, and in another (in Piedmont) to underlie the Subapennine beds. The first example of these was pointed out in 1829 by M. Desnoyers, who ascertained that the sand and marl of marine origin called faluns, near Tours, in the basin of the Loire, full of sea-shells and corals, rested upon a lacustrine formation, which constitutes the uppermost subdivision of the Parisian group, extending continuously throughout a great table-land intervening between the basin of the Seine and that of the Loire. The other example occurs in Italy, where strata containing many fossils similar to those of Bordeaux were observed by Bonelli and others in the environs of Turin, subjacent to strata belonging to the Subapennine group of Brocchi.

VALUE OF TESTACEAN FOSSILS IN CLASSIFICATION.

It will be observed that in the foregoing allusions to organic remains, the testacea or the shell-bearing mollusca are selected as the most useful and convenient class for the purposes of general classification. In the first place, they are more universally distributed through strata of every age than any other organic bodies. Those families of fossils which are of rare and casual occurrence are absolutely of no avail in establishing a chronological arrangement. If we have plants alone in one group of strata and the bones of mammalia in another, we can draw no conclusion respecting the affinity or discordance of the organic beings of the two epochs compared; and the same may be said if we have plants and vertebrated animals in one series and only shells in another. Although corals are more abundant, in a fossil state, than plants, reptiles, or fish, they are still rare when contrasted with shells, because they are more dependent for their well-being on the constant clearness of the water, and are, therefore, less likely to be included in rocks which endure in consequence of their thickness and the copiousness of sediment which prevailed when they originated. The utility of the testacea is, moreover, enhanced by the circumstance that some forms are proper to the sea, others to the land, and others to fresh water. Rivers scarcely ever fail to carry down into their deltas some land-shells, together with species which are at once fluviatile and lacustrine. By this means we learn what terrestrial, fresh-water, and marine species coexisted at particular eras of the past: and having thus identified strata formed in seas with others which originated contemporaneously in inland lakes, we are then enabled to advance a step farther, and show that certain quadrupeds or aquatic plants, found fossil in lacustrine formations, inhabited the globe at the same period when certain fish, reptiles, and zoophytes lived in the ocean.

Among other characters of the molluscous animals, which render them extremely valuable in settling chronological questions in geology, may be mentioned, first, the wide geographical range of many species; and, secondly, what is probably a consequence of the former, the great duration of species in this class, for they appear to have surpassed in longevity the greater number of the mammalia and fish. Had each species inhabited a very limited space, it could never, when imbedded in strata, have enabled the geologist to identify deposits at distant points; or had they each lasted but for a brief period, they could have thrown no light on the connection of rocks placed far from each other in the chronological, or, as it is often termed, vertical series.

CLASSIFICATION OF TERTIARY STRATA.

Many authors have divided the European Tertiary strata into three groups— lower, middle, and upper; the lower comprising the oldest formations of Paris and London before mentioned; the middle those of Bordeaux and Touraine; and the upper all those newer than the middle group.

In the first edition of the Principles of Geology, I divided the whole of the Tertiary formations into four groups, characterised by the percentage of recent shells which they contained. The lower tertiary strata of London and Paris were thought by M. Deshayes to contain only 3 1/2 per cent of recent species, and were termed Eocene. The middle tertiary of the Loire and Gironde had, according to the specific determinations of the same conchologist, 17 per cent, and formed the Miocene division. The Subapennine beds contained 35 to 50 per cent, and were termed Older Pliocene, while still more recent beds in Sicily, which had from 90 to 95 per cent of species identical with those now living, were called Newer Pliocene. The first of the above terms, Eocene, is derived from eos, dawn, and cainos, recent, because the fossil shells of this period contain an extremely small proportion of living species, which may be looked upon as indicating the dawn of the existing state of the testaceous fauna, no recent species having been detected in the older or secondary rocks.

The term Miocene (from meion, less, and cainos, recent) is intended to express a minor proportion of recent species (of testacea), the term Pliocene (from pleion, more, and cainos, recent) a comparative plurality of the same. It may assist the memory of students to remind them, that the MI-ocene contain a MI-nor proportion, and PL-iocene a comparative PL-urality of recent species; and that the greater number of recent species always implies the more modern origin of the strata.

It has sometimes been objected to this nomenclature that certain species of infusoria found in the chalk are still existing, and, on the other hand, the Miocene and Older Pliocene deposits often contain the remains of mammalia, reptiles, and fish, exclusively of extinct species. But the reader must bear in mind that the terms Eocene, Miocene, and Pliocene were originally invented with reference purely to conchological data, and in that sense have always been and are still used by me.

Since the year 1830 the number of known shells, both recent and fossil, has largely increased, and their identification has been more accurately determined. Hence some modifications have been required in the classifications founded on less perfect materials. The Eocene, Miocene, and Pliocene periods have been made to comprehend certain sets of strata of which the fossils do not always conform strictly in the proportion of recent to extinct species with the definitions first given by me, or which are implied in the etymology of those terms.

CHAPTER X.
RECENT AND POST-PLIOCENE PERIODS.

Recent and Post-pliocene Periods.
Terms defined.
Formations of the Recent Period.

Modern littoral Deposits containing Works of Art near Naples.
Danish Peat and Shell-mounds.
Swiss Lake-dwellings.
Periods of Stone, Bronze, and Iron.
Post-pliocene Formations.
Coexistence of Man with extinct Mammalia.
Reindeer Period of South of France.
Alluvial Deposits of Paleolithic Age.
Higher and Lower-level Valley-gravels.
Loess or Inundation-mud of the Nile, Rhine, etc.
Origin of Caverns.
Remains of Man and extinct Quadrupeds in Cavern Deposits.
Cave of Kirkdale.
Australian Cave-breccias.
Geographical Relationship of the Provinces of living Vertebrata and those of extinct Post-pliocene Species.
Extinct struthious Birds of New Zealand.
Climate of the Post-pliocene Period.
Comparative Longevity of Species in the Mammalia and Testacea.
Teeth of Recent and Post-pliocene Mammalia.

We have seen in the last chapter that the uppermost or newest strata are called Post-tertiary, as being more modern than the Tertiary. It will also be observed that the Post-tertiary formations are divided into two subordinate groups: the Recent, and Post-pliocene. In the former, or the Recent, the mammalia as well as the shells are identical with species now living: whereas in the Post-pliocene, the shells being all of living forms, a part, and often a considerable part, of the mammalia belonged to extinct species. To this nomenclature it may be objected that the term Post-pliocene should in strictness include all geological monuments posterior in date to the Pliocene; but when I have occasion to speak of the whole collectively, I shall call them Post-tertiary, and reserve the term Post-pliocene for the older Post-tertiary formations, while the Upper or newer ones will be called "Recent."

Cases will occur where it may be scarcely possible to draw the boundary line between the Recent and Post-pliocene deposits; and we must expect these difficulties to increase rather than diminish with every advance in our knowledge, and in proportion as gaps are filled up in the series of records.

RECENT PERIOD.

It was stated in the sixth chapter, when I treated of denudation, that the dry land, or that part of the earth's surface which is not covered by the waters of lakes or seas, is generally wasting away by the incessant action of rain and rivers, and in some cases by the undermining and removing power of waves and tides on the sea-coast. But the rate of waste is very unequal, since the level and gently sloping lands, where they are protected by a continuous covering of vegetation, escape nearly all wear and tear, so that they may remain for ages in a stationary condition, while the removal of matter is constantly widening and deepening the intervening ravines and valleys.

The materials, both fine and coarse, carried down annually by rivers from the higher regions to the lower, and deposited in successive strata in the basins of seas and lakes, must be of enormous volume. We are always liable to underrate their magnitude, because the accumulation of strata is going on out of sight.

There are, however, causes at work which, in the course of centuries, tend to render visible these modern formations, whether of marine or lacustrine origin. For a large portion of the earth's crust is always undergoing a change of level, some areas rising and others sinking at the rate of a few inches, or a few feet, perhaps sometimes yards, in a century; so that spaces which were once subaqueous are gradually converted into land, and others which were high and dry become submerged. In consequence of such movements we find in certain regions, as in Cashmere, for example, where the mountains are often shaken by earthquakes, deposits which were formed in lakes in the historical period, but through which rivers have now cut deep and wide channels. In lacustrine strata thus intersected, works of art and fresh-water shells are seen. In other districts on the borders of the sea, usually at very moderate elevations above its level, raised beaches occur, or marine littoral deposits, such as those in which, on the borders of the Bay of Baiae, near Naples, the well-known temple of Serapis was imbedded. In that case the date of the monument buried in the marine strata is ascertainable, but in many other instances the exact age of the remains of human workmanship is uncertain, as in the estuary of the Clyde at Glasgow, where many canoes have been exhumed, with other works of art, all assignable to some part of the Recent Period.

DANISH PEAT AND SHELL-MOUNDS OR KITCHEN-MIDDENS.

Sometimes we obtain evidence, without the aid of a change of level, of events which took place in pre-historic times. The combined labours, for example, of the antiquary, zoologist, and botanist have brought to light many monuments of the early inhabitants buried in peat-mosses in Denmark. Their geological age is determined by the fact that, not only the contemporaneous fresh-water and land shells, but all the quadrupeds, found in the peat, agree specifically with those now inhabiting the same districts, or which are known to have been indigenous in Denmark within the memory of man. In the lower beds of peat (a deposit varying from 20 to 30 feet in thickness), weapons of stone accompany trunks of the Scotch fir, Pinus sylvestris. This peat may be referred to that part of the stone period for which Sir John Lubbock proposed the name of "Neolithic" in contradistinction to a still older era, termed by him "Paleolithic," and which will be described in the sequel. (Sir John Lubbock Pre-historic Times page 3 1865.) In the higher portions of the same Danish bogs, bronze implements are associated with trunks and acorns of the common oak. It appears that the pine has never been a native of Denmark in historical times, and it seems to have given place to the oak about the time when articles and instruments of bronze superseded those of stone. It also appears that, at a still later period, the oak itself became scarce, and was nearly supplanted by the beech, a tree which now flourishes luxuriantly in Denmark. Again, at the still later epoch when the beech-tree abounded, tools of iron were introduced, and were gradually substituted for those of bronze.

On the coasts of the Danish islands in the Baltic, certain mounds, called in those countries "Kjokken-modding," or "kitchen-middens," occur, consisting chiefly of the castaway shells of the oyster, cockle, periwinkle, and other eatable kinds of molluscs. The mounds are from three to ten feet high, and from 100 to 1000 feet in their longest diameter. They greatly resemble heaps of shells formed by the Red Indians of North America along the eastern shores of the United States. In the old refuse-heaps, recently studied by the Danish antiquaries and naturalists with great skill and diligence, no implements of metal have ever been detected. All the knives, hatchets, and other tools, are of stone, horn, bone, or wood. With them are often intermixed fragments of rude pottery, charcoal and cinders, and the bones of quadrupeds on which the rude people fed. These bones belong to wild species still living in Europe, though some of them, like the beaver, have long been extirpated in Denmark. The only animal which they seem to have domesticated was the dog.

As there is an entire absence of metallic tools, these refuse-heaps are referred to the Neolithic division of the age of stone, which immediately preceded in Denmark the age of bronze. It appears that a race more advanced in civilisation, armed with weapons of that mixed metal, invaded Scandinavia, and ousted the aborigines.

LACUSTRINE HABITATIONS OF SWITZERLAND.

In Switzerland a different class of monuments, illustrating the successive ages of stone, bronze, and iron, has been of late years investigated with great success, and especially since 1854, in which year Dr. F. Keller explored near the shore at Meilen, in the bottom of the lake of Zurich, the ruins of an old village, originally built on numerous wooden piles, driven, at some unknown period, into the muddy bed of the lake. Since then a great many other localities, more than a hundred and fifty in all, have been detected of similar pile-dwellings, situated near the borders of the Swiss lakes, at points where the depth of water does not exceed 15 feet. (Bulletin de la Societie Vaudoise des Sciences Nat. tome 6 Lausanne 1860; and Antiquity of Man by the author chapter 2.) The superficial mud in such cases is filled with various articles, many hundreds of them being often dredged up from a very limited area. Thousands of piles, decayed at their upper extremities, are often met with still firmly fixed in the mud.

As the ages of stone, bronze, and iron merely indicate successive stages of civilisation, they may all have coexisted at once in different parts of the globe, and even in contiguous regions, among nations having little intercourse with each other. To make out, therefore, a distinct chronological series of monuments is only possible when our observations are confined to a limited district, such as Switzerland.

The relative antiquity of the pile-dwellings, which belong respectively to the ages of stone and bronze, is clearly illustrated by the associations of the tools with certain groups of animal remains. Where the tools are of stone, the castaway bones which served for the food of the ancient people are those of deer, the wild boar, and wild ox, which abounded when society was in the hunter state. But the bones of the later or bronze epoch were chiefly those of the domestic ox, goat, and pig, indicating progress in civilisation. Some villages of the stone age are of later date than others, and exhibit signs of an improved state of the arts. Among their relics are discovered carbonised grains of wheat and barley, and pieces of bread, proving that the cultivation of cereals had begun. In the same settlements, also, cloth, made of woven flax and straw, has been detected.

The pottery of the bronze age in Switzerland is of a finer texture, and more elegant in form, than that of the age of stone. At Nidau, on the lake of Bienne, articles of iron have also been discovered, so that this settlement was evidently not abandoned till that metal had come into use.

At La Thene, in the northern angle of the lake of Neufchatel, a great many articles of iron have been obtained, which in form and ornamentation are entirely different both from those of the bronze period and from those used by the Romans. Gaulish and Celtic coins have also been found there by MM. Schwab and Desor. They agree in character with remains, including many iron swords, which have been found at Tiefenau, near Berne, in ground supposed to have been a battle-field; and their date appears to have been anterior to the great Roman invasion of Northern Europe, though perhaps not long before that event. (Sir J. Lubbock's Lecture, Royal Institution February 27, 1863.) Coins, which sometimes occur in deposits of the age of iron, have never yet been found in formations of the ages of bronze or stone.

The period of bronze must have been one of foreign commerce, as tin, which enters into this metallic mixture in the proportion of about ten per cent to the copper, was obtained by the ancients chiefly from Cornwall. (Diodorus 5, 21, 22 and Sir H. James Note on Block of Tin dredged up in Falmouth Harbour. Royal Institution of Cornwall 1863.) Very few human bones of the bronze period have been met with in the Danish peat, or in the Swiss lake-dwellings, and this scarcity is generally attributed by archaeologists to the custom of burning the dead, which prevailed in the age of bronze.

POST-PLIOCENE PERIOD.

From the foregoing observations we may infer that the ages of iron and bronze in Northern and Central Europe were preceded by a stone age, the Neolithic, referable to that division of the post-tertiary epoch which I have called Recent, when the mammalia as well as the other organic remains accompanying the stone implements were of living species. But memorials have of late been brought to light of a still older age of stone, for which, as above stated, the name Paleolithic has been proposed, when man was contemporary in Europe with the elephant and rhinoceros, and various other animals, of which many of the most conspicuous have long since died out.

REINDEER PERIOD IN SOUTH OF FRANCE.

In the larger number of the caves of Europe, as for example in those of England, Belgium, Germany, and many parts of France, the animal remains agree specifically with the fauna of this oldest division of the age of stone, or that to which belongs the drift of Amiens and Abbeville presently to be mentioned, containing flint implements of a very antique type. But there are some caves in the departments of Dordogne, Aude, and other parts of the south of France, which are believed by M. Lartet to be of intermediate date between the Paleolithic and Neolithic periods. To this intermediate era M. Lartet gave, in 1863, the name of the "reindeer period," because vast quantities of the bones and horns of that deer have been met with in the French caverns. In some cases separate plates of molars of the mammoth, and several teeth of the great Irish deer, Cervus megaceros, and of the cave-lion, Felis spelaea, have been found mixed up with cut and carved bones of reindeer. On one of these sculptured bones in the cave of Perigord, a rude representation of the mammoth, with its long curved tusks and covering of wool, occurs, which is regarded by M. Lartet as placing beyond all doubt the fact that the early inhabitants of these caves must have seen this species of elephant still living in France. The presence of the marmot, as well as the reindeer and some other northern animals, in these caverns seems to imply a colder climate than that of the Swiss lake-dwellings, in which no remains of reindeer have as yet been discovered. The absence of this last in the old lacustrine habitations of Switzerland is the more significant, because in a cave in the neighbourhood of the lake of Geneva, namely, that of Mont Saleve, bones of the reindeer occur with flint implements similar to those of the caverns of Dordogne and Perigord.

The state of the arts, as exemplified by the instruments found in these caverns of the reindeer period, is somewhat more advanced than that which characterises the tools of the Amiens drift, but is nevertheless more rude than that of the Swiss lake-dwellings. No metallic articles occur, and the stone hatchets are not ground after the fashion of celts; the needles of bone are shaped in a workmanlike style, having their eyes drilled with consummate skill.

The formations above alluded to, which are as yet but imperfectly known, may be classed as belonging to the close of the Paleolithic era, of the monuments of which I am now about to treat.

ALLUVIAL DEPOSITS OF THE PALEOLITHIC AGE.

(FIGURE 87. Recent and Post-pliocene alluvial deposits. 1. Peat of the recent period. 2. Gravel of modern river. 2'. Loam of brick-earth (loess) of same age as 2, formed by inundations of the river. 3. Lower-level valley-gravel with extinct mammalia (Post-pliocene). 3'. Loam of same age. 4. Higher-level valley-gravel (Post-pliocene). 4'. Loam of same age. 5. Upland gravel of various kinds and periods, consisting in some places of unstratified boulder clay or glacial drift. 6. Older rocks.)

The alluvial and marine deposits of the Paleolithic age, the earliest to which any vestiges of man have yet been traced back, belong to a time when the physical geography of Europe differed in a marked degree from that now prevailing. In the Neolithic period, the valleys and rivers coincided almost entirely with those by which the present drainage of the land is effected, and the peat-mosses were the same as those now growing. The situation of the shell- mounds and lake-dwellings above alluded to is such as to imply that the topography of the districts where they are observed has not subsequently undergone any material alteration. Whereas we no sooner examine the Post- pliocene formations, in which the remains of so many extinct mammalia are found, than we at once perceive a more decided discrepancy between the former and present outline of the surface. Since those deposits originated, changes of considerable magnitude have been effected in the depth and width of many valleys, as also in the direction of the superficial and subterranean drainage, and, as is manifest near the sea-coast, in the relative position of land and water. In Figure 87 an ideal section is given, illustrating the different position which the Recent and Post-pliocene alluvial deposits occupy in many European valleys.

The peat, No. 1, has been formed in a low part of the modern alluvial plain, in parts of which gravel No. 2 of the recent period is seen. Over this gravel the loam or fine sediment 2' has in many places been deposited by the river during floods which covered nearly the whole alluvial plain.

No. 3 represents an older alluvium, composed of sand and gravel, formed before the valley had been excavated to its present depth. It contains the remains of fluviatile shells of living species associated with the bones of mammalia, in part of recent, and in part of extinct species. Among the latter, the mammoth (E. primigenius) and the Siberian rhinoceros (R. tichorhinus) are the most common in Europe. No. 3' is a remnant of the loam or brick-earth by which No. 3 was overspread. No. 4 is a still older and more elevated terrace, similar in its composition and organic remains to No. 3, and covered in like manner with its inundation-mud, 4'. Sometimes the valley-gravels of older date are entirely missing, or there is only one, and occasionally there are more than two, marking as many successive stages in the excavation of the valley. They usually occur at heights varying from 10 to 100 feet, sometimes on the right and sometimes on the left side of the existing river-plain, but rarely in great strength on exactly opposite sides of the valley.

Among the genera of extinct quadrupeds most frequently met with in England, France, Germany, and other parts of Europe, are the elephant, rhinoceros, hippopotamus, horse, great Irish deer, bear, tiger, and hyaena. In the peat, No. 1 (Figure 87), and in the more modern gravel and silt (No. 2), works of art of the ages of iron and bronze, and of the later or Neolithic stone period, already described, are met with. In the more ancient or Paleolithic gravels, 3 and 4, there have been found of late years in several valleys in France and England— as, for example, in those of the Seine and Somme, and of the Thames and Ouse, near Bedford— stone implements of a rude type, showing that man coexisted in those districts with the mammoth and other extinct quadrupeds of the genera above enumerated. In 1847, M. Boucher de Perthes observed in an ancient alluvium at Abbeville, in Picardy, the bones of extinct mammalia associated in such a manner with flint implements of a rude type as to lead him to infer that both the organic remains and the works of art were referable to one and the same period. This inference was soon after confirmed by Mr. Prestwich, who found in 1859 a flint tool in situ in the same stratum at Amiens that contained the remains of extinct mammalia.

The flint implements found at Abbeville and Amiens are most of them considered to be hatchets and spear-heads, and are different from those commonly called "celts." These celts, so often found in the recent formations, have a more regular oblong shape, the result of grinding, by which also a sharp edge has been given to them. The Abbeville tools found in gravel at different levels, as in Nos. 3 and 4, Figure 87, in which bones of the elephant, rhinoceros, and other extinct mammalia occur, are always unground, having evidently been brought into their present form simply by the chipping off of fragments of flint by repeated blows, such as could be given by a stone hammer.

Some of them are oval, others of a spear-headed form, no two exactly alike, and yet the greater number of each kind are obviously fashioned after the same general pattern. Their outer surface is often white, the original black flint having been discoloured and bleached by exposure to the air, or by the action of acids, as they lay in the gravel. They are most commonly stained of the same ochreous colour as the flints of the gravel in which they are imbedded. Occasionally their antiquity is indicated not only by their colour but by superficial incrustations of carbonate of lime, or by dendrites formed of oxide of iron and manganese. The edges also of most of them are worn, sometimes by having been used as tools, or sometimes by having been rolled in the old river's bed. They are met with not only in the lower-level gravels, as in No. 3, Figure 87, but also in No. 4, or the higher gravels, as at St. Acheul, in the suburbs of Amiens, where the old alluvium lies at an elevation of about 100 feet above the level of the river Somme. At both levels fluviatile and land-shells are met with in the loam as well as in the gravel, but there are no marine shells associated, except at Abbeville, in the lowest part of the gravel, near the sea, and a few feet only above the present high-water mark. Here with fossil shells of living species are mingled the bones of Elephas primigenius and E. antiquus, Rhinoceros tichorhinus, Hippopotamus, Felis spelaea, Hyaena spelaea, reindeer, and many others, the bones accompanying the flint implements in such a manner as to show that both were buried in the old alluvium at the same period.

Nearly the entire skeleton of a rhinoceros was found at one point, namely, in the Menchecourt drift at Abbeville, the bones being in such juxtaposition as to show that the cartilage must have held them together at the time of their inhumation.

The general absence here and elsewhere of human bones from gravel and sand in which flint tools are discovered, may in some degree be due to the present limited extent of our researches. But it may also be presumed that when a hunter population, always scanty in numbers, ranged over this region, they were too wary to allow themselves to be overtaken by the floods which swept away many herbivorous animals from the low river-plains where they may have been pasturing or sleeping. Beasts of prey prowling about the same alluvial flats in search of food may also have been surprised more readily than the human tenant of the same region, to whom the signs of a coming tempest were better known.

INUNDATION-MUD OF RIVERS.— BRICK-EARTH.— FLUVIATILE LOAM, OR LOESS.

As a general rule, the fluviatile alluvia of different ages (Nos. 2, 3, 4, Figure 87) are severally made up of coarse materials in their lower portions, and of fine silt or loam in their upper parts. For rivers are constantly shifting their position in the valley-plain, encroaching gradually on one bank, near which there is deep water, and deserting the other or opposite side, where the channel is growing shallower, being destined eventually to be converted into land. Where the current runs strongest, coarse gravel is swept along, and where its velocity is slackened, first sand, and then only the finest mud, is thrown down. A thin film of this fine sediment is spread, during floods, over a wide area, on one, or sometimes on both sides, of the main stream, often reaching as far as the base of the bluffs or higher grounds which bound the valley. Of such a description are the well-known annual deposits of the Nile, to which Egypt owes its fertility. So thin are they, that the aggregate amount accumulated in a century is said rarely to exceed five inches, although in the course of thousands of years it has attained a vast thickness, the bottom not having been reached by borings extending to a depth of 60 feet towards the central parts of the valley. Everywhere it consists of the same homogeneous mud, destitute of stratification— the only signs of successive accumulation being where the Nile has silted up its channel, or where the blown sands of the Libyan desert have invaded the plain, and given rise to alternate layers of sand and mud.

In European river-loams we occasionally observe isolated pebbles and angular pieces of stone which have been floated by ice to the places where they now occur; but no such coarse materials are met with in the plains of Egypt.

In some parts of the valley of the Rhine the accumulation of similar loam, called in Germany "loess," has taken place on an enormous scale. Its colour is yellowish-grey, and very homogeneous; and Professor Bischoff has ascertained, by analysis, that it agrees in composition with the mud of the Nile. Although for the most part unstratified, it betrays in some places marks of stratification, especially where it contains calcareous concretions, or in its lower part where it rests on subjacent gravel and sand which alternate with each other near the junction. About a sixth part of the whole mass is composed of carbonate of lime, and there is usually an intermixture of fine quartzose and micaceous sand.

(FIGURE 88. Succinea elongata.)

Although this loam of the Rhine is unsolidified, it usually terminates where it has been undermined by running water in a vertical cliff, from the face of which shells of terrestrial, fresh-water and amphibious mollusks project in relief. These shells do not imply the permanent sojourn of a body of fresh water on the spot, for the most aquatic of them, the Succinea, inhabits marshes and wet grassy meadows. The Succinea elongata (or S. oblongata), Figure 88, is very characteristic both of the loess of the Rhine and of some other European river- loams.

(FIGURE 89. Pupa muscorum (Linn.).)
(FIGURE 90. Helix hispida (Linn.) (plebeia).)

Among the land-shells of the Rhenish loess, Helix hispida, Figure 90, and Pupa muscorum, Figure 89, are very common. Both the terrestrial and aquatic shells are of most fragile and delicate structure, and yet they are almost invariably perfect and uninjured. They must have been broken to pieces had they been swept along by a violent inundation. Even the colour of some of the land-shells, as that of Helix nemoralis, is occasionally preserved.

In parts of the valley of the Rhine, between Bingen and Basle, the fluviatile loam or loess now under consideration is several hundred feet thick, and contains here and there throughout that thickness land and amphibious shells. As it is seen in masses fringing both sides of the great plain, and as occasionally remnants of it occur in the centre of the valley, forming hills several hundred feet in height, it seems necessary to suppose, first, a time when it slowly accumulated; and secondly, a later period, when large portions of it were removed, or when the original valley, which had been partially filled up with it, was re-excavated.

Such changes may have been brought about by a great movement of oscillation, consisting first of a general depression of the land, and then of a gradual re- elevation of the same. The amount of continental depression which first took place in the interior, must be imagined to have exceeded that of the region near the sea, in which case the higher part of the great valley would have its alluvial plain gradually raised by an accumulation of sediment, which would only cease when the subsidence of the land was at an end. If the direction of the movement was then reversed, and, during the re-elevation of the continent, the inland region nearest the mountains should rise more rapidly than that near the coast, the river would acquire a denuding power sufficient to enable it to sweep away gradually nearly all the loam and gravel with which parts of its basin had been filled up. Terraces and hillocks of mud and sand would then alone remain to attest the various levels at which the river had thrown down and afterwards removed alluvial matter.

CAVERN DEPOSITS CONTAINING HUMAN REMAINS AND BONES OF EXTINCT ANIMALS.

In England, and in almost all countries where limestone rocks abound, caverns are found, usually consisting of cavities of large dimensions, connected together by low, narrow, and sometimes torturous galleries or tunnels. These subterranean vaults are usually filled in part with mud, pebbles, and breccia, in which bones occur belonging to the same assemblage of animals as those characterising the Post-pliocene alluvia above described. Some of these bones are referable to extinct and others to living species, and they are occasionally intermingled, as in the valley-gravels, with implements of one or other of the great divisions of the stone age, and these are not unfrequently accompanied by human bones, which are much more common in cavern deposits than in valley- alluvium.

Each suite of caverns, and the passages by which they communicate the one with the other, afford memorials to the geologist of successive phases through which they must have passed. First, there was a period when the carbonate of lime was carried out gradually by springs; secondly, an era when engulfed rivers or occasional floods swept organic and inorganic debris into the subterranean hollows previously formed; and thirdly, there were such changes in the configuration of the region as caused the engulfed rivers to be turned into new channels, and springs to be dried up, after which the cave-mud, breccia, gravel, and fossil bones would bear the same kind of relation to the existing drainage of the country as the older valley-drifts with their extinct mammalian remains and works of art bear to the present rivers and alluvial plains.

The quarrying away of large masses of Carboniferous and Devonian limestone, near Liege, in Belgium, has afforded the geologist magnificent sections of some of these caverns, and the former communication of cavities in the interior of the rocks with the old surface of the country by means of vertical or oblique fissures, has been demonstrated in places where it would not otherwise have been suspected, so completely have the upper extremities of these fissures been concealed by superficial drift, while their lower ends, which extended into the roofs of the caves, are masked by stalactitic incrustations.

The origin of the stalactite is thus explained by the eminent chemist Liebig. Mould or humus, being acted on by moisture and air, evolves carbonic acid, which is dissolved by rain. The rain-water, thus impregnated, permeates the porous limestone, dissolves a portion of it, and afterwards, when the excess of carbonic acid evaporates in the caverns, parts with the calcareous matter, and forms stalactite. Even while caverns are still liable to be occasionally flooded such calcareous incrustations accumulate, but it is generally when they are no longer in the line of drainage that a solid floor of hard stalagmite is formed on the bottom.

The late Dr. Schmerling examined forty caves near Liege, and found in all of them the remains of the same fauna, comprising the mammoth, tichorhine rhinoceros, cave-bear, cave-hyaena, cave-lion, and many others, some of extinct and some of living species, and in all of them flint implements. In four or five caves only parts of human skeletons were met with, comprising sometimes skulls with a few other bones, sometimes nearly every part of the skeleton except the skull. In one of the caves, that of Engihoul, where Schmerling had found the remains of at least three human individuals, they were mingled in such a manner with bones of extinct mammalia, as to leave no doubt on his mind (in 1833) of man having co-existed with them.

In 1860, Professor Malaise, of Liege, explored with me this same cave of Engihoul, and beneath a hard floor of stalagmite we found mud full of bones of extinct and recent animals, such as Schmerling had described, and my companion, persevering in his researches after I had returned to England, extracted from the same deposit two human lower jaw-bones retaining their teeth. The skulls from these Belgian caverns display no marked deviation from the normal European type of the present day.

The careful investigations carried on by Dr. Falconer, Mr. Pengelly, and others, in the Brixham cave near Torquay, in 1858, demonstrated that flint knives were there imbedded in such a manner in loam underlying a floor of stalagmite as to prove that man had been an inhabitant of that region when the cave-bear and other members of the ancient post-pliocene fauna were also in existence.

The absence of gnawed bones had led Dr. Schmerling to infer that none of the Belgian caves which he explored had served as the dens of wild beasts; but there are many caves in Germany and England which have certainly been so inhabited, especially by the extinct hyaena and bear.

A fine example of a hyaena's den was afforded by the cave of Kirkdale, so well described by the late Dr. Buckland in his Reliquiae Diluvianae. In that cave, above twenty-five miles north-north-east of York, the remains of about 300 hyaenas, belonging to individuals of every age, were detected. The species (Hyaena spelaea) has been considered by palaeontologists as extinct; it was larger than the fierce Hyaena crocuta of South Africa, which it closely resembled, and of which it is regarded by Mr. Boyd Dawkins as a variety. Dr. Buckland, after carefully examining the spot, proved that the hyaenas must have lived there; a fact attested by the quantity of their dung, which, as in the case of the living hyaena, is of nearly the same composition as bone, and almost as durable. In the cave were found the remains of the ox, young elephant, hippopotamus, rhinoceros, horse, bear, wolf, hare, water-rat, and several birds. All the bones have the appearance of having been broken and gnawed by the teeth of the hyaenas; and they occur confusedly mixed in loam or mud, or dispersed through a crust of stalagmite which covers it. In these and many other cases it is supposed that portions of herbivorous quadrupeds have been dragged into caverns by beasts of prey, and have served as their food— an opinion quite consistent with the known habits of the living hyaena.

AUSTRALIAN CAVE-BRECCIAS.

Ossiferous breccias are not confined to Europe, but occur in all parts of the globe; and those discovered in fissures and caverns in Australia correspond closely in character with what has been called the bony breccia of the Mediterranean, in which the fragments of bone and rock are firmly bound together by a red ochreous cement.

Some of these caves were examined by the late Sir T. Mitchell in the Wellington Valley, about 210 miles west of Sidney, on the river Bell, one of the principal sources of the Macquarie, and on the Macquarie itself. The caverns often branch off in different directions through the rock, widening and contracting their dimensions, and the roofs and floors are covered with stalactite. The bones are often broken, but do not seem to be water-worn. In some places they lie imbedded in loose earth, but they are usually included in a breccia.

The remains belong to marsupial animals. Among the most abundant are those of the kangaroo, of which there are four species, while others belong to the genera Phascolomys, the wombat; Dasyurus, the ursine opossum; Phalangista, the vulpine opossum; and Hypsiprymnus, the kangaroo-rat.

(FIGURE 91. Part of lower jaw of Macropus atlas. Owen. A young individual of an extinct species. a. Permanent false molar, in the alveolus.)

(FIGURE 92. Lower jaw of largest living species of kangaroo. (Macropus major.))

In the fossils above enumerated, several species are larger than the largest living ones of the same genera now known in Australia. Figure 91 of the right side of a lower jaw of a kangaroo (Macropus atlas, Owen) will at once be seen to exceed in magnitude the corresponding part of the largest living kangaroo, which is represented in Figure 92. In both these specimens part of the substance of the jaw has been broken open, so as to show the permanent false molar (a, Figure 91), concealed in the socket. From the fact of this molar not having been cut, we learn that the individual was young, and had not shed its first teeth.

The reader will observe that all these extinct quadrupeds of Australia belong to the marsupial family, or, in other words, that they are referable to the same peculiar type of organisation which now distinguishes the Australian mammalia from those of other parts of the globe. This fact is one of many pointing to a general law deducible from the fossil vertebrate and invertebrate animals of times immediately antecedent to our own, namely, that the present geographical distribution of organic FORMS dates back to a period anterior to the origin of existing SPECIES; in other words, the limitation of particular genera or families of quadrupeds, mollusca, etc., to certain existing provinces of land and sea, began before the larger part of the species now contemporary with man had been introduced into the earth.

Professor Owen, in his excellent "History of British Fossil Mammals," has called attention to this law, remarking that the fossil quadrupeds of Europe and Asia differ from those of Australia or South America. We do not find, for example, in the Europaeo-Asiatic province fossil kangaroos, or armadillos, but the elephant, rhinoceros, horse, bear, hyaena, beaver, hare, mole, and others, which still characterise the same continent.

In like manner, in the Pampas of South America the skeletons of Megatherium, Megalonyx, Glyptodon, Mylodon, Toxodon, Macrauchenia, and other extinct forms, are analogous to the living sloth, armadillo, cavy, capybara, and llama. The fossil quadrumana, also associated with some of these forms in the Brazilian caves, belong to the Platyrrhine family of monkeys, now peculiar to South America. That the extinct fauna of Buenos Ayres and Brazil was very modern has been shown by its relation to deposits of marine shells, agreeing with those now inhabiting the Atlantic.

The law of geographical relationship above alluded to, between the living vertebrata of every great zoological province and the fossils of the period immediately antecedent, even where the fossil species are extinct, is by no means confined to the mammalia. New Zealand, when first examined by Europeans, was found to contain no indigenous land quadrupeds, no kangaroos, or opossums, like Australia; but a wingless bird abounded there, the smallest living representative of the ostrich family, called the Kiwi by the natives (Apteryx). In the fossils of the Post-pliocene period in this same island, there is the like absence of kangaroos, opossums, wombats, and the rest; but in their place a prodigious number of well-preserved specimens of gigantic birds of the struthious order, called by Owen Dinornis and Palapteryx, which are entombed in superficial deposits. These genera comprehended many species, some of which were four, some seven, others nine, and others eleven feet in height! It seems doubtful whether any contemporary mammalia shared the land with this population of gigantic feathered bipeds.

Mr. Darwin, when describing the recent and fossil mammalia of South America, has dwelt much on the wonderful relationship of the extinct to the living types in that part of the world, inferring from such geographical phenomena that the existing species are all related to the extinct ones which preceded them by a bond of common descent.

CLIMATE OF THE POST-PLIOCENE PERIOD.

The evidence as to the climate of Europe during this epoch is somewhat conflicting. The fluviatile and land-shells are all of existing species, but their geographical range has not always been the same as at present. Some, for example, which then lived in Britain are now only found in Norway and Finland, probably implying that the Post-pliocene climate of Britain was colder, especially in the winter. So also the reindeer and the musk-ox (Ovibos moschatus), now inhabitants of the Arctic regions, occur fossil in the valleys of the Thames and Avon, and also in France and Germany, accompanied in most places by the mammoth and the woolly rhinoceros. At Grays in Essex, on the other hand, another species both of elephant and rhinoceros occurs, together with a hippopotamus and the Cyrena fluminalis, a shell now extinct in Europe but still an inhabitant of the Nile and some Asiatic rivers. With it occurs the Unio littoralis, now living in the Seine and Loire. In the valley of the Somme flint tools have been found associated with Hippopotamus major and Cyrena fluminalis in the lower-level Post-pliocene gravels; while in the higher-level (and more ancient) gravels similar tools are more abundant, and are associated with the bones of the mammoth and other Post-pliocene quadrupeds indicative of a colder climate.

It is possible that we may here have evidence of summer and winter migrations rather than of a general change of temperature. Instead of imagining that the hippopotamus lived all the year round with the musk-ox and lemming, we may rather suppose that the apparently conflicting evidence may be due to the place of our observations being near the boundary line of a northern and southern fauna, either of which may have advanced or receded during comparatively slight and temporary fluctuations of climate. There may then have been a continuous land communication between England and the North of Siberia, as well as in an opposite direction with Africa, then united to Southern Europe.

In drift at Fisherton, near Salisbury, thirty feet above the river Wiley, the Greenland lemming and a new species of the Arctic genus Spermophilus have been found, along with the mammoth, reindeer, cave-hyaena, and other mammalia suited to a cold climate. A flint implement was taken out from beneath the bones of the mammoth. In a higher and older deposit in the vicinity, flint tools like those of Amiens have been discovered. Nearly all the known Post-pliocene quadrupeds have now been found accompanying flint knives or hatchets in such a way as to imply their coexistence with man; and we have thus the concurrent testimony of several classes of geological facts to the vast antiquity of the human race. In the first place, the disappearance of a great variety of species of wild animals from every part of a wide continent must have required a vast period for its accomplishment; yet this took place while man existed upon the earth, and was completed before that early period when the Danish shell-mounds were formed or the oldest of the Swiss lake-dwellings constructed. Secondly, the deepening and widening of valleys, indicated by the position of the river gravels at various heights, implies an amount of change of which that which has occurred during the historical period forms a scarcely perceptible part. Thirdly, the change in the course of rivers which once flowed through caves now removed from any line of drainage, and the formation of solid floors of stalagmite, must have required a great lapse of time. Lastly, ages must have been required to change the climate of wide regions to such an extent as completely to alter the geographical distribution of many mammalia as well as land and fresh-water shells. The 3000 or 4000 years of the historical period does not furnish us with any appreciable measure for calculating the number of centuries which would suffice for such a series of changes, which are by no means of a local character, but have operated over a considerable part of Europe.

RELATIVE LONGEVITY OF SPECIES IN THE MAMMALIA AND TESTACEA.

I called attention in 1830 to the fact, which had not at that time attracted notice, that the association in the Post-pliocene deposits of shells, exclusively of living species, with many extinct quadrupeds betokened a longevity of species in the testacea far exceeding that in the mammalia. (Principles of Geology 1st edition volume 3 page 140.) Subsequent researches seem to show that this greater duration of the same specific forms in the class mollusca is dependent on a still more general law, namely, that the lower the grade of animals, or the greater the simplicity of their structure, the more persistent are they in general in their specific characters throughout vast periods of time. Not only have the invertebrata, as shown by geological data, altered at a less rapid rate than the vertebrata, but if we take one of the classes of the former, as for example the mollusca, we find those of more simple structure to have varied at a slower rate than those of a higher and more complex organisation; the Brachiopoda, for example, more slowly than the lamellibranchiate bivalves, while the latter have been more persistent than the univalves, whether gasteropoda or cephalopoda. In like manner the specific identity of the characters of the foraminifera, which are among the lowest types of the invertebrata, has outlasted that of the mollusca in an equally decided manner.

TEETH OF POST-PLIOCENE MAMMALIA.

To those who have never studied comparative anatomy, it may seem scarcely credible that a single bone taken from any part of the skeleton may enable a skilful osteologist to distinguish, in many cases, the genus, and sometimes the species, of quadrupeds to which it belonged. Although few geologists can aspire to such knowledge, which must be the result of long practice and study, they will nevertheless derive great advantage from learning, what is comparatively an easy task, to distinguish the principal divisions of the mammalia by the forms and characters of their teeth.

Figures 93 through 105 represent the teeth of some of the more common species and genera found in alluvial and cavern deposits.

(FIGURE 93. Elephas primigenius (or Mammoth); molar of upper jaw, right side; one-third of natural size. Post-pliocene. a. Grinding surface. b. Side view.)

(FIGURE 94. Elephas antiquus, Falconer. Penultimate molar, one-third of natural size. Post-pliocene and Pliocene.)

(FIGURE 95. Elephas meridionalis, Nesti. Penultimate molar, one-third of natural size. Post-pliocene and Pliocene.)

(FIGURE 96. Rhinoceros leptorhinus, Cuvier— Rhin. megarhinus, Christol; fossil from fresh-water beds of Grays, Essex; penultimate molar, lower jaw, left side; two-thirds of natural size. Post-pliocene and Newer Pliocene.)

(FIGURE 97. Rhinoceros tichorhinus; penultimate molar, lower jaw, left side; two-thirds of natural size. Post-pliocene.)

(FIGURE 98. Hippopotamus; from cave near Palermo; molar tooth; two-thirds of natural size. Post-pliocene.)

(FIGURE 99. Horse. Equus caballus, L. (common horse); from the shell-marl, Forfarshire; second molar, lower jaw. Recent. a. Grinding surface, two-thirds natural size. b. Side view of same, half natural size.)

(FIGURE 100. Deer. Moose (Cervus alces, L.); recent; molar of upper jaw. a. Grinding surface. b. Side view, two-thirds of natural size.)

(FIGURE 101. Ox. Ox, common, from shell-marl, Forfarshire; true molar, upper jaw; two-thirds natural size. Recent. c. Grinding surface. d. Side view, fangs uppermost.)

(FIGURE 102. Bear. a. Canine tooth or tusk of bear (Ursus spelaeus); from cave near Liege. b. Molar of left side, upper jaw; one-third of natural size. Post-pliocene.)

(FIGURE 103. Tiger. c. Canine tooth of tiger (Felis tigris); recent. d. Outside view of posterior molar, lower jaw: one-third of natural size. Recent.)

(FIGURE 104. Hyaena spelaea, Goldf. (variety of H. crocuta); lower jaw. Kent's Hole, Torquay, Devonshire; one-third natural size. Post-pliocene.)

(FIGURE 105. Teeth of a new species of Arvicola, field-mouse; from the Norwich Crag. Newer Pliocene. a. Grinding surface. b. Side view of the same. c. Natural size of a and b.)

On comparing the grinding surfaces of the corresponding molars of the three species of elephants, Figures 93, 94, 95 it will be seen that the folds of enamel are most numerous in the mammoth, fewer and wider, or more open, in E. antiquus; and most open and fewest in E. meridionalis. It will be also seen that the enamel in the molar of the Rhinoceros tichorhinus (Figure 97), is much thicker than in that of the Rhinoceros leptorhinus (Figure 96).

CHAPTER XI.

POST-PLIOCENE PERIOD, CONTINUED.— GLACIAL CONDITIONS. (As to the former excess of cold, whether brought about by modifications in the height and distribution of the land or by altered astronomical conditions, see Principles volume 1 10th edition 1867 chapters 12 and 13 "Vicissitudes of Climate.")

Geographical Distribution, Form, and Characters of Glacial Drift.
Fundamental Rocks, polished, grooved, and scratched.
Abrading and striating Action of Glaciers.
Moraines, Erratic Blocks, and "Roches Moutonnees."
Alpine Blocks on the Jura.
Continental Ice of Greenland.
Ancient Centres of the Dispersion of Erratics.
Transportation of Drift by floating Icebergs.
Bed of the Sea furrowed and polished by the running aground of floating Ice-islands.

CHARACTER AND DISTRIBUTION OF GLACIAL DRIFT.

In speaking of the loose transported matter commonly found on the surface of the land in all parts of the globe, I alluded to the exceptional character of what has been called the boulder formation in the temperate and Arctic latitudes of the northern hemisphere. The peculiarity of its form in Europe north of the 50th, and in North America north of the 40th parallel of latitude, is now universally attributed to the action of ice, and the difference of opinion respecting it is now chiefly restricted to the question whether land-ice or floating icebergs have played the chief part in its distribution. It is wanting in the warmer and equatorial regions, and reappears when we examine the lands which lie south of the 40th and 50th parallels in the southern hemisphere, as, for example, in Patagonia, Tierra del Fuego, and New Zealand. It consists of sand and clay, sometimes stratified, but often wholly devoid of stratification for a depth of 50, 100, or even a greater number of feet. To this unstratified form of the deposit the name of TILL has long been applied in Scotland. It generally contains a mixture of angular and rounded fragments of rock, some of large size, having occasionally one or more of their sides flattened and smoothed, or even highly polished. The smoothed surfaces usually exhibit many scratches parallel to each other, one set of which often crosses an older set. The till is almost everywhere wholly devoid of organic remains, except those washed into it from older formations, though in some places it contains marine shells, usually of northern or Arctic species, and frequently in a fragmentary state. The bulk of the till has usually been derived from the grinding down into mud of rocks in the immediate neighbourhood, so that it is red in a region of Red Sandstone, as in Strathmore in Forfarshire; grey or black in a district of coal and bituminous shale, as around Edinburgh; and white in a chalk country, as in parts of Norfolk and Denmark. The stony fragments dispersed irregularly through the till usually belong, especially in mountainous countries, to rocks found in some part of the same hydrographical basin; but there are regions where the whole of the boulder clay has come from a distance, and huge blocks, or "erratics," as they have been called, many feet in diameter, have not unfrequently travelled hundreds of miles from their point of departure, or from the parent rocks from which they have evidently been detached. These are commonly angular, and have often one or more of their sides polished and furrowed.

The rock on which the boulder formation reposes, if it consists of granite, gneiss, marble, or other hard stone, capable of permanently retaining any superficial markings which may have been imprinted upon it, is usually smoothed or polished, like the erratics above described, and exhibits parallel striae and furrows having a determinate direction. This direction, both in Europe and North America, agrees generally in a marked manner with the course taken by the erratic blocks in the same district. The boulder clay, when it was first studied, seemed in many of its characters so singular and anomalous, that geologists despaired of ever being able to interpret the phenomena by reference to causes now in action. In those exceptional cases where marine shells of the same date as the boulder clay were found, nearly all of them were recognised as living species— a fact conspiring with the superficial position of the drift to indicate a comparatively modern origin.

The term "diluvium" was for a time the most popular name of the boulder formation, because it was referred by many to the deluge of Noah, while others retained the name as expressive of their opinion that a series of diluvial waves raised by hurricanes and storms, or by earthquakes, or by the sudden upheaval of land from the bed of the sea, had swept over the continents, carrying with them vast masses of mud and heavy stones, and forcing these stones over rocky surfaces so as to polish and imprint upon them long furrows and striae. But geologists were not long in seeing that the boulder formation was characteristic of high latitudes, and that on the whole the size and number of erratic blocks increases as we travel towards the Arctic regions. They could not fail to be struck with the contrast which the countries bordering the Baltic presented when compared with those surrounding the Mediterranean. The multitude of travelled blocks and striated rocks in the one region, and the absence of such appearances in the other, were too obvious to be overlooked. Even the great development of the boulder formation, with large erratics so far south as the Alps, offered an exception to the general rule favourable to the hypothesis that there was some intimate connection between it and accumulations of snow and ice.

TRANSPORTING AND ABRADING POWER OF GLACIERS.

(FIGURE 106. Limestone, polished, furrowed, and scratched by the glacier of Rosenlau in Switzerland. (Agassiz.) a a. White streaks or scratches, caused by small grains of flint frozen into the ice. b b. Furrows.)

I have described elsewhere ("Principles" volume 1 chapter 16 1867) the manner in which the snow of the Alpine heights is prevented from accumulating indefinitely in thickness by the constant descent of a large portion of it by gravitation. Becoming converted into ice it forms what are termed glaciers, which glide down the principal valleys. On their surface are seen mounds of rubbish or large heaps of sand and mud, with angular fragments of rock which fall from the steep slopes or precipices bounding the glaciers. When a glacier, thus laden, descends so far as to reach a region about 3500 feet above the level of the sea, the warmth of the air is such that it melts rapidly in summer, and all the mud, sand, and pieces of rock are slowly deposited at its lower end, forming a confused heap of unstratified rubbish called a MORAINE, and resembling the TILL before described.

Besides the blocks thus carried down on the top of the glacier, many fall through fissures in the ice to the bottom, where some of them become firmly frozen into the mass, and are pushed along the base of the glacier, abrading, polishing, and grooving the rocky floor below, as a diamond cuts glass, or as emery-powder polishes steel. The striae which are made, and the deep grooves which are scooped out by this action, are rectilinear and parallel to an extent never seen in those produced on loose stones or rocks, where shingle is hurried along by a torrent, or by the waves on a sea-beach. In addition to these polished, striated, and grooved surfaces of rock, another mark of the former action of a glacier is the "roche moutonnee." Projecting eminences of rock so called have been smoothed and worn into the shape of flattened domes by the glacier as it passed over them. They have been traced in the Alps to great heights above the present glaciers, and to great horizontal distances beyond them.

ALPINE BLOCKS ON THE JURA.

The moraines, erratics, polished surfaces, domes, and striae, above described, are observed in the great valley of Switzerland, fifty miles broad; and almost everywhere on the Jura, a chain which lies to the north of this valley. The average height of the Jura is about one-third that of the Alps, and it is now entirely destitute of glaciers; yet it presents almost everywhere similar moraines, and the same polished and grooved surfaces. The erratics, moreover, which cover it, present a phenomenon which has astonished and perplexed the geologist for more than half a century. No conclusion can be more incontestable than that these angular blocks of granite, gneiss, and other crystalline formations came from the Alps, and that they have been brought for a distance of fifty miles and upward across one of the widest and deepest valleys in the world; so that they are now lodged on a chain composed of limestone and other formations, altogether distinct from those of the Alps. Their great size and angularity, after a journey of so many leagues, has justly excited wonder; for hundreds of them are as large as cottages; and one in particular, composed of gneiss, celebrated under the name of Pierre a Bot, rests on the side of a hill about 900 feet above the lake of Neufchatel, and is no less than 40 feet in diameter.

In the year 1821, M. Venetz first announced his opinion that the Alpine glaciers must formerly have extended far beyond their present limits, and the proofs appealed to by him in confirmation of this doctrine were acknowledged by all subsequent observers, and greatly strengthened by new observations and arguments. M. Charpentier supposed that when the glaciers extended continuously from the Alps to the Jura, the former mountains were 2000 or 3000 feet higher than at present. Other writers, on the contrary, conjectured that the whole country had been submerged, and the moraines and erratic blocks transported on floating icebergs; but a careful study of the distribution of the travelled masses, and the total absence of marine shells from the old glacial drift of Switzerland, have entirely disproved this last hypothesis. In addition to the many evidences of the action of ice in the northern parts of Europe which we have already mentioned, there occur here and there in some of these countries, what are wanting in Switzerland, deposits of marine fossil shells, which exhibit so arctic a character that they must have led the geologist to infer the former prevalence of a much colder climate, even had he not encountered so many accompanying signs of ice-action. The same marine shells demonstrate the submergence of large areas in Scandinavia and the British Isles, during the glacial cold.

A characteristic feature of the deposits under consideration in all these countries is the occurrence of large erratic blocks, and sometimes of moraine matter, in situations remote from lofty mountains, and separated from the nearest points where the parent rocks appear at the surface by great intervening valleys, or arms of the sea. We also often observe striae and furrows, as in Norway, Sweden, and Scotland, which deviate from the direction which they ought to follow if they had been connected with the present line of drainage, and they, therefore, imply the prevalence of a very distinct condition of things at the time when the cold was most intense. The actual state of North Greenland seems to afford the best explanation of such abnormal glacial markings.

GREENLAND CONTINENTAL ICE.

Greenland is a vast unexplored continent, buried under one continuous and colossal mass of ice that is always moving seaward, a very small part of it in an easterly direction, and all the rest westward, or towards Baffin's Bay. All the minor ridges and valleys are levelled and concealed under a general covering of snow, but here and there some steep mountains protrude abruptly from the icy slope, and a few superficial lines of stones or moraines are visible at certain seasons, when no snow has fallen for many months, and when evaporation, promoted by the wind and sun, has caused much of the upper snow to disappear. The height of this continent is unknown, but it must be very great, as the most elevated lands of the outskirts, which are described as comparatively low, attain altitudes of 4000 to 6000 feet. The icy slope gradually lowers itself towards the outskirts, and then terminates abruptly in a mass about 2000 feet in thickness, the great discharge of ice taking place through certain large friths, which, at their upper ends, are usually about four miles across. Down these friths the ice is protruded in huge masses, several miles wide, which continue their course— grating along the rocky bottom like ordinary glaciers long after they have reached the salt water. When at last they arrive at parts of Baffin's Bay deep enough to buoy up icebergs from 1000 to 1500 feet in vertical thickness, broken masses of them float off, carrying with them on their surface not only fine mud and sand but large stones. These fragments of rock are often polished and scored on one or more sides, and as the ice melts, they drop down to the bottom of the sea, where large quantities of mud are deposited, and this muddy bottom is inhabited by many mollusca.

Although the direction of the ice-streams in Greenland may coincide in the main with that which separate glaciers would take if there were no more ice than there is now in the Swiss Alps, yet the striation of the surface of the rocks on an ice-clad continent would, on the whole, vary considerably in its minor details from that which would be imprinted on rocks constituting a region of separate glaciers. For where there is a universal covering of ice there will be a general outward movement from the higher and more central regions towards the circumference and lower country, and this movement will be, to a certain extent, independent of the minor inequalities of hill and valley, when these are all reduced to one level by the snow. The moving ice may sometimes cross even at right angles deep narrow ravines, or the crests of buried ridges, on which last it may afterwards seem strange to detect glacial striae and polishing after the liquefaction of the snow and ice has taken place.

Rink mentions that in North Greenland powerful springs of clayey water escape in winter from under the ice, where it descends to "the outskirts," and where, as already stated, it is often 2000 feet thick— a fact showing how much grinding action is going on upon the surface of the subjacent rocks. I also learn from Dr. Torell that there are large areas in the outskirts, now no longer covered with permanent snow or glaciers, which exhibit on their surface unmistakable signs of ancient ice-action, so that, vast as is the power now exerted by ice in Greenland, it must once have operated on a still grander scale. The land, though now very elevated, may perhaps have been formerly much higher. It is well-known that the south coast of Greenland, from latitude 60 degrees to about 70 degrees north, has for the last four centuries been sinking at the rate of several feet in a century. By this means a surface of rock, well scored and polished by ice, is now slowly subsiding beneath the sea, and is becoming strewed over, as the icebergs melt, with impalpable mud and smoothed and scratched stones. It is not precisely known how far north this downward movement extends.

DRIFT CARRIED BY ICEBERGS.

An account was given so long ago as the year 1822, by Scoresby, of icebergs seen by him in the Arctic seas drifting along in latitudes 69 and 70 degrees north, which rose above the surface from 100 to 200 feet, and some of which measured a mile in circumference. Many of them were loaded with beds of earth and rock, of such thickness that the weight was conjectured to be from 50,000 to 100,000 tons. A similar transportation of rocks is known to be in progress in the southern hemisphere, where boulders included in ice are far more frequent than in the north. One of these icebergs was encountered in 1839, in mid-ocean, in the antarctic regions, many hundred miles from any known land, sailing northward, with a large erratic block firmly frozen into it. Many of them, carefully measured by the officers of the French exploring expedition of the Astrolabe, were between 100 and 225 feet high above water, and from two to five miles in length. Captain d'Urville ascertained one of them which he saw floating in the Southern Ocean to be 13 miles long and 100 feet high, with walls perfectly vertical. The submerged portions of such islands must, according to the weight of ice relatively to sea-water, be from six to eight times more considerable than the part which is visible, so that when they are once fairly set in motion, the mechanical force which they might exert against any obstacle standing in their way would be prodigious.

We learn, therefore, from a study both of the arctic and antarctic regions, that a great extent of land may be entirely covered throughout the whole year by snow and ice, from the summits of the loftiest mountains to the sea-coast, and may yet send down angular erratics to the ocean. We may also conclude that such land will become in the course of ages almost everywhere scored and polished like the rocks which underlie a glacier. The discharge of ice into the surrounding sea will take place principally through the main valleys, although these are hidden from our sight. Erratic blocks and moraine matter will be dispersed somewhat irregularly after reaching the sea, for not only will prevailing winds and marine currents govern the distribution of the drift, but the shape of the submerged area will have its influence; inasmuch as floating ice, laden with stones, will pass freely through deep water, while it will run a ground where there are reefs and shallows. Some icebergs in Baffin's Bay have been seen stranded on a bottom 1000 or even 1500 feet deep. In the course of ages such a sea-bed may become densely covered with transported matter, from which some of the adjoining greater depths may be free. If, as in West Greenland, the land is slowly sinking, a large extent of the bottom of the ocean will consist of rock polished and striated by land-ice, and then overspread by mud and boulders detached from melting bergs.

The mud, sand, and boulders thus let fall in still water must be exactly like the moraines of terrestrial glaciers, devoid of stratification and organic remains. But occasionally, on the outer side of such packs of stranded bergs, the waves and currents may cause the detached earthy and stony materials to be sorted according to size and weight before they reach the bottom, and to acquire a stratified arrangement.

I have already alluded to the large quantity of ice, containing great blocks of stone, which is sometimes seen floating far from land, in the southern or Antarctic seas. After the emergence, therefore, of such a submarine area, the superficial detritus will have no necessary relation to the hills, valleys, and river-plains over which it will be scattered. Many a water-shed may intervene between the starting-point of each erratic or pebble and its final resting- place, and the only means of discovering the country from which it took its departure will consist in a careful comparison of its mineral or fossil contents with those of the parent rocks.

CHAPTER XII.
POST-PLIOCENE PERIOD, CONTINUED.— GLACIAL CONDITIONS, CONCLUDED.

Glaciation of Scandinavia and Russia.
Glaciation of Scotland.
Mammoth in Scotch Till.
Marine Shells in Scotch Glacial Drift.
Their Arctic Character.
Rarity of Organic Remains in Glacial Deposits.
Contorted Strata in Drift.
Glaciation of Wales, England, and Ireland.
Marine Shells of Moel Tryfaen.
Erratics near Chichester.
Glacial Formations of North America.
Many Species of Testacea and Quadrupeds survived the Glacial Cold.
Connection of the Predominance of Lakes with Glacial Action.
Action of Ice in preventing the silting up of Lake-basins.

Absence of Lakes in the Caucasus.
Equatorial Lakes of Africa.
GLACIATION OF SCANDINAVIA AND RUSSIA.
In large tracts of Norway and Sweden, where there have been no glaciers in historical times, the signs of ice-action have been traced as high as 6000 feet above the level of the sea. These signs consist chiefly of polished and furrowed rock-surfaces, of moraines and erratic blocks. The direction of the erratics, like that of the furrows, has usually been conformable to the course of the principal valleys; but the lines of both sometimes radiate outward in all directions from the highest land, in a manner which is only explicable by the hypothesis above alluded to of a general envelope of continental ice, like that of Greenland (Chapter 11.) Some of the far-transported blocks have been carried from the central parts of Scandinavia towards the Polar regions; others southward to Denmark; some south-westward, to the coast of Norfolk in England; others south-eastward, to Germany, Poland, and Russia.

In the immediate neighbourhood of Upsala, in Sweden, I had observed, in 1834, a ridge of stratified sand and gravel, in the midst of which occurs a layer of marl, evidently formed originally at the bottom of the Baltic, by the slow growth of the mussel, cockle, and other marine shells of living species, intermixed with some proper to fresh water. The marine shells are all of dwarfish size, like those now inhabiting the brackish waters of the Baltic; and the marl, in which many of them are imbedded, is now raised more than 100 feet above the level of the Gulf of Bothnia. Upon the top of this ridge repose several huge erratics, consisting of gneiss for the most part unrounded, from nine to sixteen feet in diameter, and which must have been brought into their present position since the time when the neighbouring gulf was already characterised by its peculiar fauna. Here, therefore, we have proof that the transport of erratics continued to take place, not merely when the sea was inhabited by the existing testacea, but when the north of Europe had already assumed that remarkable feature of its physical geography which separates the Baltic from the North Sea, and causes the Gulf of Bothnia to have only one- fourth of the saltness belonging to the ocean. In Denmark, also, recent shells have been found in stratified beds, closely associated with the boulder clay.

GLACIATION OF SCOTLAND.
Mr. T.F. Jamieson, in 1858, adduced a great body of facts to prove that the Grampians once sent down glaciers from the central regions in all directions towards the sea. "The glacial grooves," he observed, "radiate outward from the central heights towards all points of the compass, though they do not always strictly conform to the actual shape and contour of the minor valleys and ridges."

These facts and other characteristics of the Scotch drift lead us to the inference that when the glacial cold first set in, Scotland stood higher above the sea than at present, and was covered for the most part with snow and ice, as Greenland is now. This sheet of land-ice sliding down to lower levels, ground down and polished the subjacent rocks, sweeping off nearly all superficial deposits of older date, and leaving only till and boulders in their place. To this continental state succeeded a period of depression and partial submergence. The sea advanced over the lower lands, and Scotland was converted into an archipelago, some marine sand with shells being spread over the bottom of the sea. On this sand a great mass of boulder clay usually quite devoid of fossils was accumulated. Lastly, the land re-emerged from the water, and, reaching a level somewhat above its present height, became connected with the continent of Europe, glaciers being formed once more in the higher regions, though the ice probably never regained its former extension. (Jamieson Quarterly Geological Journal 1860 volume 16 page 370.) After all these changes, there were some minor oscillations in the level of the land, on which, although they have had important geographical consequences, separating Ireland from England, for example, and England from the Continent, we need not here enlarge.

MAMMOTH IN SCOTCH TILL.
Almost all remains of the terrestrial fauna of the Continent which preceded the period of submergence have been lost; but a few patches of estuarine and fresh- water formations escaped denudation by submergence. To these belong the peaty clay from which

several mammoths' tusks and horns of reindeer were obtained at Kilmaurs, in Ayrshire as long ago as 1816. Mr. Bryce in 1865 ascertained that the fresh-water formation containing these fossils rests on carboniferous sandstone, and is covered, first by a bed of marine sand with arctic shells, and then with a great mass of till with glaciated boulders. (Bryce Quarterly Geological Journal volume 21 page 217 1865.) Still more recent explorations in the neighbourhood of Kilmaurs have shown that the fresh-water formation contains the seed of the pond-weed Potamogeton and the aquatic Ranunculus; and Mr. Young of the Glasgow Museum washed the mud adhering to the reindeer horns of Kilmaurs and that which filled the cracks of the associated elephants' tusks, and detected in these fossils (which had been in the Glasgow Museum for half a century) abundance of the same seeds.

All doubts, therefore, as to the true position of the remains of the mammoth, a fossil so rare in Scotland, have been set at rest, and it serves to prove that part of the ancient continent sank beneath the sea at a period of great cold, as the shells of the overlying sand attest. The incumbent till or boulder clay is about 40 feet thick, but it often attains much greater thickness in the same part of Scotland.

MARINE SHELLS OF SCOTCH DRIFT.

(FIGURE 107. Astarte borealis, Chem.; (A. arctica, Moll. A. compressa, Mont.)

(FIGURE 108. Leda lanceolata (oblonga), Sowerby.)

(FIGURE 109. Saxicava rugosa, Penn.)

(FIGURE 110. Pecten islandicus, Moll. Northern shell common in the drift of the Clyde, in Scotland.)

(FIGURE 111. Natica clausa, Bred. Northern shell common in the drift of the Clyde, in Scotland.)

(FIGURE 112. Trophon clathratum, Linne. Northern shell common in the drift of the Clyde, in Scotland.)

(FIGURE 113. Leda truncata. a. Exterior of left valve. b. Interior of same.)

(FIGURE 114. Tellina calcarea, Chem. (Tellina proxima, Brown.) a. Outside of left valve. b. Interior of same.)

The greatest height to which marine shells have yet been traced in this boulder clay is at Airdie, in Lanarkshire, ten miles east of Glasgow, 524 feet above the level of the sea. At that spot they were found imbedded in stratified clays with till above and below them. There appears no doubt that the overlying deposit was true glacial till, as some boulders of granite were observed in it, which must have come from distances of sixty miles at the least.

The shells figured in Figures 107 to 112 are only a few out of a large assemblage of living species, which, taken as a whole, bear testimony to conditions far more arctic than those now prevailing in the Scottish seas. But a group of marine shells, indicating a still greater excess of cold, has been brought to light since 1860 by the Reverend Thomas Brown, from glacial drift or clay on the borders of the estuaries of the Forth and Tay. This clay occurs at Elie, in Fife, and at Errol, in Perthshire; and has already afforded about 35 shells, all of living species, and now inhabitants of arctic regions, such as Leda truncata, Tellina proxima (see Figures 113 and 114), Pecten Groenlandicus, Crenella laevigata, Crenella nigra, and others, some of them first brought by Captain Sir E. Parry from the coast of Melville Island, latitude 76 degrees north. These were all identified in 1863 by Dr. Torell, who had just returned from a survey of the seas around Spitzbergen, where he had collected no less than 150 species of mollusca, living chiefly on a bottom of fine mud derived from the moraines of melting glaciers which there protrude into the sea. He informed me that the fossil fauna of this Scotch glacial deposit exhibits not only the species but also the peculiar varieties of mollusca now characteristic of very high latitudes. Their large size implies that they formerly enjoyed a colder, or, what was to them a more genial climate, than that now prevailing in the latitude where the fossils occur. Marine shells have also been found in the glacial drift of Caithness and Aberdeenshire at heights of 250 feet, and in Banff of 350 feet, and stratified drift continuous with the above ascends to heights of 500 feet. Already 75 species are enumerated from Caithness, and the same number from Aberdeenshire and Banff, and in both cases all but six are arctic species.

I formerly suggested that the absence of all signs of organic life in the Scotch drift might be connected with the severity of the cold, and also in some places with the depth of the sea during the period of extreme submergence; but my faith in such an hypothesis has been shaken by modern investigations, an exuberance of life having been observed both in arctic and antarctic seas of great depth, and where floating ice abounds. The difficulty, moreover, of accounting for the entire dearth of marine shells in till is removed when once we have adopted the theory of this boulder clay being the product of land-ice. For glaciers coming down from a continental ice-sheet like that which covers Greenland may fill friths many hundred feet below the sea-level, and even invade parts of a bay a thousand feet deep, before they find water enough to float off their terminal portions in the form of icebergs. In such a case till without marine shells may first accumulate, and then, if the climate becomes warmer and the ice melts, a marine deposit may be superimposed on the till without any change of level being required.

Another curious phenomenon bearing on this subject was styled by the late Hugh Miller the "striated pavements" of the boulder clay. Where portions of the till have been removed by the sea on the shores of the Forth, or in the interior by railway cuttings, the boulders imbedded in what remains of the drift are seen to have been all subjected to a process of abrasion and striation, the striae and furrows being parallel and persistent across them all, exactly as if a glacier or iceberg had passed over them and scored them in a manner similar to that so often undergone by the solid rocks below the glacial drift. It is possible, as Mr. Geikie conjectures, that this second striation of the boulders may be referable to floating ice. (Geikie Transactions of the Geological Society of Glasgow volume 1 part 2 page 68 1863.)

CONTORTED STRATA IN DRIFT.

(FIGURE 115. Section of contorted drift overlying till, seen on left bank of South Esk, near Cortachie, in 1840. Height of section seen, from a to d, about 50 feet. a, b. Gravel and sand. f, g. Contorted drift. Till.)

In Scotland the till is often covered with stratified gravel, sand, and clay, the beds of which are sometimes horizontal and sometimes contorted for a thickness of several feet. Such contortions are not uncommon in Forfarshire, where I observed them, among other places, in a vertical cutting made in 1840 near the left bank of the South Esk, east of the bridge of Cortachie. The convolutions of the beds of fine and coarse sand, gravel, and loam, extend through a thickness of no less than 25 feet vertical, or from b to c, Figure 115, the horizontal stratification being resumed very abruptly at a short distance, as to the right of f, g. The overlying coarse gravel and sand, a, is in some places horizontal, in others it exhibits cross bedding, and does not partake of the disturbances which the strata b, c, have undergone. The underlying till is exposed for a depth of about 20 feet; and we may infer from sections in the neighbourhood that it is considerably thicker.

In some cases I have seen fragments of stratified clays and sands, bent in like manner, in the middle of a great mass of till. Mr. Trimmer has suggested, in explanation of such phenomena, the intercalation in the glacial period of large irregular masses of snow or ice between layers of sand and gravel. Some of the cliffs near Behring's Straits, in which the remains of elephants occur, consist of ice mixed with mud and stones; and Middendorf describes the occurrence in Siberia of masses of ice, found at various depths from the surface after digging through drift. Whenever the intercalation of snow and ice with drift, whether stratified or unstratified, has taken place, the melting of the ice will cause such a failure of support as may give rise to flexures, and sometimes to the most complicated foldings. But in many cases the strata may have been bent and deranged by the mechanical pressure of an advancing glacier, or by the sideway thrust of huge islands of ice running aground against sandbanks; in which case, the position of the beds forming the foundation of the banks may not be at all disturbed by the shock.

There are indeed many signs in Scotland of the action of floating ice, as might have been expected where proofs of submergence in the Glacial Period are not wanting. Among these are the occurrence of large erratic blocks, frequently in clusters at or near the tops of hills or ridges, places which may have formed islets or shallows in the sea where floating ice would mostly ground and discharge its cargo on melting. Glaciers or land-ice would, on the contrary, chiefly discharge their cargoes at the bottom of valleys. Traces of an earlier and independent glaciation have also been observed in some regions where the striation, apparently produced by ice proceeding from the north-west, is not explicable by the radiation of land-ice from a central mountainous region. (Milne Home Transactions of the Royal Society Edinburgh volume 25 1868-9.)

GLACIATION OF WALES AND ENGLAND.

The mountains of North Wales were recognised, in 1842, by Dr. Buckland, as having been an independent centre of the dispersion of erratics— great glaciers, long since extinct, having radiated from the Snowdonian heights in Carnarvonshire, through seven principal valleys towards as many points of the compass, carrying with them large stony fragments, and grooving the subjacent rocks in as many directions.

Besides this evidence of land-glaciers, Mr. Trimmer had previously, in 1831, detected the signs of a great submergence in Wales in the Post-pliocene period. He had observed stratified drift, from which he obtained about a dozen species of marine shells, near the summit of Moel Tryfaen, a hill 1400 feet high, on the south side of the Menai Straits. I had an opportunity of examining in the summer of 1863, together with the Reverend W.S. Symonds, a long and deep cutting made through this drift by the Alexandra Mining Company in search of slates. At the top of the hill above-mentioned we saw a stratified mass of incoherent sand and gravel 35 feet thick, from which no less than 54 species of mollusca, besides three characteristic arctic varieties— in all 57 forms— have been obtained by Mr. Darbishire. They belong without exception to species still living in British or more northern seas; eleven of them being exclusively arctic, four common to the arctic and British seas, and a large proportion of the remainder having a northward range, or, if found at all in the southern seas of Britain, being comparatively less abundant. In the lowest beds of the drift were large heavy boulders of far-transported rocks, glacially polished and scratched on more than one side. Underneath the whole we saw the edges of vertical slates exposed to view, which here, like the rocks in other parts of Wales, both at greater and less elevations, exhibit beneath the drift unequivocal marks of prolonged glaciation. The whole deposit has much the appearance of an accumulation in shallow water or on a beach, and it probably acquired its thickness during the gradual subsidence of the coast— an hypothesis which would require us to ascribe to it a high antiquity, since we must allow time, first for its sinking, and then for its re-elevation.

The height reached by these fossil shells on Moel Tryfaen is no less than 1300 feet— a most important fact when we consider how very few instances we have on record beyond the limits of Wales, whether in Europe or North America, of marine shells having been found in glacial drift at half the height above indicated. A marine molluscous fauna, however, agreeing in character with that of Moel Tryfaen, and comprising as many species, has been found in drift at Macclesfield and other places in central England, sometimes reaching an elevation of 1200 feet.

Professor Ramsay estimated the probable amount of submergence during some part of the glacial period at about 2300 feet; for he was unable to distinguish the superficial sands and gravel which reached that high elevation from the drift which, at Moel Tryfaen and at lower points, contains shells of living species. The evidence of the marine origin of the highest drift is no doubt inconclusive in the absence of shells, so great is the resemblance of the gravel and sand of a sea beach and of a river's bed, when organic remains are wanting; but, on the other hand, when we consider the general rarity of shells in drift which we know to be of marine origin, we can not suppose that, in the shelly sands of Moel Tryfaen, we have hit upon the exact uppermost limit of marine deposition, or, in other words, a precise measure of the submergence of the land beneath the sea since the glacial period.

We are gradually obtaining proofs of the larger part of England, north of a line drawn from the mouth of the Thames to the Bristol Channel, having been under the sea and traversed by floating ice since the commencement of the glacial epoch. Among recent observations illustrative of this point, I may allude to the discovery, by Mr. J.F. Bateman, near Blackpool, in Lancashire, fifty miles from the sea, and at the height of 568 feet above its level, of till containing rounded and angular stones and marine shells, such as Turritella communis, Purpura lapillus, Cardium edule, and others, among which Trophon clathratum (=Fusus Bamffius), though still surviving in North British seas, indicates a cold climate.

ERRATICS NEAR CHICHESTER.

The most southern memorials of ice-action and of a Post-pliocene fauna in Great Britain is on the coast of the county of Sussex, about 25 miles west of Brighton, and 15 south of Chichester. A marine deposit exposed between high and low tide occurs on both sides of the promontory called Selsea Bill, in which Mr. Godwin-Austen found thirty-eight species of shells, and the number has since been raised to seventy.

This assemblage is interesting because on the whole, while all the species are recent, they have a somewhat more southern aspect than those of the present British Channel. It is true that about forty of them range from British to high northern latitudes; but several of them, as, for example, Lutraria rugosa and Pecten polymorphous, which are abundant, are not known at present to range farther north than the coast of Portugal, and seem to indicate a warmer temperature than now prevails on the coast where we find them fossil. What renders this curious is the fact that the sandy loam in which they occur is overlaid by yellow clayey gravel with large erratic blocks which must have been drifted into their present position by ice when the climate had become much colder. These transported fragments of granite, syenite, and greenstone, as well as of Devonian and Silurian rocks, may have come from the coast of Normandy and Brittany, and are many of them of such large size that we must suppose them to have been drifted into their present site by coast-ice. I measured one of granite, at Pagham, 21 feet in circumference. In the gravel of this drift with erratics are a few littoral shells of living species, indicating an ancient coast-line.

GLACIAL FORMATIONS OF NORTH AMERICA.

In the western hemisphere, both in Canada and as far south as the 40th and even 38th parallel of latitude in the United States, we meet with a repetition of all the peculiarities which distinguish the European boulder formation. Fragments of rock have travelled for great distances, especially from north to south: the surface of the subjacent rock is smoothed, striated, and fluted; unstratified mud or TILL containing boulders is associated with strata of loam, sand, and clay, usually devoid of fossils. Where shells are present, they are of species still living in northern seas, and not a few of them identical with those belonging to European drift, including most of those already given in Figures 107 to 112. The fauna also of the glacial epoch in North America is less rich in species than that now inhabiting the adjacent sea, whether in the Gulf of St. Lawrence, or off the shores of Maine, or in the Bay of Massachusetts.

The extension on the American continent of the range of erratics during the Post-pliocene period to lower latitudes than they reached in Europe, agrees well with the present southward deflection of the isothermal lines, or rather the lines of equal winter temperature. It seems that formerly, as now, a more extreme climate and a more abundant supply of ice prevailed on the western side of the Atlantic. Another resemblance between the distribution of the drift fossils in Europe and North America has yet to be pointed out. In Canada and the United States, as in Europe, the marine shells are generally confined to very moderate elevations above the sea (between 100 and 700 feet), while the erratic blocks and the grooved and polished surfaces of rock extend to elevations of several thousand feet.

I have already mentioned that in Europe several quadrupeds of living, as well as extinct, species were common to pre-glacial and post-glacial times. In like manner there is reason to suppose that in North America much of the ancient mammalian fauna, together with nearly all the invertebrata, lived through the ages of intense cold. That in the United States the Mastodon giganteus was very abundant after the drift period, is evident from the fact that entire skeletons of this animal are met with in bogs and lacustrine deposits occupying hollows in the glacial drift. They sometimes occur in the bottom even of small ponds recently drained by the agriculturist for the sake of the shell-marl. In 1845 no less than six skeletons of the same species of Mastodon were found in Warren county, New Jersey, six feet below the surface, by a farmer who was digging out the rich mud from a small pond which he had drained. Five of these skeletons were lying together, and a large part of the bones crumbled to pieces as soon as they were exposed to the air.

It would be rash, however, to infer from such data that these quadrupeds were mired in MODERN times, unless we use that term strictly in a geological sense. I have shown that there is a fluviatile deposit in the valley of the Niagara, containing shells of the genera Melania, Lymnea, Planorbis, Velvata, Cyclaz, Unio, Helix, etc., all of recent species, from which the bones of the great Mastodon have been taken in a very perfect state. Yet the whole excavation of the ravine, for many miles below the Falls, has been slowly effected since that fluviatile deposit was thrown down. Other extinct animals accompany the Mastodon giganteus in the post-glacial deposits of the United States, and this, taken with the fact that so few of the mollusca, even of the commencement of the cold period, differ from species now living is important, as refuting the hypothesis, for which some have contended, that the intensity of the glacial cold annihilated all the species in temperate and arctic latitudes.

CONNECTION OF THE PREDOMINANCE OF LAKES WITH GLACIAL ACTION.

It was first pointed out by Professor Ramsay in 1862, that lakes are exceedingly numerous in those countries where erratics, striated blocks, and other signs of ice-action abound; and that they are comparatively rare in tropical and sub- tropical regions. Generally in countries where the winter cold is intense, such as Canada, Scandinavia, and Finland, even the plains and lowlands are thickly strewn with innumerable ponds and small lakes, together with some others of a larger size; while in more temperate regions, such as Great Britain, Central and Southern Europe, the United States, and New Zealand, lake districts occur in all such mountainous tracts as can be proved to have been glaciated in times comparatively modern or since the geographical configuration of the surface bore a considerable resemblance to that now prevailing. In the same countries, beyond the glaciated regions, lakes abruptly cease, and in warmer and tropical countries are either entirely absent, or consist, as in equatorial Africa, of large sheets of water unaccompanied so far as we yet know by numerous smaller ponds and tarns.

The southern limits of the lake districts of the Northern Hemisphere are found at about 40 degrees N. latitude on the American continent, and about 50 degrees in Europe, or where the Alps intervene four degrees farther south. A large proportion of the smaller lakes are dammed up by barriers of unstratified drift, having the exact character of the moraines of glaciers, and are termed by geologists "morainic," but some of them are true rock-basins, and would hold water even if all the loose drift now resting on their margins were removed.

In a paper read before the Geological Society of London in 1862, Professor Ramsay maintained that the first formation of most existing lakes took place during the glacial epoch, and was due, not to elevation or subsidence, but to actual erosion of their basins by glaciers. M. Mortillet in the same year advanced the theory that after the Alpine lake-basins had been filled up with loose fluviatile deposits, they were re-excavated by the great glaciers which passed down the valleys at the time of the greatest cold, a doctrine which would attribute to moving ice almost as great a capacity of erosion as that which assumed that the original basins were scooped out of solid rock by glaciers. It is impossible to deny that the mere geographical distribution of lakes points to the intimate connection of their origin with the abundance of ice during a former excess of cold, but how far the erosive action of moving ice has been the sole or even the principal cause of lake-basins, is a question still open to discussion.

The lakes of Switzerland and the north of Italy are some of them twenty and thirty miles in length, and so deep that their bottoms are in some cases from 1000 to 2000 feet beneath the level of the sea. It is admitted on all hands that they were once filled with ice, and as the existing glaciers polish and grind down, as before stated, the surface of the rocks, we are prepared to find that every lake-basin in countries once covered by ice should bear the marks of superficial glaciation, and also that the ice during its advance and retreat should have left behind it much transported matter as well as some evidence of its having enlarged the pre-existing cavity. But much more than this is demanded by the advocates of glacial erosion. They suggest that as the old extinct glaciers were several thousand feet thick, they were able in some places gradually to scoop out of the solid rock cavities twenty or thirty miles in length, and as in the case of Lago Maggiore from a thousand to two thousand six hundred feet below the previous level of the river-channel, and also that the ice had the power to remove from the cavity formed by its grinding action all the materials of the missing rocks. A constant supply, it is argued, of fine mud issues from the termination of every glacier in the stream which is produced by the melting of the ice, and this result of friction is exhibited both during winter and summer, affording evidence of the continual deepening and widening of the valleys through which glaciers pass. As the fine mud is carried away by a river from the deep pool which is formed from the base of every cataract, so it seems to be imagined that lake-basins may be gradually emptied of the mud formed by abrasion during the glacial period.

I am by no means disposed to object to this theory on the ground of the insufficiency of the time during which the extreme cold endured, but we must carefully consider whether that same time is not so vast as to make it probable that other forces, besides the motion of glaciers, must have cooperated in converting some parts of the ancient valley courses into lake-basins. They who have formed the most exalted conceptions of the erosive energy of moving ice do not deny that during the period termed "Glacial" there have been movements of the earth's crust sufficient to produce oscillations of level in Europe amounting to 1000 feet or more in both directions. M. Charpentier, indeed, attributed some of the principal changes of climate in Switzerland, during the glacial period, to a depression of the central Alps to the extent of 3000 feet, and Swiss geologists have long been accustomed to attribute their lake basins, in part, to those convulsions by which the shape and course of the valleys may have been modified. Our experience, in the lifetime of the present generation, of the changes of level witnessed in New Zealand during great earthquakes is entirely opposed to the notion that the movements, whether upward or downward, are uniform in amount or direction throughout areas of indefinite extent. On the contrary, the land has been permanently raised in one region several feet or yards, and the rise has been found gradually to die out, so as to be imperceptible at a distance of twenty miles, and in some areas is even exchanged for a simultaneous downward movement of several feet.

But, it is asked, if such inequality of movement can have contributed towards the production of lake basins, does it not leave unexplained the comparative rarity of lakes in tropical and subtropical countries. In reply to this question it may be observed that in our endeavour to estimate the effects of subterranean movements in modifying the superficial geography of a country we must remember that each convulsion effects a very slight change. If it interferes with the drainage, whether by raising the lower or sinking the higher portion of a hydrographical basin, the upheaval or depression will only amount to a few feet at a time, and there may be an interval of years or centuries before any further movement takes place in the same region. In the mean time an incipient lake if produced may be filled up with sediment, and the recently-formed barrier will then be cut through by the river, whereas in a country where glacial conditions prevail no such obliteration of the temporary lake-basin would take place; for however deep it became by repeated sinking of the upper or rising of the lower extremity, being always filled with ice it might remain, throughout the greater part of its extent, free from sediment or drift until the ice melted at the close of the glacial period.

One of the most serious objections to the exclusive origin by ice-erosion of wide and deep lake-basins arises from their capricious distribution, as for example in Piedmont, both to the eastward and westward of Turin, where great lakes are wanting (Antiquity of Man page 313.), although some of the largest extinct glaciers descending from Mont Blanc and Monte Rosa came down from the Alps, leaving their gigantic moraines in the low country. Here, therefore, we might have expected to find lakes of the first magnitude rivalling the contiguous Lago Maggiore in importance.

A still more striking illustration of the same absence of lakes where large glaciers abound is afforded by the Caucasus, a chain more than 300 miles long, and the loftiest peaks of which attain heights from 16,000 to 18,000 feet. This greatest altitude is reached by Elbruz, a mountain in latitude 43 degrees north three degrees south of Mont Blanc, but on the other hand 3000 feet higher. The present Caucasian glaciers are equal or superior in dimensions to those of Switzerland, and like them give rise occasionally to temporary lakes by obstructing the course of rivers, and causing great floods when the icy barriers give way. Mr. Freshfield, a careful observer, writing in 1869, says: "A total absence of lakes on both sides of the chains is the most marked feature. Not only are there no great subalpine sheets of water, like Como or Geneva, but mountain tarns, such as the Dauben See on the Gemmi, or the Klonthal See near Glarus, are equally wanting." (Travels in Central Caucasus 1869 page 452.) The same author states on the authority of the eminent Swiss geologist, Mons. E.

Favre, who also explored the Caucasus in 1868, that moraines of great height and huge erratics of granite and other rocks "justify the assertion that the present glaciers of the Caucasus, like those of the Alps, are only the shadows of their former selves."

It seems safe to assume that the chain of lakes, of which the Albert Nyanza forms one in equatorial Africa, was due to causes other than glacial. Yet if we could imagine a glacial period to visit that region filling the lakes with ice and scoring the rocks which form their sides and bottoms, we should be unable to decide how much the capacity of the basins had been enlarged and the surface modified by glacial erosion. The same may be true of the Lago Maggiore and Lake Superior, although the present basins of both of them afford abundant superficial markings due to ice-action.

But to whatever combination of causes we attribute the great Alpine lakes one thing is clear, namely, that they are, geologically speaking, of modern origin. Every one must admit that the upper valley of the Rhone has been chiefly caused by fluviatile denudation, and it is obvious that the quantity of matter removed from that valley previous to the glacial period would have been amply sufficient to fill up with sediment the basin of the Lake of Geneva, supposing it to have been in existence, even if its capacity had been many times greater than it is now. (See Principles volume 1 page 420 10th edition 1867.)

On the whole, it appears to me, in accordance with the views of Professor Ramsay, M. Mortillet, Mr. Geikie, and others, that the abrading action of ice has formed some mountain tarns and many morainic lakes; but when it is a question of the origin of larger and deeper lakes, like those of Switzerland or the north of Italy, or inland fresh-water seas, like those of Canada, it will probably be found that ice has played a subordinate part in comparison with those movements by which changes of level in the earth's crust are gradually brought about.

TERTIARY OR CAINOZOIC PERIOD.
CHAPTER XIII.
PLIOCENE PERIOD.

Glacial Formations of Pliocene Age.
Bridlington Beds.
Glacial Drifts of Ireland.
Drift of Norfolk Cliffs.
Cromer Forest-bed.
Aldeby and Chillesford Beds.
Norwich Crag.
Older Pliocene Strata.
Red Crag of Suffolk.
Coprolitic Bed of Red Crag.
White or Coralline Crag.
Relative Age, Origin, and Climate of the Crag Deposits.
Antwerp Crag.
Newer Pliocene Strata of Sicily.
Newer Pliocene Strata of the Upper Val d'Arno.
Older Pliocene of Italy.
Subapennine Strata.
Older Pliocene Flora of Italy.

It will be seen in the description given in the last chapter of the Post- pliocene formations of the British Isles that they comprise a large proportion of those commonly termed glacial, characterised by shells which, although referable to living species, usually indicate a colder climate than that now belonging to the latitudes where they occur fossil. But in parts of England, more especially in Yorkshire, Norfolk, and Suffolk, there are superficial formations of clay with glaciated boulders, and of sand and pebbles, containing occasional, though rare, patches of shells, in which the marine fauna begins to depart from that now inhabiting the neighbouring sea, and comprises some species of mollusca not yet known as living, as well as extinct varieties of others, entitling us to class them as Newer Pliocene, although belonging to the close of that period and chronologically on the verge of the later or Post-pliocene epoch.

BRIDLINGTON DRIFT.

To this era belongs the well-known locality of Bridlington, near the mouth of the Humber, in Yorkshire, where about seventy species or well-marked varieties of shells have been found on the coast, near the sea-level, in a bed of sand several feet thick resting on glacial clay with much chalk debris, and covered by a deposit of purple clay with glaciated boulders. More than a third of the species in this drift are now inhabitants of arctic regions, none of them extending southward to the British seas; which is the more remarkable as Bridlington is situated in latitude 54 degrees north. Fifteen species are British and Arctic, a very few belong to those species which range south of our British seas. Five species or well-marked varieties are not known living, namely, the variety of Astarte borealis (called A. Withami); A. mutabilis; the sinistral form of Tritonium carinatum, Cardita analis, and Tellina obliqua, Figure 120. Mr. Searles Wood also inclines to consider Nucula Cobboldiae, Figure 119, now absent from the European seas and the Atlantic, as specifically distinct from a closely-allied shell now living in the seas surrounding Vancouver's Island, which some conchologists regard as a variety. Tellina obliqua also approaches very near to a shell now living in Japan.

GLACIAL DRIFT OF IRELAND.

Marine drift containing the last-mentioned Nucula and other glacial shells reaches a height of from 1000 to 1200 feet in the county of Wexford, south of Dublin. More than eighty species have already been obtained from this formation, of which two, Conovulus pyramidalis and Nassa monensis, are not known as living; while Turritella incrassata and Cypraea lucida no longer inhabit the British seas, but occur in the Mediterranean. The great elevation of these shells, and the still greater height to which the surface of the rocks in the mountainous regions of Ireland have been smoothed and striated by ice-action, has led geologists to the opinion that that island, like the greater part of England and Scotland, after having been united with the continent of Europe, from whence it received the plants and animals now inhabiting it, was in great part submerged. The conversion of this and other parts of Great Britain into an archipelago was followed by a re-elevation of land and a second continental period. After all these changes the final separation of Ireland from Great Britain took place, and this event has been supposed to have preceded the opening of the straits of Dover. (See Antiquity of Man chapter 14.)

DRIFT OF NORFOLK CLIFFS.

(FIGURE 116. Tellina balthica (T. solidula).)

There are deposits of boulder clay and till in the Norfolk cliffs principally made up of the waste of white chalk and flints which, in the opinion of Mr. Searles Wood, jun., and others, are older than the Bridlington drift, and contain a larger proportion of shells common to the Norwich and Red Crag, including a certain number of extinct forms, but also abounding in Tellina balthica (T. solidula, Figure 116), which is found fossil at Bridlington, and living in our British seas, but wanting in all the formations, even the newest, afterwards to be described as Crag. As the greater part of these drifts are barren of organic remains, their classification is at present a matter of great uncertainty.

They can nowhere be so advantageously studied as on the coast between Happisburgh and Cromer. Here we may see vertical cliffs, sometimes 300 feet and more in height, exposed for a distance of fifty miles, at the base of which the chalk with flints crops out in nearly horizontal strata. Beds of gravel and sand repose on this undisturbed chalk. They are often strangely contorted, and envelop huge masses or erratics of chalk with layers of vertical flint. I measured one of these fragments in 1839 at Sherringham, and found it to be eighty feet in its longest diameter. It has been since entirely removed by the waves of the sea. In the floor of the chalk beneath it the layers of flint were horizontal. Such erratics have evidently been moved bodily from their original site, probably by the same glacial action which has polished and striated some of the accompanying granitic and other boulders, occasionally six feet in diameter, which are imbedded in the drift.

CROMER FOREST-BED.

Intervening between these glacial formations and the subjacent chalk lies what has been called the Cromer Forest-bed. This buried forest has been traced from Cromer to near Kessingland, a distance of more than forty miles, being exposed at certain seasons between high and low water mark. It is the remains of an old land and estuarine deposit, containing the submerged stumps of trees standing erect with their roots in the ancient soil. Associated with the stumps and overlying them, are lignite beds with fresh-water shells of recent species, and laminated clay without fossils. Through the lignite and forest-bed are scattered cones of the Scotch and spruce firs with the seeds of recent plants, and the bones of at least twenty species of terrestrial mammalia. Among these are two species of elephant, E. meridionalis, Nesti, and E. antiquus, the former found in the Newer Pliocene beds of the Val d'Arno, near Florence. In the same bed occur Hippopotamus major, Rhinoceros etruscus, both of them also Val d'Arno species, many species of deer considered by Mr. Boyd Dawkins to be characteristic of warmer countries, and also a horse, beaver, and field-mouse. Half of these mammalia are extinct, and the rest still survive in Europe. The vegetation taken alone does not imply a temperature higher than that now prevailing in the British Isles. There must have been a subsidence of the forest to the amount of 400 or 500 feet, and a re-elevation of the same to an equal extent in order to allow the ancient surface of the chalk or covering of soil, on which the forest grew, to be first covered with several hundred feet of drift, and then upheaved so that the trees should reach their present level. Although the relative antiquity of the forest-bed to the overlying glacial till is clear, there is some difference of opinion as to its relation to the crag presently to be described.

CHILLESFORD AND ALDEBY BEDS.

(FIGURE 117. Natica helicoides, Johnson.)

It is in the counties of Norfolk, Suffolk, and Essex, that we obtain our most valuable information respecting the British Pliocene strata, whether newer or older. They have obtained in those counties the provincial name of "Crag," applied particularly to masses of shelly sand which have long been used in agriculture to fertilise soils deficient in calcareous matter. At Chillesford, between Woodbridge and Aldborough in Suffolk, and Aldeby, near Beccles, in the same county, there occur stratified deposits, apparently older than any of the preceding drifts of Yorkshire, Norfolk, and Suffolk. They are composed at Chillesford of yellow sands and clays, with much mica, forming horizontal beds about twenty feet thick. Messrs. Prestwich and Searles Wood, senior, who first described these beds, point out that the shells indicate on the whole a colder climate than the Red Crag; two-thirds of them being characteristic of high latitudes. Among these are Cardium Groenlandicum, Leda limatula, Tritonium carinatum, and Scalaria Groenlandica. In the upper part of the laminated clays a skeleton of a whale was found associated with casts of the characteristic shells, Nucula Cobboldiae and Tellina obliqua, already referred to as no longer inhabiting our seas, and as being extinct varieties if not species. The same shells occur in a perfect state in the lower part of the formation. Natica helicoides (Figure 117) is an example of a species formerly known only as fossil, but which has now been found living in our seas.

At Aldeby, where beds occur decidedly similar in mineral character as well as fossil remains, Messrs. Crowfoot and Dowson have now obtained sixty-six species of mollusca, comprising the Chillesford species and some others. Of these about nine-tenths are recent. They are in a perfect state, clearly indicating a cold climate; as two-thirds of them are now met with in arctic regions. As a rule, the lamellibranchiate molluscs have both valves united, and many of them, such as Mya arenaria, stand with the siphonal end upward, as when in a living state. Tellina balthica, before mentioned (Figure 116) as so characteristic of the glacial beds, including the drift of Bridlington, has not yet been found in deposits of Chillesford and Aldeby age, whether at Sudbourn, East Bavent, Horstead, Coltishall, Burgh, or in the highest beds overlying the Norwich Crag proper at Bramerton and Thorpe.

NORWICH OR FLUVIO-MARINE CRAG.

(FIGURE 118. Mastodon arvernensis, third milk molar, left side, upper jaw; grinding surface, natural size. Norwich Crag, Postwick, also found in Red Crag, see below.)

The beds above alluded to ought, perhaps, to be regarded as beds of passage between the glacial formations and those called from a provincial name "Crag," the newest member of which has been commonly called the "Norwich Crag." It is chiefly seen in the neighbourhood of Norwich, and consists of beds of incoherent sand, loam, and gravel, which are exposed to view on both banks of the Yare, as at Bramerton and Thorpe. As they contain a mixture of marine, land, and fresh-water shells, with bones of fish and mammalia, it is clear that these beds have been accumulated at the bottom of a sea near the mouth of a river. They form patches rarely exceeding twenty feet in thickness, resting on white chalk. At their junction with the chalk there invariably intervenes a bed called the "Stone-bed," composed of unrolled chalk-flints, commonly of large size, mingled with the remains of a land fauna comprising Mastodon arvernensis, Elephas meridionalis, and an extinct species of deer. The mastodon, which is a species characteristic of the Pliocene strata of Italy and France, is the most abundant fossil, and one not found in the Cromer forest before mentioned. When these flints, probably long exposed in the atmosphere, became submerged, they were covered with barnacles, and the surface of the chalk became perforated by the Pholas crispata, each fossil shell still remaining at the bottom of its cylindrical cavity, now filled up with loose sand from the incumbent crag. This species of Pholas still exists, and drills the rocks between high and low water on the British coast. The name of "Fluvio-marine" has often been given to this formation, as no less than twenty species of land and fresh-water shells have been found in it. They are all of living species; at least only one univalve, Paludina lenta, has any, and that a very doubtful, claim to be regarded as extinct.

(FIGURE 119. Nucula Cobboldiae.)
(FIGURE 120. Tellina obliqua.)

Of the marine shells, 124 in number, about 18 per cent are extinct, according to the latest estimate given me by Mr. Searles Wood; but, for reasons presently to be mentioned, this percentage must be only regarded as provisional. It must also be borne in mind that the proportion of recent shells would be augmented if the uppermost beds at Bramerton, near Norwich, which belong to the most modern or Chillesford division of the Crag, had been included, as they were formerly, by Mr. Woodward and myself, in the Norwich series. Arctic shells, which formed so large a proportion in the Chillesford and Aldeby beds, are more rare in the Norwich Crag, though many northern species— such as Rhynchonella psittacea, Scalaria Groenlandica, Astarte borealis, Panopaea Norvegia, and others— still occur. The Nucula Cobboldiae and Tellina obliqua, Figures 119 and 120, before mentioned, are frequent in these beds, as are also Littorina littorea, Cardium edule, and Turritella communis, of our seas, proving the littoral origin of the beds.

OLDER PLIOCENE STRATA.
RED CRAG.

(FIGURE 121. Section through (left) sea, Red Crag, London Clay and Chalk (right).)

Among the English Pliocene beds the next in antiquity is the Red Crag, which often rests immediately on the London Clay, as in the county of Essex, illustrated in Figure 121.

It is chiefly in the county of Suffolk that it is found, rarely exceeding twenty feet in thickness, and sometimes overlying another Pliocene deposit, the Coralline Crag, to be mentioned in the sequel. It has yielded— exclusive of 25 species regarded by Mr. Wood as derivative— 256 species of mollusca, of which 65, or 25 per cent, are extinct. Thus, apart from its order of superposition, its greater antiquity than the Norwich and glacial beds, already described, is proved by the greater departure from the fauna of our seas. It may also be observed that in most of the deposits of this Red Crag, the northern forms of the Norwich Crag, and of such glacial formations as Bridlington, are less numerous, while those having a more southern aspect begin to make their appearance. Both the quartzose sand, of which it chiefly consists, and the included shells, are most commonly distinguished by a deep ferruginous or ochreous colour, whence its name. The shells are often rolled, sometimes comminuted, and the beds have much the appearance of having been shifting sand-banks, like those now forming on the Dogger-bank, in the sea, sixty miles east of the coast of Northumberland. Cross stratification is almost always present, the planes of the strata being sometimes directed towards one point of the compass, sometimes to the opposite, in beds immediately overlying. That such a structure is not deceptive or due to any subsequent concretionary rearrangement of particles, or to mere bands of colour produced by the iron, is proved by each bed being made up of flat pieces of shell which lie parallel to the planes of the smaller strata.

(FIGURE 122. Purpura tetragona, Sowerby; natural size.)

(FIGURE 123. Voluta Lamberti, Sowerby. Variety characteristic of Suffolk Crag. Pliocene.)

(FIGURE 124. Voluta Lamberti, young individual, Cor. and Red Crag.)

It has long been suspected that the different patches of Red Crag are not all of the same age, although their chronological relation can not be decided by superposition. Separate masses are characterised by shells specifically distinct or greatly varying in relative abundance, in a manner implying that the deposits containing them were separated by intervals of time. At Butley, Tunstall, Sudbourn, and in the Red Crag of Chillesford, the mollusca appear to assume their most modern aspect when the climate was colder than when the earliest deposits of the same period were formed. At Butley, Nucula Cobboldiae, so common in the Norwich and certain glacial beds, is found, and Purpura tetragona (Figure 122) is very abundant. On the other hand, at Walton-on-the-Naze, in Essex, we seem to have an exhibition of the oldest phase of the Red Crag; and a warmer climate seems indicated, not only by the absence of many northern forms, but also by the abundance of some now living in the British seas and the Mediterranean. Voluta Lamberti (see Figures 123 and 124), an extinct form, which seems to have flourished chiefly in the antecedent Coralline Crag period, is still represented here by individuals of every age.

(FIGURE 125. Trophon antiquum, Muller. (Fusus contrarius) half natural size.)

The reversed whelk (Figure 125) is common at Walton, where the dextral form of that shell is unknown. Here also we find most frequently specimens of lamellibranchiate molluscs, with both the valves united, showing that they belonged to this sea of the Upper Crag, and were not washed in from an older bed, such as the Coralline, in which case the ligament would not have held together the valves in strata so often showing signs of the boisterous action of the waves. No less than forty species of lamellibranchiate molluscs, with double valves, have been collected by Mr. Bell from the various localities of the Red Crag.

At and near the base of the Red Crag is a loose bed of brown nodules, first noticed by Professor Henslow as containing a large percentage of earthy phosphates. This bed of coprolites (as it is called, because they were originally supposed to be the faeces of animals) does not always occur at one level, but is generally in largest quantity at the junction of the Crag and the underlying formation. In thickness it usually varies from six to eighteen inches, and in some rare cases amounts to many feet. It has been much used in agriculture for manure, as not only the nodules, but many of the separate bones associated with them, are largely impregnated with phosphate of lime, of which there is sometimes as much as sixty per cent. They are not unfrequently covered with barnacles, showing that they were not formed as concretions in the stratum where they now lie buried, but had been previously consolidated. The phosphatic nodules often collect fossil crabs and fishes from the London Clay, together with the teeth of gigantic sharks. In the same bed have been found many ear-bones of whales, and the teeth of Mastodon arvernensis, Rhinoceros Schleiermacheri, Tapirus priscus, and Hipparion (a quadruped of the horse family), and antlers of a stag, Cervus anoceros. Organic remains also of the older chalk and Lias are met with, showing how great was the denudation of previous formations during the Pliocene period. As the older White Crag, presently to be mentioned, contains similar phosphatic nodules near its base, those of the Red Crag may be partly derived from this source.

WHITE OR CORALLINE CRAG.

The lower or Coralline Crag is of very limited extent, ranging over an area about twenty miles in length, and three or four in breadth, between the rivers Stour and Alde, in Suffolk. It is generally calcareous and marly— often a mass of comminuted shells, and the remains of bryozoa (or polyzoa), passing occasionally into a soft building-stone. (Ehrenberg proposed in 1831 the term Bryozoum, or "Moss-animal," for the molluscous or ascidian form of polyp, characterised by having two openings to the digestive sack, as in Eschara, Flustra, Retepora, and other zoophytes popularly included in the corals, but now classed by naturalists as mollusca. The term Polyzoum, synonymous with Bryozoum, was, it seems, proposed in 1830, or the year before, by Mr. J.O. Thompson.) At Sudbourn and Gedgrave, near Orford, this building-stone has been largely quarried. At some places in the neighbourhood the softer mass is divided by thin flags of hard limestone, and bryozoa placed in the upright position in which they grew. From the abundance of these coralloid mollusca the lowest or White Crag obtained its popular name, but true corals, as now defined, or zoantharia, are very rare in this formation.

The Coralline Crag rarely, if ever, attains a thickness of thirty feet in any one section. Mr. Prestwich imagines that if the beds found at different localities were united in the probable order of their succession, they might exceed eighty feet in thickness, but Mr. Searles Wood does not believe in the possibility of establishing such a chronological succession by aid of the organic remains, and questions whether proof could be obtained of more than forty feet. I was unable to come to any satisfactory opinion on the subject, although at Orford, especially at Gedgrave, in the neighbourhood of that place, I saw many sections in pits, where this crag is cut through. These pits are so unconnected, and of such limited extent, that no continuous section of any length can be obtained, so that speculations as to the thickness of the whole deposit must be very vague. At the base of the formation at Sutton a bed of phosphatic nodules, very similar to that before alluded to in the Red Crag, with remains of mammalia, has been met with.

(FIGURE 126. Section near Woodbridge, in Suffolk.
Through Sutton (left), Shottisham Creek, Ramsholt (right) and R. Deben.
a. Red Crag.
b. Coralline Crag.
c. London Clay.)

Whenever the Red and Coralline Crag occur in the same district, the Red Crag lies uppermost; and in some cases, as in the section represented in Figure 126, which I had an opportunity of seeing exposed to view in 1839, it is clear that the older deposit, or Coralline Crag, b, had suffered denudation, before the newer formation, a, was thrown down upon it. At D there was not only seen a distinct cliff, eight or ten feet high, of Coralline Crag, running in a direction N.E. and S.W., against which the Red Crag abuts with its horizontal layers, but this cliff occasionally overhangs. The rock composing it is drilled everywhere by Pholades, the holes which they perforated having been afterwards filled with sand, and covered over when the newer beds were thrown down. The older formation is shown by its fossils to have accumulated in a deeper sea, and contains none of those littoral forms such as the limpet, Patella, found in the Red Crag. So great an amount of denudation could scarcely take place, in such incoherent materials, without some of the fossils of the inferior beds becoming mixed up with the overlying crag, so that considerable difficulty must be occasionally experienced by the palaeontologist in deciding which species belong severally to each group.

(FIGURE 127. Fascicularia aurantium, Milne Edwards. Family, Tubuliporidae, of same author. Bryozoan of extinct genus, from the inferior or Coralline Crag, Suffolk. a. Exterior. b. Vertical section of interior. c. Portion of exterior magnified. d. Portion of interior magnified, showing that it is made up of long, thin, straight tubes, united in conical bundles.)

(FIGURE 128. Astarte omalii, laj.; species common to Upper and lower crag.)

Mr. Searles Wood estimates the total number of marine testaceous mollusca of the Coralline Crag at 350, of which 110 are not known as living, being in the proportion of thirty-one per cent extinct. No less than 130 species of bryozoa have been found in the Coralline Crag, and some belong to genera unknown in the living creation, and of a very peculiar structure; as, for example, that represented in Figure 127, which is one of several species having a globular form. Among the testacea the genus Astarte (see Figure 128) is largely represented, no less than fourteen species being known, and many of these being rich in individuals. There is an absence of genera peculiar to hot climates, such as Conus, Oliva, Fasciolaria, Crassatella, and others. The absence also of large cowries (Cyprea), those found belonging exclusively to the section Trivia, is remarkable. The large volute, called Voluta Lamberti (Figure 123), may seem an exception; but it differs in form from the volutes of the torrid zone, and, like the living Voluta Magellanica, must have been fitted for an extra-tropical climate.

(FIGURE 129. Lingula Dumortieri, Nyst; Suffolk and Antwerp Crag.)
(FIGURE 130. Pyrula reticulata, Lam.; Coralline Crag, Ramsholt.)

(FIGURE 131. Temnechinus excavatus, Forbes; Temnopleurus excavatus, Wood; Coralline Crag, Ramsholt.)

The occurrence of a species of Lingula at Sutton (see Figure 129) is worthy of remark, as these Brachiopoda seem now confined to more equatorial latitudes; and the same may be said still more decidedly of a species of Pyrula, supposed by Mr. Wood to be identical with P. reticulata (Figure 130), now living in the Indian Ocean. A genus also of echinoderms, called by Professor Forbes Temnechinus (Figure 131), occurs in the Red and Coralline Crag of Suffolk, and until lately was unknown in a living state, but it has been brought to light as an existing form by the deep-sea dredgings, both of the United States survey, off Florida, at a depth of from 180 to 480 feet, and more recently (1869), in the British seas, during the explorations of the "Porcupine."

CLIMATE OF THE CRAG DEPOSITS.

One of the most interesting conclusions deduced from a careful comparison of the shells of the British Pliocene strata and the fauna of our present seas has been pointed out by Professor E. Forbes. It appears that, during the Glacial period, a period intermediate, as we have seen, between that of the Crag and our own time, many shells, previously established in the temperate zone, retreated southward to avoid an uncongenial climate, and they have been found fossil in the Newer Pliocene strata of Sicily, Southern Italy, and the Grecian Archipelago, where they may have enjoyed, during the era of floating icebergs, a climate resembling that now prevailing in higher European latitudes. (E. Forbes Mem. Geological Survey of Great Britain volume 1 page 386.) The Professor gave a list of fifty shells which inhabited the British seas while the Coralline and Red Crag were forming, and which, though now living in our seas, were wanting, as far as was then known, in the glacial deposits. Some few of these species have subsequently been found in the glacial drift, but the general conclusion of Forbes remains unshaken.

The transport of blocks by ice, when the Red Crag was being deposited, appears to me evident from the large size of some huge, irregular, quite unrounded chalk flints, retaining their white coating, and 2 feet long by 18 inches broad, in beds worked for phosphatic nodules at Foxhall, four miles south-east of Ipswich. These must have been tranquilly drifted to the spot by floating ice. Mr. Prestwich also mentions the occurrence of a large block of porphyry in the base of the Coralline Crag at Sutton, which would imply that the ice-action had begun in our seas even in this older period. The cold seems to have gone on increasing from the time of the Coralline to that of the Norwich Crag, and became more and more severe, not perhaps without some oscillations of temperature, until it reached its maximum in what has been called the Glacial period, or at the close of the Newer Pliocene, and in the Post-pliocene periods.

RELATION OF THE FAUNA OF THE CRAG TO THAT OF THE RECENT SEAS.

By far the greater number of the recent marine species occurring in the several Crag formations are still inhabitants of the British seas; but even these differ considerably in their relative abundance, some of the commonest of the Crag shells being now extremely scarce— as, for example, Buccinum Dalei— while others, rarely met with in a fossil state, are now very common, as Murex erinaceus and Cardium echinatum. Some of the species also, the identity of which with the living would not be disputed by any conchologist, are nevertheless distinguishable as varieties, whether by slight deviations in form or a difference in average dimensions. Since Mr. Searles Wood first described the marine testacea of the Crags, the additions made to that fossil fauna have not been considerable, whereas we have made in the same period immense progress in our knowledge of the living testacea of the British and arctic seas, and of the Mediterranean. By this means the naturalist has been enabled to identify with existing species many forms previously supposed to be extinct.

In the forthcoming supplement to the invaluable monograph communicated by Mr. Wood to the Palaeontographical Society, in which he has completed his figures and descriptions of the British crag shells of every age, list will be found of all the fossil shells, of which a summary is given in the table below.

TABLE OF NUMBER OF KNOWN SPECIES OF MARINE TESTACEA IN THE CRAG.

COLUMN 1: KNOWN SPECIES. COLUMN 2: TOTAL NUMBER OF KNOWN SPECIES. COLUMN 3: NUMBER OF SPECIES NOT KNOWN AS LIVING.
CHILLESFORD AND ALDEBY BEDS:
Bivalves: 61 : 4.
Univalves: 33 : 5.
Brachiopods: 0 : 0.
PERCENTAGE OF SHELLS NOT KNOWN AS LIVING : 9.5.
NORWICH OR FLUVIO-MARINE CRAG:
Bivalves: 61 : 10.
Univalves: 64 : 12.
Brachiopods: 1 : 0.
PERCENTAGE OF SHELLS NOT KNOWN AS LIVING : 17.5.
RED CRAG (Exclusive of many derivative shells):
Bivalves: 128 : 31.
Univalves: 127 : 33.
Brachiopods: 1 : 1.
PERCENTAGE OF SHELLS NOT KNOWN AS LIVING : 25.0.
CORALLINE CRAG:
Bivalves: 161 : 47.
Univalves: 184 : 60
Brachiopods: 5 : 3
PERCENTAGE OF SHELLS NOT KNOWN AS LIVING : 31.5

To begin with the uppermost or Chillesford beds, it will be seen that about 9 per cent only are extinct, or not known as living, whereas in the Norwich, which succeeds in the descending order, seventeen in a hundred are extinct. Formerly, when the Norwich or Fluvio-marine Crag was spoken of, both these formations were included under the same head, for both at Bramerton and Thorpe, the chief localities where the Norwich Crag was studied, an overlying deposit occurs referable to the Chillesford age. If now the two were fused together as of old, their shells would, according to Mr. Wood, yield a percentage of fifteen in a hundred of species extinct or not known as living.

To come next to the Red Crag, the reader will observe that a percentage of 25 is given of shells unknown as living, and this increases to 31 in the antecedent Coralline Crag. But the gap between these two stages of our Pliocene deposits is really wider than these numbers would indicate, for several reasons. In the first place, the Coralline Crag is more strictly the product of a single period, the Red Crag, as we have seen, consisting of separate and independent patches, slightly varying in age, of which the newest is probably not much anterior to the Norwich Crag. Secondly, there was a great change of conditions, both as to the depth of the sea and climate, between the periods of the Coralline and Red Crag, causing the fauna in each to differ far more widely than would appear from the above numerical results.

The value of the analysis given in the above table of the shells of the Red and Coralline Crags is in no small degree enhanced by the fact that they were all either collected by Mr. Wood himself, or obtained by him direct from their discoverers, so that he was enabled in each case to test their authenticity, and as far as possible to avoid those errors which arise from confounding together shells belonging to the sea of a newer deposit, and those washed into it from a formation of older date. The danger of this confusion may be conceived when we remember that the number of species rejected from the Red Crag as derivative by Mr. Wood is no less than 25. Some geologists have held that on the same grounds it is necessary to exclude as spurious some of the species found in the Norwich Crag proper; but Mr. Wood does not entertain this view, believing that the spurious shells which have sometimes found their way into the lists of this crag have been introduced by want of care from strata of Red Crag.

There can be no doubt, on the other hand, that conchologists have occasionally rejected from the Red and Norwich Crags, as derivative, shells which really belonged to the seas of those periods, because they were extinct or unknown as living, which in their eyes afforded sufficient ground for suspecting them to be intruders. The derivative origin of a species may sometimes be indicated by the extreme scarcity of the individuals, their colour, and worn condition; whereas an opposite conclusion may be arrived at by the integrity of the shells, especially when they are of delicate and tender structure, or their abundance, and, in the case of the lamellibranchiata, by their being held together by the ligament, which often happens when the shells have been so broken that little more than the hinges of the two valves are preserved. As to the univalves, I have seen from a pit of Red Crag, near Woodbridge, a large individual of the extinct Voluta Lamberti, seven inches in length, of which the lip, then perfect, had in former stages of its growth been frequently broken, and as often repaired. It had evidently lived in the sea of the Red Crag, where it had been exposed to rough usage, and sustained injuries like those which the reversed whelk, Trophon antiquum, so characteristic of the same formation, often exhibits. Additional proofs, however, have lately been obtained by Mr. Searles Wood that this shell had not died out in the era of the Red Crag by the discovery of the same fossil near Southwold, in beds of the later Norwich Crag.

ANTWERP CRAG.
Strata of the same age as the Red and Coralline Crag of Suffolk have been long known in the country round Antwerp, and on the banks of the Scheldt, below that city; and the lowest division, or Black Crag, there found, is shown by the shells to be somewhat more ancient than any of our British series, and probably forms the first links of a downward passage from the strata of the Pliocene to those of the Upper Miocene period.

NEWER PLIOCENE STRATA OF SICILY.
(FIGURE 132. Murex vaginatus, Phil.)

At several points north of Catania, on the eastern sea-coast of Sicily— as at Aci-Castello, for example, Trezza, and Nizzeti— marine strata, associated with volcanic tuffs and basaltic lavas, are seen, which belong to a period when the first igneous eruptions of Mount Etna were taking place in a shallow bay of the Mediterranean. They contain numerous fossil shells, and out of 142 species that have been collected all but eleven are identical with species now living. Some few of these eleven shells may possibly still linger in the depths of the Mediterranean, like Murex vaginatus, see Figure 132. The last-mentioned shell had already become rare when the associated marine and volcanic strata above alluded to were formed. On the whole, the modern character of the testaceous fauna under consideration is expressed not only by the small proportion of extinct species, but by the relative number of individuals by which most of the other species are represented, for the proportion agrees with that observed in the present fauna of the Mediterranean. The rarity of individuals in the extinct species is such as to imply that they were already on the point of dying out, having flourished chiefly in the earlier Pliocene times, when the Subapennine strata were in progress.

Yet since the accumulation of these Newer Pliocene sands and clays, the whole cone of Etna, 11,000 feet in height and about 90 miles in circumference at its base, has been slowly built up; an operation requiring many tens of thousands of years for its accomplishment, and to estimate the magnitude of which it is necessary to study in detail the internal structure of the mountain, and to see the proofs of its double axis, or the evidence of the lavas of the present great centre of eruption having gradually overwhelmed and enveloped a more ancient cone, situated 3 1/2 miles to the east of the present one. (See a Memoir on the Lavas and Mode of Origin of Mount Etna by the Author in Philosophical Transactions 1858.)

It appears that while Etna was increasing in bulk by a series of eruptions, its whole mass, comprising the foundations of subaqueous origin above alluded to, was undergoing a slow upheaval, by which those marine strata were raised to the height of 1200 feet above the sea, as seen at Catera, and perhaps to greater heights, for we can not trace their extension westward, owing to the dense and continuous covering of modern lava under which they are buried. During the gradual rise of these Newer Pliocene formations (consisting of clays, sands, and basalts) other strata of Post-pliocene date, marine as well as fluviatile, accumulated round the base of the mountain, and these, in their turn, partook of the upward movement, so that several inland cliffs and terraces at low levels, due partly to the action of the sea and partly to the river Simeto, originated in succession. Fossil remains of the elephant, and other extinct quadrupeds, have been found in these Post-Pliocene strata, associated with recent shells.

There is probably no part of Europe where the Newer Pliocene formations enter so largely into the structure of the earth's crust, or rise to such heights above the level of the sea, as Sicily. They cover nearly half the island, and near its centre, at Castrogiovanni, reach an elevation of 3000 feet. They consist principally of two divisions, the upper calcareous and the lower argillaceous, both of which may be seen at Syracuse, Girgenti, and Castrogiovanni. According to Philippi, to whom we are indebted for the best account of the tertiary shells of this island, thirty-five species out of one hundred and twenty-four obtained from the beds in central Sicily are extinct.

A geologist, accustomed to see nearly all the Newer Pliocene formations in the north of Europe occupying low grounds and very incoherent in texture, is naturally surprised to behold formations of the same age so solid and stony, of such thickness, and attaining so great an elevation above the level of the sea. The upper or calcareous member of this group in Sicily consists in some places of a yellowish-white stone, like the Calcaire Grossier of Paris; in others, of a rock nearly as compact as marble. Its aggregate thickness amounts sometimes to 700 or 800 feet. It usually occurs in regular horizontal beds, and is occasionally intersected by deep valleys, such as those of Sortino and Pentalica, in which are numerous caverns. The fossils are in every stage of preservation, from shells retaining portions of their animal matter and colour to others which are mere casts. The limestone passes downward into a sandstone and conglomerate, below which is clay and blue marl, from which perfect shells and corals may be disengaged. The clay sometimes alternates with yellow sand.

South of the plain of Catania is a region in which the tertiary beds are intermixed with volcanic matter, which has been for the most part the product of submarine eruptions. It appears that, while the clay, sand, and yellow limestone before mentioned were in course of deposition at the bottom of the sea, volcanoes burst out beneath the waters, like that of Graham Island, in 1831, and these explosions recurred again and again at distant intervals of time. Volcanic ashes and sand were showered down and spread by the waves and currents so as to form strata of tuff, which are found intercalated between beds of limestone and clay containing marine shells, the thickness of the whole mass exceeding 2000 feet. The fissures through which the lava rose may be seen in many places, forming what are called DIKES.

(FIGURE 133. Pecten jacobaeus; half natural size.)

No shell is more conspicuous in these Sicilian strata than the great scallop, Pecten jacobaeus (Figure 133), now so common in the neighbouring seas. The more we reflect on the preponderating number of this and other recent shells, the more we are surprised at the great thickness, solidity, and height above the sea of the rocky masses in which they are entombed, and the vast amount of geographical change which has taken place since their origin. It must be remembered that, before they began to emerge, the uppermost strata of the whole must have been deposited under water. In order, therefore, to form a just conception of their antiquity, we must first examine singly the innumerable minute parts of which the whole is made up, the successive beds of shells, corals, volcanic ashes, conglomerates, and sheets of lava; and we must afterwards contemplate the time required for the gradual upheaval of the rocks, and the excavation of the valleys. The historical period seems scarcely to form an appreciable unit in this computation, for we find ancient Greek temples, like those of Girgenti (Agrigentum), built of the modern limestone of which we are speaking, and resting on a hill composed of the same; the site having remained to all appearances unaltered since the Greeks first colonised the island.

It follows, from the modern geological date of these rocks, that the fauna and flora of a large part of Sicily are of higher antiquity than the country itself. The greater part of the island has been raised above the sea since the epoch of existing species, and the animals and plants now inhabiting it must have migrated from adjacent countries, with whose productions the species are now identical. The average duration of species would seem to be so great that they are destined to outlive many important changes in the configuration of the earth's surface, and hence the necessity for those innumerable contrivances by which they are enabled to extend their range to new lands as they are formed, and to escape from those which sink beneath the sea.

NEWER PLIOCENE STRATA OF THE UPPER VAL D'ARNO.

When we ascend the Arno for about ten miles above Florence, we arrive at a deep narrow valley called the Upper Val d'Arno, which appears once to have been a lake, at a time when the valley below Florence was an arm of the sea. The horizontal lacustrine strata of this upper basin are twelve miles long and two broad. The depression which they fill has been excavated out of Eocene and Cretaceous rocks, which form everywhere the sides of the valley in highly inclined stratification. The thickness of the more modern and unconformable beds is about 750 feet, of which the upper 200 feet consist of Newer Pliocene strata, while the lower are Older Pliocene. The newer series are made up of sands and a conglomerate called "sansino." Among the imbedded fossil mammalia are Mastodon arvernensis, Elephas meridionalis, Rhinoceros etruscus, Hippopotamus major, and remains of the genera bear, hyaena, and felis, nearly all of which occur in the Cromer forest-bed (see Chapter 13).

In the same upper strata are found, according to M. Gaudin, the leaves and cones of Glyptostrobus europaeus, a plant closely allied to G. heterophyllus, now inhabiting the north of China and Japan. This conifer had a wide range in time, having been traced back to the Lower Miocene strata of Switzerland, and being common at Oeningen in the Upper Miocene, as we shall see in the sequel (Chapter 14.)

OLDER PLIOCENE OF ITALY.— SUBAPENNINE STRATA.

The Apennines, it is well-known, are composed chiefly of Secondary or Mesozoic rocks, forming a chain which branches off from the Ligurian Alps and passes down the middle of the Italian peninsula. At the foot of these mountains, on the side both of the Adriatic and the Mediterranean, are found a series of tertiary strata, which form, for the most part, a line of low hills occupying the space between the older chain and the sea. Brocchi was the first Italian geologist who described this newer group in detail, giving it the name of the Subapennine. Though chiefly composed of Older Pliocene strata, it belongs, nevertheless, in part, both to older and newer members of the tertiary series. The strata, for example, of the Superga, near Turin, are Miocene; those of Asti and Parma Older Pliocene, as is the blue marl of Sienna; while the shells of the incumbent yellow sand of the same territory approach more nearly to the recent fauna of the Mediterranean, and may be Newer Pliocene.

We have seen that most of the fossil shells of the Older Pliocene strata of Suffolk which are of recent species are identical with testacea now living in British seas, yet some of them belong to Mediterranean species, and a few even of the genera are those of warmer climates. We might therefore expect, in studying the fossils of corresponding age in countries bordering the Mediterranean, to find among them some species and genera of warmer latitudes. Accordingly, in the marls belonging to this period at Asti, Parma, Sienna, and parts of the Tuscan and Roman territories, we observe the genera Conus, Cypraea, Strombus, Pyrula, Mitra, Fasciolaria, Sigaretus, Delphinula, Ancillaria, Oliva, Terebellum, Terebra, Perna, Plicatula, and Corbis, some characteristic of tropical seas, others represented by species more numerous or of larger size than those now proper to the Mediterranean.

OLDER PLIOCENE FLORA OF ITALY.

(FIGURE 134. Oreodaphne Heerii.
Leaf half natural size. (Feuilles fossiles de la Toscane.))

I have already alluded to the Newer Pliocene deposits of the Upper Val d'Arno above Florence, and stated that below those sands and conglomerates, containing the remains of the Elephas meridionalis and other associated quadrupeds, lie an older horizontal and conformable series of beds, which may be classed as Older Pliocene. They consist of blue clays with some subordinate layers of lignite, and exhibit a richer flora than the overlying Newer Pliocene beds, and one receding farther from the existing vegetation of Europe. They also comprise more species common to the antecedent Miocene period. Among the genera of flowering plants, M. Gaudin enumerates pine, oak, evergreen oak, plum, plane, alder, elm, fig, laurel, maple, walnut, birch, buckthorn, hickory, sumach, sarsaparilla, sassafras, cinnamon, Glyptostrobus, Taxodium, Sequoia, Persea, Oreodaphne (Figure 134), Cassia, and

Psoralea, and some others. This assemblage of plants indicates a warm climate, but not so subtropical an one as that of the Upper Miocene period, which will presently be considered.

(FIGURE 135. Liquidambar europaeum, var. trilobatum, A. Br. (sometimes four-lobed, and more commonly five-lobed). a. Leaf, half natural size. b. Part of same, natural size. c. Fruit, natural size. d. Seed, natural size. Oeningen.)

M. Gaudin, jointly with the Marquis Strozzi, has thrown much light on the botany of beds of the same age in another part of Tuscany, at a place called Montajone, between the rivers Elsa and Evola, where, among other plants, is found the Oreodaphne Heerii, Gaud. (See Figure 134), which is probably only a variety of Oreodaphne foetens, or the laurel called the Til in Madeira, where, as in the Canaries, it constitutes a large portion of the native woods, but can not now endure the climate of Europe. In the fossil specimens the same glands or protuberances are preserved (see Figure 134) as those which are seen in the axils of the primary veins of the leaves in the recent Til. (Contributions a la Flore fossile Italienne. Gaudin and Strozzi. Plate 11 Figure 3. Gaudin page 22.) Another plant also indicating a warmer climate is the Liquidambar europaeum, Brong. (see Figure 135), a species nearly allied to L. styracifluum, L., which flourishes in most places in the Southern States of North America, on the borders of the Gulf of Mexico.

CHAPTER XIV.
MIOCENE PERIOD.— UPPER MIOCENE.

Upper Miocene Strata of France.— faluns of Touraine.
Tropical Climate implied by Testacea.
Proportion of recent Species of Shells.
faluns more ancient than the Suffolk Crag.
Upper Miocene of Bordeaux and the South of France.
Upper Miocene of Oeningen, in Switzerland.
Plants of the Upper Fresh-water Molasse.
Fossil Fruit and Flowers as well as Leaves.
Insects of the Upper Molasse.
Middle or Marine Molasse of Switzerland.
Upper Miocene Beds of the Bolderberg, in Belgium.
Vienna Basin.
Upper Miocene of Italy and Greece.
Upper Miocene of India; Siwalik Hills.
Older Pliocene and Miocene of the United States.

UPPER MIOCENE STRATA OF FRANCE.— FALUNS OF TOURAINE.

The strata which we meet with next in the descending order are those called by many geologists "Middle Tertiary," for which in 1833 I proposed the name of Miocene, selecting the "faluns" of the valley of the Loire, in France, as my example or type. I shall now call these falunian deposits Upper Miocene, to distinguish them from others to which the name of Lower Miocene will be given.

No British strata have a distinct claim to be regarded as Upper Miocene, and as the Lower Miocene are also but feebly represented in the British Isles, we must refer to foreign examples in illustration of this important period in the earth's history. The term "faluns" is given provincially by French agriculturists to shelly sand and marl spread over the land in Touraine, just as similar shelly deposits were formerly much used in Suffolk to fertilise the soil, before the coprolitic or phosphatic nodules came into use. Isolated masses of such faluns occur from near the mouth of the Loire, in the neighbourhood of Nantes, to as far inland as a district south of Tours. They are also found at Pontlevoy, on the Cher, about seventy miles above the junction of that river with the Loire, and thirty miles south-east of Tours. Deposits of the same age also appear under new mineral conditions near the towns of Dinan and Rennes, in Brittany. I have visited all the localities above enumerated, and found the beds on the Loire to consist principally of sand and marl, in which are shells and corals, some entire, some rolled, and others in minute fragments. In certain districts, as at Doue, in the Department of Maine and Loire, ten miles south- west of Saumur, they form a soft building-stone, chiefly composed of an aggregate of broken shells, bryozoa, corals, and echinoderms, united by a calcareous cement; the whole mass being very like the Coralline

Crag near Aldborough, and Sudbourn in Suffolk. The scattered patches of faluns are of slight thickness, rarely exceeding fifty feet; and between the district called Sologne and the sea they repose on a great variety of older rocks; being seen to rest successively upon gneiss, clay-slate, various secondary formations, including the chalk; and, lastly, upon the upper fresh-water limestone of the Parisian tertiary series, which, as before mentioned (Chapter 9), stretches continuously from the basin of the Seine to that of the Loire.

(FIGURE 136. Dinotherium giganteum, Kaup.)

At some points, as at Louans, south of Tours, the shells are stained of a ferruginous colour, not unlike that of the Red Crag of Suffolk. The species are, for the most part, marine, but a few of them belong to land and fluviatile genera. Among the former, Helix turonensis (Figure 38, Chapter 3) is the most abundant. Remains of terrestrial quadrupeds are here and there intermixed, belonging to the genera Dinotherium (Figure 136), Mastodon, Rhinoceros, Hippopotamus, Chaeropotamus, Dichobune, Deer, and others, and these are accompanied by cetacea, such as the Lamantin, Morse, Sea-calf, and Dolphin, all of extinct species.

The fossil testacea of the faluns of the Loire imply, according to the late Edward Forbes, that the beds were formed partly on the shore itself at the level of low water, and partly at very moderate depths, not exceeding ten fathoms below that level. The molluscan fauna is, on the whole, much more littoral than that of the Pliocene Red and Coralline Crag of Suffolk, and implies a shallower sea. It is, moreover, contrasted with the Suffolk Crag by the indications it affords of an extra-European climate. Thus it contains seven species of Cypraea, some larger than any existing cowry of the Mediterranean, several species of Oliva, Ancillaria, Mitra, Terebra, Pyrula, Fasciolaria, and Conus. Of the cones there are no less than eight species, some very large, whereas the only European cone now living is of diminutive size. The genus Nerita, and many others, are also represented by individuals of a type now characteristic of equatorial seas, and wholly unlike any Mediterranean forms. These proofs of a more elevated temperature seem to imply the higher antiquity of the faluns as compared with the Suffolk Crag, and are in perfect accordance with the fact of the smaller proportion of testacea of recent species found in the faluns.

Out of 290 species of shells, collected by myself in 1840 at Pontlevoy, Louans, Bossee, and other villages twenty miles south of Tours, and at Savigne, about fifteen miles north-west of that place, seventy-two only could be identified with recent species, which is in the proportion of twenty-five per cent. A large number of the 290 species are common to all the localities, those peculiar to each not being more numerous than we might expect to find in different bays of the same sea.

The total number of species of testaceous mollusca from the faluns in my possession is 302, of which forty-five only, or fourteen per cent, were found by Mr. Wood to be common to the Suffolk Crag. The number of corals, including bryozoa and zoantharia, obtained by me at Doue and other localities before adverted to, amounts to forty-three, as determined by Mr. Lonsdale, of which seven (one of them a zoantharian) agree specifically with those of the Suffolk Crag. Some of the genera occurring fossil in Touraine, as the corals Astrea and Dendrophyllia, and the bryozoan Lunulites, have not been found in European seas north of the Mediterranean; nevertheless, the zoantharia of the faluns do not seem to indicate, on the whole, so warm a climate as would be inferred from the shells.

It was stated that, on comparing about 300 species of Touraine shells with about 450 from the Suffolk Crag, forty-five only were found to be common to both, which is in the proportion of only fifteen per cent. The same small amount of agreement is found in the corals also. I formerly endeavoured to reconcile this marked difference in species with the supposed co-existence of the two faunas, by imagining them to have severally belonged to distinct zoological provinces or two seas, the one opening to the north and the other to the south, with a barrier of land between them, like the Isthmus of Suez, now separating the Red Sea and the Mediterranean. But I now abandon that idea for several reasons; among others, because I succeeded in 1841 in tracing the Crag fauna southward in Normandy to within seventy miles of the Falunian type, near Dinan, yet found that both assemblages of fossils retained their distinctive characters, showing no signs of any blending of species or transition of climate.

The principal grounds, however, for referring the English Crag to the older Pliocene and the French faluns to the Upper Miocene epochs, consist in the predominance of fossil shells in the British strata identifiable with species not only still living, but which are now inhabitants of neighbouring seas, while the accompanying extinct species are of genera such as characterise Europe. In the faluns, on the contrary, the recent species are in a decided minority; and most of them are now inhabitants of the Mediterranean, the coast of Africa, and the Indian Ocean; in a word, less northern in character, and pointing to the prevalence of a warmer climate. They indicate a state of things receding farther from the present condition of Central Europe in physical geography and climate, and doubtless, therefore, receding farther from our era in time.

(FIGURE 137. Voluta Lamberti, Sowerby. Variety characteristic of Faluns of Touraine. Miocene.)

Among the conspicuous fossils common to the faluns of the Loire and the Suffolk Crag is a variety of the Voluta Lamberti, a shell already alluded to (Figure 123). The specimens of this shell which I have myself collected in Touraine, or have seen in museums, are thicker and heavier than British individuals of the same species, and shorter in proportion to their width, and have the folds on the columella less oblique, as represented in Figure 137.

UPPER MIOCENE OF BORDEAUX AND THE SOUTH OF FRANCE.

A great extent of country between the Pyrenees and the Gironde is overspread by tertiary deposits of various ages, and chiefly of Miocene date. Some of these, near Bordeaux, coincide in age with the faluns of Touraine, already mentioned, but many of the species of shells are peculiar to the south. The succession of beds in the basin of the Gironde implies several oscillations of level by which the same wide area was alternately converted into sea and land and into brackish-water lagoons, and finally into fresh-water ponds and lakes.

Among the fresh-water strata of this age near the base of the Pyrenees are marls, limestones and sands, in which the eminent comparative anatomist, M. Lartet, has obtained a great number of fossil mammalia common to the faluns of the Loire and the Upper Miocene beds of Switzerland, such as Dinotherium giganteum and Mastodon angustidens; also the bones of quadrumana, or of the ape and monkey tribe, which were discovered in 1837, the first of that order of quadrupeds detected in Europe. They were found near Auch, in the Department of Gers, in latitude 43 degrees 39' N. About forty miles west of Toulouse. They were referred by MM. Lartet and Blainville to a genus closely allied to the Gibbon, to which they gave the name of Pliopithecus. Subsequently, in 1856, M. Lartet described another species of the same family of long-armed apes (Hylobates), which he obtained from strata of the same age at Saint-Gaudens, in the Haute Garonne. The fossil remains of this animal consisted of a portion of a lower jaw with teeth and the shaft of a humerus. It is supposed to have been a tree-climbing frugivorous ape, equalling man in stature. As the trunks of oaks are common in the lignite beds in which it lay, it has received the generic name of Dryopithecus. The angle formed by the ascending ramus of the jaw and the alveolar border is less open, and therefore more like the human subject, than in the Chimpanzee, and what is still more remarkable, the fossil, a young but adult individual, had all its milk teeth replaced by the second set, while its last true molar (or wisdom-tooth) was still undeveloped, or only existed as a germ in the jaw-bone. In the mode, therefore, of the succession of its teeth (which, as in all the old-World apes, exactly agree in number with those in man) it differed from the Gorilla and Chimpanzee, and corresponded with the human species.

UPPER MIOCENE BEDS OF OENINGEN, IN SWITZERLAND.

The faluns of the Loire first served, as already stated, as the type of the Miocene formations in Europe. They yielded a plentiful harvest of marine fossil shells and corals, but were entirely barren of plants and insects. In Switzerland, on the other hand, deposits of the same age have been discovered, remarkable for their botanical and entomological treasures. We are indebted to Professor Heer, of Zurich, for the description, restoration, and classification of several hundred species and varieties of these fossil plants, the whole of which he has illustrated by excellent figures in his "Flora Tertiaria Helvetiae." This great work, and those of Adolphe Brongniart, Unger, Goppert and others, show that this class of fossils is beginning to play the same important part in the classification of the tertiary strata containing lignite or brown coal as an older flora has long played in enabling us to understand the ancient coal or carboniferous formation. No small skepticism has always prevailed among botanists as to whether the leaves alone and the wood of plants could ever afford sufficient data for determining even genera and families in the vegetable kingdom. In truth, before such remains could be rendered available a new science had to be created. It was necessary to study the outlines, nervation, and microscopic structure of the leaves, with a degree of care which had never been called for in the classification of living plants, where the flower and fruit afforded characters so much more definite and satisfactory. As geologists, we can not be too grateful to those who, instead of despairing when so difficult a task was presented to them, or being discouraged when men of the highest scientific attainments treated the fossil leaves as worthless, entered with full faith and enthusiasm into this new and unexplored field. That they should frequently have fallen into errors was unavoidable, but it is remarkable, especially if we inquire into the history of Professor Heer's researches, how often early conjectures as to the genus and family founded on the leaves alone were afterwards confirmed when fuller information was obtained. As examples to be found on comparing Heer's earlier and later works, I may instance the chestnut, elm, maple, cinnamon, magnolia, buckbean or Menyanthes, vine, buckthorn (Rhamnus), Andromeda and Myrica, and among the conifers Sequoia and Taxodium. In all these cases the plants were first recognised by their leaves, and the accuracy of the determination was afterwards confirmed when the fruit, and in some instances both fruit and flower, were found attached to the same stem as the leaves.

But let us suppose that no fruit, seed, or flower had ever been met with in a fossil state, we should still have been indebted to the persevering labours of botanical palaeontologists for one of the grandest scientific discoveries for which the present century is remarkable— namely, the proofs now established of the prevalence of a mild climate and a rich arborescent flora in the arctic regions in that Miocene epoch of which we are now entering. It may be useful if I endeavour to give the reader in a few words some idea of the nature of the evidence of these important conclusions, to show how far they may be safely based on fossil leaves alone. When we begin by studying the fossils of the Newer Pliocene deposits, such as those of the Upper Val d'Arno, before alluded to, we perceive that the fossil foliage agrees almost entirely with the trees and shrubs of a modern European forest. In the plants of the Older Pliocene strata of the same region we observe a larger proportion of species and genera which, although they may agree with well-known Asiatic or other foreign types, are at present wanting in Italy. If we then examine the Miocene formations of the same country, exotic forms become more abundant, especially the palms, whether they belong to the European or American fan-palms, Chamaerops and Sabal, or to the more tropical family of the date-palms or Phoenicites, which last are conspicuous in the Lower Miocene beds of Central Europe. Although we have not found the fruit or flower of these palms in a fossil state, the leaves are so characteristic that no one doubts the family to which they belong, or hesitates to accept them as indications of a warm and sub-tropical climate.

When the Miocene formations are traced to the northward of the 50th degree of latitude, the fossil palms fail us, but the greater proportion of the leaves, whether identical with those of existing European trees or of forms now unknown in Europe, which had accompanied the Miocene palms, still continue to characterise rocks of the same age, until we meet with them not only in Iceland, but in Greenland, in latitude 70 degrees N., and in Spitzbergen, latitude 78 degrees 56', or within about 11 degrees of the pole, and under circumstances which clearly show them to have been indigenous in those regions, and not to have been drifted from the south (see Chapter 15). Not only, therefore, has the botanist afforded the geologist much palaeontological assistance in identifying distinct tertiary formations in distant places by his power of accurately discriminating the forms, veining, and microscopic structure of leaves or wood, but, independently of that exact knowledge derivable from the organs of fructification, we are indebted to him for one of the most novel, unexpected results of modern scientific inquiry.

The Miocene formations of Switzerland have been called MOLASSE, a term derived from the French MOL, and applied to a SOFT, incoherent, greenish sandstone, occupying the country between the Alps and the Jura. This molasse comprises three divisions, of which the middle one is marine, and being closely related by its shells to the faluns of Touraine, may be classed as Upper Miocene. The two others are fresh-water, the upper of which may be also grouped with the faluns, while the lower must be referred to the Lower Miocene, as defined in the next chapter.

UPPER FRESH-WATER MOLASSE.

This formation is best seen at Oeningen, in the valley of the Rhine, between Constance and Schaffhausen, a locality celebrated for having produced in the year 1700 the supposed human skeleton called by Scheuchzer "homo diluvii testis," a fossil afterwards demonstrated by Cuvier to be a reptile, or aquatic salamander, of larger dimensions than even its great living representative, the salamander of Japan.

The Oeningen strata consist of a series of marls and limestones, many of them thinly laminated, and which appear to have slowly accumulated in a lake probably fed by springs holding carbonate of lime in solution. The elliptical area over which this fresh-water formation has been traced extends, according to Sir Roderick Murchison, for a distance of ten miles east and west from Berlingen, on the right bank of the river to Wangen, and to Oeningen, near Stein, on the left bank. The organic remains have been chiefly derived from two quarries, the lower of which is about 550 feet above the level of the Lake of Constance, while the upper quarry is 150 feet higher. In this last, a section thirty feet deep displays a great succession of beds, most of them splitting into slabs and some into very thin laminae. Twenty-one beds are enumerated by Professor Heer, the uppermost a bluish-grey marl seven feet thick, with organic remains, resting on a limestone with fossil plants, including leaves of poplar, cinnamon, and pond- weed (Potamogeton), together with some insects; while in the bed No. 4, below, is a bituminous rock, in which the Mastodon tapiroides, a characteristic Upper Miocene quadruped, has been met with. The 5th bed, two or three inches thick, contains fossil fish, e.g., Leuciscus (roach), and the larvae of dragon-flies, with plants such as the elm (Ulmus), and the aquatic Chara. Below these are other plant-beds; and then, in No. 9, the stone in which the great salamander (Andrias Scheuchzeri) and some fish were found. Below this other strata occur with fish, tortoises, the great salamander before alluded to, fresh-water mussels, and plants. In No. 16 the fossil fox of Oeningen, galecynus Oeningensis, Owen, was obtained by Sir R. Murchison. To this succeed other beds with mammalia (Lagomys), reptiles, (Emys), fish, and plants, such as walnut, maple, and poplar. In the 19th bed are numerous fish, insects, and plants, below which are marls of a blue indigo colour.

In the lower quarry eleven beds are mentioned, in which, as in the upper, both land and fresh-water plants and many insects occur. In the 6th, reckoning from the top, many plants have been obtained, such as Liquidambar, Daphnogene, Podogonium, and Ulmus, together with tortoises, besides the bones and teeth of a ruminant quadruped, named by H. von Meyer Palaeomeryx eminens. No. 9 is called the insect-bed, a layer only a few inches thick, which, when exposed to the frost, splits into leaves as thin as paper. In these thin laminae plants such as Liquidambar, Daphnogene, and Glyptostrobus, occur, with innumerable insects in a wonderful state of preservation, usually found singly. Below this is an indigo- blue marl, like that at the bottom of the higher quarry, resting on yellow marl ascertained to be at least thirty feet thick.

(FIGURE 138. Cinnamomum polymorphum, Ad. Brong. Upper and Lower Miocene. a. Leaf. b. Flower, natural size; Heer Plate 93 Figure 28. c. Ripe fruit of Cinnamomum polymorphum, from Oeningen; Heer, Plate 94 Figure 14. d. Fruit of recent Cinnamomum camphorum of Japan; Heer, Plate 152 Figure 18.)

All the above fossil-bearing strata were evidently formed with extreme slowness. Although the fossiliferous beds are, in the aggregate, no more than a few yards in thickness, and have only been examined in the small area comprised in the two quarries just alluded to, they give us an insight into the state of animal and vegetable life in part of the Upper Miocene period, such as no other region in the world has elsewhere supplied. In the year 1859, Professor Heer had already determined no less than 475 species of plants and more than 800 insects from these Oeningen beds. He supposes that a river entering a lake floated into it some of the leaves and land insects, together with the carcasses of quadrupeds, among others a great Mastodon. Occasionally, during tempests, twigs and even boughs of trees with their leaves were torn off and carried for some distance so as to reach the lake. Springs, containing carbonate of lime, seem at some points to have supplied calcareous matter in solution, giving origin locally to a kind of travertin, in which organic bodies sinking to the bottom became hermetically sealed up. The laminae, says Heer, which immediately succeed each other were not all formed at the same season, for it can be shown that, when some of them originated, certain plants were in flower, whereas, when the next of these layers was produced, the same plants had ripened their fruit. This inference is confirmed by independent proofs derived from insects. The principal insect-bed is rarely two inches thick, and is composed, says Heer, of about 250 leaf-like laminae, some of which were deposited in the spring, when the Cinnamomum polymorphum (Figure 138) was in flower, others in summer, when winged ants were numerous, and when the poplar and willow had matured their seed; others, again, in autumn, when the same Cinnamomum polymorphum (Figure 138) was in fruit, as well as the liquidambar, oak, clematis, and many other plants. The ancient lake seems to have had a belt of poplars and willows round its borders, countless leaves of which were imbedded in mud, and together with them, at some points, a species of reed, Arundo, which was very common.

One of the most characteristic shrubs is a papilionaceous and leguminous plant of an extinct genus, called by Heer Podogonium, of which two species are known. Entire twigs have been found with flowers, and always without leaves, as the flowers evidently came out, as in the poplar and willow tribe, before any leaves made their appearance. Other specimens have been obtained with ripe fruits accompanied by leaves, which resemble those of the tamarind, to which it was evidently allied, being of the family Caesalpineae, now proper to warmer regions.

(FIGURE 139. Acer trilobatum, normal form; Heer, Flora Tert. Helv. Plate 114 Figure 2. Size 1/2 diam. (Part only of the long stalk of the original fossil specimen is here given). Upper Miocene, Oeningen; also found in Lower Miocene of Switzerland.)

(FIGURE 140. Acer trilobatum. a. Abnormal variety of leaf; Heer, Plate 110 Figure 16. b. Flower and bracts, normal form; Heer, Plate 111 Figure 21. c. Half a seed-vessel; Heer, Plate 111 Figure 5.)

(FIGURE 141. Platanus aceroides, Gopp.; Heer, Plate 88 Figures 5-8. Size 2/3 diam. Upper Miocene, Oeningen. a. Leaf. b. The core of a bundle of pericarps. c. Single fruit or pericarp, natural size.)

The Upper Miocene flora of Oeningen is peculiarly important, in consequence of the number of genera of which not merely the leaves, but, as in the case of the Podogonium just mentioned, the fruit also and even the flower are known. Thus there are nineteen species of maple, ten of which have already been found with fruit. Although in no one region of the globe do so many maples now flourish, we need not suspect Professor Heer of having made too many species in this genus when we consider the manner in which he has dealt with one of them, Acer trilobatum, Figures 139 and 140. Of this plant the number of marked varieties figured and named is very great, and no less than three of them had been considered as distinct species by other botanists, while six of the others might have laid claim, with nearly equal propriety, to a like distinction. The common form, called Acer trilobatum, Figure 139, may be taken as a normal representative of the Oeningen fossil, and Figure 140, as one of the most divergent varieties, having almost four lobes in the leaf instead of three.

(FIGURE 142. Smilax sagittifera; Heer, Plate 30 Figure 7. Size 1/2 diameter. a. Leaf. b. Flower magnified, one of the six petals wanting at d. Upper Miocene, Oeningen. c. Smilax obtusifolia; Heer, Plate 30 Figure 9; natural size. Upper Miocene, Oeningen.)

(FIGURE 143. Fruit of the fossil and recent species of Hakea, a genus of Proteaceae. a. Leaf of fossil species, Hakea salicina. Upper Miocene, Oeningen; Heer Plate 97 Figure 29. 1/3 diameter. b. Impression of woody fruit of same, showing thick stalk. 2/3 diameter. c. Seed of same, natural size. d. Fruit of living Australian species, Hakea saligna, R. Brown. 1/2 diameter. e. Seed of same, natural size.)

Among the conspicuous genera which abounded in the Miocene period in Europe is the plane-tree, Platanus, the fossil species being considered by Heer to come nearer to the American P. occidentalis than to P. orientalis of Greece and Asia Minor. In some of the fossil specimens the male flowers are preserved. Among other points of resemblance with the living plane-trees, as we see them in the parks and squares of London, fossil fragments of the trunk are met with, having pieces of their bark peeling off.

The vine of Oeningen, Vitis teutonica, Ad. Brong, is of a North American type. Both the leaves and seeds have been found at Oeningen, and bunches of compressed grapes of the same species have been met with in the brown coal of Wetteravia in Germany. No less than eight species of smilax, a monocotyledonous genus, occur at Oeningen and in other Upper Miocene localities, the flowers of some of them, as well as the leaves, being preserved; as in the case of the very common fossil, S. sagittifera, Figure 142, a.

Leaves of plants supposed to belong to the order Proteaceae have been obtained partly from Oeningen and partly from the lacustrine formation of the same age at Locle in the Jura. They have been referred to the genera Banksia, Grevillea, Hakea, and Persoonia. Of Hakea there is the impression of a supposed seed- vessel, with its characteristic thick stalk and seeds, but as the fruit is without structure, and has not yet been found attached to the same stem as the leaf, the proof is incomplete.

To whatever family the foliage hitherto regarded as proteaceous by many able palaeontologists may eventually be shown to belong, we must be careful not to question their affinity to that order of plants on those geographical considerations which have influenced some botanists. The nearest living Proteaceae now feel the in Abyssinia in latitude 20 degrees N., but the greatest number are confined to the Cape and Australia. The ancestors, however, of the Oeningen fossils ought not to be looked for in such distant regions, but from that European land which in Lower Miocene times bore trees with similar foliage, and these had doubtless an Eocene source, for cones admitted by all botanists to be proteaceous have been met with in one division of that older Tertiary group (see Figure 206 Chapter 16). The source of these last, again, must not be sought in the antipodes, for in the white chalk of Aix-la-Chapelle leaves like those of Grevillea and other proteaceous genera have been found in abundance, and, as we shall see in Chapter 17, in a most perfect state of preservation. All geologists agree that the distribution of the Cretaceous land and sea had scarcely any connection with the present geography of the globe.

(FIGURE 144. Glyptostrobus Europaeus.
Branch with ripe fruit; Heer, Plate 20 Figure 1. Upper Miocene, Oeningen.)

In the same beds with the supposed Proteaceae there occurs at Locle a fan-palm of the American type Sabal (for genus see Figure 151), a genus which ranges throughout the low country near the sea from the Carolinas to Florida and Louisiana. Among the Coniferae of Upper Miocene age is found a deciduous cypress nearly allied to the Taxodium distichum of North America, and a Glyptostrobus (Figure 144), very like the Japanese G. heterophyllus, now common in our shrubberies.

Before the appearance of Heer's work on the Miocene Flora of Switzerland, Unger and Goppert had already pointed out the large proportion of living North American genera which distinguished the vegetation of the Miocene period in Central Europe. Next in number, says Heer, to these American forms at Oeningen the European genera preponderate, the Asiatic ranking in the third, the African in the fourth, and the Australian in the fifth degree. The American forms are more numerous than in the Italian Pliocene flora, and the whole vegetation indicates a warmer climate than the Pliocene, though not so high a temperature as that of the older or Lower Miocene period.

The conclusions drawn from the insects are for the most part in perfect harmony with those derived from the plants, but they have a somewhat less tropical and less American aspect, the South European types being more numerous. On the whole, the insect fauna is richer than that now inhabiting any part of Europe. No less than 844 species are reckoned by Heer from the Oeningen beds alone, the number of specimens which he has examined being 5080. The entire list of Swiss species from the Upper and Lower Miocene together amount to 1322. Almost all the living families of Coleoptera are represented, but, as we might have anticipated from the preponderance of arborescent and ligneous plants, the wood-eating beetles play the most conspicuous part, the Buprestidae and other long-horned beetles being particularly abundant.

(FIGURE 145. Harpactor maculipes, Heer. Upper Miocene, Oeningen.)

The patterns and some remains of the colours both of Coleoptera and Hemiptera are preserved at Oeningen, as, for example in Harpactor (Figure 145), in which the antennae, one of the eyes, and the legs and wings are retained. The characters, indeed, of many of the insects are so well defined as to incline us to believe that if this class of the invertebrata were not so rare and local, they might be more useful than even the plants and shells in settling chronological points in geology.

MIDDLE OR MARINE MOLASSE (UPPER MIOCENE) OF SWITZERLAND.

It was before stated that the Miocene formation of Switzerland consisted of, first, the upper fresh-water molasse, comprising the lacustrine marls of Oeningen; secondly, the marine molasse, corresponding in age to the faluns of Touraine; and thirdly, the lower fresh-water molasse. Some of the beds of the marine or middle series reach a height of 2470 feet above the sea. A large number of the shells are common to the faluns of Touraine, the Vienna basin, and other Upper Miocene localities. The terrestrial plants play a subordinate part in the fossiliferous beds, yet more than ninety of them are enumerated by Heer as belonging to this falunian division, and of these more than half are common to subjacent Lower Miocene beds, while a proportion of about forty-five in one hundred are common to the overlying Oeningen flora. Twenty-six of the ninety-two species are peculiar.

UPPER MIOCENE OF THE BOLDERBERG, IN BELGIUM.

(FIGURE 146. Oliva Dufresnii, Bast. Bolderberg, Belgium; natural size. a. Front view. b. Back view.)

In a small hill or ridge called the Bolderberg, which I visited in 1851, situated near Hasselt, about forty miles E.N.E. of Brussels, strata of sand and gravel occur, to which M. Dumont first called attention as appearing to constitute a northern representative of the faluns of Touraine. On the whole, they are very distinct in their fossils from the two upper divisions of the Antwerp Crag before mentioned (Chapter 13), and contain shells of the genera Oliva, Conus, Ancillaria, Pleurotoma, and Cancellaria in abundance. The most common shell is an Olive (Figure 146), called by Nyst Oliva Dufresnii; and constituting, as M. Bosquet observes, a smaller and shorter variety of the Bordeaux species.

So far as the shells of the Bolderberg are known, the proportion of recent species agrees with that in the faluns of Touraine, and the climate must have been warmer than that of the Coralline Crag of England.

UPPER MIOCENE BEDS OF THE VIENNA BASIN.

In South Germany the general resemblance of the shells of the Vienna tertiary basin with those of the faluns of Touraine has long been acknowledged. In the late Dr. Hornes's excellent work on the fossil mollusca of that formation, we see accurate figures of many shells, clearly of the same species as those found in the falunian sands of Touraine.

According to Professor Suess, the most ancient and purely marine of the Miocene strata in this basin consist of sands, conglomerates, limestones, and clays, and they are inclined inward, or from the borders of the trough towards the centre, their outcropping edges rising much higher than the newer beds, whether Miocene or Pliocene, which overlie them, and which occupy a smaller area at an inferior elevation above the sea. M. Hornes has described no less than 500 species of gasteropods, of which he identifies one-fifth with living species of the Mediterranean, Indian, or African seas, but the proportion of existing species among the lamellibranchiate bivalves exceeds this average. Among many univalves agreeing with those of Africa on the eastern side of the Atlantic are Cypraea sanguinolenta, Buccinum lyratum, and Oliva flammulata. In the lowest marine beds of the Vienna basin the remains of several mammalia have been found, and among them a species of Dinotherium, a Mastodon of the Trilophodon family, a Rhinoceros (allied to R. megarhinus, Christol), also an animal of the hog tribe, Listriodon, von Meyer, and a carnivorous animal of the canine family. The Helix turonensis (Figure 38 Chapter 3), the most common land shell of the French faluns, accompanies the above land animals. In a higher member of the Vienna Miocene series are found Dinotherium giganteum (Figure 136 Chapter 14), Mastodon longirostris, Rhinoceros Schleiermacheri, Acerotherium incisivum, and Hippotherium gracile, all of them equally characteristic of an Upper Miocene deposit occurring at Eppelsheim, in Hesse Darmstadt; a locality also remarkable as having furnished in latitude 49 degrees 50 north the bone of a large ape of the Gibbon kind, the most northerly example yet discovered of a quadrumanous animal.

(FIGURE 147. Amphistegina Hauerina, d'Orbigny. Upper Miocene strata, Vienna.)

M. Alcide d'Orbigny has shown that the foraminifera of the Vienna basin differ alike from the Eocene and Pliocene species, and agree with those of the faluns, so far as the latter are known. Among the Vienna foraminifera, the genus Amphistegina (Figure 147) is very characteristic, and is supposed by d'Archiac to take the same place among the Rhizopods of the Upper Miocene era which the Nummulites occupy in the Eocene period.

The flora of the Vienna basin exhibits some species which have a general range through the whole Miocene period, such as Cinnamomum polymorphum (Figure 138 Chapter 14), and C. Scheuchzeri, also Planera Richardi, Mich., Liquidambar europaeum (Figure 135 Chapter 13) Juglans bilinica, Cassia ambigua, and C. lignitum. Among the plants common to the Upper Miocene beds of Oeningen, in Switzerland, are Platanus aceroides (Figure 141 Chapter 14), Myrica vindobonensis, and others.

UPPER MIOCENE STRATA OF ITALY.

We are indebted to Signor Michelotti for a valuable work on the Miocene shells of Northern Italy. Those found in the hill called the Superga, near Turin, have long been known to correspond in age with the faluns of Touraine, and they contain so many species common to the Upper Miocene strata of Bordeaux as to lead to the conclusion that there was a free communication between the northern part of the Mediterranean and the Bay of Biscay in the Upper Miocene period.

UPPER MIOCENE FORMATIONS OF GREECE.

At Pikerme, near Athens, MM. Wagner and Roth have described a deposit in which they found the remains of the genera Mastodon, Dinotherium, Hipparion, two species of Giraffe, Antelope, and others, some living and some extinct. With them were also associated fossil bones of the Semnopithecus, showing that here, as in the south of France, the quadrumana were characteristic of this period. The whole fauna attests the former extension of a vast expanse of grassy plains where we have now the broken and mountainous country of Greece; plains, which were probably united with Asia Minor, spreading over the area where the deep Aegean Sea and its numerous islands are now situated. We are indebted to M. Gaudry, who visited Pikerme, for a treatise on these fossil bones, showing how many data they contribute to the theory of a transition from the mammalia of the Upper Miocene through the Pliocene and Post-pliocene forms to those of living genera and species.

UPPER MIOCENE OF INDIA. SIWALIK HILLS.

The Siwalik Hills lie at the southern foot of the Himalayan chain, rising to the height of 2000 and 3000 feet. Between the Jumna and the Ganges they consist of inclined strata of sandstone, shingle, clay, and marl. We are indebted to the indefatigable researches of Dr. Falconer and Sir Proby Cautley, continued for fifteen years, for the discovery in these marls and sandstones of a great variety of fossil mammalia and reptiles, together with many freshwater shells. Out of fifteen species of shells of the genera Paludina, Melania, Ampullaria, and Unio, all are extinct or unknown species with the exception of four, which are still inhabitants of Indian rivers. Such a proportion of living to extinct mollusca agrees well with the usual character of an Upper Miocene or Falunian fauna, as observed in Touraine, or in the basin of Vienna and elsewhere.

The genera of mammalia point in the same direction. One of them, of the genus Chalicotherium (or Anisodon of Lartet), is a pachyderm intermediate between the Rhinoceros and Anoplothere, and characteristic of the Upper Miocene strata of Eppelsheim, and of the south of France. With it occurs also an extinct form of Hippopotamus, called Hexaprotodon, and a species of Hippotherium and pig, also two species of Mastodon, two of elephant, and three other elephantine proboscidians; none of them agreeing with any fossil forms of Europe, and being intermediate between the genera Elephas and Mastodon, constituting the sub-genus Stegodon of Falconer. With these are associated a monkey, allied to the Semnopithecus entellus, now living in the Himalaya, and many ruminants. Among these last, besides the giraffe, camel, antelope, stag, and others, we find a remarkable new type, the Sivatherium, like a gigantic four-horned deer. There are also new forms of carnivora, both feline and canine, the Machairodus among the former, also hyaenas, and a subursine form called the Hyaenarctos, and a genus allied to the otter (Enhydriodon), of formidable size.

The giraffe, camel, and a large ostrich may be cited as proofs that there were formerly extensive plains where now a steep chain of hills, with deep ravines, runs for many hundred miles east and west. Among the accompanying reptiles are several crocodiles, some of huge dimensions, and one not distinguishable, says Dr. Falconer, from a species now living in the Ganges (C. Gangeticus); and there is still another saurian which the same anatomist has identified with a species now inhabiting India. There was also an extinct species of tortoise of gigantic proportions (Colossochelys Atlas), the curved shell of which was twelve feet three inches long and eight feet in diameter, the entire length of the animal being estimated at eighteen feet, and its probable height seven feet.

Numerous fossils of the Siwalik type have also been found in Perim Island, in the Gulf of Cambay, and among these a species of Dinotherium, a genus so characteristic of the Upper Miocene period in Europe.

OLDER PLIOCENE AND MIOCENE FORMATIONS IN THE UNITED STATES.

Between the Alleghany Mountains, formed of older rocks, and the Atlantic, there intervenes, in the United States, a low region occupied principally by beds of marl, clay, and sand, consisting of the cretaceous and tertiary formations, and chiefly of the latter. The general elevation of this plain bordering the Atlantic does not exceed 100 feet, although it is sometimes several hundred feet high. Its width in the middle and southern states is very commonly from 100 to 150 miles. It consists, in the South, as in Georgia, Alabama, and South Carolina, almost exclusively of Eocene deposits; but in North Carolina, Maryland, Virginia, Delaware, more modern strata predominate, of the age of the English Crag and faluns of Touraine. (Proceedings of the Geological Society volume 4 part 3 1845 page 547.)

(FIGURE 148. Fulgur canaliculatus. Maryland.)

(FIGURE 149. Fusus quadricostatus, Say. Maryland.)

In the Virginian sands, we find in great abundance a species of Astarte (A. undulata, Conrad), which resembles closely, and may possibly be a variety of, one of the commonest fossils of the Suffolk Crag (A. Omalii); the other shells also, of the genera Natica, Fissurella, Artemis, Lucina, Chama, Pectunculus, and Pecten, are analagous to shells both of the English Crag and French faluns, although the species are almost all distinct. Out of 147 of these American fossils I could only find thirteen species common to Europe, and these occur partly in the Suffolk Crag, and partly in the faluns of Touraine; but it is an important characteristic of the American group, that it not only contains many peculiar extinct forms, such as Fusus quadricostatus, Say (see Figure 149), and Venus tridacnoides, abundant in these same formations, but also some shells which, like Fulgur carica of Say and F. canaliculatus (see Figure 148), Calyptraea costata, Venus mercenaria, Lam., Modiola glandula, Totten, and Pecten magellanicus, Lam., are recent species, yet of forms now confined to the western side of the Atlantic— a fact implying that some traces of the beginning of the present geographical distribution of mollusca date back to a period as remote as that of the Miocene strata.

Of ten species of corals which I procured on the banks of the James River, one agrees generically with a coral now living on the coast of the United States. Mr. Lonsdale regarded these corals as indicating a temperature exceeding that of the Mediterranean, and the shells would lead to similar conclusions. Those occurring on the James River are in the 37th degree of N. latitude, while the French faluns are in the 47th; yet the forms of the American fossils would scarcely imply so warm a climate as must have prevailed in France when the Miocene strata of Touraine originated.

(FIGURE 150. Astrangia lineata, Lonsdale. Syn. Anthophyllum lineatum. Williamsburg, Virginia.)

Among the remains of fish in these post-eocene strata of the United States are several large teeth of the shark family, not distinguishable specifically from fossils of the faluns of Touraine.

CHAPTER XV.
LOWER MIOCENE (OLIGOCENE OF BEYRICH).

Lower Miocene Strata of France.
Line between Miocene and Eocene.
Lacustrine Strata of Auvergne.
Fossil mammalia of the Limagne d'Auvergne.
Lower Molasse of Switzerland.
Dense Conglomerates and Proofs of Subsidence.
Flora of the Lower Molasse.
American Character of the Flora.
Theory of a Miocene Atlantis.
Lower Miocene of Belgium.
Rupelian Clay of Hermsdorf near Berlin.
Mayence Basin.
Lower Miocene of Croatia.
Oligocene Strata of Beyrich.
Lower Miocene of Italy.
Lower Miocene of England.
Hempstead Beds.
Bovey Tracey Lignites in Devonshire.
Isle of Mull Leaf-Beds.
Arctic Miocene Flora.
Disco Island.
Lower Miocene of United States.
Fossils of Nebraska.

LINE BETWEEN MIOCENE AND EOCENE FORMATIONS.

The marine faluns of the valley of the Loire have been already described as resting in some places on a fresh-water tertiary limestone, fragments of which have been broken off and rolled on the shores and in the bed of the Miocene sea. Such pebbles are frequent at Pontlevoy on the Cher, with hollows drilled in them in which the perforating marine shells of the Falunian period still remain. Such a mode of superposition implies an interval of time between the origin of the fresh-water limestone and its submergence beneath the waters of the Upper Miocene sea. The limestone in question forms a part of the formation called the Calcaire de la Beauce, which constitutes a large table-land between the basins of the Loire and the Seine. It is associated with marls and other deposits, such as may have been formed in marshes and shallow lakes in the newest part of a great delta. Beds of flint, continuous or in nodules, accumulated in these lakes, and aquatic plants called Charae, left their stems and seed-vessels imbedded both in the marl and flint, together with fresh-water and land shells. Some of the siliceous rocks of this formation are used extensively for mill-stones. The flat summits or platforms of the hills round Paris, and large areas in the forest of Fontainebleau, as well as the Plateau de la Beauce, already alluded to, are chiefly composed of these fresh-water strata. Next to these in the descending order are marine sands and sandstone, commonly called the Gres de Fontainebleau, from which a considerable number of shells, very distinct from those of the faluns, have been obtained at Etampes, south of Paris, and at Montmartre and other hills in Paris itself, or in its suburbs. At the bottom of these sands a green clay occurs, containing a small oyster, Ostrea cyathula, Lam., which, although of slight thickness, is spread over a wide area. This clay rests immediately on the Paris gypsum, or that series of beds of gypsum and gypseous marl from which Cuvier first obtained several species of Palaeotherium and other extinct mammalia. (Bulletin 1856 Journal volume 12 page 768.)

At this junction of the clay and the gypsum the majority of French geologists have always drawn the line between the Middle and Lower Tertiary, or between the Miocene and Eocene formations, regarding the Fontainebleau sands and the Ostrea cyathula clay as the base of the Miocene, and the gypsum, with its mammalia, as the top of the Eocene group. I formerly dissented from this division, but I now find that I must admit it to be the only one which will agree with the distribution of the Miocene mammalia, while even the mollusca of the Fontainebleau sands, which were formerly supposed to present at preponderance of affinities to an Eocene fauna, have since been shown to agree more closely with the fossils of certain deposits always regarded as Middle Tertiary at Mayence and in Belgium. In fact, we are now arriving at that stage of progress when the line, wherever it be drawn between Miocene and Eocene, will be an arbitrary one, or one of mere convenience, as I shall have an opportunity of showing when the Upper Eocene formations in the Isle of Wight are described in the sixteenth chapter.

LOWER MIOCENE OF CENTRAL FRANCE.

Lacustrine strata, belonging, for the most part, to the same Miocene system as the Calcaire de la Beauce, are again met with farther south in Auvergne, Cantal, and Velay. They appear to be the monuments of ancient lakes, which, like some of those now existing in Switzerland, once occupied the depressions in a mountainous region, and have been each fed by one or more rivers and torrents. The country where they occur is almost entirely composed of granite and different varieties of granitic schist, with here and there a few patches of Secondary strata, much dislocated, and which have suffered great denudation. There are also some vast piles of volcanic matter, the greater part of which is newer than the fresh-water strata, and is sometimes seen to rest upon them, while a small part has evidently been of contemporaneous origin. Of these igneous rocks I shall treat more particularly in the sequel.

The study of these regions possesses a peculiar interest very distinct in kind from that derivable from the investigation either of the Parisian or English Tertiary areas. For we are presented in Auvergne with the evidence of a series of events of astonishing magnitude and grandeur, by which the original form and features of the country have been greatly changed, yet never so far obliterated but that they may still, in part at least, be restored in imagination. Great lakes have disappeared— lofty mountains have been formed, by the reiterated emission of lava, preceded and followed by showers of sand and scoriae— deep valleys have been subsequently furrowed out through masses of lacustrine and volcanic origin— at a still later date, new cones have been thrown up in these valleys— new lakes have been formed by the damming up of rivers— and more than one assemblage of quadrupeds, birds, and plants, Eocene, Miocene, and Pliocene, have followed in succession; yet the region has preserved from first to last its geographical identity; and we can still recall to our thoughts its external condition and physical structure before these wonderful vicissitudes began, or while a part only of the whole had been completed. There was first a period when the spacious lakes, of which we still may trace the boundaries, lay at the foot of mountains of moderate elevation, unbroken by the bold peaks and precipices of Mont Dor, and unadorned by the picturesque outline of the Puy de Dome, or of the volcanic cones and craters now covering the granitic platform. During this earlier scene of repose deltas were slowly formed; beds of marl and sand, several hundred feet thick, deposited; siliceous and calcareous rocks precipitated from the waters of mineral springs; shells and insects imbedded, together with the remains of the crocodile and tortoise, the eggs and bones of water-birds, and the skeletons of quadrupeds, most of them of genera and species characteristic of the Miocene period. To this tranquil condition of the surface succeeded the era of volcanic eruptions, when the lakes were drained, and when the fertility of the mountainous district was probably enhanced by the igneous matter ejected from below, and poured down upon the more sterile granite. During these eruptions, which appear to have taken place towards the close of the Miocene epoch, and which continued during the Pliocene, various assemblages of quadrupeds successively inhabited the district, among which are found the genera mastodon, rhinoceros, elephant, tapir, hippopotamus, together with the ox, various kinds of deer, the bear, hyaena, and many beasts of prey which ranged the forest or pastured on the plain, and were occasionally overtaken by a fall of burning cinders, or buried in flows of mud, such as accompany volcanic eruptions. Lastly, these quadrupeds became extinct, and gave place in their turn to the species now existing. There are no signs, during the whole time required for this series of events, of the sea having intervened, nor of any denudation which may not have been accomplished by currents in the different lakes, or by rivers and floods accompanying repeated earthquakes, or subterranean movements, during which the levels of the district have in some places been materially modified, and perhaps the whole upraised relatively to the surrounding parts of France.

AUVERGNE.

The most northern of the fresh-water groups is situated in the valley-plain of the Allier, which lies within the department of the Puy de Dome, being the tract which went formerly by the name of the Limagne d'Auvergne. The average breadth of this tract is about twenty miles; and it is for the most part composed of nearly horizontal strata of sand, sandstone, calcareous marl, clay, and limestone, none of which observe a fixed and invariable order of superposition. The ancient borders of the lake wherein the fresh-water strata were accumulated may generally be traced with precision, the granite and other ancient rocks rising up boldly from the level country. The actual junction, however, of the lacustrine beds and the granite is rarely seen, as a small valley usually intervenes between them. The fresh-water strata may sometimes be seen to retain their horizontality within a very slight distance of the border-rocks, while in some places they are inclined, and in few instances vertical. The principal divisions into which the lacustrine series may be separated are the following:— first, Sandstone, grit, and conglomerate, including red marl and red sandstone; secondly, Green and white foliated marls; thirdly, Limestone, or travertin, often oolitic in structure; fourthly, Gypseous marls.

The relations of these different groups can not be learnt by the study of any one section; and the geologist who sets out with the expectation of finding a fixed order of succession may perhaps complain that the different parts of the basin give contradictory results. The arenaceous division, the marls, and the limestone may all be seen in some places to alternate with each other; yet it can by no means be affirmed that there is no order of arrangement. The sands, sandstone, and conglomerate constitute in general a littoral group; the foliated white and green marl, a contemporaneous central deposit more than 700 feet thick, and thinly foliated, a character which often arises from the innumerable thin shells or carapace valves shed by the small crustacean called Cypris in the ancient lakes of Auvergne; and lastly the limestone is for the most part subordinate to the newer portions of both the above formations.

It seems that, when the ancient lake of the Limagne first began to be filled with sediment, no volcanic action had yet produced lava and scoriae on any part of the surface of Auvergne. No pebbles, therefore, of lava were transported into the lake— no fragments of volcanic rocks imbedded in the conglomerate. But at a later period, when a considerable thickness of sandstone and marl had accumulated, eruptions broke out, and lava and tuff were deposited, at some spots, alternately with the lacustrine strata. It is not improbable that cold and thermal springs, holding different mineral ingredients in solution, became more numerous during the successive convulsions attending this development of volcanic agency, and thus deposits of carbonate and sulphate of lime, silex, and other minerals were produced. Hence these minerals predominate in the uppermost strata. The subterranean movements may then have continued until they altered the relative levels of the country, and caused the waters of the lakes to be drained off, and the further accumulation of regular fresh-water strata to cease.

LOWER MIOCENE MAMMALIA OF THE LIMAGNE.

It is scarcely possible to determine the age of the oldest part of the fresh- water series of the Limagne, large masses both of the sandy and marly strata being devoid of fossils. Some of the lowest beds may be of Upper Eocene date, although, according to M. Pomel, only one bone of a Palaeotherium has been discovered in Auvergne. But in Velay, in strata containing some species of fossil mammalia common to the Limagne, no less than four species of Palaeothere have been found by M. Aymard, and one of these is generally supposed to be identical with Palaeotherium magnum, an undoubted Upper Eocene fossil, of the Paris gypsum, the other three being peculiar.

Not a few of the other mammalia of the Limagne belong undoubtedly to genera and species elsewhere proper to the Lower Miocene. Thus, for example, the Cainotherium of Bravard, a genus not far removed from the Anoplotherium, is represented by several species, one of which, as I learn from Mr. Waterhouse, agrees with Microtherium Renggeri of the Mayence basin. In like manner, the Amphitragulus elegans of Pomel, an Auvergne fossil, is identified by Waterhouse with Dorcatherium nanum of Kaup, a Rhenish species from Weissenau, near Mayence. A small species, also, of rodent, of the genus Titanomys of H. von Meyer, is common to the Lower Miocene of Mayence and the Limagne d'Auvergne, and there are many other points of agreement which the discordance of nomenclature tends to conceal. A remarkable carnivorous genus, the Hyaenodon of Laizer, is represented by more than one species. The same genus has also been found in the Upper Eocene marls of Hordwell Cliff, Hampshire, just below the level of the Bembridge Limestone, and therefore a formation older than the Gypsum of Paris. Several species of opossum (Didelphis) are met with in the same strata of the Limagne. The total number of mammalia enumerated by M. Pomel as appertaining to the Lower Miocene fauna of the Limagne and Velay falls little short of a hundred, and with them are associated some large crocodiles and tortoises, and some Ophidian and Batrachian reptiles.

LOWER MOLASSE OF SWITZERLAND.

The two upper divisions of the Swiss Molasse— the one fresh-water, the other marine— have already been described in the preceding chapter. I shall now proceed to treat of the third division, which is of Lower Miocene age. Nearly the whole of this Lower Molasse is fresh-water, yet some of the inferior beds contain a mixture of marine and fluviatile shells, the Cerithium margaritaceum, a well-known Lower Miocene fossil, being one of the marine species. Notwithstanding, therefore, that some of these Lower Miocene strata consist of old shingle-beds several thousand feet in thickness, as in the Rigi, near Lucerne, and in the Speer, near Wesen, mountains 5000 and 7000 feet above the sea, the deposition of the whole series must have begun at or below the sea-level.

The conglomerates, as might be expected, are often very unequal in thickness, in closely adjoining districts, since in a littoral formation accumulations of pebbles would swell out in certain places where rivers entered the sea, and would thin out to comparatively small dimensions where no streams or only small ones came down to the coast. For ages, in spite of a gradual depression of the land and adjacent sea-bottom, the rivers continued to cover the sinking area with their deltas; until finally, the subsidence being in excess, the sea of the Middle Molasse gained upon the land, and marine beds were thrown down over the dense mass of fresh-water and brackish-water deposit, called the Lower Molasse, which had previously accumulated.

FLORA OF THE LOWER MOLASSE.

In part of the Swiss Molasse, which belongs exclusively to the Lower Miocene period, the number of plants has been estimated at more than 500 species, somewhat exceeding those which were before enumerated as occurring in the two upper divisions. The Swiss Lower Miocene may best be studied on the northern borders of the Lake of Geneva, between Lausanne and Vevay, where the contiguous villages of Monod and Rivaz are situated. The strata there, which I have myself examined, consist of alternations of conglomerate, sandstone, and finely laminated marls with fossil plants. A small stream falls in a succession of cascades over the harder beds of pudding-stone, which resist, while the sandstone and plant-bearing shales and marls give way. From the latter no less than 193 species of plants have been obtained by the exertions of MM. Heer and Gaudin, and they are considered to afford a true type of the vegetation of the Lower Miocene formations of Switzerland— a vegetation departing farther in its character from that now flourishing in Europe than any of the higher members of the series before alluded to, and yet displaying so much affinity to the flora of Oeningen as to make it natural for the botanist to refer the whole to one and the same Miocene period. There are, indeed, no less than 81 species of these Older Miocene plants which pass up into the flora of Oeningen.

This fact is important as bearing on the propriety of classing the Lower Molasse of Switzerland as belonging to the Miocene rather than to the latter part of the Eocene period. There are, indeed, so many types among the fossils, both specific and generic, which have a wide range through the whole of the Molasse, that a unity of character is thereby stamped on the whole flora, in spite of the contrast between the plants of the uppermost and lowest formations, or between Oeningen and Monod. The proofs of a warmer climate, and the excess of arborescent over herbaceous plants, and of evergreen trees over deciduous species, are characters common to the whole flora, but which are intensified as we descend to the inferior deposits.

(FIGURE 151. Sabal major, Unger sp. Vevay. Lower Miocene; Heer, Plate 41.)

Nearly all the plants at Monod are contained in three layers of marl separated by two of soft sandstone. The thickness of the marls is ten feet, and vegetable matter predominates so much in some layers as to form an imperfect lignite. One bed is filled with large leaves of a species of fig (Ficus populina), and of a hornbeam (Carpinus grandis), the strength of the wind having probably been great when they were blown into the lake; whereas another contiguous layer contains almost exclusively smaller leaves, indicating, apparently, a diminished strength in the wind. Some of the upper beds at Monod abound in leaves of Proteaceae, Cyperaceae, and ferns, while in some of the lower ones Sequoia, Cinnamomum, and Sparganium are common. In one bed of sandstone the trunk of a large palm-tree was found unaccompanied by other fossils, and near Vevay, in the same series of Lower Miocene strata, the leaves of a palm of the genus Sabal (Figure 151), a genus now proper to America, were obtained.

Among other genera of the same class is a Flabellaria occurring near Lausanne, and a magnificent Phoenicites allied to the date palm. When these plants flourished the climate must have been much hotter than now. The Alps were no doubt much lower, and the palms now found fossil in strata elevated 2000 feet above the sea grew nearly at the sea-level, as is demonstrated by the brackish-water character of some of the beds into which they were carried by winds or rivers from the adjoining coast.

(FIGURE 152. Banksia. a. Fruit of fossil Banksia. b. Leaf of Banksia Deekiana.)

In the same plant-bearing deposits of the Lower Molasse in Switzerland leaves have been found which have been ascribed to the order Proteaceae already spoken of as well represented in the Oeningen beds (see Chapter 14). The Proteas and other plants of this family now flourish at the Cape of Good Hope; while the Banksias, and a set of genera distinct from those of Africa, grow most luxuriantly in the southern and temperate parts of Australia. They were probably inhabitants, says Heer, of dry hilly ground, and the stiff leathery character of their leaves must have been favourable to their preservation, allowing them to float on a river for great distances without being injured, and then to sink, when water-logged, to the bottom. It has been objected that the fruit of the Proteaceae is of so tough and enduring a texture that it ought to have been more commonly met with; but in the first place we must not forget the numerous cones found in the Eocene strata of Sheppey, which all admit to be proteaceous and to belong to at least two species (see Chapter 14). Secondly, besides the fruit of Hakea before mentioned (Chapter 14), Heer found associated with fossil leaves, having the exact form and nervation of Banksia, fruit precisely such as may have come from a cone of that plant, and lately he has received another similar fruit from the Lower Miocene strata of Lucerne. They may have fallen out of a decayed cone in the same way as often happens to the seeds of the spruce fir, Pinus abies, found scattered over the ground in our woods. It is a known fact that among the living Proteaceae the cones are very firmly attached to the branches, so that the seeds drop out without the cone itself falling to the ground, and this may perhaps be the reason why, in some instances in which fossil seeds have been found, no traces of the cone have been observed.

(FIGURE 153. Sequoia Langsdorfii. Ad. Brong., 1/3 natural size. Rivaz, near Lausanne; Heer, Plate 21 Figure 4. Upper and Lower Miocene and Lower Pliocene, Val d'Arno.
a. Branch with leaves.
b. Young cone.)

Among the Coniferae the Sequoia here figured is common at Rivaz, and is one of the most universal plants in the Lowest Miocene of Switzerland, while it also characterises the Miocene Brown Coals of Germany and certain beds of the Val d'Arno, which I have called Older Pliocene, Chapter 13.

(FIGURE 154. Lastraea stiriaca, Unger; Heer's Flora, Plate 143 Figure 8. Natural size. Lower and Upper Miocene, Switzerland. a. Specimen from Monod, showing the position of the sori on the middle of the tertiary nerves. b. More common appearance, where the sori remain and the nerves are obliterated.)

Among the ferns met with in profusion at Monod is the Lastraea stiriaca, Unger, which has a wide range in the Miocene period from strata of the age of Oeningen to the lowest part of the Swiss Molasse. In some specimens, as shown in Figure 154, the fructification is distinctly seen.

(FIGURE 155. Cinnamomum Rossmassleri, Heer. Daphnogene cinnamomifolia, Unger.

Upper and Lower Miocene, Switzerland and Germany.)

Among the laurels several species of Cinnamomum are very conspicuous. Besides the C. polymorphum, before figured, Chapter 14, another species also ranges from the Lower to the Upper Molasse of Switzerland, and is very characteristic of different deposits of Brown Coal in Germany. It has been called Cinnamomum Rossmassleri by Heer (see Figure 155). The leaves are easily recognised as having two side veins, which run up uninterruptedly to their point.

AMERICAN CHARACTER OF THE FLORA.

If we consider not merely the number of species but those plants which constitute the mass of the Lower Miocene vegetation, we find the European part of the fossil flora very much less prominent than in the Oeningen beds, while the foreground is occupied by American forms, by evergreen oaks, maples, poplars, planes, Liquidambar, Robinia, Sequoia, Taxodium, and ternate-leaved pines. There is also a much greater fusion of the characters now belonging to distinct botanical provinces than in the Upper Miocene flora, and we shall find this fusion still more strikingly exemplified as we go back to the antecedent Eocene and Cretaceous periods.

Professor Heer has advocated the doctrine, first advanced by Unger to explain the large number of American genera in the Miocene flora of Europe, that the present basin of the Atlantic was occupied by land over which the Miocene flora could pass freely. But other able botanists have shown that it is far more probable that the American plants came from the east and not from the west, and instead of reaching Europe by the shortest route over an imaginary Atlantis, migrated in an opposite direction, crossing the whole of Asia.

ARCTIC MIOCENE FLORA.

But when we indulge in speculations as to the geographical origin of the Miocene plants of Central Europe, we must take into account the discoveries recently made of a rich terrestrial flora having flourished in the Arctic Regions in the Miocene period from which many species may have migrated from a common centre so as to reach the present continents of Europe, Asia, and America. Professor Heer has examined the various collections of fossil plants that have been obtained in North Greenland (latitude 70 degrees), Iceland, Spitzbergen, and other parts of the Arctic regions, and has determined that they are of Miocene age and indicate a temperate climate. (Heer "Miocene baltische Flora" and "Fossil-flora von Alaska" 1869.) Including the collections recently brought from Greenland by Mr. Whymper, the Arctic Miocene flora now comprises 194 species, and that of Greenland 137 species, of which 46, or exactly one-third, are identical with plants found in the Miocene beds of Central Europe. Considerably more than half the number are trees, which is the more remarkable since, at the present day, trees do not exist in any part of Greenland even 10 degrees farther south.

More than thirty species of Coniferae have been found, including several Sequoias (allied to the gigantic Wellingtonia of California), with species of Thujopsis and Salisburia now peculiar to Japan. There are also beeches, oaks, planes, poplars, maples, walnuts, limes, and even a magnolia, two cones of which have recently been obtained, proving that this splendid evergreen not only lived but ripened its fruit within the Arctic circle. Many of the limes, planes, and oaks were large-leaved species, and both flowers and fruit, besides immense quantities of leaves, are in many cases preserved. Among the shrubs were many evergreens, as Andromeda, and two extinct genera, Daphnogene and M'Clintockia, with fine leathery leaves, together with hazel, blackthorn, holly, logwood, and hawthorn. A species of Zamia (Zamites) grew in the swamps, with Potamogeton, Sparganium, and Menyanthes, while ivy and vines twined around the forest trees and broad-leaved ferns grew beneath their shade. Even in Spitzbergen, as far north as latitude 78 degrees 56', no less than ninety-five species of fossil plants have been obtained, including Taxodium of two species, hazel, poplar, alder, beech, plane-tree, and lime. Such a vigorous growth of trees within 12 degrees of the pole, where now a dwarf willow and a few herbaceous plants form the only vegetation, and where the ground is covered with almost perpetual snow and ice, is truly remarkable.

The identity of so many of the fossils with Miocene species of Central Europe and Italy not only proves that the climate of Greenland was much warmer than it is now, but also renders it probable that a much more uniform climate prevailed over the entire northern hemisphere. This is also indicated by the whole character of the Upper Miocene flora of Central Europe, which does not necessitate a mean temperature very much greater than exists at present, if we suppose such absence of winter cold as is proper to insular climates. Professor Heer believes that the mean temperature of North Greenland must have been at least 30 degrees higher than at present, while an addition of 10 degrees to the mean temperature of Central Europe would probably be as much as was required. The chief locality where this wonderful flora is preserved is at Atanekerdluk in North Greenland (latitude 70 degrees), on a hill at an elevation of about 1200 feet above the sea. There is here a considerable succession of sedimentary strata pierced by volcanic rocks. Fossil plants occur in all the beds, and the erect trunks as thick as a man's body which are sometimes found, together with the abundance of specimens of flowers and fruit in good preservation, sufficiently prove that the plants grew where they are now found. At Disco island and other localities on the same part of the coast, good coal is abundant, interstratified with beds of sandstone, in some of which fossil plants have also been found, similar to those at Atanekerdluk.

LOWER MIOCENE, BELGIUM.

(FIGURE 156. Leda (Nucula) Deshayesiana, Nyst.)

(FIGURE 157. Vanessa pluto; natural size. Lower Miocene, Radaboj, Croatia.)

The Upper Miocene Bolderberg beds, mentioned in Chapter 14, rest on a Lower Miocene formation called the Rupelian of Dumont. This formation is best seen at the villages of Rupelmonde and Boom, ten miles south of Antwerp, on the banks of the Scheldt and near the junction with it of a small stream called the Rupel. A stiff clay abounding in fossils is extensively worked at the above localities for making tiles. It attains a thickness of about 100 feet, and though very different in age, much resembles in mineral character the "London clay," containing, like it, septaria or concretions of argillaceous limestone traversed by cracks in the interior, which are filled with calc-spar. The shells, referable to about forty species, have been described by MM. Nyst and De Koninck. Among them Leda (or Nucula) Deshayesiana (see Figure 156) is by far the most abundant; a fossil unknown as yet in the English tertiary strata, but when young much resembling Leda amygdaloides of the London Clay proper (see Figure 213 Chapter 16). Among other characteristic shells are Pecten Hoeninghausii, and a species of Cassidaria, and several of the genus Pleurotoma. Not a few of these testacea agree with English Eocene species, such as Actaeon simulatus, Sowb, Cancellaria evulsa, Brander, Corbula pisum (Figure 157), and Nautilus (Aturia) ziczac. They are accompanied by many teeth of sharks, as Lamna contortidens, Ag., Oxyrhinaxiphodon, Ag., Carcharodon angustidens (see Figure 196 Chapter 16), Ag., and other fish, some of them common to the Middle Eocene strata.

KLEYN SPAWEN BEDS.

The succession of the Lower Miocene strata of Belgium can be best studied in the environs of Kleyn Spawen, a village situated about seven miles west of Maestricht, in the old province of Limburg in Belgium. In that region, about 200 species of testacea, marine and fresh-water, have been obtained, with many foraminifera and remains of fish. In none of the Belgian Lower Miocene strata could I find any nummulites; and M. d'Archiac had previously observed that these foraminifera characterise his "Lower Tertiary Series," as contrasted with the Middle, and they therefore serve as a good test of age between Eocene and Miocene, at least in Belgium and the North of France. (D'Archiac Monograph pages 79, 100.) Between the Bolderberg beds and the Rupelian clay there is a great gap in Belgium, which seems, according to M. Beyrich, to be filled up in the North of Germany by what he calls the Sternberg beds, and which, had Dumont found them in Belgium, he might probably have termed Upper Rupelian.

LOWER MIOCENE OF GERMANY.
RUPELIAN CLAY OF HERMSDORF, NEAR BERLIN.

Professor Beyrich has described a mass of clay, used for making tiles, within seven miles of the gates of Berlin, near the village of Hermsdorf, rising up from beneath the sands with which that country is chiefly overspread. This clay is more than forty feet thick, of a dark bluish-grey colour, and, like that of Rupelmonde, contains septaria. Among other shells, the Leda Deshayesiana, before mentioned (Figure 156), abounds, together with many species of Pleurotoma, Voluta, etc., a certain proportion of the fossils being identical in species with those of Rupelmonde.

MAYENCE BASIN.

An elaborate description has been published by Dr. F. Sandberger of the Mayence tertiary area, which occupies a tract from five to twelve miles in breadth, extending for a great distance along the left bank of the Rhine from Mayence to the neighbourhood of Manheim, and which is also found to the east, north, and south-west of Frankfort. M. De Koninck, of Liege, first pointed out to me that the purely marine portion of the deposit contained many species of shells common to the Kleyn Spawen beds, and to the clay of Rupelmonde, near Antwerp. Among these he mentioned Cassidaria depressa, Tritonium argutum, Brander (T. flandricum, De Koninck), Tornatella simulata, Aporrhais Sowbyi, Leda Deshayesiana (Figure 156), Corbula pisum, (Figure 158) and others.

LOWER MIOCENE BEDS OF CROATIA.

The Brown Coal of Radaboj, near Angram in Croatia, not far from the borders of Styria, is covered, says Von Buch, by beds containing the marine shells of the Vienna basin, or, in other words, by Upper Miocene or Falunian strata. They appear to correspond in age to the Mayence basin, or to the Rupelian strata of Belgium. They have yielded more than 200 species of fossil plants, described by the late Professor Unger. These plants are well preserved in a hard marlstone, and contain several palms; among them the Sabal, Figure 151, and another genus allied to the date-palm Phoenicites spectabilis. The only abundant plant among the Radaboj fossils which is characteristic of the Upper Miocene period is the Populus mutabilis, whereas no less than fifty of the Radaboj species are common to the more ancient flora of the Lower Molasse of Switzerland.

The insect fauna is very rich, and, like the plants, indicates a more tropical climate than do the fossils of Oeningen presently to be mentioned. There are ten species of Termites, or white ants, some of gigantic size, and large dragon- flies with speckled wings, like those of the Southern States in North America; there are also grasshoppers of considerable size, and even the Lepidoptera are not unrepresented. In one instance, the pattern of a butterfly's wing has escaped obliteration in the marl-stone of Radaboj; and when we reflect on the remoteness of the time from which it has been faithfully transmitted to us, this fact may inspire the reader with some confidence as to the reliable nature of the characters which other insects of a more durable texture, such as the beetles, may afford for specific determination. The Vanessa above figured retains, says Heer, some of its colours, and corresponds with Vanessa Hadena of India.

Professor Beyrich has made known to us the existence of a long succession of marine strata in North Germany, which lead by an almost gradual transition from beds of Upper Miocene age to others of the age of the base of the Lower Miocene. Although some of the German lignites called Brown Coal belong to the upper parts of this series, the most important of them are of Lower Miocene date, as, for example, those of the Siebengebirge, near Bonn, which are associated with volcanic rocks.

Professor Beyrich confines the term "Miocene" to those strata which agree in age with the faluns of Touraine, and he has proposed the term "Oligocene" for those older formations called Lower Miocene in this work.

LOWER MIOCENE OF ITALY.

In the hills of which the Superga forms a part there is a great series of Tertiary strata which pass downward into the Lower Miocene. Even in the Superga itself there are some fossil plants which, according to Heer, have never been found in Switzerland so high as the marine Molasse, such as Banksia longifolia, and Carpinus grandis. In several parts of the Ligurian Apennines, as at Dego and Carcare, the Lower Miocene appears, containing some nummulites, and at Cadibona, north of Savona, fresh-water strata of the same age occur, with dense beds of lignite inclosing remains of the Anthracotherium magnum and

Anthracotherium minimum, besides other mammalia enumerated by Gastaldi. In these beds a great number of the Lower Miocene plants of Switzerland have been discovered.

LOWER MIOCENE OF ENGLAND— HEMPSTEAD BEDS.

We have already stated that the Upper Miocene formation is nowhere represented in the British Isles; but strata referable to the Lower Miocene period are found both in England, Scotland, and Ireland. In the Hampshire basin these occupy a very small superficial area, having been discovered by the late Edward Forbes at Hempstead near Yarmouth, in the northern part of the Isle of Wight, where they are 170 feet thick, and rich in characteristic marine shells. They overlie the uppermost of an extensive series of Eocene deposits of marine, brackish, and fresh-water formations, which rest on the Chalk and terminate upward in strata corresponding in age to the Paris gypsum, and containing the same extinct genera of quadrupeds, Palaeotherium, Anoplotherium, and others which Cuvier first described. The following is the succession of these Lower Miocene strata, most of them exposed in a cliff east of Yarmouth:

(FIGURE 158. Corbula pisum. Hempstead Beds, Isle of Wight.)
(FIGURE 159. Cyrena semistriata. Hempstead Beds.)

1. The uppermost or Corbula beds, consisting of marine sands and clays, contain Voluta Rathieri, a characteristic Lower Miocene shell; Corbula pisum (Figure 158), a species common to the Upper Eocene clay of Barton; Cyrena semistriata (Figure 159), several Cerithia, and other shells peculiar to this series.

(FIGURE 160. Cerithium plicatum, Lam., Hempstead.)
(FIGURE 161. Cerithium elegans. Hempstead.)
(FIGURE 162. Rissoa Chastelii, Nyst, sp. Hempstead, Isle of Wight.)
(FIGURE 163. Paludina lenta. Hempstead Bed.)

2. Next below are fresh-water and estuary marls and carbonaceous clays in the brackish-water portion of which are found abundantly Cerithium plicatum, Lam. (Figure 160), Cerithium elegans (Figure 161), and Cerithium tricinctum; also Rissoa Chastelii (Figure 162), a very common Kleyn Spawen shell, and which occurs in each of the four subdivisions of the Hempstead series down to its base, where it passes into the Bembridge beds. In the fresh-water portion of the same beds Paludina lenta (Figure 163) occurs; a shell identified by some conchologists with a species now living, Paludina unicolor; also several species of Lymneus, Planorbis, and Unio.

3. The next series, or middle fresh-water and estuary marls, are distinguished by the presence of Melania fasciata, Paludina lenta, and clays with Cypris; the lowest bed contains Cyrena semistriata (Figure 159), mingled with Cerithia and a panopaea.

4. The lower fresh-water and estuary marls contain Melania costata, Sowerby, Melanopsis, etc. The bottom bed is carbonaceous, and called the "Black band," in which Rissoa Chastelii (Figure 162), before alluded to, is common. This bed contains a mixture of Hempstead shells with those of the underlying Upper Eocene or Bembridge series. The mammalia, among which is Hyopotamus bovinus, differ, so far as they are known, from those of the Bembridge beds. Among the plants, Professor Heer has recognised four species common to the lignite of Bovey Tracey, a Lower Miocene formation presently to be described: namely, Sequoia Couttsiae, Heer; Andromeda reticulata, Ettings.; Nelumbium (Nymphoea) doris, Heer; and Carpolithes Websteri, Brong. (Pengelly, preface to The Lignite Formation of Bovey Tracey page 17. London 1863.) The seed-vessels of Chara medicaginula, Brong, and Chara helicteres are characteristic of the Hempstead beds generally.

The Hyopotamus belongs to the hog tribe, or the same family as the Anthracotherium, of which seven species, varying in size from the hippopotamus to the wild boar, have been found in Italy and other part of Europe associated with the lignites of the Lower Miocene period.

LIGNITES AND CLAYS OF BOVEY TRACEY, DEVONSHIRE.

Surrounded by the granite and other rocks of the Dartmoor hills in Devonshire, is a formation of clay, sand, and lignite, long known to geologists as the Bovey Coal formation, respecting the age of which, until the year 1861, opinions were very unsettled. This deposit is situated at Bovey Tracey, a village distant eleven miles from Exeter in a south-west, and about as far from Torquay in a north-west direction. The strata extend over a plain nine miles long, and they consist of the materials of decomposed and worn-down granite and vegetable matter, and have evidently filled up an ancient hollow or lake-like expansion of the valleys of the Bovey and Teign.

The lignite is of bad quality for economical purposes, as there is a great admixture in it of iron pyrites, and it emits a sulphurous odour, but it has been successfully applied to the baking of pottery, for which some of the fine clays are well adapted. Mr. Pengelly has confirmed Sir H. De la Beche's opinion that much of the upper portion of this old lacustrine formation has been removed by denudation. (Philosophical Transactions 1863. Paper by W. Pengelly F.R.S. and Dr. Oswald Heer.)

At the surface is a dense covering of clay and gravel with angular stones probably of the Post-pliocene period, for in the clay are three species of willow and the dwarf birch, Betula nana, indicating a climate colder than that of Devonshire at the present day.

Below this are Lower Miocene strata about 300 feet in thickness, in the upper part of which are twenty-six beds of lignite, clay, and sand, and at their base a ferruginous quartzose sand, varying in thickness from two to twenty-seven feet. Below this sand are forty-five beds of alternating lignite and clay. No shells or bones of mammalia, and no insect, with the exception of one fragment of a beetle (Buprestis); in a word, no organic remains, except plants, have as yet been found. These plants occur in fourteen of the beds— namely, in two of the clays, and the rest in the lignites. One of the beds is a perfect mat of the debris of a coniferous tree, called by Heer Sequoia Couttsiae, intermixed with leaves of ferns. The same Sequoia (before mentioned as a Hempstead fossil) is spread through all parts of the formation, its cones, and seeds, and branches of every age being preserved. It is a species supplying a link between Sequoia Langsdorfii (see Figure 153) and Sequoia Sternbergi, the widely spread fossil representatives of the two living trees Sequoia sempervirens and Sequoia gigantea (or Wellingtonia), both now confined to California. Another bed is full of the large rhizomes of ferns, while two others are rich in dicotyledonous leaves. In all, Professor Heer enumerates forty-nine species of plants, twenty of which are common to the Miocene beds of the Continent, a majority of them being characteristic of the Lower Miocene. The new species, also of Bovey, are allied to plants of the older Miocene deposits of Switzerland, Germany, and other Continental countries. The grape-stones of two species of vine occur in the clays, and leaves of the fig and seeds of a water-lily. The oak and laurel have supplied many leaves. Of the triple-nerved laurels several are referred to Cinnamomum. There are leaves also of a palm of which the genus is not determined. Leaves also of proteaceous forms, like some of the Continental fossils before mentioned, occur, and ferns like the well-known Lastraea stiriaca (Figure 154), displaying at Bovey, as in Switzerland, its fructification.

The croziers of some of the young ferns are very perfect, and were at first mistaken by collectors for shells of the genus Planorbis. On the whole, the vegetation of Bovey implies the existence of a sub-tropical climate in Devonshire, in the Lower Miocene period.

SCOTLAND: ISLE OF MULL.

In the sea-cliffs forming the headland of Ardtun, on the west coast of Mull, in the Hebrides, several bands of tertiary strata containing leaves of dicotyledonous plants were discovered in 1851 by the Duke of Argyll. (Quarterly Geological Journal 1851 page 19.) From his description it appears that there are three leaf-beds, varying in thickness from 1 1/2 to 5 1/2 feet, which are interstratified with volcanic tuff and trap, the whole mass being about 130 feet in thickness. A sheet of basalt 40 feet thick covers the whole; and another columnar bed of the same rock, ten feet thick, is exposed at the bottom of the cliff. One of the leaf-beds consists of a compressed mass of leaves unaccompanied by any stems, as if they had been blown into a marsh where a species of Equisetum grew, of which the remains are plentifully imbedded in clay.

It is supposed by the Duke of Argyll that this formation was accumulated in a shallow lake or marsh in the neighbourhood of a volcano, which emitted showers of ashes and streams of lava. The tufaceous envelope of the fossils may have fallen into the lake from the air as volcanic dust, or have been washed down into it as mud from the adjoining land. Even without the aid of organic remains we might have decided that the deposit was newer than the chalk, for chalk- flints containing cretaceous fossils were detected by the duke in the principal mass of volcanic ashes or tuff. (Quarterly Geological Journal 1851 page 90.)

The late Edward Forbes observed that some of the plants of this formation resembled those of Croatia, described by Unger, and his opinion has been confirmed by Professor Heer, who found that the conifer most prevalent was the Sequoia Langsdorfii (Figure 153), also Corylus grossedentata, a Lower Miocene species of Switzerland and of Menat in Auvergne. There is likewise a plane-tree, the leaves of which seem to agree with those of Platanus aceroides (Figure 141 Chapter 14), and a fern which is as yet peculiar to Mull, Filicites hebridica, Forbes.

These interesting discoveries in Mull led geologists to suspect that the basalt of Antrim, in Ireland, and of the celebrated Giant's Causeway, might be of the same age. The volcanic rocks that overlie the chalk, and some of the strata associated with and interstratified between masses of basalt, contain leaves of dicotyledonous plants, somewhat imperfect, but resembling the beech, oak, and plane, and also some coniferae of the genera pine and Sequoia. The general dearth of strata in the British Isles, intermediate in age between the formation of the Eocene and Pliocene periods, may arise, says Professor Forbes, from the extent of dry land which prevailed in that vast interval of time. If land predominated, the only monuments we are likely ever to find of Miocene date are those of lacustrine and volcanic origin, such as the Bovey Coal in Devonshire, the Ardtun beds in Mull, or the lignites and associated basalts in Antrim.

LOWER MIOCENE, UNITED STATES: NEBRASKA.

In the territory of Nebraska, on the Upper Missouri, near the Platte River, latitude 42 degrees N., a tertiary formation occurs, consisting of white limestone, marls, and siliceous clay, described by Dr. D. Dale Owen (David Dale Owen Geological Survey of Wisconsin etc. Philadelphia 1852.), in which many bones of extinct quadrupeds, and of chelonians of land or fresh-water forms, are met with. Among these, Dr. Leidy describes a gigantic quadruped, called by him Titanotherium, nearly allied to the Palaeotherium, but larger than any of the species found in the Paris gypsum. With these are several species of the genus Oreodon, Leidy, uniting the characters of pachyderms and ruminants also; Eucrotaphus, another new genus of the same mixed character; two species of rhinoceros of the sub-genus Acerotherium, a Lower Miocene form of Europe before mentioned; two species of Archaeotherium, a pachyderm allied to Chaeropotamus and Hyracotherium; also Paebrotherium, an extinct ruminant allied to Dorcatherium, Kaup; also Agriochoerus, of Leidy, a ruminant allied to Merycopotamus of Falconer and Cautley; and, lastly, a large carnivorous animal of the genus Machairodus, the most ancient example of which in Europe occurs in the Lower Miocene strata of Auvergne, but of which some species are found in Pliocene deposits. The turtles are referred to the genus Testudo, but have some affinity to Emys. On the whole, the Nebraska formation is probably newer than the Paris gypsum, and referable to the Lower Miocene period, as above defined.

CHAPTER XVI.
EOCENE FORMATIONS.

Eocene Areas of North of Europe.
Table of English and French Eocene Strata.
Upper Eocene of England.
Bembridge Beds.
Osborne or St. Helen's Beds.
Headon Series.
Fossils of the Barton Sands and Clays.
Middle Eocene of England.
Shells, Nummulites, Fish and Reptiles of the Bracklesham Beds and Bagshot Sands.
Plants of Alum Bay and Bournemouth.
Lower Eocene of England.
London Clay Fossils.
Woolwich and Reading Beds formerly called "Plastic Clay."
Fluviatile Beds underlying Deep-sea Strata.
Thanet Sands.
Upper Eocene Strata of France.
Gypseous Series of Montmartre and Extinct Quadrupeds.

Fossil Footprints in Paris Gypsum.
Imperfection of the Record.
Calcaire Silicieux.
Gres de Beauchamp.
Calcaire Grossier.
Miliolite Limestone.
Soissonnais Sands.
Lower Eocene of France.
Nummulitic Formations of Europe, Africa, and Asia.
Eocene Strata in the United States.
Gigantic Cetacean.

EOCENE AREAS OF THE NORTH OF EUROPE.

(FIGURE 164. Map of the principal Eocene areas of North-western Europe, showing:
Shaded dotted: Hypogene rocks and strata older than the Devonian.
Shaded horizontal lines: Eocene formations.
NB.— the space left blank is occupied by fossiliferous formations from the Devonian to the chalk inclusive.)

The strata next in order in the descending series are those which I term Eocene.

In the map in Figure 164, the position of several Eocene areas in the north of Europe is pointed out. When this map was constructed I classed as the newer part of the Eocene those Tertiary strata which have been described in the last chapter as Lower Miocene, and to which M. Beyrich has given the name of Oligocene. None of these occur in the London Basin, and they occupy in that of Hampshire, as we have seen in Chapter 15, too insignificant a superficial area to be noticed in a map on this scale. They fill a larger space in the Paris Basin between the Seine and the Loire, and constitute also part of the northern limits of the area of the Netherlands which are shaded in the map.

TABLE 16.1. TABLE OF ENGLISH AND FRENCH EOCENE STRATA.
COLUMN 1: NAME OF STRATA.
COLUMN 2: ENGLISH SUBDIVISIONS.
COLUMN 3: FRENCH EQUIVALENTS.
UPPER EOCENE.
A.1: Bembridge series, Isle of Wight: Gypseous series of Montmartre.
A.2: Osborne or St. Helen's series, Isle of Wight: Calcaire silicieux, or Travertin Inferieur.
A.3: Headon series, Isle of Wight: Calcaire silicieux, or Travertin Inferieur.
A.4: Barton series. Sands and clays of Barton Cliff, Hants: Gres de Beauchamp, or Sables Moyens.
MIDDLE EOCENE.
B.1: Bracklesham series: Calcaire Grossier.
B.2: Alum Bay and Bournemouth beds: Wanting in France?
B.2: Wanting in England?: Soissonnais Sands, or Lits Coquilliers.
LOWER EOCENE.
C.1: London Clay: Argile de Londres, Cassel, near Dunkirk.
C.2: Woolwich and Reading series: Argile plastique and lignite.
C.3: Thanet sands: Sables de Bracheux.

It is in the northern part of the Isle of Wight that we have the uppermost beds of the true Eocene best exhibited— namely, those which correspond in their fossils with the celebrated gypsum of the Paris basin before alluded to in Chapter 15 (see Table 16.1). That gypsum has been selected by almost all Continental geologists as affording the best line of demarkation between the Middle and Lower Tertiary, or, in other words, between the Lower Miocene and Eocene formations.

In reference to Table 16.1 I may observe, that the correlation of the French and English subdivisions here laid down is often a matter of great doubt and difficulty, notwithstanding their geographical proximity. This arises from various circumstances, partly from the former prevalence of marine conditions in one basin simultaneously with fluviatile or lacustrine in the other, and sometimes from the existence of land in one area causing a break or absence of all records during a period when deposits may have been in progress in the other basin. As bearing on this subject, it may be stated that we have unquestionable evidence of oscillations of level shown by the superposition of salt or brackish-water strata to fluviatile beds; and those of deep-sea origin to strata formed in shallow water. Even if the upward and downward movements were uniform in amount and direction, which is very improbable, their effect in producing the conversion of sea into land or land into sea would be different, according to the previous shape and varying elevation of the land and bottom of the sea. Lastly, denudation, marine and subaerial, has frequently caused the absence of deposits in one basin of corresponding age to those in the other, and this destructive agency has been more than ordinarily effective on account of the loose and unconsolidated nature of the sands and clays.

UPPER EOCENE OF ENGLAND.
BEMBRIDGE SERIES, A.1.

These beds are about 120 feet thick, and, as stated in Chapter 15, lie immediately under the Hempstead beds, near Yarmouth, in the Isle of Wight, being conformable with those Lower Miocene strata. They consist of marls, clays, and limestones of fresh-water, brackish, and marine origin. Some of the most abundant shells, as Cyrena semistriata var., and Paludina lenta, Figure 163 Chapter 15, are common to this and to the overlying Hempstead series; but the majority of the species are distinct. The following are the subdivisions described by the late Professor Forbes:

(FIGURE 165. Melania turritissima, Forbes. Bembridge.)

a. Upper marls, distinguished by the abundance of Melania turritissima, Forbes (Figure 165).

(FIGURE 166. Fragment of carapace of Trionyx. Bembridge Beds, Isle of Wight.)

b. Lower marls, characterised by Cerithium mutabile, Cyrena pulchra, etc., and by the remains of Trionyx (see Figure 166).

c. Green marls, often abounding in a peculiar species of oyster, and accompanied by Cerithium, Mytilus, Arca, nucula, etc.)

(FIGURE 167. Bulimus ellipticus, Sowerby. Bembridge Limestone. 1/2 natural size.)
(FIGURE 168. Helix occlusa, Edwards. Bembridge Limestone, Isle of Wight.)
(FIGURE 169. Paludina orbicularis. Bembridge.)
(FIGURE 170. Planorbis discus, Edwards. Bembridge. 1/2 diameter.)
(FIGURE 171. Lymnea longiscata, Brand. Natural size.)
(FIGURE 172. Chara tuberculata, seed-vessel. Bembridge Limestone, Isle of Wight.)

d. Bembridge limestones, compact cream-coloured limestones alternating with shales and marls, in all of which land-shells are common, especially at Sconce, near Yarmouth, as described by Mr. F. Edwards. The Bulimus ellipticus, Figure 167, and Helix occlusa, Figure 168, are among its best known land-shells. Paludina orbicularis, Figure 169, is also of frequent occurrence. One of the bands is filled with a little globular Paludina. Among the fresh-water pulmonifera, Lymnea longiscata (Figure 171) and Planorbis discus (Figure 170) are the most generally distributed: the latter represents or takes the place of the Planorbis euomphalus (see Figure 175) of the more ancient Headon series. Chara tuberculata (Figure 172) is the characteristic Bembridge gyrogonite or seed-vessel.

(FIGURE 173. Anoplotherium commune. Binstead, Isle of Wight.
Lower molar tooth, natural size.)
(FIGURE 174. Palaeotherium magnum, Cuvier.)
(FIGURE 175. Planorbis euomphalus, Sowerby. Headon Hill. 1/2 diameter.)

From this formation on the shores of Whitecliff Bay, Dr. Mantell obtained a fine specimen of a fan palm, Flabellaria Lamanonis, Brong., a plant first obtained from beds of corresponding age in the suburbs of Paris. The well-known building-stone of Binstead, near Ryde, a limestone with numerous hollows caused by Cyrenae which have disappeared and left the moulds of their shells, belongs to this subdivision of the Bembridge series. In the same Binstead stone Mr. Pratt and the Reverend Darwin Fox first discovered the remains of mammalia characteristic of the gypseous series of Paris, as Palaeotherium magnum (Figure

174), Palaeotherium medium, Palaeotherium minus, Palaeotherium minimum, Palaeotherium curtum, Palaeotherium crassum; also Anoplotherium commune (Figure 173), Anoplotherium secundarium, Dichobune cervinum, and Chaeropotamus Cuvieri. The Palaeothere above alluded to resembled the living tapir in the form of the head, and in having a short proboscis, but its molar teeth were more like those of the rhinoceros. Palaeotherium magnum was of the size of a horse, three or four feet high. The woodcut, Figure 174, is one of the restorations which Cuvier attempted of the outline of the living animal, derived from the study of the entire skeleton. As the vertical range of particular species of quadrupeds, so far as our knowledge extends, is far more limited than that of the testacea, the occurrence of so many species at Binstead, agreeing with fossils of the Paris gypsum, strengthens the evidence derived from shells and plants of the synchronism of the two formations.

OSBORNE OR ST. HELEN'S SERIES, A.2.

This group is of fresh and brackish-water origin, and very variable in mineral character and thickness. Near Ryde, it supplies a freestone much used for building, and called by Professor Forbes the Nettlestone grit. In one part ripple-marked flagstones occur, and rocks with fucoidal markings. The Osborne beds are distinguished by peculiar species of Paludina, Melania, and Melanopsis, as also of Cypris and the seeds of Chara.

HEADON SERIES A.3.

These beds are seen both in Whitecliff Bay, Headon Hill, and Alum Bay, or at the east and west extremities of the Isle of Wight. The upper and lower portions are fresh-water, and the middle of mixed origin, sometimes brackish and marine. Everywhere Planorbis euomphalus, Figure 175, characterises the fresh-water deposits, just as the allied form, Planorbis discus, Figure 170, does the Bembridge limestone. The brackish-water beds contain Potamomya plana, Cerithium mutabile, and Potamides cinctus (Figure 37 Chapter 3), and the marine beds Venus (or Cytherea) incrassata, a species common to the Limburg beds and Gres de Fontainebleau, or the Lower Miocene series. The prevalence of salt-water remains is most conspicuous in some of the central parts of the formation.

(FIGURE 176. Helix labyrinthica, Say. Headon Hill, Isle of Wight; and Hordwell Cliff, Hants— also recent.)

(FIGURE 177. Neritina concava, Sowerby. Headon series.)

(FIGURE 178. Lymnea caudata, Edw. Headon series.)

(FIGURE 179. Cerithium concavum, Sowerby. Headon series.)

Among the shells which are widely distributed through the Headon series are Neritina concava (Figure 177), Lymnea caudata (Figure 178), and Cerithium concavum (Figure 179). Helix labyrinthica, Say (Figure 176), a land-shell now inhabiting the United States, was discovered in this series by Mr. Searles Wood in Hordwell Cliff. It is also met with in Headon Hill, in the same beds. At Sconce, in the Isle of Wight, it occurs in the Bembridge series, and affords a rare example of an Eocene fossil of a species still living, though, as usual in such cases, having no local connection with the actual geographical range of the species. The lower and middle portion of the Headon series is also met with in Hordwell Cliff (or Hordle, as it is often spelt), near Lymington, Hants. Among the shells which abound in this cliff are Paludina lenta and various species of Lymnea, Planorbis, Melania, Cyclas, Unio, Potamomya, Dreissena, etc.

Among the chelonians we find a species of Emys, and no less than six species of Trionyx; among the saurians an alligator and a crocodile; among the ophidians two species of land-snakes (Paleryx, Owen); and among the fish Sir P. Egerton and Mr. Wood have found the jaws, teeth, and hard shining scales of the genus Lepidosteus, or bony pike of the American rivers. This same genus of fresh-water ganoids has also been met with in the Hempstead beds in the Isle of Wight. The bones of several birds have been obtained from Hordwell, and the remains of quadrupeds of the genera Palaeotherium (Palaeotherium minus), Anoplotherium, Anthracotherium, Dichodon, Dichobune, Spalacodon, and Hyaenodon. The latter offers, I believe, the oldest known example of a true carnivorous animal in the series of British fossils, although I attach very little theoretical importance to the fact, because herbivorous species are those most easily met with in a fossil state in all save cavern deposits. In another point of view, however, this fauna deserves notice. Its geological position is considerably lower than that of the Bembridge or Montmartre beds, from which it differs almost as much in species as it does from the still more ancient fauna of the Lower Eocene beds to be mentioned in the sequel. It therefore teaches us what a grand succession of distinct assemblages of mammalia flourished on the earth during the Eocene period.

Many of the marine shells of the brackish-water beds of the above series, both in the Isle of Wight and Hordwell Cliff, are common to the underlying Barton Clay: and, on the other hand, there are some fresh-water shells, such as Cyrena obovata, which are common to the Bembridge beds, notwithstanding the intervention of the St. Helen's series. The white and green marls of the Headon series, and some of the accompanying limestones, often resemble the Eocene strata of France in mineral character and colour in so striking a manner as to suggest the idea that the sediment was derived from the same region or produced contemporaneously under very similar geographical circumstances.

(FIGURE 180. Solenastraea cellulosa, Duncan. Brockenhurst.)

At Brockenhurst, near Lyndhurst, in the New Forest, marine strata have recently been found containing fifty-nine shells, of which many have been described by Mr. Edwards. These beds rest on the Lower Headon, and are considered as the equivalent of the middle part of the Headon series, many of the shells being common to the brackish-water or Middle Headon beds of Colwell and Whitecliff Bays, such as Cancellaria muricata, Sowerby, Fusus labiatus, Sowerby, etc. In these beds at Brockenhurst, corals, ably described by Dr. Duncan, have recently been found in abundance and perfection; see Figure 180, Solenastraea cellulosa.

Baron von Konen has pointed out that no less than forty-six out of the fifty-nine Brockenhurst shells, or a proportion of 78 per cent, agree with species occurring in Dumont's Lower Tongrian formation in Belgium. (Quarterly Geological Journal volume 20 page 97 1864.) This being the case, we might fairly expect that if we had a marine equivalent of the Bembridge series or of the contemporaneous Paris gypsum, we should find it to contain a still greater number of shells common to the Tongrian beds of Belgium, but the exact correlation of these fresh-water groups of France, Belgium, and Britain has not yet been fully made out. It is possible that the Tongrian of Dumont may be newer than the Bembridge series, and therefore referable to the Lower Miocene. If ever the whole series should be complete, we must be prepared to find the marine equivalent of the Bembridge beds, or the uppermost Eocene, passing by imperceptible shades into the inferior beds of the overlying Miocene strata.

Among the fossils found in the Middle Headon are Cytherea incrassata and Cerithium plicatum (Figure 160 Chapter 15). These shells, especially the latter, are very characteristic of the Lower Miocene, and their occurrence in the Headon series has been cited as an objection to the line proposed to be drawn between Miocene and Eocene. But if we were to attach importance to such occasional passages, we should soon find that no lines of division could be drawn anywhere, for in the present state of our knowledge of the Tertiary series there will always be species common to beds above and below our boundary-lines.

BARTON SERIES (SANDS AND CLAYS), A.4 TABLE 16.1.)

(FIGURE 181. Chama squamosa, Eichw. Barton.)

Both in the Isle of Wight, and in Hordwell Cliff, Hants, the Headon beds, above-mentioned, rest on white sands usually devoid of fossils, and used in the Isle of Wight for making glass. In one of these sands Dr. Wright found Chama squamosa, a Barton Clay shell, in great plenty, and certain impressions of marine shells have been found in sands supposed to be of the same age in Whitecliff Bay. These sands have been called Upper Bagshot in the maps of our Government Survey, but this identification of a fossiliferous series in the Isle of Wight with an unfossiliferous formation in the London Basin can scarcely be depended upon. The Barton Clay, which immediately underlies these sands, is seen vertical in Alum Bay, Isle of Wight, and nearly horizontal in the cliffs of the mainland near Lymington. This clay, together with the Bracklesham beds, presently to be described, has been termed Middle Bagshot by the Survey. In Barton Cliff, where it attains a thickness of about 300 feet, it is rich in marine fossils.

It was formerly confounded with the London Clay, an older Eocene deposit of very similar mineral character, to be mentioned below, which contains many shells in common, but not more than one-fourth of the whole. In other words, there are known at present 247 species in the London Clay and 321 in that of Barton, and only 70 common to the two formations. Fifty-six of these have been found in the intermediate Bracklesham beds, and the reappearance of the other 14 may imply a return of similar conditions, whether of temperature or depth or of a muddy argillaceous bottom, common to the two periods of the London and Barton Clays. According to M. Hebert, the most characteristic Barton Clay fossils correspond to those of the Gres de Beauchamp, or Sables Moyens, of the Paris Basin, but it also contains many common to the older Calcaire Grossier.

SHELLS OF THE BARTON CLAY.
(FIGURE 182. Mitra scabra, Sowerby.)
(FIGURE 183. Voluta ambigua, Sol.)
(FIGURE 184. Typhis pungens, Brand.)
(FIGURE 185. Voluta athleta, Sol. Barton and Bracklesham.)
(FIGURE 186. Terebellum fusiforme, Lam. Barton and Bracklesham.)
(FIGURE 187. Terebellum sopita, Brand.)
(FIGURE 188. Cardita sulcata, Brand. Barton.)
(FIGURE 189. Crassatella sulcata, Sowerby. Bracklesham and Barton.)
(FIGURE 190. Nummulites variolaria, Lam. Var. of Nummulites radiata, Sowerby. Middle Eocene, Bracklesham Bay.
a. Natural size.
b. Magnified.)

Certain foraminifera called Nummulites begin, when we study the Tertiary formations in a descending order, to make their first appearance in these beds. A small species called Nummulites variolaria, Figure 190, is found both on the Hampshire coast and in beds of the same age in Whitecliff Bay, in the Isle of Wight. Several marine shells, such as Corbula pisum (Figure 158), are common to the Barton beds and the Hempstead or Lower Miocene series, and a still greater number, as before stated, are common to the Headon series.

MIDDLE EOCENE, ENGLAND.
BRACKLESHAM BEDS AND BAGSHOT SANDS (B.1, TABLE 16.1).
(FIGURE 191. Cardita (Venericardia) planicosta, Lam.)
(FIGURE 192. Nummulites (Nummularia) laevigata. Bracklesham. Dixon's Fossils of Sussex, Plate 8. a. Section of nummulite. b. Group, with an individual showing the exterior of the shell.)

Beneath the Barton Clay we find in the north of the Isle of Wight, both in Alum and Whitecliff Bays, a great series of various coloured sands and clays for the most part unfossiliferous, and probably of estuarine origin. As some of these beds contain Cardita planicosta (Figure 191) they have been identified with the marine beds much richer in fossils seen in the coast section in Bracklesham Bay near Chichester in Sussex, where the strata consist chiefly of green clayey sands with some lignite. Among the Bracklesham fossils besides the Cardita, the huge Cerithium giganteum is seen, so conspicuous in the Calcaire Grossier of Paris, where it is sometimes two feet in length. The Nummulites laevigata (see Figure 192), so characteristic of the lower beds of the Calcaire Grossier in France, where it sometimes forms stony layers, as near Compiegne, is very common in these beds, together with Nummulites scabra and Nummulites variolaria. Out of 193 species of testacea procured from the Bagshot and Bracklesham beds in England, 126 occur in the Calcaire Grossier in France. It was clearly, therefore, coeval with that part of the Parisian series more nearly than with any other.

(FIGURE 193. Palaeophis typhoeus, Owen; an Eocene sea-serpent. Bracklesham. a, b. Vertebra, with long neural spine preserved. c. Two vertebrae articulated together.)
(FIGURE 194. Defensive spine of Ostracion. Bracklesham.)
(FIGURE 195. Dental plates of Myliobates Edwardsi. Bracklesham Bay. Dixon's Fossils of Sussex, Plate 8.)

According to tables compiled from the best authorities by Mr. Etheridge, the number of mollusca now known from the Bracklesham beds in Great Britain is 393, of which no less than 240 are peculiar to this subdivision of the British Eocene series, while 70 are common to the Older London Clay, and 140 to the Newer Barton Clay. The volutes and cowries of this formation, as well as the lunulites and corals, favour the idea of a warm climate having prevailed, which is borne out by the discovery of a serpent, Palaeophis typhoeus (see Figure 193), exceeding, according to Professor Owen, twenty feet in length, and allied in its osteology to the Boa, Python, Coluber, and Hydrus. The compressed form and diminutive size of certain caudal vertebrae indicate so much analogy with Hydrus as to induce Professor Owen to pronounce this extinct ophidian to have been marine. (Palaeontological Society Monograph Reptiles part 2 page 61.) Among the companions of the sea-snake of Bracklesham was an extinct crocodile (Gavialis Dixoni, Owen), and numerous fish, such as now frequent the seas of warm latitudes, as the Ostracion of the family Balistidae, of which a dorsal spine is figured (see Figure 194), and gigantic rays of the genus Myliobates (see Figure 195).

(FIGURE 196. Carcharodon angustidens, Agassiz.)
(FIGURE 197. Otodus obliquus, Agassiz.)
(FIGURE 198. Lamna elegans, Agassiz.)
(FIGURE 199. Galeocerdo latidens, Agassiz.)

The teeth of sharks also, of the genera Carcharodon, Otodus, Lamna, Galeocerdo, and others, are abundant. (See Figures 196, 197, 198, 199.)

MARINE SHELLS OF BRACKLESHAM BEDS.
ALUM BAY AND BOURNEMOUTH BEDS. (LOWER BAGSHOT OF ENGLISH SURVEY), B.2, TABLE 16.1.)

(FIGURE 200. Pleurotoma attenuata, Sowerby.)
(FIGURE 201. Voluta Selseiensis, Edwards.)
(FIGURE 202. Turritella multisulcata, Lam.)
(FIGURE 203. Lucina serrata, Sowerby. Magnified.)
(FIGURE 204. Conus deperditus, Brug.)

To that great series of sands and clays which intervene between the equivalents of the Bracklesham Beds and the London Clay or Lower Eocene, our Government Survey has given the name of the Lower Bagshot sands, for they are supposed to agree in age with the inferior unfossiliferous sands of the country round Bagshot in the London Basin. This part of the series is finely exposed in the vertical beds of Alum bay, in the Isle of Wight, and east and west of Bournemouth, on the south coast of Hampshire. In some of the close and white compact clays of this locality, there are not only dicotyledonous leaves, but numerous fronds of ferns allied to Gleichenia which are well preserved with their fruit.

None of the beds are of great horizontal extent, and there is much cross-stratification in the sands, and in some places black carbonaceous seams and lignite. In the midst of these leaf-beds in Studland Bay, Purbeck shells of the genus Unio attest the fresh-water origin of the white clay.

No less than forty species of plants are mentioned by MM. de la Harpe and Gaudin from this formation in Hampshire, among which the Proteaceae (Dryandra, etc.) and the fig tribe are abundant, as well as the cinnamon and several other laurineae, with some papilionaceous plants. On the whole, they remind the botanist of the types of subtropical India and Australia. (Heer Climat et Vegetation du Pays Tertiaire page 172.)

Heer has mentioned several species which are common to this Alum Bay flora and that of Monte Bolca, near Verona, so celebrated for its fossil fish, and where the strata contain nummulites and other Middle Eocene fossils. He has particularly alluded to Aralia primigenia (of which genus a fruit has since been found by Mr. Mitchell at Bournemouth), Daphnogene Veronensis, and Ficus granadilla, as among the species common to and characteristic of the Isle of Wight and Italian Eocene beds; and he observes that in the flora of this period these forms of a temperate climate which constitute a marked feature in the European Miocene formations, such as the willow, poplar, birch, alder, elm, hornbeam, oak, fir, and pine, are wanting. The American types are also absent, or much more feebly represented than in the Miocene period, although fine specimens of the fan-palm (Sabal) have been found in these Eocene clays at Studland. The number of exotic forms which are common to the Eocene and Miocene strata of Europe, like those to be alluded to in the sequel which are common to the Eocene and Cretaceous fauna, demonstrate the remoteness of the times in which the geographical distribution of living plants originated. A great majority of the Eocene genera have disappeared from our temperate climates, but not the whole of them; and they must all have exerted some influence on the assemblages of species which succeeded them. Many of these last occurring in the Upper Miocene are indeed so closely allied to the flora now surviving as to make it questionable, even in the opinion of naturalists opposed to the doctrine of transmutation, whether they are not genealogically related the one to the other.

LOWER EOCENE FORMATIONS, ENGLAND.
LONDON CLAY (C.1, TABLE 16.1).

This formation underlies the preceding, and sometimes attains a thickness of 500 feet. It consists of tenacious brown and bluish-grey clay, with layers of concretions called septaria, which abound chiefly in the brown clay, and are obtained in sufficient numbers from sea-cliffs near Harwich, and from shoals off the coast of Essex and the Isle of Sheppey, to be used for making Roman cement. The total number of British fossil mollusca known at present (January, 1870) in this formation are 254, of which 166 are peculiar, or not found in other Eocene beds in this country. The principal localities of fossils in the London clay are Highgate Hill, near London, the Island of Sheppey at the mouth of the Thames, and Bognor on the Sussex coast. Out of 133 fossil shells, Mr. Prestwich found only 20 to be common to the Calcaire Grossier (from which 600 species have been obtained), while 33 are common to the "Lits Coquilliers" (see below), in which 200 species are known in France.

In the Island of Sheppey near the mouth of the Thames, the thickness of the London Clay is estimated by Mr. Prestwich to be more than 500 feet, and it is in the uppermost 50 feet that a great number of fossil fruits were obtained, being chiefly found on the beach when the sea has washed away the clay of the rapidly wasting cliffs.

(FIGURE 205. Nipadites ellipticus, Bowerbank. Fossil fruit of palm, from Sheppey.)

Mr. Bowerbank, in a valuable publication on these fossil fruits and seeds, has described no less than thirteen fruits of palms of the recent type Nipa, now only found in the Molucca and Philippine Islands, and in Bengal (see Figure 205). In the delta of the Ganges, Dr. Hooker observed the large nuts of Nipa fruticans floating in such numbers in the various arms of that great river, as to obstruct the paddle-wheels of steamboats. These plants are allied to the cocoanut tribe on the one side, and on the other to the Pandanus, or screw-pine. There are also met with three species of Anona, or custard-apple; and cucurbitaceous fruits (of the gourd and melon family), and fruits of various species of Acacia.

Besides fir-cones or fruit of true Coniferae there are cones of Proteaceae in abundance, and the celebrated botanist the late Robert Brown pointed out the affinity of these to the New Holland types Petrophila and Isopogon. Of the first there are about fifty, and of the second thirty described species now living in Australia.

(FIGURE 206. Eocene Proteaceous Fruit.
Petrophiloides Richardsoni. London Clay, Sheppey. Natural size.
a. Cone.
b. Section of cone showing the position of the seeds.)

Ettingshausen remarked in 1851 that five of the fossil species from Sheppey, named by Bowerbank (Fossil Fruits and Seeds of London Clay Plates 9 and 10.) were specimens of the same fruit (see Figure 206), in different states of preservation; and Mr. Carruthers, having examined the original specimens now in the British Museum, tells me that all these cones from Sheppey may be reduced to two species, which have an undoubted affinity to the two existing Australian genera above mentioned, although their perfect identity in structure can not be made out.

The contiguity of land may be inferred not only from these vegetable productions, but also from the teeth and bones of crocodiles and turtles, since these creatures, as Dean Conybeare remarked, must have resorted to some shore to lay their eggs. Of turtles there were numerous species referred to extinct genera. These are, for the most part, not equal in size to the largest living tropical turtles. A sea-snake, which must have been thirteen feet long, of the genus Palaeophis before mentioned, has also been described by Professor Owen from Sheppey, of a different species from that of Bracklesham, and called Palaeophis toliapicus. A true crocodile, also, Crocodilus toliapicus, and another saurian more nearly allied to the gavial, accompany the above fossils; also the relics of several birds and quadrupeds. One of these last belongs to the new genus Hyracotherium of Owen, of the hog tribe, allied to Chaeropotamus, another is a Lophiodon; a third a pachyderm called Coryphodon eocaenus by Owen, larger than any existing tapir. All these animals seem to have inhabited the banks of the great river which floated down the Sheppey fruits. They imply the existence of a mammiferous fauna antecedent to the period when nummulites flourished in Europe and Asia, and therefore before the Alps, Pyrenees, and other mountain-chains now forming the backbones of great continents, were raised from the deep; nay, even before a part of the constituent rocky masses now entering into the central ridges of these chains had been deposited in the sea.

SHELLS OF THE LONDON CLAY.

(FIGURE 207. Voluta nodosa, Sowerby. Highgate.)

(FIGURE 208. Phorus extensus, Sowerby. Highgate.)

(FIGURE 209. Rostellaria (Hippocrenes) ampla, Brander. 1/3 of natural size; also found in the Barton clay.)

(FIGURE 210. Nautilus centralis, Sowerby. Highgate.)

(FIGURE 211. Aturia ziczac, Bronn. Syn. Nautilus ziczac, Sowerby. London clay. Sheppey.)

(FIGURE 212. Belosepia sepioidea, De Blainv. London clay. Sheppey.)

(FIGURE 213. Leda amygdaloides, Sowerby. Highgate.)

(FIGURE 214. Cyptodon (Axinus) angulatum, Sowerby. London clay. Hornsey.)

(FIGURE 215. Astropecten crispatus, E. Forbes. Sheppey.)

The marine shells of the London Clay confirm the inference derivable from the plants and reptiles in favour of a high temperature. Thus many species of Conus and Voluta occur, a large Cypraea, C. oviformis, a very large Rostellaria (Figure 209), a species of Cancellaria, six species of Nautilus (Figure 211), besides other Cephalopoda of extinct genera, one of the most remarkable of which is the Belosepia (Figure 212). Among many characteristic bivalve shells are Leda amygdaloides (Figure 213) and Cryptodon angulatum (Figure 214), and among the Radiata a star-fish, Astropecten (Figure 215).

These fossils are accompanied by a sword-fish (Tetrapterus priscus, Agassiz), about eight feet long, and a saw-fish (Pristis bisulcatus, Agassiz), about ten feet in length; genera now foreign to the British seas. On the whole, about eighty species of fish have been described by M. Agassiz from these beds of Sheppey, and they indicate, in his opinion, a warm climate.

In the lower part of the London clay at Kyson, a few miles east of Woodbridge, the remains of mammalia have been detected. Some of these have been referred by Professor Owen to an opossum, and others to the genus Hyracotherium. The teeth of this last-mentioned pachyderm were at first, in 1840, supposed to belong to a monkey, an opinion afterwards abandoned by Owen when more ample materials for comparison were obtained.

WOOLWICH AND READING SERIES (C.2, TABLE 16.1.)

This formation was formerly called the Plastic Clay, as it agrees with a similar clay used in pottery which occupies the same position in the French series, and it has been used for the like purposes in England. (Prestwich Quarterly Geological Journal volume 10.)

No formations can be more dissimilar, on the whole, in mineral character than the Eocene deposits of England and Paris; those of our own island being almost exclusively of mechanical origin— accumulations of mud, sand, and pebbles; while in the neighbourhood of Paris we find a great succession of strata composed of limestones, some of them siliceous, and of crystalline gypsum and siliceous sandstone, and sometimes of pure flint used for millstones. Hence it is often impossible, as before stated, to institute an exact comparison between the various members of the English and French series, and to settle their respective ages. But in regard to the division which we have now under consideration, whether we study it in the basins of London, Hampshire, or Paris, we recognise as a general rule the same mineral character, the beds consisting over a large area of mottled clays and sand, with lignite, and with some strata of well-rolled flint pebbles, derived from the chalk, varying in size, but occasionally several inches in diameter. These strata may be seen in the Isle of Wight in contact with the chalk, or in the London basin, at Reading, Blackheath, and Woolwich. In some of the lowest of them, banks of oysters are observed, consisting of Ostrea bellovacina, so common in France in the same relative position. In these beds at Bromley, Dr. Buckland found a large pebble to which five full-grown oysters were affixed, in such a manner as to show that they had commenced their first growth upon it, and remained attached to it through life.

(FIGURE 216. Cyrena cuneiformis, Sowerby. Natural size. Woolwich clays.)

(FIGURE 217. Melania (Melanatria) inquinata, Des. Syn. Cerithium melanoides, Sowerby. Woolwich clays.)

In several places, as at Woolwich on the Thames, at Newhaven in Sussex, and elsewhere, a mixture of marine and fresh-water testacea distinguishes this member of the series. Among the latter, Cyrena cuneiformis (see Figure 216) and Melania inquinata (see Figure 217) are very common, as in beds of corresponding age in France. They clearly indicate points where rivers entered the Eocene sea. Usually there is a mixture of brackish, fresh-water, and marine shells, and sometimes, as at Woolwich, proofs of the river and the sea having successively prevailed on the same spot. At New Charlton, in the suburbs of Woolwich, Mr. de la Condamine discovered in 1849, and pointed out to me, a layer of sand associated with well-rounded flint pebbles in which numerous individuals of the Cyrena tellinella were seen standing endwise with both their valves united, the siphonal extremity of each shell being uppermost, as would happen if the mollusks had died in their natural position. I have described a bank of sandy mud, in the delta of the Alabama River at Mobile, on the borders of the Gulf of Mexico, where in 1846 I dug out at low tide specimens of living species of Cyrena and of a Gnathodon, which were similarly placed with their shells erect, or in a posture which enables the animal to protrude its siphon upward, and draw in or reject water at pleasure. (Second Visit to the United States volume 2 page 104.) The water at Mobile is usually fresh, but sometimes brackish. At Woolwich a body of river-water must have flowed permanently into the sea where the Cyrenae lived, and they may have been killed suddenly by an influx of pure salt-water, which invaded the spot when the river was low, or when a subsidence of land took place. Traced in one direction, or eastward towards Herne Bay, the Woolwich beds assume more and more of a marine character; while in an opposite, or south-western direction, they become, as near Chelsea and other places, more fresh-water, and contain Unio, Paludina, and layers of lignite, so that the land drained by the ancient river seems clearly to have been to the south-west of the present site of the metropolis.

FLUVIATILE BEDS UNDERLYING DEEP-SEA STRATA.

Before the minds of geologists had become familiar with the theory of the gradual sinking of land, and its conversion into sea at different periods, and the consequent change from shallow to deep water, the fluviatile and littoral character of this inferior group appeared strange and anomalous. After passing through hundreds of feet of London clay, proved by its fossils to have been deposited in deep salt-water, we arrive at beds of fluviatile origin, and associated with them masses of shingle, attaining at Blackheath, near London, a thickness of 50 feet. These shingle banks are probably of marine origin, but they indicate the proximity of land, and the existence of a shore where the flints of the chalk were rolled into sand and pebbles, and spread over a wide space. We have, therefore, first, as before stated, evidence of oscillations of level during the accumulation of the Woolwich series, then of a great submergence, which allowed a marine deposit 500 thick to be laid over the antecedent beds of fresh and brackish water origin.

THANET SANDS (C.3 TABLE 16.1).

The Woolwich or plastic clay above described may often be seen in the Hampshire basin in actual contact with the chalk, constituting in such places the lowest member of the British Eocene series. But at other points another formation of marine origin, characterised by a somewhat different assemblage of organic remains, has been shown by Mr. Prestwich to intervene between the chalk and the Woolwich series. For these beds he has proposed the name of "Thanet Sands," because they are well seen in the Isle of Thanet, in the northern part of Kent, and on the sea-coast between Herne Bay and the Reculvers, where they consist of sands with a few concretionary masses of sandstone, and contain, among other fossils, Pholadomya cuneata, Cyprina morrisii, Corbula longirostris, Scalaria Bowerbankii, etc. The greatest thickness of these beds is 90 feet.

UPPER EOCENE FORMATIONS OF FRANCE.

The tertiary formations in the neighbourhood of Paris consist of a series of marine and fresh-water strata, alternating with each other, and filling up a depression in the chalk. The area which they occupy has been called the Paris Basin, and is about 180 miles in its greatest length from north to south, and about 90 miles in breadth from east to west. MM. Cuvier and Brongniart attempted, in 1810, to distinguish five different groups, comprising three fresh-water and two marine, which were supposed to imply that the waters of the ocean, and of rivers and lakes, had been by turns admitted into and excluded from the same area. Investigations since made in the Hampshire and London basins have rather tended to confirm these views, at least so far as to show that since the commencement of the Eocene period there have been great movements of the bed of the sea, and of the adjoining lands, and that the superposition of deep-sea to shallow-water deposits (the London Clay, for example, to the Woolwich beds) can only be explained by referring to such movements. It appears, notwithstanding, from the researches of M. Constant Prevost, that some of the minor alternations and intermixtures of fresh-water and marine deposits, in the Paris basin, may be accounted for without such changes of level, by imagining both to have been simultaneously in progress, in the same bay of the same sea, or a gulf into which many rivers entered.

GYPSEOUS SERIES OF MONTMARTRE (A.1, TABLE 16.1).

To enlarge on the numerous subdivisions of the Parisian strata would lead me beyond my present limits; I shall therefore give some examples only of the most important formations. Beneath the Gres de Fontainebleau, belonging to the Lower Miocene period, as before stated, we find, in the neighbourhood of Paris, a series of white and green marls, with subordinate beds of gypsum. These are most largely developed in the central parts of the Paris basin, and, among other places, in the hill of Montmartre, where its fossils were first studied by Cuvier.

The gypsum quarried there for the manufacture of plaster of Paris occurs as a granular crystalline rock, and, together with the associated marls, contains land and fluviatile shells, together with the bones and skeletons of birds and quadrupeds. Several land-plants are also met with, among which are fine specimens of the fan-palm or palmetto tribe (Flabellaria). The remains also of fresh-water fish, and of crocodiles and other reptiles, occur in the gypsum. The skeletons of mammalia are usually isolated, often entire, the most delicate extremities being preserved; as if the carcasses, clothed with their flesh and skin, had been floated down soon after death, and while they were still swollen by the gases generated by their first decomposition. The few accompanying shells are of those light kinds which frequently float on the surface of rivers, together with wood.

In this formation the relics of about fifty species of quadrupeds, including the genera Palaeotherium (see Figure 174), Anoplotherium (see Figure 218), and others, have been found, all extinct, and nearly four-fifths of them belonging to the Perissodactyle or odd-toed division of the order Pachydermata, which now contains only four living genera, namely, rhinoceros, tapir, horse, and hyrax. With them a few carnivorous animals are associated, among which are the Hyaenodon dasyuroides, a species of dog, Canis Parisiensis, and a weasel, Cynodon Parisiensis. Of the Rodentia are found a squirrel; of the Cheiroptera, a bat; while the Marsupalia (an order now confined to America, Australia, and some contiguous islands) are represented by an opossum.

Of birds, about ten species have been ascertained, the skeletons of some of which are entire. None of them are referable to existing species. (Cuvier, Oss. Foss. tome 3 page 255.) The same remark, according to MM. Cuvier and Agassiz, applies both to the reptiles and fish. Among the last are crocodiles and tortoises of the genera Emys and Trionyx.

(FIGURE 218. Xiphodon gracile, or Anoplotherium gracile, Cuvier. Restored outline.)

The tribe of land quadrupeds most abundant in this formation is such as now inhabits alluvial plains and marshes, and the banks of rivers and lakes, a class most exposed to suffer by river inundations. Among these were several species of Palaeotherium, a genus before alluded to. These were associated with the Anoplotherium, a tribe intermediate between pachyderms and ruminants. One of the three divisions of this family was called by Cuvier Xiphodon. Their forms were slender and elegant, and one, named Xiphodon gracile (Figure 218), was about the size of the chamois; and Cuvier inferred from the skeleton that it was as light, graceful, and agile as the gazelle.

FOSSIL FOOTPRINTS.

There are three superimposed masses of gypsum in the neighbourhood of Paris, separated by intervening deposits of laminated marl. In the uppermost of the three, in the valley of Montmorency, M. Desnoyers discovered in 1859 many footprints of animals occurring at no less than six different levels. (Sur des Empreintes de Pas d'Animaux par M. J. Desnoyers. Compte rendu de l'Institut 1859.) The gypsum to which they belong varies from thirty to fifty feet in thickness, and is that which has yielded to the naturalist the largest number of bones and skeletons of mammalia, birds, and reptiles. I visited the quarries, soon after the discovery was made known, with M. Desnoyers, who also showed me large slabs in the Museum at Paris, where, on the upper planes of stratification, the indented foot-marks were seen, while corresponding casts in relief appeared on the lower surfaces of the strata of gypsum which were immediately superimposed. A thin film of marl, which before it was dried and condensed by pressure must have represented a much thicker layer of soft mud, intervened between the beds of solid gypsum. On this mud the animals had trodden, and made impressions which had penetrated to the gypseous mass below, then evidently unconsolidated. Tracks of the Anoplotherium with its bisulcate hoof, and the trilobed footprints of Palaeotherium, were seen of different sizes, corresponding to those of several species of these genera which Cuvier had reconstructed, while in the same beds were foot-marks of carnivorous mammalia. The tracks also of fluviatile, lacustrine, and terrestrial tortoises (Emys, Trionyx, etc.) were discovered, also those of crocodiles, iguanas, geckos, and great batrachians, and the footprints of a huge bird, apparently a wader, of the size of the gastornis, to be mentioned in the sequel. There were likewise the impressions of the feet of other creatures, some of them clearly distinguishable from any of the fifty extinct types of mammalia of which the bones have been found in the Paris gypsum. The whole assemblage, says Desnoyers, indicate the shores of a lake, or several small lakes communicating with each other, on the borders of which many species of pachyderms wandered, and beasts of prey which occasionally devoured them. The tooth-marks of these last had been detected by palaeontologists long before on the bones and skulls of Paleotheres entombed in the gypsum.

IMPERFECTION OF THE RECORD.

These foot-marks have revealed to us new and unexpected proofs that the air-breathing fauna of the Upper Eocene period in Europe far surpassed in the number and variety of its species the largest estimate which had previously been formed of it. We may now feel sure that the mammalia, reptiles, and birds which have left portions of their skeletons as memorials of their existence in the solid gypsum constituted but a part of the then living creation. Similar inferences may be drawn from the study of the whole succession of geological records. In each district the monuments of periods embracing thousands, and probably in some instances hundreds of thousands of years, are totally wanting. Even in the volumes which are extant the greater number of the pages are missing in any given region, and where they are found they contain but few and casual entries of the physical events or living beings of the times to which they relate. It may also be remarked that the subordinate formations met with in two neighbouring countries, such as France and England (the minor Tertiary groups above enumerated), commonly classed as equivalents and referred to corresponding periods, may nevertheless have been by no means strictly coincident in date. Though called contemporaneous, it is probable that they were often separated by intervals of many thousands of years. We may compare them to double stars, which appear single to the naked eye because seen from a vast distance in space, and which really belong to one and the same stellar system, though occupying places in space extremely remote if estimated by our ordinary standard of terrestrial measurements.

CALCAIRE SILICIEUX, OR TRAVERTIN INFERIEUR (A.2 AND 3 TABLE 16.1).

This compact siliceous limestone extends over a wide area. It resembles a precipitate from the waters of mineral springs, and is often traversed by small empty sinuous cavities. It is, for the most part, devoid of organic remains, but in some places contains fresh-water and land species, and never any marine fossils. The calcaire silicieux and the calcaire grossier usually occupy distinct parts of the Paris basin, the one attaining its fullest development in those places where the other is of slight thickness. They are described by some writers as alternating with each other towards the centre of the basin, as at Sergy and Osny.

The gypsum, with its associated marls before described, is in greatest force towards the centre of the basin, where the calcaire grossier and calcaire silicieux are less fully developed.

GRES DE BEAUCHAMP, OR SABLES MOYENS (A.4 TABLE 16.1).

In some parts of the Paris basin, sands and marls, called the Gres de Beauchamp, or Sables moyens, divide the gypseous beds from the calcaire grossier proper. These sands, in which a small nummulite (N. variolaria) is very abundant, contain more than 300 species of marine shells, many of them peculiar, but others common to the next division.

MIDDLE EOCENE FORMATIONS OF FRANCE.
CALCAIRE GROSSIER, UPPER AND MIDDLE (B.1 TABLE 16.1).

The upper division of this group consists in great part of beds of compact, fragile limestone, with some intercalated green marls. The shells in some parts are a mixture of Cerithium, Cyclostoma, and Corbula; in others Limnea, Cerithium, Paludina, etc. In the latter, the bones of reptiles and mammalia, Palaeotherium and Lophiodon, have been found. The middle division, or calcaire grossier proper, consists of a coarse limestone, often passing into sand. It contains the greater number of the fossil shells which characterise the Paris basin. No less than 400 distinct species have been procured from a single spot near Grignon, where they are imbedded in a calcareous sand, chiefly formed of comminuted shells, in which, nevertheless, individuals in a perfect state of preservation, both of marine, terrestrial, and fresh-water species, are mingled together. Some of the marine shells may have lived on the spot; but the Cyclostoma and Limnea, being land and fresh-water shells, must have been brought thither by rivers and currents, and the quantity of triturated shells implies considerable movement in the waters.

Nothing is more striking in this assemblage of fossil testacea than the great proportion of species referable to the genus Cerithium (Figures 160 and 161 Chapter 15). There occur no less than 137 species of this genus in the Paris basin, and almost all of them in the calcaire grossier. Most of the living Cerithia inhabit the sea near the mouths of rivers, where the waters are brackish; so that their abundance in the marine strata now under consideration is in harmony with the hypothesis that the Paris basin formed a gulf into which several rivers flowed.

EOCENE FORAMINIFERA.

(FIGURE 219. Calcarina rarispina, Desh. a. Natural size. b. Magnified.)
(FIGURE 220. Spirolina stenostoma, Desh. a. Natural size. b. Magnified.)
(FIGURE 221. Triloculina inflata, Desh. a. Natural size. b. Magnified.)

In some parts of the calcaire grossier round Paris, certain beds occur of a stone used in building, and called by the French geologists "Miliolite limestone." It is almost entirely made up of millions of microscopic shells, of the size of minute grains of sand, which all belong to the class Foraminifera. Figures of some of these are given in Figures 219 to 221. As this miliolitic stone never occurs in the Faluns, or Upper Miocene strata of Brittany and Touraine, it often furnishes the geologist with a useful criterion for distinguishing the detached Eocene and Upper Miocene formations scattered over those and other adjoining provinces. The discovery of the remains of Palaeotherium and other mammalia in some of the upper beds of the calcaire grossier shows that these land animals began to exist before the deposition of the overlying gypseous series had commenced.

LOWER CALCAIRE GROSSIER, OR GLAUCONIE GROSSIERE (B.1 TABLE 16.1).

The lower part of the calcaire grossier, which often contains much green earth, is characterised at Auvers, near Pontoise, to the north of Paris, and still more in the environs of Compiegne, by the abundance of nummulites, consisting chiefly of N. laevigata, N. scabra, and N. Lamarcki, which constitute a large proportion of some of the stony strata, though these same foraminifera are wanting in beds of similar age in the immediate environs of Paris.

SOISSONNAIS SANDS, OR LITS COQUILLIERS (B.2 TABLE 16.1).

(FIGURE 222. Nerita conoidea, Lam. Syn. N. Schmidelliana, Chemnitz.)

Below the preceding formation, shelly sands are seen, of considerable thickness, especially at Cuisse-Lamotte, near Compiegne, and other localities in the Soissonnais, about fifty miles N.E. of Paris, from which about 300 species of shells have been obtained, many of them common to the calcaire grossier and the Bracklesham beds of England, and many peculiar. The Nummulites planulata is very abundant, and the most characteristic shell is the Nerita conoidea, Lam., a fossil which has a very wide geographical range; for, as M. d'Archiac remarks, it accompanies the nummulitic formation from Europe to India, having been found in Cutch, near the mouths of the Indus, associated with Nummulites scabra. No less than 33 shells of this group are said to be identical with shells of the London clay proper, yet, after visiting Cuisse-Lamotte and other localities of the "Sables inferieurs" of Archiac, I agree with Mr. Prestwich, that the latter are probably newer than the London clay, and perhaps older than the Bracklesham beds of England. The London clay seems to be unrepresented in the Paris basin, unless partially so, by these sands. (d'Archiac Bulletin tome 10 and Prestwich Quarterly Geological Journal 1847 page 377.)

LOWER EOCENE FORMATIONS OF FRANCE.
ARGILE PLASTIQUE (C.2 TABLE 16.1).

At the base of the tertiary system in France are extensive deposits of sands, with occasional beds of clay used for pottery, and called "argile plastique." Fossil oysters (Ostrea bellovacina) abound in some places, and in others there is a mixture of fluviatile shells, such as Cyrena cuneiformis (Figure 216), Melania inquinata (Figure 217), and others, frequently met with in beds occupying the same position in the London Basin. Layers of lignite also accompany the inferior clays and sands.

Immediately upon the chalk at the bottom of all the tertiary strata in France there generally is a conglomerate or breccia of rolled and angular chalk-flints, cemented by siliceous sand. These beds appear to be of littoral origin, and imply the previous emergence of the chalk, and its waste by denudation. In the year 1855, the tibia and femur of a large bird equalling at least the ostrich in size were found at Meudon, near Paris, at the base of the Plastic clay. This bird, to which the name of Gastornis Parisiensis has been assigned, appears, from the Memoirs of MM. Hebert, Lartet, and Owen, to belong to an extinct genus. Professor Owen refers it to the class of wading land birds rather than to an aquatic species. (Quarterly Geological Journal volume 12 page 204 1856.)

That a formation so much explored for economical purposes as the Argile plastique around Paris, and the clays and sands of corresponding age near London, should never have afforded any vestige of a feathered biped previously to the year 1855, shows what diligent search and what skill in osteological interpretation are required before the existence of birds of remote ages can be established.

SABLES DE BRACHEUX (C.3 TABLE 16.1).

The marine sands called the Sables de Bracheux (a place near Beauvais), are considered by M. Hebert to be older than the Lignites and Plastic clay, and to coincide in age with the Thanet Sands of England. At La Fere, in the Department of Aisne, in a deposit of this age, a fossil skull has been found of a quadruped called by Blainville Arctocyon primaevus, and supposed by him to be related both to the bear and to the Kinkajou (Cercoleptes). This creature appears to be the oldest known tertiary mammifer.

NUMMULITIC FORMATIONS OF EUROPE, ASIA, ETC.

Of all the rocks of the Eocene period, no formations are of such great geographical importance as the Upper and Middle Eocene, as above defined, assuming that the older tertiary formation, commonly called nummulitic, is correctly ascribed to this group. It appears that of more than fifty species of these foraminifera described by D'Archiac, one or two species only are found in other tertiary formations whether of older or newer date. Nummulites intermedia, a Middle Eocene form, ascends into the Lower Miocene, but it seems doubtful whether any species descends to the level of the London clay, still less to the Argile plastique or Woolwich beds. Separate groups of strata are often characterised by distinct species of nummulite; thus the beds between the lower Miocene and the lower Eocene may be divided into three sections, distinguished by three different species of nummulites, N. variolaria in the upper, N. laevigata in the middle, and N. planulata in the lower beds. The nummulitic limestone of the Swiss Alps rises to more than 10,000 feet above the level of the sea, and attains here and in other mountain chains a thickness of several thousand feet. It may be said to play a far more conspicuous part than any other tertiary group in the solid framework of the earth's crust, whether in Europe, Asia, or Africa. It occurs in Algeria and Morocco, and has been traced from Egypt, where it was largely quarried of old for the building of the Pyramids, into Asia Minor, and across Persia by Bagdad to the mouths of the Indus. It has been observed not only in Cutch, but in the mountain ranges which separate Scinde from Persia, and which form the passes leading to Caboul; and it has been followed still farther eastward into India, as far as eastern Bengal and the frontiers of China.

(FIGURE 223. Nummulites Puschi, D'Archiac. Peyrehorade, Pyrenees. a. External surface of one of the nummulites, of which longitudinal sections are seen in the limestone. b. Transverse section of same.)

Dr. T. Thompson found nummulites at an elevation of no less than 16,500 feet above the level of the sea, in Western Thibet. One of the species, which I myself found very abundant on the flanks of the Pyrenees, in a compact crystalline marble (Figure 223) is called by M. D'Archiac Nummulites Puschi. The same is also very common in rocks of the same age in the Carpathians. In many distant countries, in Cutch, for example, some of the same shells, such as Nerita conoidea (Figure 222), accompany the nummulites, as in France. The opinion of many observers, that the Nummulitic formation belongs partly to the cretaceous era, seems chiefly to have arisen from confounding an allied genus, Orbitoides, with the true Nummulite.

When we have once arrived at the conviction that the nummulitic formation occupies a middle and upper place in the Eocene series, we are struck with the comparatively modern date to which some of the greatest revolutions in the physical geography of Europe, Asia, and Northern Africa must be referred. All the mountain-chains, such as the Alps, Pyrenees, Carpathians, and Himalayas, into the composition of whose central and loftiest parts the nummulitic strata enter bodily, could have had no existence till after the Middle Eocene period. During that period the sea prevailed where these chains now rise, for nummulites and their accompanying testacea were unquestionably inhabitants of salt water. Before these events, comprising the conversion of a wide area from a sea to a continent, England had been peopled, as I before pointed out, by various quadrupeds, by herbivorous pachyderms, by insectivorous bats, and by opossums.

Almost all the volcanoes which preserve any remains of their original form, or from the craters of which lava streams can be traced, are more modern than the Eocene fauna now under consideration; and besides these superficial monuments of the action of heat, Plutonic influences have worked vast changes in the texture of rocks within the same period. Some members of the nummulitic and overlying tertiary strata called flysch have actually been converted in the central Alps into crystalline rocks, and transformed into marble, quartz-rock, micha-schist, and gneiss. (Murchison Quarterly Journal of Geological Society volume 5 and Lyell volume 6 1850 Anniversary Address.)

EOCENE STRATA IN THE UNITED STATES.

In North America the Eocene formations occupy a large area bordering the Atlantic, which increases in breadth and importance as it is traced southward from Delaware and Maryland to Georgia and Alabama. They also occur in Louisiana and other States both east and west of the valley of the Mississippi. At Claiborne, in Alabama, no less than 400 species of marine shells, with many echinoderms and teeth of fish, characterise one member of this system. Among the shells, the Cardita planicosta, before mentioned (Figure 191), is in abundance; and this fossil and some others identical with European species, or very nearly allied to them, make it highly probable that the Claiborne beds agree in age with the central or Bracklesham group of England, and with the calcaire grossiere of Paris. (See paper by the Author Quarterly Journal of Geological Society volume 4 page 12 and Second Visit to the United States volume 2 page 59.)

Higher in the series is a remarkable calcareous rock, formerly called "the nummulite limestone," from the great number of discoid bodies resembling nummulites which it contains, fossils now referred by A. d'Orbigny to the genus Orbitoides, which has been demonstrated by Dr. Carpenter to belong to the foraminifera. (Quarterly Journal of Geological Society volume 6 page 32.) That naturalist, moreover, is of opinion that the Orbitoides alluded to (O. Mantelli) is of the same species as one found in Cutch, in the Middle Eocene or nummulitic formation of India.

Above the orbitoidal limestone is a white limestone, sometimes soft and argillaceous, but in parts very compact and calcareous. It contains several peculiar corals, and a large Nautilus allied to N. ziczac; also in its upper bed a gigantic cetacean, called Zeuglodon by Owen. (See Memoir by R.W. Gibbes Journal of Academy of Natural Science Philadelphia volume 1 1847.)

The colossal bones of this cetacean are so plentiful in the interior of Clarke County, Alabama, as to be characteristic of the formation. The vertebral column of one skeleton found by Dr. Buckley at a spot visited by me, extended to the length of nearly seventy feet, and not far off part of another backbone nearly fifty feet long was dug up. I obtained evidence, during a short excursion, of so many localities of this fossil animal within a distance of ten miles, as to lead me to conclude that they must have belonged to at least forty distinct individuals.

(FIGURE 224. Zeuglodon cetoides, Owen. Basilosaurus, Harlan.
Molar tooth, natural size.)

(FIGURE 225. Zeuglodon cetoides, Owen. Basilosaurus, Harlan.
Vertebra, reduced.)

Professor Owen first pointed out that this huge animal was not reptilian, since each tooth was furnished with double roots (Figure 224), implanted in corresponding double sockets; and his opinion of the cetacean nature of the fossil was afterwards confirmed by Dr. Wyman and Dr. R.W. Gibbes. That it was an extinct mammal of the whale tribe has since been placed beyond all doubt by discovery of the entire skull of another fossil species of the same family, having the double occipital condyles only met with in mammals, and the convoluted tympanic bones which are characteristic of cetaceans.

CHAPTER XVII.
UPPER CRETACEOUS GROUP.

Lapse of Time between Cretaceous and Eocene Periods.
Table of successive Cretaceous Formations.
Maestricht Beds.
Pisolitic Limestone of France.
Chalk of Faxoe.
Geographical Extent and Origin of the White Chalk.
Chalky Matter now forming in the Bed of the Atlantic.
Marked Difference between the Cretaceous and existing Fauna.
Chalk-flints.
Pot-stones of Horstead.
Vitreous Sponges in the Chalk.
Isolated Blocks of Foreign Rocks in the White Chalk supposed to be ice-borne.
Distinctness of Mineral Character in contemporaneous Rocks of the Cretaceous Epoch.
Fossils of the White Chalk.
Lower White Chalk without Flints.
Chalk Marl and its Fossils.
Chloritic Series or Upper Greensand.

Coprolite Bed near Cambridge.
Fossils of the Chloritic Series.
Gault.
Connection between Upper and Lower Cretaceous Strata.
Blackdown Beds.
Flora of the Upper Cretaceous Period.
Hippurite Limestone.
Cretaceous Rocks in the United States.

We have treated in the preceding chapters of the Tertiary or Cainozoic strata, and have next to speak of the Secondary or Mesozoic formations. The uppermost of these last is commonly called the chalk or the cretaceous formation, from creta, the latin name for that remarkable white earthy limestone, which constitutes an upper member of the group in those parts of Europe where it was first studied. The marked discordance in the fossils of the tertiary, as compared with the cretaceous formations, has long induced many geologists to suspect that an indefinite series of ages elapsed between the respective periods of their origin. Measured, indeed, by such a standard, that is to say, by the amount of change in the Fauna and Flora of the earth effected in the interval, the time between the Cretaceous and Eocene may have been as great as that between the Eocene and Recent periods, to the history of which the last seven chapters have been devoted. Several deposits have been met with here and there, in the course of the last half century, of an age intermediate between the white chalk and the plastic clays and sands of the Paris and London districts, monuments which have the same kind of interest to a geologist which certain medieval records excite when we study the history of nations. For both of them throw light on ages of darkness, preceded and followed by others of which the annals are comparatively well-known to us. But these newly-discovered records do not fill up the wide gap, some of them being closely allied to the Eocene, and others to the Cretaceous type, while none appear as yet to possess so distinct and characteristic a fauna as may entitle them to hold an independent place in the great chronological series.

Among the formations alluded to, the Thanet Sands of Prestwich have been sufficiently described in the last chapter, and classed as Lower Eocene. To the same tertiary series belong the Belgian formations, called by Professor Dumont, Landenian. On the other hand, the Maestricht and Faxoe limestones are very closely connected with the chalk, to which also the Pisolitic limestone of France is referable.

CLASSIFICATION OF THE CRETACEOUS ROCKS.
TABLE 17.1.
UPPER CRETACEOUS OR CHALK PERIOD.

1. Maestricht Beds and Faxoe Limestone. 2. Upper White Chalk, with flints. 3. Lower White Chalk, without flints. 4. Chalk Marl. 5. Chloritic series (or Upper Greensand). 6. Gault.

LOWER CRETACEOUS OR NEOCOMIAN.

1. Marine: Upper Neocomian, see Chapter 18. Fresh-water: Wealden Beds (upper part). 2. Marine: Middle Neocomian, see Chapter 18. Fresh-water: Wealden Beds (upper part). 3. Marine: Lower Neocomian, see Chapter 18. Fresh-water: Wealden Beds (upper part).

The cretaceous group has generally been divided into an Upper and a Lower series, the Upper called familiarly THE CHALK, and the Lower THE GREENSAND; the one deriving its name from the predominance of white earthy limestone and marl, of which it consists in a great part of France and England, the other or lower series from the plentiful mixture of green or chloritic grains contained in some of the sands and cherts of which it largely consists in the same countries. But these mineral characters often fail, even when we attempt to follow out the same continuous subdivisions throughout a small portion of the north of Europe, and are worse than valueless when we desire to apply them to more distant regions. It is only by aid of the organic remains which characterise the successive marine subdivisions of the formation that we are able to recognise in remote countries, such as the south of Europe or North America, the formations which were there contemporaneously in progress. To the English student of geology it will be sufficient to begin by enumerating those groups which characterise the series in this country and others immediately contiguous, alluding but slightly to those of more distant regions. In Table 17.1 it will be seen that I have used the term Neocomian for that commonly called "Lower Greensand;" as this latter term is peculiarly objectionable, since the green grains are an exception to the rule in many of the members of this group even in districts where it was first studied and named.

MAESTRICHT BEDS.

(FIGURE 226. Belemnitella mucronata, Maestricht, Faxoe, and White Chalk. a. Entire specimen, showing vascular impression on outer surface, and characteristic slit. b. Section of same, showing place of phragmocone. (For particulars of structure see Chapter 18.))

On the banks of the Meuse, at Maestricht, reposing on ordinary white chalk with flints, we find an upper calcareous formation about 100 feet thick, the fossils of which are, on the whole, very peculiar, and all distinct from tertiary species. Some few are of species common to the inferior white chalk, among which may be mentioned Belemnitella mucronata (Figure 226) and Pecten quadricostatus, a shell regarded by many as a mere variety of Pecten quinquecostatus (see Figure 270). Besides the Belemnite there are other genera, such as Baculites and Hamites, never found in strata newer than the cretaceous, but frequently met with in these Maestricht beds. On the other hand, Voluta, Fasciolaria, and other genera of univalve shells, usually met with only in tertiary strata, occur.

The upper part of the rock, about 20 feet thick, as seen in St. Peter's Mount, in the suburbs of Maestricht, abounds in corals and Bryozoa, often detachable from the matrix; and these beds are succeeded by a soft yellowish limestone 50 feet thick, extensively quarried from time immemorial for building. The stone below is whiter, and contains occasional nodules of grey chert or chalcedony.

(FIGURE 227. Mosasaurus Camperi. Original more than three feet long.)

(FIGURE 228. Hemipneustes radiatus, Ag. Spatangus radiatus, Lam. Chalk of Maestricht and white chalk.)

M. Bosquet, with whom I examined this formation (August, 1850), pointed out to me a layer of chalk from two to four inches thick, containing green earth and numerous encrinital stems, which forms the line of demarkation between the strata containing the fossils peculiar to Maestricht and the white chalk below. The latter is distinguished by regular layers of black flint in nodules, and by several shells, such as Terebratula carnea (see Figure 246), wholly wanting in beds higher than the green band. Some of the organic remains, however, for which St. Peter's Mount is celebrated, occur both above and below that parting layer, and, among others, the great marine reptile called Mosasaurus (see Figure 227), a saurian supposed to have been 24 feet in length, of which the entire skull and a great part of the skeleton have been found. Such remains are chiefly met with in the soft freestone, the principal member of the Maestricht beds. Among the fossils common to the Maestricht and white chalk may be instanced the echinoderm, Figure 228.

I saw proofs of the previous denudation of the white chalk exhibited in the lower bed of the Maestricht formation in Belgium, about 30 miles S.W. of Maestricht, at the village of Jendrain, where the base of the newer deposit consisted chiefly of a layer of well-rolled, black chalk-flint pebbles, in the midst of which perfect specimens of Thecidea papillata and Belemnitella mucronata are imbedded. To a geologist accustomed in England to regard rolled pebbles of chalk-flint as a common and distinctive feature of tertiary beds of different ages, it is a new and surprising phenomenon to behold strata made up of such materials, and yet to feel no doubt that they were accumulated in a sea in which the belemnite and other cretaceous mollusca flourished.

PISOLITIC LIMESTONE OF FRANCE.

Geologists were for many years at variance respecting the chronological relations of this rock, which is met with in the neighbourhood of Paris, and at places north, south, east, and west of that metropolis, as between Vertus and Laversines, Meudon and Montereau. By many able palaeontologists the species of fossils, more than fifty in number, were declared to be more Eocene in their appearance than Cretaceous. But M. Hebert found in this formation at Montereau, near Paris, the Pecten quadricostatus, a well-known Cretaceous species, together with some other fossils common to the Maestricht chalk and to the Baculite limestone of the Cotentin, in Normandy. He therefore, as well as M. Alcide d'Orbigny, who had carefully studied the fossils, came to the opinion that it was an upper member of the Cretaceous group. It is usually in the form of a coarse yellowish or whitish limestone, and the total thickness of the series of beds already known is about 100 feet. Its geographical range, according to M. Hebert, is not less than 45 leagues from east to west, and 35 from north to south. Within these limits it occurs in small patches only, resting unconformably on the white chalk.

(FIGURE 229. Portion of Baculites Faujasii. Maestricht and Faxoe beds and white chalk.)

(FIGURE 230. Nautilus Danicus, Schl. Faxoe, Denmark.)

The Nautilus Danicus, Figure 230, and two or three other species found in this rock, are frequent in that of Faxoe, in Denmark, but as yet no Ammonites, Hamites, Scaphites, Turrilites, Baculites, or Hippurites have been met with. The proportion of peculiar species, many of them of tertiary aspect, is confessedly large; and great aqueous erosion suffered by the white chalk, before the pisolitic limestone was formed, affords an additional indication of the two deposits being widely separated in time. The pisolitic formation, therefore, may eventually prove to be somewhat more intermediate in date between the secondary and tertiary epochs than the Maestricht rock.

CHALK OF FAXOE.

In the island of Seeland, in Denmark, the newest member of the chalk series, seen in the sea-cliffs at Stevensklint resting on white chalk with flints, is a yellow limestone, a portion of which, at Faxoe, where it is used as a building stone, is composed of corals, even more conspicuously than is usually observed in recent coral reefs. It has been quarried to the depth of more than 40 feet, but its thickness is unknown. The imbedded shells are chiefly casts, many of them of univalve mollusca, which are usually very rare in the white chalk of Europe. Thus, there are two species of Cypraea, two of Oliva, two of Mitra, four of the genus Cerithium, six of Fusus, two of Trochus, one of Patella, one of Emarginula, etc.; on the whole, more than thirty univalves, spiral or patelliform. At the same time, some of the accompanying bivalve shells, echinoderms, and zoophytes, are specifically identical with fossils of the true Cretaceous series. Among the cephalopoda of Faxoe may be mentioned Baculites Faujasii (Figure 229), and Belemnitella mucronata (Figure 226), shells of the white chalk. The Nautilus Danicus (see Figure 230) is characteristic of this formation; and it also occurs in France in the calcaire pisolitique of Laversin (Department of Oise). The claws and entire skull of a small crab, Brachyurus rugosus (Schlott.), are scattered through the Faxoe stone, reminding us of similar crustaceans inclosed in the rocks of modern coral reefs. Some small portions of this coralline formation consist of white earthy chalk.

COMPOSITION, EXTENT AND ORIGIN OF THE WHITE CHALK.

(FIGURE 231. Diagrammatic section from Hertfordshire, in England, to Sens, in France.

Through London (left), Hythe, Boulogne, Valley of Bray, Paris and Sens (right).)

The highest beds of chalk in England and France consist of a pure, white, calcareous mass, usually too soft for a building-stone, but sometimes passing into a more solid state. It consists, almost purely, of carbonate of lime; the stratification is often obscure, except where rendered distinct by interstratified layers of flint, a few inches thick, occasionally in continuous beds, but oftener in nodules, and recurring at intervals generally from two to four feet distant from each other. This upper chalk is usually succeeded, in the descending order, by a great mass of white chalk without flints, below which comes the chalk marl, in which there is a slight admixture of argillaceous matter. The united thickness of the three divisions in the south of England equals, in some places, 1000 feet. The section in Figure 231 will show the manner in which the white chalk extends from England into France, covered by the tertiary strata described in former chapters, and reposing on lower cretaceous beds.

The area over which the white chalk preserves a nearly homogeneous aspect is so vast, that the earlier geologists despaired of discovering any analogous deposits of recent date. Pure chalk, of nearly uniform aspect and composition, is met with in a north-west and south-east direction, from the north of Ireland to the Crimea, a distance of about 1140 geographical miles, and in an opposite direction it extends from the south of Sweden to the south of Bordeaux, a distance of about 840 geographical miles. In Southern Russia, according to Sir R. Murchison, it is sometimes 600 feet thick, and retains the same mineral character as in France and England, with the same fossils, including Inoceramus Cuvieri, Belemnitella mucronata, and Ostrea vesicularis (Figure 251).

(Figures 232 to 236.— Organic bodies forming the ooze of the bed of the Atlantic at great depths.
(FIGURE 232. Globigerina bulloides. Calcareous Rhizopod.)
(FIGURE 233. Actinocyclus. Siliceous Diatomaceae.)
(FIGURE 234. Pinnularia. Siliceous Diatomaceae.)
(FIGURE 235. Eunotia bidens. Siliceous Diatomaceae.)
(FIGURE 236. Spicula of sponge. Siliceous sponge.))

Great light has recently been thrown upon the origin of the unconsolidated white chalk by the deep soundings made in the North Atlantic, previous to laying down, in 1858, the electric telegraph between Ireland and Newfoundland. At depths sometimes exceeding two miles, the mud forming the floor of the ocean was found, by Professor Huxley, to be almost entirely composed (more than nineteen- twentieths of the whole) of minute Rhizopods, or foraminiferous shells of the genus Globigerina, especially the species Globigerina bulloides (see Figure 232.) the organic bodies next in quantity were the siliceous shells called Polycystineae, and next to them the siliceous skeletons of plants called Diatomaceae (Figures 233, 234, 235), and occasionally some siliceous spiculae of sponges (Figure 236) were intermixed. These were connected by a mass of living gelatinous matter to which he has given the name of Bathybius, and which contains abundance of very minute bodies termed Coccoliths and Coccospheres, which have also been detected fossil in chalk.

Sir Leopold MacClintock and Dr. Wallich have ascertained that 95 per cent of the mud of a large part of the North Atlantic consists of Globigerina shells. But Captain Bullock, R.N., lately brought up from the enormous depth of 16,860 feet a white, viscid, chalky mud, wholly devoid of Globigerinae. This mud was perfectly homogeneous in composition, and contained no organic remains visible to the naked eye. Mr. Etheridge, however, has ascertained by microscopical examination that it is made up of Coccoliths, Discoliths, and other minute fossils like those of the Chalk classed by Huxley as Bathybius, when this term is used in its widest sense. This mud, more than three miles deep, was dredged up in latitude 20 degrees 19' N., longitude 4 degrees 36' E., or about midway between Madeira and the Cape of Good Hope.

The recent deep-sea dredgings in the Atlantic conducted by Dr. Wyville Thomson, Dr. Carpenter, Mr. Gwyn Jeffreys, and others, have shown that on the same white mud there sometimes flourish Mollusca, Crustacea, and Echinoderms, besides abundance of siliceous sponges, forming, on the whole, a marine fauna bearing a striking resemblance in its general character to that of the ancient chalk.

POPULAR ERROR AS TO THE GEOLOGICAL CONTINUITY OF THE CRETACEOUS PERIOD.

We must be careful, however, not to overrate the points of resemblance which the deep-sea investigations have placed in a strong light. They have been supposed by some naturalists to warrant a conclusion expressed in these words: "We are still living in the Cretaceous epoch;" a doctrine which has led to much popular delusion as to the bearing of the new facts on geological reasoning and classification. The reader should be reminded that in geology we have been in the habit of founding our great chronological divisions, not on foraminifera and sponges, nor even on echinoderms and corals, but on the remains of the most highly organised beings available to us, such as the mollusca; these being met with, as explained in Chapter 9, in stratified rocks of almost every age. In dealing with the mollusca, it is those of the highest or most specialised organisation, which afford us the best characters in proportion as their vertical range is the most limited. Thus the Cephalopoda are the most valuable, as having a more restricted range in time than the Gasteropoda; and these, again, are more characteristic of the particular stratigraphical subdivisions than are the Lamellibranchiate Bivalves, while these last, again, are more serviceable in classification than the Brachiopoda, a still lower class of shell-fish, which are the most enduring of all.

When told that the new dredgings prove that "we are still living in the Chalk Period," we naturally ask whether some cuttle-fish has been found with a Belemnite forming part of its internal framework; or have Ammonites, Baculites, Hamites, Turrilites, with four or five other Cephalopodous genera characteristic of the chalk and unknown as tertiary, been met with in the abysses of the ocean? Or, in the absence of these long-extinct forms, has a single spiral univalve, or species of Cretaceous Gasteropod, been found living? Or, to descend still lower in the scale, has some characteristic Cretaceous genus of Lamellibranchiate Bivalve, such as the Inoceramus, or Hippurite, foreign to the Tertiary seas, been proved to have survived down to our time? Or, of the numerous genera of lamellibranchiates common to the Cretaceous and Recent seas, has one species been found living? The answer to all these questions is— not one has been found. Even of the humblest shell-fish, the Brachiopods, no new species common to the Cretaceous and recent seas has yet been met with. It has been very generally admitted by conchologists that out of a hundred species of this tribe occurring fossil in the Upper Chalk— one, and one only, Terebratulina striata, is still living, being thought to be identical with Terebratula caput-serpentis. Although this identity is still questioned by some naturalists of authority, it would certainly not surprise us if another lamp-shell of equal antiquity should be met with in the deep sea.

Had it been declared that we are living in the Eocene epoch, the idea would not be so extravagant, for the great reptiles of the Upper Chalk, the Mosasaurus, Pliosaurus, and Pterodactyle, and many others, as well as so many genera of chambered univalves, had already disappeared from the earth, and the marine fauna had made a greater approach to our own by nearly the entire difference which separates it from the fauna of the Cretaceous seas. The Eocene nummulitic limestone of Egypt is a rock mainly composed, like the more ancient white chalk, of globigerine mud; and if the reader will refer to what we have said of the extent to which the nummulitic marine strata, formed originally at the bottom of the sea, now enter into the framework of mountain chains of the principal continents, he will at once perceive that the present Atlantic, Pacific, and Indian Oceans are geographical terms, which must be wholly without meaning when applied to the Eocene, and still more to the Cretaceous Period; so that to talk of the chalk having been uninterruptedly forming in the Atlantic from the Cretaceous Period to our own, is as inadmissible in a geographical as in a geological sense.

CHALK-FLINTS.

The origin of the layers of flint, whether in the form of nodules, or continuous sheets, or in veins or cracks not parallel to the stratification, has always been more difficult to explain than that of the white chalk. But here, again, the late deep-sea soundings have suggested a possible source of such mineral matter. During the cruise of the "Bulldog," already alluded to, it was ascertained that while the calcareous Globigerinae had almost exclusive possession of certain tracts of the sea-bottom, they were wholly wanting in others, as between Greenland and Labrador. According to Dr. Wallich, they may flourish in those spaces where they derive nutriment from organic and other matter, brought from the south by the warm waters of the Gulf Stream, and they may be absent where the effects of that great current are not felt. Now, in several of the spaces where the calcareous Rhizopods are wanting, certain microscopic plants, called Diatomaceae, above-mentioned (Figures 233-235), the solid parts of which are siliceous, monopolise the ground at a depth of nearly 400 fathoms, or 2400 feet.

The large quantities of silex in solution required for the formation of these plants may probably arise from the disintegration of feldspathic rocks, which are universally distributed. As more than half of their bulk is formed of siliceous earth, they may afford an endless supply of silica to all the great rivers which flow into the ocean. We may imagine that, after a lapse of many years or centuries, changes took place in the direction of the marine currents, favouring at one time a supply in the same area of siliceous, and at another of calcareous matter in excess, giving rise in the one case to a preponderance of Globigerinae, and in the other of Diatomaceae. These last, and certain sponges, may by their decomposition have furnished the silex, which, separating from the chalky mud, collected round organic bodies, or formed nodules, or filled shrinkage cracks.

POT-STONES.

(FIGURE 237. View of a chalk-pit at Horstead, near Norwich, showing the position of the pot-stones. From a drawing by Mrs. Gunn.)

A more difficult enigma is presented by the occurrence of certain huge flints, or pot-stones, as they are called in Norfolk, occurring singly, or arranged in nearly continuous columns at right angles to the ordinary and horizontal layers of small flints. I visited in the year 1825 an extensive range of quarries then open on the river Bure, near Horstead, about six miles from Norwich, which afforded a continuous section, a quarter of a mile in length, of white chalk, exposed to the depth of about twenty-six feet, and covered by a bed of gravel. The pot-stones, many of them pear-shaped, were usually about three feet in height and one foot in their transverse diameter, placed in vertical rows, like pillars, at irregular distances from each other, but usually from twenty to thirty feet apart, though sometimes nearer together, as in Figure 237. These rows did not terminate downward in any instance which I could examine, nor upward, except at the point where they were cut off abruptly by the bed of gravel. On breaking open the pot-stones, I found an internal cylindrical nucleus of pure chalk, much harder than the ordinary surrounding chalk, and not crumbling to pieces like it, when exposed to the winter's frost. At the distance of half a mile, the vertical piles of pot-stones were much farther apart from each other. Dr. Buckland has described very similar phenomena as characterising the white chalk on the north coast of Antrim, in Ireland. (Geological Transactions 1st Series volume 4 page 413.)

VITREOUS SPONGES OF THE CHALK.

These pear-shaped masses of flint often resemble in shape and size the large sponges called Neptune's Cups (Spongia patera, Hardw.), which grow in the seas of Sumatra; and if we could suppose a series of such gigantic sponges to be separated from each other, like trees in a forest, and the individuals of each successive generation to grow on the exact spot where the parent sponge died and was enveloped in calcareous mud, so that they should become piled one above the other in a vertical column, their growth keeping pace with the accumulation of the enveloping calcareous mud, a counterpart of the phenomena of the Horstead pot-stones might be obtained.

(FIGURE 238. Ventriculites radiatus, Mantell. Syn. Ocellaria radiata. D'Orbigny. White chalk.)

Professor Wyville Thomson, describing the modern soundings in 1869 off the north coast of Scotland, speaks of the ooze or chalk mud brought from a depth of about 3000 feet, and states that at one haul they obtained forty specimens of vitreous sponges buried in the mud. He suggests that the Ventriculites of the chalk were nearly allied to these sponges, and that when the silica of their spicules was removed, and was dissolved out of the calcareous matrix, it set into flint.

BOULDERS AND GROUPS OF PEBBLES IN CHALK.

The occurrence here and there, in the white chalk of the south of England, of isolated pebbles of quartz and green schist has justly excited much wonder. It was at first supposed that they had been dropped from the roots of some floating tree, by which means stones are carried to some of the small coral islands of the Pacific. But the discovery in 1857 of a group of stones in the white chalk near Croydon, the largest of which was syenite and weighed about forty pounds, accompanied by pebbles and fine sand like that of a beach, has been shown by Mr. Godwin Austen to be inexplicable except by the agency of floating ice. If we consider that icebergs now reach 40 degrees north latitude in the Atlantic, and several degrees nearer the equator in the southern hemisphere, we can the more easily believe that even during the Cretaceous epoch, assuming that the climate was milder, fragments of coast ice may have floated occasionally as far as the south of England.

DISTINCTNESS OF MINERAL CHARACTER IN CONTEMPORANEOUS ROCKS OF THE CRETACEOUS PERIOD.

But we must not imagine that because pebbles are so rare in the white chalk of England and France there are no proofs of sand, shingle, and clay having been accumulated contemporaneously even in European seas. The siliceous sandstone called "upper quader" by the Germans overlies white argillaceous chalk or "planer-kalk," a deposit resembling in composition and organic remains the chalk marl of the English series. This sandstone contains as many fossil shells common to our white chalk as could be expected in a sea-bottom formed of such different materials. It sometimes attains a thickness of 600 feet, and, by its jointed structure and vertical precipices, plays a conspicuous part in the picturesque scenery of Saxon Switzerland, near Dresden. It demonstrates that in the Cretaceous sea, as in our own, distinct mineral deposits were simultaneously in progress. The quartzose sandstone alluded to, derived from the detritus of the neighbouring granite, is absolutely devoid of carbonate of lime, yet it was formed at the distance only of four hundred miles from a sea-bottom now constituting part of France, where the purely calcareous white chalk was forming. In the North American continent, on the other hand, where the Upper Cretaceous formations are so widely developed, true white chalk, in the ordinary sense of that term, does not exist.

FOSSILS OF THE WHITE CHALK.

(FIGURE 239. Ananchytes ovatus, Leske. White chalk, upper and lower. a. Side view. b. Base of the shell, on which both the oral and anal apertures are placed; the anal being more round, and at the smaller end.)

(FIGURE 240. Micraster cor-angulum, Leske. White chalk.)

(FIGURE 241. Galerites albogalerus, Lam. White chalk.)

(FIGURE 242. Marsupites Milleri. Mant. White chalk.)

Among the fossils of the white chalk, echinoderms are very numerous; and some of the genera, like Ananchytes (see Figure 239), are exclusively cretaceous. Among the Crinoidea, the Marsupites (Figure 242) is a characteristic genus. Among the mollusca, the cephalopoda are represented by Ammonites, Baculites (Figure 229), and Belemnites (Figure 226). Although there are eight or more species of Ammonites and six of them peculiar to it, this genus is much less fully represented than in each of the other subdivisions of the Upper Cretaceous group.

(FIGURE 243. Terebratulina striata, Wahlenb. Upper white chalk.)

(FIGURE 244. Rhynchonella octoplicata, Sowerby. (Var. of R. plicatilis). Upper white chalk.

(FIGURE 245. Magas pumila, Sowerby. Upper white chalk.)

(FIGURE 246. Terebratula carnea, Sowerby. Upper white chalk.)

(FIGURE 247. Terebratula biplicata, Brocch. Upper cretaceous.)

(FIGURE 248. Crania Parisiensis, Duf. Inferior or attached valve. Upper white chalk.)

(FIGURE 249. Pecten Beaveri, Sowerby. Reduced to one-third diameter. Lower white chalk and chalk marl. Maidstone.)

(FIGURE 250. Lima spinosa, Sowerby. Syn. Spondylus spinosus. Upper white chalk.)

(FIGURE 251. Ostrea vesicularis. Syn. Gryphaea convexa. Upper chalk and upper greensand.)

Among the brachiopoda in the white chalk, the Terebratulae are very abundant (see Figures 243-247). With these are associated some forms of oyster (see Figure 251), and other bivalves (Figures 249, 250).

(FIGURE 252. Inoceramus Lamarckii. Syn. Catillus Lamarckii. White chalk (Dixon's Geology Sussex Table 28 Figure 29).)

Among the bivalve mollusca, no form marks the Cretaceous era in Europe, America, and India in a more striking manner than the extinct genus Inoceramus (Catillus of Lam.; see Figure 252), the shells of which are distinguished by a fibrous texture, and are often met with in fragments, having probably been extremely friable.

(Figures 253 to 256. Radiolites Mortoni. Mantell. Houghton, Sussex. White chalk. Diameter one-seventh natural size. On the side where the shell is thinnest, there is one external furrow and corresponding internal ridge, a, b, Figures 253, 254; but they are usually less prominent than in these figures. The upper or opercular valve is wanting.

(FIGURE 253. Two individuals deprived of their upper valves, adhering together.)

(FIGURE 254. Same seen from above.)

(FIGURE 255. Transverse section of part of the wall of the shell, magnified to show the structure.)

(FIGURE 256. Vertical section of the same.))

Of the singular family called Rudistes by Lamarck, hereafter to be mentioned as extremely characteristic of the chalk of southern Europe, a single representative only (Figure 253) has been discovered in the white chalk of England.

(FIGURE 257. Eschara disticha. White chalk. a. Natural size. b. Portion magnified.)

(FIGURE 258. Escharina oceani. a. Natural size. b. Part of the same magnified. White chalk.)

(FIGURE 259. A branching sponge in a flint, from the white chalk. From the collection of Mr. Bowerbank.)

The general absence of univalve mollusca in the white chalk is very marked. Of bryozoa there is an abundance, such as Eschara and Escharina (Figures 257, 258). These and other organic bodies, especially sponges, such as Ventriculites (Figure 238), are dispersed indifferently through the soft chalk and hard flint, and some of the flinty nodules owe their irregular forms to inclosed sponges, such as Figure 259, a, where the hollows in the exterior are caused by the branches of a sponge (Figure 259, b), seen on breaking open the flint.

(FIGURE 260. Palatal tooth of Ptychodus decurrens. Lower white chalk. Maidstone.)

(FIGURE 261. Cestracion Phillippi; recent. Port Jackson. Buckland, Bridgewater Treatise Plate 27 d.))

The remains of fishes of the Upper Cretaceous formations consist chiefly of teeth belonging to the shark family. Some of the genera are common to the Tertiary formations, and some are distinct. To the latter belongs the genus Ptychodus (Figure 260), which is allied to the living Port Jackson shark, Cestracion Phillippi, the anterior teeth of which (see Figure 261, a) are sharp and cutting, while the posterior or palatal teeth (b) are flat (Figure 260). But we meet with no bones of land-animals, nor any terrestrial or fluviatile shells, nor any plants, except sea-weeds, and here and there a piece of drift-wood. All the appearances concur in leading us to conclude that the white chalk was the product of an open sea of considerable depth.

The existence of turtles and oviparous saurians, and of a Pterodactyl or winged lizard, found in the white chalk of Maidstone, implies, no doubt, some neighbouring land; but a few small islets in mid-ocean, like Ascension, formerly so much frequented by migratory droves of turtle, might perhaps have afforded the required retreat where these creatures laid their eggs in the sand, or from which the flying species may have been blown out to sea. Of the vegetation of such islands we have scarcely any indication, but it consisted partly of cycadaceous plants; for a fragment of one of these was found by Captain Ibbetson in the Chalk Marl of the Isle of Wight, and is referred by A. Brongniart to Clathraria Lyellii, Mantell, a species common to the antecedent Wealden period. The fossil plants, however, of beds corresponding in age to the white chalk at Aix-la-Chapelle, presently to be described, like the sandy beds of Saxony, before alluded to, afford such evidence of land as to prove how vague must be any efforts of ours to restore the geography of that period.

The Pterodactyl of the Kentish chalk, above alluded to, was of gigantic dimensions, measuring 16 feet 6 inches from tip to tip of its outstretched wings. Some of its elongated bones were at first mistaken by able anatomists for those of birds; of which class no osseous remains have as yet been derived from the white chalk, although they have been found (as will be seen) in the Chloritic sand.

(FIGURE 262. Coprolites of fish, from the chalk.)

The collector of fossils from the white chalk was formerly puzzled by meeting with certain bodies which they call larch-cones, which were afterwards recognised by Dr. Buckland to be the excrement of fish (see Figure 262). They are composed in great part of phosphate of lime.

LOWER WHITE CHALK.

(FIGURE 263. Baculites anceps, Lam. Lower chalk.)

The Lower White Chalk, which is several hundred feet thick, without flints, has yielded 25 species of Ammonites, of which half are peculiar to it. The genera Baculite, Hamite, Scaphite, Turrilite, Nautilus, Belemnite, and Belemnitella, are also represented.

CHALK MARL.

(FIGURE 264. Ammonites Rhotomagensis. Chalk marl. Back and side view.)

(FIGURE 265. Turrilites costatus, Lam. Lower chalk and chalk marl. a. Section, showing the foliated border of the sutures of the chambers.)

(FIGURE 266. Scaphites aequalis. Chloritic marl and sand, Dorsetshire.)

The lower chalk without flints passes gradually downward, in the south of England, into an argillaceous limestone, "the chalk marl," already alluded to. It contains 32 species of Ammonites, seven of which are peculiar to it, while eleven pass up into the overlying lower white chalk. A. Rhotomagensis is characteristic of this formation. Among the British cephalopods of other genera may be mentioned Scaphites aequalis (Figure 266) and Turrilites costatus (Figure 265).

CHLORITIC SERIES (OR UPPER GREENSAND).

According to the old nomenclature, this subdivision of the chalk was called Upper Greensand, in order to distinguish it from those members of the Neocomian or Lower Cretaceous series below the Gault to which the name of Greensand had been applied. Besides the reasons before given for abandoning this nomenclature, it is objectionable in this instance as leading the uninitiated to suppose that the divisions thus named Upper and Lower Greensand are of co-ordinate value, instead of which the chloritic sand is quite a subordinate member of the Upper Cretaceous group, and the term Greensand has very commonly been used for the whole of the Lower Cretaceous rocks, which are almost comparable in importance to the entire Upper Cretaceous series. The higher portion of the Chloritic series in some districts has been called chloritic marl, from its consisting of a chalky marl with chloritic grains. In parts of Surrey, where calcareous matter is largely intermixed with sand, it forms a stone called malm-rock or firestone. In the cliffs of the southern coast of the Isle of Wight it contains bands of calcareous limestone with nodules of chert.

COPROLITE BED.

The so-called coprolite bed, found near Farnham, in Surrey, and near Cambridge, contains nodules of phosphate of lime in such abundance as to be largely worked for the manufacture of artificial manure. It belongs to the upper part of the Chloritic series, and is doubtless chiefly of animal origin, and may perhaps be partly coprolitic, derived from the excrement of fish and reptiles. The late Mr. Barrett discovered in it, near Cambridge, in 1858, the remains of a bird, which was rather larger than the common pigeon, and probably of the order Natatores, and which, like most of the Gull tribe, had well-developed wings. Portions of the metacarpus, metatarsus, tibia, and femur have been detected, and the determinations of Mr. Barrett have been confirmed by Professor Owen.

This phosphatic bed in the suburbs of Cambridge must have been formed partly by the denudation of pre-existing rocks, mostly of Cretaceous age. The fossil shells and bones of animals washed out of these denuded strata, now forming a layer only a few feet thick, have yielded a rich harvest to the collector. A large Rudist of the genus Radiolite, no less than two feet in height, may be seen in the Cambridge Museum, obtained from this bed. The number of reptilian remains, all apparently of Cretaceous age, is truly surprising; more than ten species of Pterodactyl, five or six of Ichthyosaurus, one of Pliosaurus, one of Dinosaurus, eight of Chelonians, besides other forms, having been recognised.

The chloritic sand is regarded by many geologists as a littoral deposit of the Chalk Ocean, and therefore contemporaneous with part of the chalk marl, and even, perhaps, with some part of the white chalk. For, as the land went on sinking, and the cretaceous sea widened its area, white mud and chloritic sand were always forming somewhere, but the line of sea-shore was perpetually shifting its position. Hence, though both sand and mud originated simultaneously, the one near the land, the other far from it, the sands in every locality where a shore became submerged might constitute the underlying deposit.

(FIGURE 267. Ostrea columba. Syn. Gryphaea columba. Chloritic sand.)

(FIGURE 268. Ostrea carinata. Chalk marl and chloritic sand. Neocomian.)

(FIGURE 269. Terebrirostra lyra, Sowerby. Chloritic sand.)

(FIGURE 270. Pecten 5-costatus. White chalk and chloritic sand. Neocomian.)

(FIGURE 271. Plagiostoma Hoperi, Sowerby. Syn. Lima Hoperi. White chalk and chloritic sand.)

Among the characteristic mollusca of the chloritic sand may be mentioned Terebrirostra lyra (Figure 269), Plagiostoma Hoperi (Figure 271), Pecten quinque-costatus (Figure 270), and Ostrea columba (Figure 267).

The Cephalopoda are abundant, among which 40 species of Ammonites are now known, 10 being peculiar to this subdivision, and the rest common to the beds immediately above or below.

GAULT.

(FIGURE 272. Ancyloceras spinigerum, d'Orb. Syn. Hamites spiniger, Sowerby. Near Folkestone. Gault.)

The lowest member of the Upper Cretaceous group, usually about 100 feet thick in the S.E. of England, is provincially termed Gault. It consists of a dark blue marl, sometimes intermixed with green sand. Many peculiar forms of cephalopoda, such as the Hamite (Figure 272), and Scaphite, with other fossils, characterise this formation, which, small as is its thickness, can be traced by its organic remains to distant parts of Europe, as, for example, to the Alps.

Twenty-one species of British Ammonites are recorded as found in the Gault, of which only eight are peculiar to it, ten being common to the overlying Chloritic series.

CONNECTION BETWEEN UPPER AND LOWER CRETACEOUS STRATA.— BLACKDOWN BEDS.

The break between the Upper and Lower Cretaceous formations will be appreciated when it is stated that, although the Neocomian contains 31 species of Ammonite, and the Gault, as we have seen, 21, there are only three of those common to both divisions. Nevertheless, we may expect the discovery in England, and still more when we extend our survey to the Continent, of beds of passage intermediate between the Upper and Lower Cretaceous. Even now the Blackdown beds in Devonshire, which rest immediately on Triassic strata, and which evidently belong to some part of the Cretaceous series, have been referred by some geologists to the Upper group, by others to the Lower or Neocomian. They resemble the Folkestone beds of the latter series in mineral character, and 59 out of 156 of their fossil mollusca are common to them; but they have also 16 species common to the Gault, and 20 to the overlying Chloritic series; and what is very important, out of seven Ammonites six are found also in the Gault and Chloritic series, only one being peculiar to the Blackdown beds.

Professor Ramsay has remarked that there is a stratigraphical break; for in Kent, Surrey, and Sussex, at those few points where there are exposures of junctions of the Gault and Neocomian, the surface of the latter has been much eroded or denuded, while to the westward of the great chalk escarpment the unconformability of the two groups is equally striking. At Blackdown this unconformability is still more marked, for though distant only 100 miles from Kent and Surrey, no formation intervenes between these beds and the Trias; all intermediate groups, such as the Lower Neocomian and Oolite, having either not been deposited or destroyed by denudation.

FLORA OF THE UPPER CRETACEOUS PERIOD.

As the Upper Cretaceous rocks of Europe are, for the most part, of purely marine origin, and formed in deep water usually far from the nearest shore, land-plants of this period, as we might naturally have anticipated, are very rarely met with. In the neighbourhood of Aix-la-Chapelle, however, an important exception occurs, for there certain white sands and laminated clays, 400 feet in thickness, contain the remains of terrestrial plants in a beautiful state of preservation. These beds are the equivalents of the white chalk and chalk marl of England, or Senonien of d'Orbigny, although the white siliceous sands of the lower beds, and the green grains in the upper part of the formation, cause it to differ in mineral character from our white chalk.

Beds of fine clay, with fossil plants, and with seams of lignite, and even perfect coal, are intercalated. Floating wood, containing perforating shells, such as Pholas and Gastrochoena, occur. There are likewise a few beds of a yellowish-brown limestone, with marine shells, which enable us to prove that the lowest and highest plant-beds belong to one group. Among these shells are Pecten quadricostatus, and several others which are common to the upper and lower part of the series, and Trigonia limbata, d'Orbigny, a shell of the white chalk. On the whole, the organic remains and the geological position of the strata prove distinctly that in the neighbourhood of Aix-la-Chapelle a gulf of the ancient Cretaceous sea was bounded by land composed of Devonian rocks. These rocks consisted of quartzose and schistose beds, the first of which supplied white sand and the other argillaceous mud to a river which entered the sea at this point, carrying down in its turbid waters much drift-wood and the leaves of plants. Occasionally, when the force of the river abated, marine shells of the genera Trigonia, Turritella, Pecten, etc., established themselves in the same area, and plants allied to Zostera and Fucus grew on the bottom.

The fossil plants of this member of the upper chalk at Aix have been diligently collected and studied by Dr. Debey, and as they afford the only example yet known of a terrestrial flora older than the Eocene, in which the great divisions of the vegetable kingdom are represented in nearly the same proportions as in our own times, they deserve particular attention. Dr. Debey estimates the number of species as amounting to more than two hundred, of which sixty-seven are cryptogamous, chiefly ferns, twenty species of which can be well determined, most of them being in fructification. The scars on the bark of one or two are supposed to indicate tree-ferns. Of thirteen genera three are still existing, namely, Gleichenia, now inhabiting the Cape of Good Hope, and New Holland; Lygodium, now spread extensively through tropical regions, but having some species which live in Japan and North America; and Asplenium, a cosmopolite form. Among the phaenogamous plants, the Conifers are abundant, the most common belonging to a genus called Cycadopteris by Debey, and hardly separable from Sequoia (or Wellingtonia), of which both the cones and branches are preserved. When I visited Aix, I found the silicified wood of this plant very plentifully dispersed through the white sands in the pits near that city. In one silicified trunk 200 rings of annual growth could be counted. Species of Araucaria like those of Australia are also found. Cycads are extremely rare, and of Monocotyledons there are but few. No palms have been recognised with certainty, but the genus Pandanus, or screw pine, has been distinctly made out. The number of the Dicotyledonous Angiosperms is the most striking feature in so ancient a flora.

(In this and subsequent remarks on fossil plants I shall often use Dr. Lindley's terms, as most familiar in this country; but as those of M. A. Brongniart are much cited, it may be useful to geologists to give a table explaining the corresponding names of groups so much spoken of in palaeontology.

COLUMN 1. BRONGNIART.
COLUMN 2. LINDLEY.
COLUMN 3. EXAMPLES.
CRYPTOGAMIC.

1. Cryptogamous amphigens, or cellular cryptogamic: Thallogens: Lichens, seaweeds, fungi.

2. Cryptogamous acrogens: Acrogens: Mosses, equisetums, ferns, lycopodiums—Lepidodendra.

PHAENEROGAMIC.

3. Dicotyledonous gymnosperms: Gymnogens: Conifers and Cycads.

4. Dicotyledonous Angiosperms: Exogens: Compositae, leguminosae, umbelliferae, cruciferae, heaths, etc. All native European trees except conifers.

5. Monocotyledons: Endogens. Palms, lilies, aloes, rushes, grasses, etc.)

Among them we find the familiar forms of the Oak, Fig, and Walnut (Quercus, Ficus, and Juglans), of the last both the nuts and leaves; also several genera of the Myrtaceae. But the predominant order is the Proteaceae, of which there are between sixty and seventy supposed species, many of extinct genera, but some referred to the following living forms—Dryandra, Grevillea, Hakea, Banksia, Persoonia— all now belonging to Australia, and Leucospermum, species of which form small bushes at the Cape.

The epidermis of the leaves of many of these Aix plants, especially of the Proteaceae, is so perfectly preserved in an envelope of fine clay, that under the microscope the stomata, or polygonal cellules, can be detected, and their peculiar arrangement is identical with that known to characterise some living Proteaceae (Grevillea, for example). Although this peculiarity of the structure of stomata is also found in plants of widely distant orders, it is, on the whole, but rarely met with, and being thus observed to characterise a foliage previously suspected to be proteaceous, it adds to the probability that the botanical evidence had been correctly interpreted.

An occasional admixture at Aix-la-Chapelle of Fucoids and Zosterites attests, like the shells, the presence of salt-water. Of insects, Dr. Debey has obtained about ten species of the families Curculionidae and Carabidae.

The resemblance of the flora of Aix-la-Chapelle to the tertiary and living floras in the proportional number of dicotyledonous angiosperms as compared to the gymnogens, is a subject of no small theoretical interest, because we can now affirm that these Aix plants flourished before the rich reptilian fauna of the secondary rocks had ceased to exist. The Ichthyosaurus, Pterodactyl, and Mosasaurus were of coeval date with the oak, the walnut, and the fig. Speculations have often been hazarded respecting a connection between the rarity of Exogens in the older rocks and a peculiar state of the atmosphere. A denser air, it was suggested, had in earlier times been alike adverse to the well-being of the higher order of flowering plants, and of the quick-breathing animals, such as mammalia and birds, while it was favourable to a cryptogamic and gymnospermous flora, and to a predominance of reptile life. But we now learn that there is no incompatibility in the co-existence of a vegetation like that of the present globe, and some of the most remarkable forms of the extinct reptiles of the age of gymnosperms.

If the passage seem at present to be somewhat sudden from the flora of the Lower or Neocomian to that of the Upper Cretaceous period, the abruptness of the change will probably disappear when we are better acquainted with the fossil vegetation of the uppermost beds of the Neocomian and that of the lowest strata of the Gault or true Cretaceous series.

HIPPURITE LIMESTONE.— DIFFERENCE BETWEEN THE CHALK OF THE NORTH AND SOUTH OF EUROPE.

(FIGURE 273. Map of part of S.W. France, from the Loire river to the Pyrenees.)

By the aid of the three tests, superposition, mineral character, and fossils, the geologist has been enabled to refer to the same Cretaceous period certain rocks in the north and south of Europe, which differ greatly both in their fossil contents and in their mineral composition and structure.

If we attempt to trace the cretaceous deposits from England and France to the countries bordering the Mediterranean, we perceive, in the first place, that in the neighbourhood of London and Paris they form one great continuous mass, the Straits of Dover being a trifling interruption, a mere valley with chalk cliffs on both sides. We then observe that the main body of the chalk which surrounds Paris stretches from Tours to near Poitiers (see Figure 273, in which the shaded part represents chalk).

Between Poitiers and La Rochelle, the space marked A on the map separates two regions of chalk. This space is occupied by the Oolite and certain other formations older than the Chalk and Neocomian, and has been supposed by M. E. de Beaumont to have formed an island in the Cretaceous sea. South of this space we again meet with rocks which we at once recognise to be cretaceous, partly from the chalky matrix and partly from the fossils being very similar to those of the white chalk of the north: especially certain species of the genera Spatangus, Ananchytes, Cidarites, Nucula, Ostrea, Gryphaea (Exogyra), Pecten, Plagiostoma (Lima), Trigonia, Catillus (Inoceramus), and Terebratula. (d'Archiac, Sur la form. Cretacee du S.-O. de la France Mem. de la Soc. Geol. de France tome 2.) But Ammonites, as M. d'Archiac observes, of which so many species are met with in the chalk of the north of France, are scarcely ever found in the southern region; while the genera Hamite, Turrilite, and Scaphite, and perhaps Belemnite, are entirely wanting.

(FIGURE 274. Radiolites radiosa, d'Orbigny. White chalk of France. b. Upper valve of same.)

(FIGURE 275. Radiolites foliaceus, d'Orbigny. Syn. Sphaerulites agarici-formis, Blainv. White chalk of France.)

(FIGURE 276. Hippurites organisans, Desmoulins. Upper chalk:— chalk marl of Pyrenees? (d'Orbigny's Palaeontologie francaise plate 533.) a. Young individual; when full grown they occur in groups adhering laterally to each other. b. Upper side of the upper valve, showing a reticulated structure in those parts, b, where the external coating is worn off. c. Upper end or opening of the lower and cylindrical valve. d. Cast of the interior of the lower conical valve.)

On the other hand, certain forms are common in the south which are rare or wholly unknown in the north of France. Among these may be mentioned many Hippurites, Sphaerulites, and other members of that great family of mollusca called Rudistes by Lamarck, to which nothing analogous has been discovered in the living creation, but which is quite characteristic of rocks of the Cretaceous era in the south of France, Spain, Sicily, Greece, and other countries bordering the Mediterranean. The species called Hippurites organisans (Figure 276) is more abundant than any other in the south of Europe; and the geologist should make himself well acquainted with the cast of the interior, d, which is often the only part preserved in many compact marbles of the Upper Cretaceous period. The flutings on the interior of the Hippurite, which are represented on the cast by smooth, rounded longitudinal ribs, and in some individuals attain a great size and length, are wholly unlike the markings on the exterior of the shell.

CRETACEOUS ROCKS IN THE UNITED STATES.

If we pass to the American continent, we find in the State of New Jersey a series of sandy and argillaceous beds wholly unlike in mineral character to our Upper Cretaceous system; which we can, nevertheless, recognise as referable, palaeontologically, to the same division.

That they were about the same age generally as the European chalk and Neocomian, was the conclusion to which Dr. Morton and Mr. Conrad came after their investigation of the fossils in 1834. The strata consist chiefly of green sand and green marl, with an overlying coralline limestone of a pale yellow colour, and the fossils, on the whole, agree most nearly with those of the Upper European series, from the Maestricht beds to the Gault inclusive. I collected sixty shells from the New Jersey deposits in 1841, five of which were identical with European species— Ostrea larva, O. vesicularis, Gryphaea costata, Pecten quinque-costatus, Belemnitella mucronata. As some of these have the greatest vertical range in Europe, they might be expected more than any others to recur in distant parts of the globe. Even where the species were different, the generic forms, such as the Baculite and certain sections of Ammonites, as also the Inoceramus (see above, Figure 252) and other bivalves, have a decidedly cretaceous aspect. Fifteen out of the sixty shells above alluded to were regarded by Professor Forbes as good geographical representatives of well-known cretaceous fossils of Europe. The correspondence, therefore, is not small, when we reflect that the part of the United States where these strata occur is between 3000 and 4000 miles distant from the chalk of Central and Northern Europe, and that there is a difference of ten degrees in the latitude of the places compared on opposite sides of the Atlantic. Fish of the genera Lamna, Galeus, and Carcharodon are common to New Jersey and the European cretaceous rocks. So also is the genus Mosasaurus among reptiles.

It appears from the labours of Dr. Newberry and others, that the Cretaceous strata of the United States east and west of the Appalachians are characterised by a flora decidedly analogous to that of Aix-la-Chapelle above-mentioned, and therefore having considerable resemblance to the vegetation of the Tertiary and Recent Periods.

CHAPTER XVIII.
LOWER CRETACEOUS OR NEOCOMIAN FORMATION.

Classification of marine and fresh-water Strata.
Upper Neocomian.
Folkestone and Hythe Beds.
Atherfield Clay.
Similarity of Conditions causing Reappearance of Species after short Intervals.
Upper Speeton Clay.
Middle Neocomian.
Tealby Series.
Middle Speeton Clay.
Lower Neocomian.
Lower Speeton Clay.

Wealden Formation.
Fresh-water Character of the Wealden.
Weald Clay.
Hastings Sands.
Punfield Beds of Purbeck, Dorsetshire.
Fossil Shells and Fish of the Wealden.
Area of the Wealden.
Flora of the Wealden.

We now come to the Lower Cretaceous Formation which was formerly called Lower Greensand, and for which it will be useful for reasons before explained (Chapter 17) to use the term "Neocomian."

TABLE 18.1. LOWER CRETACEOUS OR NEOCOMIAN GROUP. COLUMN 1: MARINE. COLUMN 2: FRESH-WATER.

1. Upper Neocomian— Greensand of Folkestone, Sandgate, and Hythe, Atherfield clay, upper part of Speeton clay: Part of Wealden beds of Kent, Surrey, Sussex, Hants, and Dorset.

2. Middle Neocomian— Punfield Marine bed, Tealby beds, middle part of Speeton clay: Part of Wealden beds of Kent, Surrey, Sussex, Hants, and Dorset.

3. Lower Neocomian— Lower part of Speeton clay: Part of Wealden beds of Kent, Surrey, Sussex, Hants, and Dorset.

In Western France, the Alps, the Carpathians, Northern Italy, and the Apennines, an extensive series of rocks has been described by Continental geologists under the name of Tithonian. These beds, which are without any marine equivalent in this country, appear completely to bridge over the interval between the Neocomian and the Oolites. They may, perhaps, as suggested by Mr. Judd, be of the same age as part of the Wealden series.

UPPER NEOCOMIAN.
FOLKSTONE AND HYTHE BEDS.

(FIGURE 277. Nautilus plicatus, Sowerby, in Fitton's Monog.)
(FIGURE 278. Ancyloceras gigas, d'Orbigny.)
(FIGURE 279. Gervillia anceps, Desh. Upper Neocomian, Surrey.)
(FIGURE 280. Trigonia caudata, Agassiz. Upper Neocomian.)
(FIGURE 281. Terebratula sella, Sowerby. Upper Neocomian, Hythe.)
(FIGURE 282. Diceras Lonsdalii. Upper Neocomian, Wilts. a. The bivalve shell. b. Cast of one of the valves enlarged.)

The sands which crop out beneath the Gault in Wiltshire, Surrey, and Sussex are sometimes in the uppermost part pure white, at others of a yellow and ferruginous colour, and some of the beds contain much green matter. At Folkestone they contain layers of calcareous matter and chert, and at Hythe, in the neighbourhood, as also at Maidstone and other parts of Kent, the limestone called Kentish Rag is intercalated. This somewhat clayey and calcareous stone forms strata two feet thick, alternating with quartzose sand. The total thickness of these Folkestone and Hythe beds is less than 300 feet, and they are seen to rest immediately on a grey clay, to which we shall presently allude as the Atherfield clay. Among the fossils of the Folkestone and Hythe beds we may mention Nautilus plicatus (Figure 277), Ancyloceras (Scaphites) gigas (Figure 278), which has been aptly described as an Ammonite more or less uncoiled; Trigonia caudata (Figure 280), Gervillia anceps (Figure 279), a bivalve genus allied to Avicula, and Terebratula sella (Figure 281). In ferruginous beds of the same age in Wiltshire is found a remarkable shell called Diceras Lonsdalii (Figure 282), which abounds in the Upper and Middle Neocomian of Southern Europe. This genus is closely allied to Chama, and the cast of the interior has been compared to the horns of a goat.

ATHERFIELD CLAY.

We mentioned before that the Folkestone and Hythe series rest on a grey clay. This clay is only of slight thickness in Kent and Surrey, but acquires great dimensions at Atherfield, in the Isle of Wight. The difference, indeed, in mineral character and thickness of the Upper Neocomian formation near Folkestone, and the corresponding beds in the south of the Isle of Wight, about 100 miles distant, is truly remarkable. In the latter place we find no limestone answering to the Kentish Rag, and the entire thickness from the bottom of the

Atherfield clay to the top of the Neocomian, instead of being less than 300 feet as in Kent, is given by the late Professor E. Forbes as 843 feet, which he divides into sixty-three strata, forming three groups. The uppermost of these consists of ferruginous sands, the second of sands and clay, and the third or lowest of a brown clay, abounding in fossils.

Pebbles of quartzose sandstone, jasper, and flinty slate, together with grains of chlorite and mica, and, as Mr. Godwin-Austen has shown, fragments and water-worn fossils of the oolitic rocks, speak plainly of the nature of the pre-existing formations, by the wearing down of which the Neocomian beds were formed. The land, consisting of such rocks, was doubtless submerged before the origin of the white chalk, a deposit which was formed in a more open sea, and in clearer waters.

(FIGURE 283. Perna Mulleti, Desh. One-eighth natural size. a. Exterior. b. Part of hinge-line of upper or right valve.)

Among the shells of the Atherfield clay the biggest and most abundant shell is the large Perna Mulleti, of which a reduced figure is given in Figure 283.

SIMILARITY OF CONDITIONS CAUSING REAPPEARANCE OF SPECIES.

Some species of mollusca and other fossils range through the whole series, while others are confined to particular subdivisions, and Forbes laid down a law which has since been found of very general application in regard to estimating the chronological relations of consecutive strata. Whenever similar conditions, he says, are repeated, the same species reappear, provided too great a lapse of time has not intervened; whereas if the length of the interval has been geologically great, the same genera will reappear represented by distinct species. Changes of depth, or of the mineral nature of the sea-bottom, the presence or absence of lime or of peroxide of iron, the occurrence of a muddy, or a sandy, or a gravelly bottom, are marked by the banishment of certain species and the predominance of others. But these differences of conditions being mineral, chemical, and local in their nature, have no necessary connection with the extinction, throughout a large area, of certain animals or plants. When the forms proper to loose sand or soft clay, or to perfectly clear water, or to a sea of moderate or great depth, recur with all the same species, we may infer that the interval of time has been, geologically speaking, small, however dense the mass of matter accumulated. But if, the genera remaining the same, the species are changed, we have entered upon a new period; and no similarity of climate, or of geographical and local conditions, can then recall the old species which a long series of destructive causes in the animate and inanimate world has gradually annihilated.

SPEETON CLAY, UPPER DIVISION.

(FIGURE 284. Ammonites Deshayesii, Leym. Upper Neocomian.)

On the coast, beneath the white chalk of Flamborough Head, in Yorkshire, an argillaceous formation crops out, called the Speeton clay, several hundred feet in thickness, the palaeontological relations of which have been ably worked out by Mr. John W. Judd, and he has shown that it is separable into three divisions, the uppermost of which, 150 feet thick, and containing 87 species of mollusca, decidedly belongs to the Atherfield clay and associated strata of Hythe and Folkestone, already described. (Judd, Speeton clay, Quarterly Geological Journal volume 24 1868 page 218.) It is characterised by the Perna Mulleti (Figure 283) and Terebratula sella (Figure 281), and by Ammonites Deshayesii (Figure 284), a well-known Hythe fossil. Fine skeletons of reptiles of the genera Pliosaurus and Teleosaurus have been obtained from this clay. At the base of this upper division of the Speeton clay there occurs a layer of large Septaria, formerly worked for the manufacture of cement. This bed is crowded with fossils, especially Ammonites, one species of which, three feet in diameter, was observed by Mr. Judd.

MIDDLE NEOCOMIAN.
TEALBY SERIES.

(FIGURE 285. Pecten cinctus, Sowerby. (P. crassitesta, Rom.) Middle Neocomian, England; Middle and Lower Neocomian, Germany. One-fifth natural size.)

(FIGURE 286. Ancyloceras (Crioceras) Duvallei, Leveille. Middle and Lower Neocomian. One-fifth natural size.)

At Tealby, a village in the Lincolnshire Wolds, there crop out beneath the white chalk some non-fossiliferous ferruginous sands about twenty-feet thick, beneath which are beds of clay and limestone, about fifty feet thick, with an interesting suite of fossils, among which are Pecten cinctus (Figure 285), from 9 to 12 inches in diameter, Ancyloceras Duvallei (Figure 286), and some forty other shells, many of them common to the Middle Speeton clay, about to be mentioned. Mr. Judd remarks that as Ammonites clypei-formis and Terebratula hippopus characterise the Middle Neocomian of the Continent, it is to this stage that the Tealby series containing the same fossils may be assigned. (Judd Quarterly Geological Journal 1867 volume 23 page 249.)

The middle division of the Speeton clay, occurring at Speeton below the cement-bed, before alluded to, is 150 feet thick, and contains about 39 species of mollusca, half of which are common to the overlying clay. Among the peculiar shells, Pecten cinctus (Figure 285) and Ancyloceras (Crioceras) Duvallei (Figure 286) occur.

LOWER NEOCOMIAN.

(FIGURE 287. Ammonites Noricus, Schloth. Lower Neocomian, Speeton.)

In the lower division of the Speeton clay, 200 feet thick, 46 species of mollusca have been found, and three divisions, each characterised by its peculiar ammonite, have been noticed by Mr. Judd. The central zone is marked by Ammonites Noricus (see Figure 287). On the Continent these beds are well-known by their corresponding fossils, the Hils clay and conglomerate of the north of Germany agreeing with the Middle and Lower Speeton, the latter of which, with the same mineral characters and fossils as in Yorkshire, is also found in the little island of Heligoland. Yellow limestone, which I have myself seen near Neuchatel, in Switzerland, represents the Lower Neocomian at Speeton.

WEALDEN FORMATION.

Beneath the Atherfield clay or Upper Neocomian of the S.E. of England, a fresh-water formation is found, called the Wealden, which, although it occupies a small horizontal area in Europe, as compared to the White Chalk and the marine Neocomian beds, is nevertheless of great geological interest, since the imbedded remains give us some insight into the nature of the terrestrial fauna and flora of the Lower Cretaceous epoch. The name of Wealden was given to this group because it was first studied in parts of Kent, Surrey, and Sussex, called the Weald; and we are indebted to Dr. Mantell for having shown, in 1822, in his "Geology of Sussex," that the whole group was of fluviatile origin. In proof of this he called attention to the entire absence of Ammonites, Belemnites, Brachiopoda, Echinodermata, Corals, and other marine fossils, so characteristic of the Cretaceous rocks above, and of the Oolitic strata below, and to the presence in the Weald of Paludinae, Melaniae, Cyrenae, and various fluviatile shells, as well as the bones of terrestrial reptiles and the trunks and leaves of land-plants.

(FIGURE 288. Section from (left) W.S.W. through Brixton bay, Isle of Wight, Solent and South Downs to E.N.E. (right). 1. Tertiary. 2. Chalk and Gault. 3. Upper Neocomian (or Lower Greensand). 4. Wealden (Weald Clay and Hastings Sands).)

The evidence of so unexpected a fact as that of a dense mass of purely fresh-water origin underlying a deep-sea deposit (a phenomenon with which we have since become familiar) was received, at first, with no small doubt and incredulity. But the relative position of the beds is unequivocal; the Weald Clay being distinctly seen to pass beneath the Atherfield Clay in various parts of Surrey, Kent, and Sussex, and to reappear in the Isle of Wight at the base of the Cretaceous series, being, no doubt, continuous far beneath the surface, as indicated by the dotted lines in Figure 288. They are also found occupying the same relative position below the chalk in the peninsula of Purbeck, Dorsetshire, where, as we shall see in the sequel, they repose on strata referable to the Upper Oolite.

WEALD CLAY.

The Upper division, or Weald Clay, is, in great part, of fresh-water origin, but in its highest portion contains beds of oysters and other marine shells which indicate fluvio-marine conditions. The uppermost beds are not only conformable, as Dr. Fitton observes, to the inferior strata of the overlying Neocomian, but of similar mineral composition. To explain this, we may suppose that, as the delta of a great river was tranquilly subsiding, so as to allow the sea to encroach upon the space previously occupied by fresh-water, the river still continued to carry down the same sediment into the sea. In confirmation of this view it may be stated that the remains of the Iguanodon Mantelli, a gigantic terrestrial reptile, very characteristic of the Wealden, has been discovered near Maidstone, in the overlying Kentish Rag, or marine limestone of the Upper Neocomian. Hence we may infer that some of the saurians which inhabited the country of the great river continued to live when part of the district had become submerged beneath the sea. Thus, in our own times, we may suppose the bones of large alligators to be frequently entombed in recent fresh-water strata in the delta of the Ganges. But if part of that delta should sink down so as to be covered by the sea, marine formations might begin to accumulate in the same space where fresh-water beds had previously been formed; and yet the Ganges might still pour down its turbid waters in the same direction, and carry seaward the carcasses of the same species of alligator, in which case their bones might be included in marine as well as in subjacent fresh-water strata.

(FIGURES 289 AND 290. Tooth of Iguanodon Mantelli.

(FIGURE 289. a, and b.)

(FIGURE 290. A. Partially worn tooth of young individual of the same. b. Crown of tooth in adult worn down. (Mantell.)))

The Iguanodon, first discovered by Dr. Mantell, was an herbivorous reptile, of which the teeth, though bearing a great analogy, in their general form and crenated edges (see Figure 289 a and b), to the modern Iguanas which now frequent the tropical woods of America and the West Indies, exhibit many important differences. It appears that they have often been worn by the process of mastication; whereas the existing herbivorous reptiles clip and gnaw off the vegetable productions on which they feed, but do not chew them. Their teeth frequently present an appearance of having been chipped off, but never, like the fossil teeth of the Iguanodon, have a flat ground surface (see Figure 290, b) resembling the grinders of herbivorous mammalia. Dr. Mantell computes that the teeth and bones of this species which passed under his examination during twenty years must have belonged to no less than seventy-one distinct individuals, varying in age and magnitude from the reptile just burst from the egg, to one of which the femur measured twenty-four inches in circumference. Yet, notwithstanding that the teeth were more numerous than any other bones, it is remarkable that it was not until the relics of all these individuals had been found, that a solitary example of part of a jaw-bone was obtained. Soon afterwards remains both of the upper and lower jaw were met with in the Hastings beds in Tilgate Forest, near Cuckfield. In the same sands at Hastings, Mr. Beckles found large tridactyle impressions which it is conjectured were made by the hind feet of this animal, on which it is ascertained that there were only three well-developed toes.

(FIGURE 291. Cypris spinigera, Fitton.)

(FIGURE 292. Weald clay with Cyprides.)

Occasionally bands of limestone, called Sussex Marble, occur in the Weald Clay, almost entirely composed of a species of Paludina, closely resembling the common P. vivipara of English rivers. Shells of the Cypris, a genus of Crustaceans mentioned in Chapter 3 as abounding in lakes and ponds, are also plentifully scattered through the clays of the Wealden, sometimes producing, like plates of mica, a thin lamination (see Figure 292).

HASTINGS SANDS.

This lower division of the Wealden consists of sand, sandstone, calciferous grit, clay, and shale; the argillaceous strata, notwithstanding the name, predominating somewhat over the arenaceous, as will be seen by reference to the following table, drawn up by Messrs. Drew and Foster, of the Geological Survey of Great Britain:

TABLE 18.1. SUBORDINATE FORMATIONS IN THE HASTINGS SAND. COLUMN 1: NAME OF SUBORDINATE FORMATION. COLUMN 2: MINERAL COMPOSITION OF THE STRATA. COLUMN 3: THICKNESS IN FEET.

Tunbridge Wells Sand: Sandstone and loam: 150.

Wadhurst Clay: Blue and brown shale and clay, with a little calc-grit: 100.

Ashdown Sand: Hard sand, with some beds of calc-grit: 160.

Ashburnham Beds: Mottled white and red clay, with some sandstone: 330.

The picturesque scenery of the "High Rocks" and other places in the neighbourhood of Tunbridge Wells is caused by the steep natural cliffs, to which a hard bed of white sand, occurring in the upper part of the Tunbridge Wells Sand, mentioned in the above table, gives rise. This bed of "rock-sand" varies in thickness from 25 to 48 feet. Large masses of it, which were by no means hard or capable of making a good building-stone, form, nevertheless, projecting rocks with perpendicular faces, and resist the degrading action of the river because, says Mr. Drew, they present a solid mass without planes of division. The calcareous sandstone and grit of Tilgate Forest, near Cuckfield, in which the remains of the Iguanodon and Hylaeosaurus were first found by Dr. Mantell, constitute an upper member of the Tunbridge Wells Sand, while the "sand-rock" of the Hastings cliffs, about 100 feet thick, is one of the lower members of the same. The reptiles, which are very abundant in this division, consist partly of saurians, referred by Owen and Mantell to eight genera, among which, besides those already enumerated, we find the Megalosaurus and Plesiosaurus. The Pterodactyl also, a flying reptile, is met with in the same strata, and many remains of Chelonians of the genera Trionyx and Emys, now confined to tropical regions.

(FIGURE 293. Lepidotus Mantelli, Agassiz. Wealden. a. Palate and teeth. b. Side view of teeth. c. Scale.)

The fishes of the Wealden are chiefly referable to the Ganoid and Placoid orders. Among them the teeth and scales of Lepidotus are most widely diffused (see Figure 293). These ganoids were allied to the Lepidosteus, or Gar-pike, of the American rivers. The whole body was covered with large rhomboidal scales, very thick, and having the exposed part coated with enamel. Most of the species of this genus are supposed to have been either river-fish, or inhabitants of the sea at the mouth of estuaries.

(FIGURE 294. Unio Valdensis, Mant. Isle of Wight and Dorsetshire; in the lower beds of the Hastings Sands. a, b.)

(FIGURE 295. Underside of slab of sandstone about one yard in diameter. Stammerham, Sussex.)

At different heights in the Hastings Sands, we find again and again slabs of sandstone with a strong ripple-mark, and between these slabs beds of clay many yards thick. In some places, as at Stammerham, Horsham, near there, are indications of this clay having been exposed so as to dry and crack before the next layer was thrown down upon it. The open cracks in the clay have served as moulds, of which casts have been taken in relief, and which are, therefore, seen on the lower surface of the sandstone (see Figure 295).

(FIGURE 296. Sphenopteris gracilis, Fitton. From the Hastings Sands near Tunbridge Wells. a. A portion of the same magnified.)

Near the same place a reddish sandstone occurs in which are innumerable traces of a fossil vegetable, apparently Sphenopteris, the stems and branches of which are disposed as if the plants were standing erect on the spot where they originally grew, the sand having been gently deposited upon and around them; and similar appearances have been remarked in other places in this formation. (Mantell Geology of S.E. of England page 244.) In the same division also of the Wealden, at Cuckfield, is a bed of gravel or conglomerate, consisting of water- worn pebbles of quartz and jasper, with rolled bones of reptiles. These must have been drifted by a current, probably in water of no great depth.

From such facts we may infer that, notwithstanding the great thickness of this division of the Wealden, the whole of it was a deposit in water of a moderate depth, and often extremely shallow. This idea may seem startling at first, yet such would be the natural consequence of a gradual and continuous sinking of the ground in an estuary or bay, into which a great river discharged its turbid waters. By each foot of subsidence, the fundamental rock would be depressed one foot farther from the surface; but the bay would not be deepened, if newly- deposited mud and sand should raise the bottom one foot. On the contrary, such new strata of sand and mud might be frequently laid dry at low water, or overgrown for a season by a vegetation proper to marshes.

PUNFIELD BEDS, BRACKISH AND MARINE.

(FIGURE 297. Vicarya Lujani, De Verneuil (Foss. de Utrillas.) Wealden, Punfield. a. Nearly perfect shell.

b. Vertical section of smaller specimen, showing continuous ridges as in Nerinaea.)

The shells of the Wealden beds belong to the genera Melanopsis, Melania, Paludina, Cyrena, Cyclas, Unio (see Figure 294), and others, which inhabit rivers or lakes; but one band has been found at Punfield, in Dorsetshire, indicating a brackish state of the water, where the genera Corbula, Mytilus, and Ostrea occur; and in some places this bed becomes purely marine, containing some well-known Neocomian fossils, among which Ammonites Deshayesii (Figure 284) may be mentioned. Others are peculiar as British, but very characteristic of the Upper and Middle Neocomian of Spain, and among these the Vicarya Lujani (Figure 297), a shell allied to Nerinea, is conspicuous.

By reference to Table 18.1 it will be seen that the Wealden beds are given as the fresh-water equivalents of the Marine Neocomian. The highest part of them in England may, for reasons just given, be regarded as Upper Neocomian, while some of the inferior portions may correspond in age to the Middle and Lower divisions of that group. In favour of this latter view, M. Marcou mentions that a fish called Asteracanthus granulosus, occurring in the Tilgate beds, is characteristic of the lowest beds of the Neocomian of the Jura, and it is well known that Corbula alata, common in the Ashburnham beds, is found also at the base of the Neocomian of the Continent.

AREA OF THE WEALDEN.

In regard to the geographical extent of the Wealden, it can not be accurately laid down, because so much of it is concealed beneath the newer marine formations. It has been traced about 320 English miles from west to east, from the coast of Dorsetshire to near Boulogne, in France; and nearly 200 miles from north-west to south-east, from Surrey and Hampshire to Vassy, in France. If the formation be continuous throughout this space, which is very doubtful, it does not follow that the whole was contemporaneous; because, in all likelihood, the physical geography of the region underwent frequent changes throughout the whole period, and the estuary may have altered its form, and even shifted its place. Dr. Dunker, of Cassel, and H. von Meyer, in an excellent monograph on the Wealdens of Hanover and Westphalia, have shown that they correspond so closely, not only in their fossils, but also in their mineral characters, with the English series, that we can scarcely hesitate to refer the whole to one great delta. Even then, the magnitude of the deposit may not exceed that of many modern rivers. Thus, the delta of the Quorra or Niger, in Africa, stretches into the interior for more than 170 miles, and occupies, it is supposed, a space of more than 300 miles along the coast, thus forming a surface of more than 25,000 square miles, or equal to about one-half of England. (Fitton Geology of Hastings page 58, who cites Lander's Travels.) Besides, we know not, in such cases, how far the fluviatile sediment and organic remains of the river and the land may be carried out from the coast, and spread over the bed of the sea. I have shown, when treating of the Mississippi, that a more ancient delta, including species of shells such as now inhabit Louisiana, has been upraised, and made to occupy a wide geographical area, while a newer delta is forming; and the possibility of such movements and their effects must not be lost sight of when we speculate on the origin of the Wealden. (See Chapter 6 and Second Visit to the United States volume 2 chapter 34.)

It may be asked where the continent was placed, from the ruins of which the Wealden strata were derived, and by the drainage of which a great river was fed. If the Wealden was gradually going downward 1000 feet or more perpendicularly, a large body of fresh-water would not continue to be poured into the sea at the same point. The adjoining land, if it participated in the movement, could not escape being submerged. But we may suppose such land to have been stationary, or even undergoing contemporaneous slow upheaval. There may have been an ascending movement in one region, and a descending one in a contiguous parallel zone of country. But even if that were the case, it is clear that finally an extensive depression took place in that part of Europe where the deep sea of the Cretaceous period was afterwards brought in.

THICKNESS OF THE WEALDEN.

In the Weald area itself, between the North and South Downs, fresh-water beds to the thickness of 1600 feet are known, the base not being reached. Probably the thickness of the whole Wealden series, as seen in Swanage Bay, can not be estimated as less than 2000 feet.

WEALDEN FLORA.

The flora of the Wealden is characterised by a great abundance of Coniferae, Cycadeae, anD Ferns, and by the absence of leaves and fruits of Dicotyledonous Angiosperms. The discovery in 1855, in the Hastings beds of the Isle of Wight, of Gyrogonites, or spore-vessels of the Chara, was the first example of that genus of plants, so common in the tertiary strata, being found in a Secondary or Mesozoic rock.

CHAPTER XIX.
JURASSIC GROUP.— PURBECK BEDS AND OOLITE.

The Purbeck Beds a Member of the Jurassic Group.
Subdivisions of that Group.
Physical Geography of the Oolite in England and France.
Upper Oolite.
Purbeck Beds.
New Genera of fossil Mammalia in the Middle Purbeck of Dorsetshire.
Dirt-bed or ancient Soil.
Fossils of the Purbeck Beds.
Portland Stone and Fossils.
Kimmeridge Clay.
Lithographic Stone of Solenhofen.
Archaeopteryx.
Middle Oolite.
Coral Rag.
Nerinaea Limestone.
Oxford Clay, Ammonites and Belemnites.
Kelloway Rock.
Lower, or Bath, Oolite.
Great Plants of the Oolite.
Oolite and Bradford Clay.
Stonesfield Slate.
Fossil Mammalia.
Fuller's Earth.
Inferior Oolite and Fossils.
Northamptonshire Slates.
Yorkshire Oolitic Coal-field.
Brora Coal.
Palaeontological Relations of the several Subdivisions of the Oolitic group.

CLASSIFICATION OF THE OOLITE.

Immediately below the Hastings Sands we find in Dorsetshire another remarkable fresh-water formation, called THE PURBECK, because it was first studied in the sea-cliffs of the peninsula of Purbeck in that county. These beds are for the most part of fresh-water origin, but the organic remains of some few intercalated beds are marine, and show that the Purbeck series has a closer affinity to the Oolitic group, of which it may be considered as the newest or uppermost member.

In England generally, and in the greater part of Europe, both the Wealden and Purbeck beds are wanting, and the marine cretaceous group is followed immediately, in the descending order, by another series called the Jurassic. In this term, the formations commonly designated as "the Oolite and Lias" are included, both being found in the Jura Mountains. The Oolite was so named because in the countries where it was first examined the limestones belonging to it had an Oolitic structure (see Chapter 3). These rocks occupy in England a zone nearly thirty miles in average breadth, which extends across the island, from Yorkshire in the north-east, to Dorsetshire in the south-west. Their mineral characters are not uniform throughout this region; but the following are the names of the principal subdivisions observed in the central and south- eastern parts of England.

TABLE 19.1. OOLITE.

UPPER OOLITE: a. Purbeck beds. b. Portland stone and sand. c. Kimmeridge clay.
MIDDLE OOLITE: d. Coral rag. e. Oxford clay, and Kelloway rock.
LOWER OOLITE: f. Cornbrash and Forest marble. g. Great Oolite and Stonesfield slate. h. Fuller's earth. i. Inferior Oolite.

The Upper Oolitic system of the Table 19.1 has usually the Kimmeridge clay for its base; the Middle Oolitic system, the Oxford clay. The Lower system reposes on the Lias, an argillo-calcareous formation, which some include in the Lower Oolite, but which will be treated of separately in the next chapter. Many of these subdivisions are distinguished by peculiar organic remains; and, though varying in thickness, may be traced in certain directions for great distances, especially if we compare the part of England to which the above-mentioned type refers with the north-east of France and the Jura Mountains adjoining. In that country, distant above 400 geographical miles, the analogy to the accepted English type, notwithstanding the thinness or occasional absence of the clays, is more perfect than in Yorkshire or Normandy.

PHYSICAL GEOGRAPHY.

The alternation, on a grand scale, of distinct formations of clay and limestone has caused the oolitic and liassic series to give rise to some marked features in the physical outline of parts of England and France. Wide valleys can usually be traced throughout the long bands of country where the argillaceous strata crop out; and between these valleys the limestones are observed, forming ranges of hills or more elevated grounds. These ranges terminate abruptly on the side on which the several clays rise up from beneath the calcareous strata.

(FIGURE 298. Section through Lias (left), Lower Oolite, Oxford Clay, Middle Oolite, Kim. Clay. Upper Oolite. Gault, Chalk and London Clay (right).)

Figure 298 will give the reader an idea of the configuration of the surface now alluded to, such as may be seen in passing from London to Cheltenham, or in other parallel lines, from east to west, in the southern part of England. It has been necessary, however, in this drawing, greatly to exaggerate the inclination of the beds, and the height of the several formations, as compared to their horizontal extent. It will be remarked, that the lines of steep slope, or escarpment, face towards the west in the great calcareous eminences formed by the chalk and the Upper, Middle, and Lower Oolites; and at the base of which we have respectively the Gault, Kimmeridge clay, Oxford clay, and Lias. This last forms, generally, a broad vale at the foot of the escarpment of inferior Oolite, but where it acquires considerable thickness, and contains solid beds of marlstone, it occupies the lower part of the escarpment.

The external outline of the country which the geologist observes in travelling eastward from Paris to Metz, is precisely analogous, and is caused by a similar succession of rocks intervening between the tertiary strata and the Lias; with this difference, however, that the escarpments of Chalk, Upper, Middle, and Lower Oolites face towards the east instead of the west. It is evident, therefore, that the denuding causes (see Chapter 6) have acted similarly over an area several hundred miles in diameter, removing the softer clays more extensively than the limestones, and causing these last to form steep slopes or escarpments wherever the harder calcareous rock was based upon a more yielding and destructible formation.

UPPER OOLITE.
PURBECK BEDS.

These strata, which we class as the uppermost member of the Oolite, are of limited geographical extent in Europe, as already stated, but they acquire importance when we consider the succession of three distinct sets of fossil remains which they contain. Such repeated changes in organic life must have reference to the history of a vast lapse of ages. The Purbeck beds are finely exposed to view in Durdlestone Bay, near Swanage, Dorsetshire, and at Lulworth Cove and the neighbouring bays between Weymouth and Swanage. At Meup's Bay, in particular, Professor E. Forbes examined minutely, in 1850, the organic remains of this group, displayed in a continuous sea-cliff section, and it appears from his researches that the Upper, Middle, and Lower Purbecks are each marked by peculiar species of organic remains, these again being different, so far as a comparison has yet been instituted, from the fossils of the overlying Hastings Sands and Weald Clay.

UPPER PURBECK.

(FIGURE 299. Cyprides from the Upper Purbeck. a. Cypris gibbosa, E. Forbes. b. Cypris tuberculata, E. Forbes. c. Cypris leguminella, E. Forbes.)

The highest of the three divisions is purely fresh-water, the strata, about fifty feet in thickness, containing shells of the genera Paludina, Physa, Limnaea, Planorbis, Valvata, Cyclas, and Unio, with Cyprides and fish. All the species seem peculiar, and among these the Cyprides are very abundant and characteristic (see Figure 299, a, b, c.)

The stone called "Purbeck Marble," formerly much used in ornamental architecture in the old English cathedrals of the southern counties, is exclusively procured from this division.

MIDDLE PURBECK.

Next in succession is the Middle Purbeck, about thirty feet thick, the uppermost part of which consists of fresh-water limestone, with cyprides, turtles, and fish, of different species from those in the preceding strata. Below the limestone are brackish-water beds full of Cyrena, and traversed by bands abounding in Corbula and Melania. These are based on a purely marine deposit, with Pecten, Modiola, Avicula, and Thracia. Below this, again, come limestones and shales, partly of brackish and partly of fresh-water origin, in which many fish, especially species of Lepidotus and Microdon radiatus, are found, and a crocodilian reptile named Macrorhynchus. Among the mollusks, a remarkable ribbed Melania, of the section Chilina, occurs.

(FIGURE 300. Ostrea distorta, Sowerby. Cinder-bed. Middle Purbeck.)

(FIGURE 301. Hemicidaris Purbeckensis, E. Forbes. Middle Purbeck.)

(FIGURE 302. Cyprides from the Middle Purbecks. a. Cypris striato-punctata, E. Forbes. b. Cypris fasciculata, E. Forbes. c. Cypris granulata, Sowerby.)

(FIGURE 303. Physa Bristovii, E. Forbes. Middle Purbeck.)

Immediately below is a great and conspicuous stratum, twelve feet thick, formed of a vast accumulation of shells of Ostrea distorta (Figure 300), long familiar to geologists under the local name of "Cinder-bed." In the uppermost part of this bed Professor Forbes discovered the first echinoderm (Figure 301) as yet known in the Purbeck series, a species of Hemicidaris, a genus characteristic of the Oolitic period, and scarcely, if at all, distinguishable from a previously known Oolitic fossil. It was accompanied by a species of Perna. Below the Cinder-bed fresh-water strata are again seen, filled in many places with species of Cypris (Figure 302, a, b, c), and with Valvata, Paludina, Planorbis, Limnaea, Physa (Figure 303), and Cyclas, all different from any occurring higher in the series. It will be seen that Cypris fasciculata (Figure 302, b) has tubercles at the end only of each valve, a character by which it can be immediately recognised. In fact, these minute crustaceans, almost as frequent in some of the shales as plates of mica in a micaceous sandstone, enable geologists at once to identify the Middle Purbeck in places far from the Dorsetshire cliffs, as, for example, in the Vale of Wardour in Wiltshire. Thick beds of chert occur in the Middle Purbeck filled with mollusca and cyprides of the genera already enumerated, in a beautiful state of preservation, often converted into chalcedony. Among these Professor Forbes met with gyrogonites (the spore-vessels of Chara), plants never until 1851 discovered in rocks older than the Eocene. About twenty feet below the "Cinder-bed" is a stratum two or three inches thick, in which fossil mammalia presently to be mentioned occur, and beneath this a thin band of greenish shales, with marine shells and impressions of leaves like those of a large Zostera, forming the base of the Middle Purbeck.

FOSSIL MAMMALIA OF THE MIDDLE PURBECK.

In 1852, after alluding to the discovery of numerous insects and air-breathing mollusca in the Purbeck strata, I remarked that, although no mammalia had then been found, "it was too soon to infer their non-existence on mere negative evidence." (Elements of Geology 4th edition.) Only two years after this remark was in print, Mr. W.R. Brodie found in the Middle Purbeck, about twenty feet below the "Cinder-bed" above alluded to, in Durdlestone Bay, portions of several small jaws with teeth, which Professor Owen recognised as belonging to a small mammifer of the insectivorous class, more closely allied in its dentition to the Amphitherium (or Thylacotherium) than to any existing type.

Four years later (in 1856) the remains of several other species of warm-blooded quadrupeds were exhumed by Mr. S.H. Beckles, F.R.S., from the same thin bed of marl near the base of the Middle Purbeck. In this marly stratum many reptiles, several insects, and some fresh-water shells of the genera Paludina, Planorbis, and Cyclas, were found.

Mr. Beckles had determined thoroughly to explore the thin layer of calcareous mud from which in the suburbs of Swanage the bones of the Spalacotherium had already been obtained, and in three weeks he brought to light from an area forty feet long and ten wide, and from a layer the average thickness of which was only five inches, portions of the skeletons of six new species of mammalia, as interpreted by Dr. Falconer, who first examined them. Before these interesting inquiries were brought to a close, the joint labours of Professor Owen and Dr. Falconer had made it clear that twelve or more species of mammalia characterised this portion of the Middle Purbeck, most of them insectivorous or predaceous, varying in size from that of a mole to that of the common polecat, Mustela putorius. While the majority had the character of insectivorous marsupials, Dr. Falconer selected one as differing widely from the rest, and pointed out that in certain characters it was allied to the living Kangaroo-rat, or Hypsiprymnus, ten species of which now inhabit the prairies and scrub-jungle of Australia, feeding on plants, and gnawing scratched-up roots. A striking peculiarity of their dentition, one in which they differ from all other quadrupeds, consists in their having a single large pre-molar, the enamel of which is furrowed with vertical grooves, usually seven in number.

(FIGURE 304. Pre-molar of the recent Australian Hypsiprymnus Gaimardi, showing 7 grooves, at right angles to the length of the jaw, magnified 3 1/2 diameters.)

(FIGURE 305. Third and largest pre-molar (lower jaw) of Plagiaulax Becklesii, magnified 5 1/2 diameters, showing 7 diagonal grooves.)

(FIGURE 306. Plagiaulex Becklesii, Falconer. Middle Purbeck. Right ramus of lower jaw, magnified two diameters. a. Incisor. b, c. Line of vertical fracture behind the pre-molars. pm. Three pre-molars, the third and last (much larger than the other two taken together) being divided by a crack. m. Sockets of two missing molars.)

The largest pre-molar (see Figure 305) in the fossil genus exhibits in like manner seven parallel grooves, producing by their termination a similar serrated edge in the crown; but their direction is diagonal— a distinction, says Dr. Falconer, which is "trivial, not typical." As these oblique furrows form so marked a character of the majority of the teeth, Dr. Falconer gave to the fossil the generic name of Plagiaulax. The shape and relative size of the incisor, a, Figure 306, exhibit a no less striking similarity to Hypsiprymnus. Nevertheless, the more sudden upward curve of this incisor, as well as other characters of the jaw, indicate a great deviation in the form of Plagiaulax from that of the living kangaroo-rats.

There are two fossil specimens of lower jaws of this genus evidently referable to two distinct species extremely unequal in size and otherwise distinguishable. The Plagiaulax Becklesii (Figure 306) was about as big as the English squirrel or the flying phalanger of Australia (Petaurus Australis, Waterhouse). The smaller fossil, having only half the linear dimensions of the other, was probably only one-twelfth of its bulk. It is of peculiar geological interest, because, as shown by Dr. Falconer, its two back molars bear a decided resemblance to those of the Triassic Microlestes (Figure 389 Chapter 19), the most ancient of known mammalia, of which an account will be given in Chapter 21.

Up to 1857 all the mammalian remains discovered in secondary rocks had consisted solely of single branches of the lower jaw, but in that year Mr. Beckles obtained the upper portion of a skull, and on the same slab the lower jaw of another quadruped with eight molars, a large canine, and a broad and thick incisor. It has been named Triconodon from its bicuspid teeth, and is supposed to have been a small insectivorous marsupial, about the size of a hedgehog. Other jaws have since been found indicating a larger species of the same genus.

Professor Owen has proposed the name of Galestes for the largest of the mammalia discovered in 1858 in Purbeck, equalling the polecat (Mustela putorius) in size. It is supposed to have been predaceous and marsupial.

Between forty and fifty pieces or sides of lower jaws with teeth have been found in oolitic strata in Purbeck; only five upper maxillaries, together with one portion of a separate cranium, occur at Stonesfield, and it is remarkable that with these there were no examples in Purbeck of an entire skeleton, nor of any considerable number of bones in juxtaposition. In several portions of the matrix there were detached bones, often much decomposed, and fragments of others apparently mammalian; but if all of them were restored, they would scarcely suffice to complete the five skeletons to which the five upper maxillaries above alluded to belonged. As the average number of pieces in each mammalian skeleton is about 250, there must be many thousands of missing bones; and when we endeavour to account for their absence, we are almost tempted to indulge in speculations like those once suggested to me by Dr. Buckland, when he tried to solve the enigma in reference to Stonesfield; "The corpses," he said, "of drowned animals, when they float in a river, distended by gases during putrefaction, have often their lower jaw hanging loose, and sometimes it has dropped off. The rest of the body may then be drifted elsewhere, and sometimes may be swallowed entire by a predaceous reptile or fish, such as an ichthyosaur or a shark."

As all the above-mentioned Purbeck marsupials, belonging to eight or nine genera and to about fourteen species, insectivorous, predaceous, and herbivorous, have been obtained from an area less than 500 square yards in extent, and from a single stratum no more than a few inches thick, we may safely conclude that the whole lived together in the same region, and in all likelihood they constituted a mere fraction of the mammalia which inhabited the lands drained by one river and its tributaries. They afford the first positive proof as yet obtained of the co-existence of a varied fauna of the highest class of vertebrata with that ample development of reptile life which marks all the periods from the Trias to the Lower Cretaceous inclusive, and with a gymnospermous flora, or that state of the vegetable kingdom when cycads and conifers predominated over all kinds of plants, except the ferns, so far, at least, as our present imperfect knowledge of fossil botany entitles us to speak.

TABLE 19.2. NUMBER AND DISTRIBUTION OF ALL THE KNOWN SPECIES OF FOSSIL MAMMALIA FROM STRATA OLDER THAN THE PARIS GYPSUM, OR THAN THE BEMBRIDGE SERIES OF THE ISLE OF WIGHT.

TERTIARY:

Headon Series and beds between the Paris Gypsum and the Gres de Beauchamp: 14: 10 English, 4 French.

Barton Clay and Sables de Beauchamp: 0.

Bagshot Beds, Calcaire Grossier, and Upper Soissonnais of Cuisse-Lamotte: 20: 16 French, 1 English, 3 United States (I allude to several Zeuglodons found in Alabama, and referred by some zoologists to three species.)

London Clay, including the Kyson Sand: 7 English.

Plastic Clay and Lignite: 9: 7 French, 2 English.

Sables de Bracheux: 1 French.

Thanet Sands and Lower Landenien of Belgium: 0.

SECONDARY:

Maestricht Chalk: 0.

White Chalk: 0.

Chalk Marl: 0.

Chloritic Series (Upper Greensand): 0.

Gault: 0.

Neocomian (Lower Greensand): 0.

Wealden: 0.

Upper Purbeck Oolite : 0.

Middle Purbeck Oolite : 14 Swanage.

Lower Purbeck Oolite: 0.

Portland Oolite: 0.

Kimmeridge Clay: 0.

Coral Rag: 0.

Oxford Clay: 0.

Great Oolite: 4 Stonesfield.

Inferior Oolite: 0.
Lias: 0.
Upper Trias: 4 Wurtemberg, Somersetshire. N. Carolina.
Middle Trias: 0.
Lower Trias: 0.
PRIMARY.
Permian: 0.
Carboniferous : 0.
Devonian: 0.
Silurian: 0.
Cambrian: 0.
Laruentian: 0.

Table 19.2 will enable the reader to see at a glance how conspicuous a part, numerically considered, the mammalian species of the Middle Purbeck now play when compared with those of other formations more ancient than the Paris gypsum, and, at the same time, it will help him to appreciate the enormous hiatus in the history of fossil mammalia which at present occurs between the Eocene and Purbeck periods, and between the latter and the Stonesfield Oolite, and between this again and the Trias.

The Sables de Bracheux, enumerated in the Tertiary division of the table, supposed by Mr. Prestwich to be somewhat newer than the Thanet Sands, and by M. Hebert to be of about that age, have yielded at La Fere the Arctocyon (Palaeocyon) primaevus, the oldest known tertiary mammal.

It is worthy of notice, that in the Hastings Sands there are certain layers of clay and sandstone in which numerous footprints of quadrupeds have been found by Mr. Beckles, and traced by him in the same set of rocks through Sussex and the Isle of Wight. They appear to belong to three or four species of reptiles, and no one of them to any warm-blooded quadruped. They ought, therefore, to serve as a warning to us, when we fail in like manner to detect mammalian footprints in older rocks (such as the New Red Sandstone), to refrain from inferring that quadrupeds, other than reptilian, did not exist or pre-exist.

But the most instructive lesson read to us by the Purbeck strata consists in this: They are all, with the exception of a few intercalated brackish and marine layers, of fresh-water origin; they are 160 feet in thickness, have been well searched by skillful collectors, and by the late Edward Forbes in particular, who studied them for months consecutively. They have been numbered, and the contents of each stratum recorded separately, by the officers of the Geological Survey of Great Britain. They have been divided into three distinct groups by Forbes, each characterised by the same genera of pulmoniferous mollusca and cyprides, these genera being represented in each group by different species; they have yielded insects of many orders, and the fruits of several plants; and lastly, they contain "dirt-beds," or old terrestrial surfaces and vegetable soils at different levels, in some of which erect trunks and stumps of cycads and conifers, with their roots still attached to them, are preserved. Yet when the geologist inquires if any land-animals of a higher grade than reptiles lived during any one of these three periods, the rocks are all silent, save one thin layer a few inches in thickness; and this single page of the earth's history has suddenly revealed to us in a few weeks the memorials of so many species of fossil mammalia, that they already outnumber those of many a subdivision of the tertiary series, and far surpass those of all the other secondary rocks put together!

LOWER PURBECK.

(FIGURE 307. Cyprides from the Lower Purbeck. a. Cypris Purbeckensis, Forbes. b. Same magnified. c. Cypris punctata, Forbes. d, e. Two views magnified of the same.)

Beneath the thin marine band mentioned above as the base of the Middle Purbeck, some purely fresh-water marls occur, containing species of Cypris (Figure 307 a, c), Valvata, and Limnaea, different from those of the Middle Purbeck. This is the beginning of the inferior division, which is about 80 feet thick. Below the marls are seen, at Meup's Bay, more than thirty feet of brackish-water strata, abounding in a species of Serpula, allied to, if not identical with, Serpula coacervites, found in beds of the same age in Hanover. There are also shells of the genus Rissoa (of the subgenus Hydrobia), and a little Cardium of the subgenus

Protocardium, in these marine beds, together with Cypris. Some of the cypris-bearing shales are strangely contorted and broken up, at the west end of the Isle of Purbeck. The great dirt-bed or vegetable soil containing the roots and stools of Cycadeae, which I shall presently describe, underlies these marls, and rests upon the lowest fresh-water limestone, a rock about eight feet thick, containing Cyclas, Valvata, and Limnaea, of the same species as those of the uppermost part of the Lower Purbeck, or above the dirt-bed. The fresh-water limestone in its turn rests upon the top beds of the Portland stone, which, although it contains purely marine remains, often consists of a rock undistinguishable in mineral character from the Lowest Purbeck limestone.

DIRT-BED OR ANCIENT SURFACE-SOIL.

(FIGURE 308. Mantellia nidiformis, Brongniart. The upper part shows the woody stem, the lower part the bases of the leaves.)

The most remarkable of all the varied succession of beds enumerated in the above list is that called by the quarrymen "the dirt," or "black dirt," which was evidently an ancient vegetable soil. It is from 12 to 18 inches thick, is of a dark brown or black colour, and contains a large proportion of earthy lignite. Through it are dispersed rounded and sub-angular fragments of stone, from 3 to 9 inches in diameter, in such numbers that it almost deserves the name of gravel. I also saw in 1866, in Portland, a smaller dirt-bed six feet below the principal one, six inches thick, consisting of brown earth with upright Cycads of the same species, Mantellia nidiformis, as those found in the upper bed, but no Coniferae. The weight of the incumbent strata squeezing down the compressible dirt-bed has caused the Cycads to assume that form which has led the quarrymen to call them "petrified birds' nests," which suggested to Brongniart the specific name of nidiformis. I am indebted to Mr. Carruthers for Figure 308 of one of these Purbeck specimens, in which the original cylindrical figure has been less distorted than usual by pressure.

Many silicified trunks of coniferous trees, and the remains of plants allied to Zamia and Cycas, are buried in this dirt-bed, and must have become fossil on the spots where they grew. The stumps of the trees stand erect for a height of from one to three feet, and even in one instance to six feet, with their roots attached to the soil at about the same distances from one another as the trees in a modern forest. The carbonaceous matter is most abundant immediately around the stumps, and round the remains of fossil Cycadeae.

(FIGURE 309. Section in Isle of Portland, Dorset. (Buckland and De la Beche.)showing layers (from top to bottom): Fresh-water calcareous slate: Dirt-bed and ancient forest: Lowest fresh-water beds of the Lower Purbeck: and Portland stone, marine.)

Besides the upright stumps above mentioned, the dirt-bed contains the stems of silicified trees laid prostrate. These are partly sunk into the black earth, and partly enveloped by a calcareous slate which covers the dirt-bed. The fragments of the prostrate trees are rarely more than three or four feet in length; but by joining many of them together, trunks have been restored, having a length from the root to the branches of from 20 to 23 feet, the stems being undivided for 17 or 20 feet, and then forked. The diameter of these near the root is about one foot; but I measured one myself, in 1866, which was 3 1/2 feet in diameter, said by the quarrymen to be unusually large. Root-shaped cavities were observed by Professor Henslow to descend from the bottom of the dirt-bed into the subjacent fresh-water stone, which, though now solid, must have been in a soft and penetrable state when the trees grew. The thin layers of calcareous slate (Figure 309) were evidently deposited tranquilly, and would have been horizontal but for the protrusion of the stumps of the trees, around the top of each of which they form hemispherical concretions.

(FIGURE 310. Section of cliff east of Lulworth Cove. (Buckland and De la Beche.) showing layers (from top to bottom): Fresh-water calcareous slate: Dirt-bed, with stools of trees: Fresh-water: Portland stone, marine.)

The dirt-bed is by no means confined to the island of Portland, where it has been most carefully studied, but is seen in the same relative position in the cliffs east of Lulworth Cove, in Dorsetshire, where, as the strata have been disturbed, and are now inclined at an angle of 45 degrees, the stumps of the trees are also inclined at the same angle in an opposite direction— a beautiful illustration of a change in the position of beds originally horizontal (see Figure 310).

From the facts above described we may infer, first, that those beds of the Upper Oolite, called "the Portland," which are full of marine shells, were overspread with fluviatile mud, which became dry land, and covered by a forest, throughout a portion of the space now occupied by the south of England, the climate being such as to permit the growth of the Zamia and Cycas. Secondly. This land at length sank down and was submerged with its forests beneath a body of fresh-water, from which sediment was thrown down enveloping fluviatile shells. Thirdly. The regular and uniform preservation of this thin bed of black earth over a distance of many miles, shows that the change from dry land to the state of a fresh-water lake or estuary, was not accompanied by any violent denudation, or rush of water, since the loose black earth, together with the trees which lay prostrate on its surface, must inevitably have been swept away had any such violent catastrophe taken place.

The forest of the dirt-bed, as before hinted, was not everywhere the first vegetation which grew in this region. Besides the lower bed containing upright Cycadeae, before mentioned, another has sometimes been found above it, which implies oscillations in the level of the same ground, and its alternate occupation by land and water more than once.

SUBDIVISIONS OF THE PURBECK.

It will be observed that the division of the Purbecks into upper, middle, and lower, was made by Professor Forbes strictly on the principle of the entire distinctness of the species of organic remains which they include. The lines of demarkation are not lines of disturbance, nor indicated by any striking physical characters or mineral changes. The features which attract the eye in the Purbecks, such as the dirt-beds, the dislocated strata at Lulworth, and the Cinder-bed, do not indicate any breaks in the distribution of organised beings. "The causes which led to a complete change of life three times during the deposition of the fresh-water and brackish strata must," says this naturalist, "be sought for, not simply in either a rapid or a sudden change of their area into land or sea, but in the great lapse of time which intervened between the epochs of deposition at certain periods during their formation."

Each dirt-bed may, no doubt, be the memorial of many thousand years or centuries, because we find that two or three feet of vegetable soil is the only monument which many a tropical forest has left of its existence ever since the ground on which it now stands was first covered with its shade. Yet, even if we imagine the fossil soils of the Lower Purbeck to represent as many ages, we need not be surprised to find that they do not constitute lines of separation between strata characterised by different zoological types. The preservation of a layer of vegetable soil, when in the act of being submerged, must be regarded as a rare exception to a general rule. It is of so perishable a nature, that it must usually be carried away by the denuding waves or currents of the sea, or by a river; and many Purbeck dirt-beds were probably formed in succession and annihilated, besides those few which now remain.

The plants of the Purbeck beds, so far as our knowledge extends at present, consist chiefly of Ferns, Coniferae, and Cycadeae (Figure 308), without any angiosperms; the whole more allied to the Oolitic than to the Cretaceous vegetation. The same affinity is indicated by the vertebrate and invertebrate animals. Mr. Brodie has found the remains of beetles and several insects of the homopterous and trichopterous orders, some of which now live on plants, while others are of such forms as hover over the surface of our present rivers.

PORTLAND OOLITE AND SAND (b, TABLE 19.1).

(FIGURE 311. Cerithium Portlandicum (=Terebra) Sowerby. a. Cast of shell known as "Portland screw." b. The shell itself.)

(FIGURE 312. Isastraea oblonga, M. Edw. and J. Haime. As seen on a polished slab of chert from the Portland Sand, Tisbury.)

(FIGURE 313. Trigonia gibbosa. 1/2 natural size. Portland Stone, Tisbury. a. The hinge.)

(FIGURE 314. Cardium dissimile. 1/4 natural size. Portland Stone.)

(FIGURE 315. Ostrea expansa. Portland Sand.)

The Portland Oolite has already been mentioned as forming in Dorsetshire the foundation on which the fresh-water limestone of the Lower Purbeck reposes (see above). It supplies the well-known building-stone of which St. Paul's and so many of the principal edifices of London are constructed. About fifty species of mollusca occur in this formation, among which are some ammonites of large size. The cast of a spiral univalve called by the quarrymen the "Portland screw" (a, Figure 311), is common; the shell of the same (b) being rarely met with. Also Trigonia gibbosa (Figure 313) and Cardium dissimile (Figure 314). This upper member rests on a dense bed of sand, called the Portland Sand, containing similar marine fossils, below which is the Kimmeridge Clay. In England these Upper Oolite formations are almost wholly confined to the southern counties. But some fragments of them occur beneath the Neocomian or Speeton Clay on the coast of Yorkshire, containing many more fossils common to the Portlandian of the Continent than does the same formation in Dorsetshire. Corals are rare in this formation, although one species is found plentifully at Tisbury, Wiltshire, in the Portland Sand, converted into flint and chert, the original calcareous matter being replaced by silex (Figure 312).

KIMMERIDGE CLAY.

The Kimmeridge Clay consists, in great part, of a bituminous shale, sometimes forming an impure coal, several hundred feet in thickness. In some places in Wiltshire it much resembles peat; and the bituminous matter may have been, in part at least, derived from the decomposition of vegetables. But as impressions of plants are rare in these shales, which contain ammonites, oysters, and other marine shells, with skeletons of fish and saurians, the bitumen may perhaps be of animal origin. Some of the saurians (Pliosaurus) in Dorsetshire are among the most gigantic of their kind.

(FIGURE 316. Cardium striatulum. Kimmeridge Clay, Hartwell.)
(FIGURE 317. Ostrea deltoidea. Kimmeridge Clay, 1/4 natural size.)
(FIGURE 318. Gryphaea (Exogyra) virgula. Kimmeridge Clay.)
(FIGURE 319. Trigonellites latus, Park, Kimmeridge Clay.)

Among the fossils, amounting to nearly 100 species, may be mentioned Cardium striatulum (Figure 316) and Ostrea deltoidea (Figure 317), the latter found in the Kimmeridge Clay throughout England and the north of France, and also in Scotland, near Brora. The Gryphaea virgula (Figure 318), also met with in the Kimmeridge Clay near Oxford, is so abundant in the Upper Oolite of parts of France as to have caused the deposit to be termed "marnes a gryphees virgules." Near Clermont, in Argonne, a few leagues from St. Menehould, where these indurated marls crop out from beneath the Gault, I have seen them, on decomposing, leave the surface of every ploughed field literally strewed over with this fossil oyster. The Trigonellites latus (Aptychus of some authors)(Figure 319) is also widely dispersed through this clay. The real nature of the shell, of which there are many species in oolitic rocks, is still a matter of conjecture. Some are of opinion that the two plates have been the gizzard of a cephalopod; others, that it may have formed a bivalve operculum of the same.

SOLENHOFEN STONE.

(FIGURE 320. Skeleton of Pterodactylus crassirostris. Oolite of Pappenheim, near Solenhofen. a. This bone, consisting of four joints, is part of the fifth or outermost digit elongated, as in bats, for the support of a wing.)

The celebrated lithographic stone of Solenhofen in Bavaria, appears to be of intermediate age between the Kimmeridge clay and the Coral Rag, presently to be described. It affords a remarkable example of the variety of fossils which may be preserved under favourable circumstances, and what delicate impressions of the tender parts of certain animals and plants may be retained where the sediment is of extreme fineness. Although the number of testacea in this slate is small, and the plants few, and those all marine, count Munster had determined no less than 237 species of fossils when I saw his collection in 1833; and among them no less than seven SPECIES of flying reptiles or pterodactyls (see Figure 320), six saurians, three tortoises, sixty species of fish, forty-six of crustacea, and twenty-six of insects. These insects, among which is a libellula, or dragon-fly, must have been blown out to sea, probably from the same land to which the pterodactyls, and other contemporaneous air-breathers, resorted.

(FIGURE 321. Tail and feather of Archaeopteryx, from Solenhofen, and tail of living bird for comparison. A. Caudal vertebrae of Archaeopteryx macrura, Owen; with impression of tail- feathers; one-fifth natural size. B. Two caudal vertebrae of same; natural size. C. Single feather, found in 1861 at Solenhofen, by Von Meyer, and called Archaeopteryx lithographica; natural size. D. Tail of recent vulture (Gyps Bengalensis) showing attachment of tail-feathers in living birds; one-quarter natural size. E. Profile of caudal vertebrae of same; one-third natural size. e, e. Direction of tail-feathers when seen in profile. f. Ploughshare bone or broad terminal joint (seen also in f, D.))

In the same slate of Solenhofen a fine example was met with in 1862 of the skeleton of a bird almost entire, and retaining even its feathers so perfect that the vanes as well as the shaft are preserved. The head was at first supposed to be wanting, but Mr. Evans detected on the slab what seems to be the impression of the cranium and beak, much resembling in size and shape that of the jay or woodcock. This valuable specimen is now in the British Museum, and has been called by Professor Owen Archaeopteryx macrura. Although anatomists agree that it is a true bird, yet they also find that in the length of the bones of the tail, and some other minor points of its anatomy, it approaches more nearly to reptiles than any known living bird. In the living representatives of the class Aves, the tail-feathers are attached to a coccygian bone, consisting of several vertebrae united together, whereas in the Archaeopteryx the tail is composed of twenty vertebrae, each of which supports a pair of quill-feathers. The first five only of the vertebrae, as seen in A, have transverse processes, the fifteen remaining ones become gradually longer and more tapering. The feathers diverge outward from them at an angle of 45 degrees.

Professor Huxley in his late memoirs on the order of reptiles called Dinosaurians, which are largely represented in all the formations, from the Neocomian to the Trias inclusive, has shown that they present in their structure many remarkable affinities to birds. But a reptile about two feet long, called Compsognathus, lately found in the Stonesfield slate, makes a much greater approximation to the class Aves than any Dinosaur, and therefore forms a closer link between the classes Aves and Reptilia than does the Archaeopteryx.

It appears doubtful whether any species of British fossil, whether of the vertebrate or invertebrate class, is common to the Oolite and Chalk. But there is no similar break or discordance as we proceed downward, and pass from one to another of the several leading members of the Jurassic group, the Upper, Middle, and Lower Oolite, and the Lias, there being often a considerable proportion of the mollusca, sometimes as much as a fourth, common to such divisions as the Upper and Middle Oolite.

MIDDLE OOLITE.
CORAL RAG.

(FIGURE 322. Thecosmilia annularis, Milne Edwards and J. Haime. Coral Rag, Steeple Ashton.)

(FIGURE 323. Thamnastraea. Coral Rag. Steeple Ashton.)

(FIGURE 324. Ostrea gregaria, Coral Rag, Steeple Ashton.)

One of the limestones of the Middle Oolite has been called the "Coral Rag," because it consists, in part, of continuous beds of petrified corals, most of them retaining the position in which they grew at the bottom of the sea. In their forms they more frequently resemble the reef-building polyparia of the Pacific than do the corals of any other member of the Oolite. They belong chiefly to the genera Thecosmilia (Figure 322), Protoseris, and Thamnastraea, and sometimes form masses of coral fifteen feet thick. In Figure 323 of a Thamnastraea from this formation, it will be seen that the cup-shaped cavities are deepest on the right-hand side, and that they grow more and more shallow, until those on the left side are nearly filled up. The last-mentioned stars are supposed to represent a perfected condition, and the others an immature state. These coralline strata extend through the calcareous hills of the north-west of Berkshire, and north of Wilts, and again recur in Yorkshire, near Scarborough. The Ostrea gregarea (Figure 324) is very characteristic of the formation in England and on the Continent.

(FIGURE 325. Nerinaea Goodhallii, Fitton. Coral Rag, Weymouth. 1/4 natural size.)

One of the limestones of the Jura, referred to the age of the English Coral Rag, has been called "Nerinaean limestone" (Calcaire a Nerinees) by M. Thirria; Nerinaea being an extinct genus of univalve shells (Figure 325) much resembling the Cerithium in external form. Figure 325 shows the curious and continuous ridges on the columnella and whorls.

OXFORD CLAY.
(FIGURE 326. Belemnites hastatus. Oxford Clay.)
(FIGURE 327. Ammonites Jason, Reinecke. (Syn. A. Elizabethae, Pratt. Oxford Clay, Christian Malford, Wiltshire.)
(FIGURE 328. Belemnites Puzosianus, d'Orbigny. B. Owenii, Pierce. Oxford Clay, Christian Malford, Wiltshire. a. Section of the shell projecting from the phragmacone. b-c. External covering to the ink-bag and phragmacone. c, d. Osselet, or that portion commonly called the belemnite. e. Conical chambered body called the phragmacone. f. Position of ink-bag beneath the shelly covering.)

The coralline limestone, or "Coral Rag," above described, and the accompanying sandy beds, called "calcareous grits," of the Middle Oolite, rest on a thick bed of clay, called the "Oxford Clay," sometimes not less than 600 feet thick. In this there are no corals, but great abundance of cephalopoda, of the genera Ammonite and Belemnite (Figures 326 and 327). In some of the finely laminated clays ammonites are very perfect, although somewhat compressed, and are frequently found with the lateral lobe extended on each side of the opening of the mouth into a horn-like projection (Figure 327). These were discovered in the cuttings of the Great Western Railway, near Chippenham, in 1841, and have been described by Mr. Pratt (Annals of Natural History November 1841).

Similar elongated processes have been also observed to extend from the shells of some Belemnites discovered by Dr. Mantell in the same clay (see Figure 328), who, by the aid of this and other specimens, has been able to throw much light on the structure of singular extinct forms of cuttle-fish. (See Philosophical Transactions 1850 page 363; also Huxley Memoirs of Geological Survey 1864; Phillips Palaeontological Society.)

KELLOWAY ROCK.

The arenaceous limestone which passes under this name is generally grouped as a member of the Oxford clay, in which it forms, in the south-west of England, lenticular masses, 8 or 10 feet thick, containing at Kelloway, in Wiltshire, numerous casts of ammonites and other shells. But in Yorkshire this calcareo-arenaceous formation thickens to about 30 feet, and constitutes the lower part of the Middle Oolite, extending inland from Scarborough in a southerly direction. The number of mollusca which it contains is, according to Mr. Etheridge, 143, of which only 34, or 23 1/2 per cent, are common to the Oxford clay proper. Of the 52 Cephalopoda, 15 (namely 13 species of ammonite, the Ancyloceras Calloviense and one Belemnite) are common to the Oxford Clay, giving a proportion of nearly 30 per cent.

LOWER OOLITE.
CORNBRASH AND FOREST MARBLE.

The upper division of this series, which is more extensive than the preceding or Middle Oolite, is called in England the Cornbrash, as being a brashy, easily broken rock, good for corn land. It consists of clays and calcareous sandstones, which pass downward into the Forest Marble, an argillaceous limestone, abounding in marine fossils. In some places, as at Bradford, this limestone is replaced by a mass of clay. The sandstones of the Forest Marble of Wiltshire are often ripple-marked and filled with fragments of broken shells and pieces of drift-wood, having evidently been formed on a coast. Rippled slabs of fissile oolite are used for roofing, and have been traced over a broad band of country from Bradford in Wilts, to Tetbury in Gloucestershire. These calcareous tile-stones are separated from each other by thin seams of clay, which have been deposited upon them, and have taken their form, preserving the undulating ridges and furrows of the sand in such complete integrity, that the impressions of small footsteps, apparently of crustaceans, which walked over the soft wet sands, are still visible. In the same stone the claws of crabs, fragments of echini, and other signs of a neighbouring beach, are observed. (P. Scrope Proceedings of the Geological Society March 1831.)

GREAT (OR BATH) OOLITE.

(FIGURE 329. Eunomia radiata, Lamouroux. (Calamophyllia, Milne Edwards.) a. Section transverse to the tubes. b. Vertical section, showing the radiation of the tubes. c. Portion of interior of tubes magnified, showing striated surface.)

Although the name of Coral Rag has been appropriated, as we have seen, to a member of the Middle Oolite before described, some portions of the Lower Oolite are equally entitled in many places to be called coralline limestones. Thus the Great Oolite near Bath contains various corals, among which the Eunomia radiata (Figure 329) is very conspicuous, single individuals forming masses several feet in diameter; and having probably required, like the large existing brain-coral (Meandrina) of the tropics, many centuries before their growth was completed.

(FIGURE 330. Apiocrinites rotundus, or Pear Encrinite; Miller. Fossil at Bradford, Wilts. a. Stem of Apiocrinites, and one of the articulations, natural size. b. Section at Bradford of Great Oolite and overlying clay, containing the fossil encrinites. (See text.) c. Three perfect individuals of Apiocrinites, represented as they grew on the surface of the Great Oolite. d. Body of the Apiocrinites rotundus. Half natural size.)

(FIGURE 331. Apiocrinus. a. Single plate of body of Apiocrinus, overgrown with serpulae and bryozoa. Natural size. Bradford Clay. b. Portion of the same magnified, showing the bryozoan Diastopora diluviana covering one of the serpulae.)

Different species of crinoids, or stone-lilies, are also common in the same rocks with corals; and, like them, must have enjoyed a firm bottom, where their base of attachment remained undisturbed for years (c, Figure 330). Such fossils, therefore, are almost confined to the limestones; but an exception occurs at Bradford, near Bath, where they are enveloped in clay sometimes 60 feet thick. In this case, however, it appears that the solid upper surface of the "Great Oolite" had supported, for a time, a thick submarine forest of these beautiful zoophytes, until the clear and still water was invaded by a current charged with mud, which threw down the stone-lilies, and broke most of their stems short off near the point of attachment. The stumps still remain in their original position; but the numerous articulations, once composing the stem, arms, and body of the encrinite, were scattered at random through the argillaceous deposit in which some now lie prostrate. These appearances are represented in the section b, Figure 330, where the darker strata represent the Bradford clay, which is however a formation of such local development that in many places it can not easily be separated from the clays of the overlying "forest-marble" and underlying "fuller's earth." The upper surface of the calcareous stone below is completely incrusted over with a continuous pavement, formed by the stony roots or attachments of the Crinoidea; and besides this evidence of the length of time they had lived on the spot, we find great numbers of single joints, or circular plates of the stem and body of the encrinite, covered over with serpulae. Now these serpulae could only have begun to grow after the death of some of the stone-lilies, parts of whose skeletons had been strewed over the floor of the ocean before the irruption of argillaceous mud. In some instances we find that, after the parasitic serpulae were full grown, they had become incrusted over with a bryozoan, called Diastopora diluviana (see b, Figure 331); and many generations of these molluscoids had succeeded each other in the pure water before they became fossil.

We may, therefore, perceive distinctly that, as the pines and cycadeous plants of the ancient "dirt-bed," or fossil forest, of the Lower Purbeck were killed by submergence under fresh water, and soon buried beneath muddy sediment, so an invasion of argillaceous matter put a sudden stop to the growth of the Bradford Encrinites, and led to their preservation in marine strata.

Such differences in the fossils as distinguish the calcareous and argillaceous deposits from each other, would be described by naturalists as arising out of a difference in the STATIONS of species; but besides these, there are variations in the fossils of the higher, middle, and lower part of the oolitic series, which must be ascribed to that great law of change in organic life by which distinct assemblages of species have been adapted, at successive geological periods, to the varying conditions of the habitable surface. In a single district it is difficult to decide how far the limitation of species to certain minor formations has been due to the local influence of STATIONS, or how far it has been caused by time or the law of variation above alluded to. But we recognise the reality of the last-mentioned influence, when we contrast the whole oolitic series of England with that of parts of the Jura, Alps, and other distant regions, where, although there is scarcely any lithological resemblance, yet some of the same fossils remain peculiar in each country to the Upper,

Middle, and Lower Oolite formations respectively. Mr. Thurmann has shown how remarkably this fact holds true in the Bernese Jura, although the argillaceous divisions, so conspicuous in England, are feebly represented there, and some entirely wanting.

(FIGURE 332. Terebratula digona, Sowerby. Natural size. Bradford Clay.)
(FIGURE 333. Purpuroidea nodulata. One-fourth natural size. Great Oolite, Minchinhampton.)
(FIGURE 334. Cylindrites acutus. Sowb. Syn. Actaeon acutus. Great Oolite, Minchinhampton.)
(FIGURE 335. Patella rugosa, Sowerby. Great Oolite.)
(FIGURE 336. Nerita costulata, Desh. Great Oolite.)
(FIGURE 337. Rimula (Emarginula) clathrata, Sowerby. Great Oolite.)

The calcareous portion of the Great Oolite consists of several shelly limestones, one of which, called the Bath Oolite, is much celebrated as a building-stone. In parts of Gloucestershire, especially near Minchinhampton, the Great Oolite, says Mr. Lycett, "must have been deposited in a shallow sea, where strong currents prevailed, for there are frequent changes in the mineral character of the deposit, and some beds exhibit false stratification. In others, heaps of broken shells are mingled with pebbles of rocks foreign to the neighbourhood, and with fragments of abraded madrepores, dicotyledonous wood, and crabs' claws. The shelly strata, also, have occasionally suffered denudation, and the removed portions have been replaced by clay." In such shallow-water beds shells of the genera Patella, Nerita, Rimula, Cylindrites are common (see Figures 334 to 337); while cephalopods are rare, and instead of ammonites and belemnites, numerous genera of carnivorous trachelipods appear. Out of 224 species of univalves obtained from the Minchinhampton beds, Mr. Lycett found no less than 50 to be carnivorous. They belong principally to the genera Buccinum, Pleurotoma, Rostellaria, Murex, Purpuroidea (Figure 333), and Fusus, and exhibit a proportion of zoophagous species not very different from that which obtains in seas of the Recent period. These zoological results are curious and unexpected, since it was imagined that we might look in vain for the carnivorous trachelipods in rocks of such high antiquity as the Great Oolite, and it was a received doctrine that they did not begin to appear in considerable numbers till the Eocene period, when those two great families of cephalopoda, the ammonites and belemnites, and a great number of other representatives of the same class of chambered shells, had become extinct.

STONESFIELD SLATE: MAMMALIA.

(FIGURE 338. Elytron of Buprestis? Stonesfield.)

The slate of Stonesfield has been shown by Mr. Lonsdale to lie at the base of the Great Oolite. (Proceedings of the Geological Society volume 1 page 414.) It is a slightly oolitic shelly limestone, forming large lenticular masses imbedded in sand only six feet thick, but very rich in organic remains. It contains some pebbles of a rock very similar to itself, and which may be portions of the deposit, broken up on a shore at low water or during storms, and redeposited. The remains of belemnites, trigoniae, and other marine shells, with fragments of wood, are common, and impressions of ferns, cycadeae, and other plants. Several insects, also, and, among the rest, the elytra or wing-covers of beetles, are perfectly preserved (see Figure 338), some of them approaching nearly to the genus Buprestis. The remains, also, of many genera of reptiles, such as Plesiosaur, Crocodile, and Pterodactyl, have been discovered in the same limestone.

But the remarkable fossils for which the Stonesfield slate is most celebrated are those referred to the mammiferous class. The student should be reminded that in all the rocks described in the preceding chapters as older than the Eocene, no bones of any land-quadruped, or of any cetacean, had been discovered until the Spalacotherium of the Purbeck beds came to light in 1854. Yet we have seen that terrestrial plants were not wanting in the Upper Cretaceous formation (see Chapter 17), and that in the Wealden there was evidence of fresh-water sediment on a large scale, containing various plants, and even ancient vegetable soils. We had also in the same Wealden many land-reptiles and winged insects, which render the absence of terrestrial quadrupeds the more striking. The want, however, of any bones of whales, seals, dolphins, and other aquatic mammalia, whether in the chalk or in the upper or middle oolite, is certainly still more remarkable.

These observations are made to prepare the reader to appreciate more justly the interest felt by every geologist in the discovery in the Stonesfield slate of no less than ten specimens of lower jaws of mammiferous quadrupeds, belonging to four different species and to three distinct genera, for which the names of Amphitherium, Phascolotherium, and Stereognathus have been adopted.

(FIGURE 339. Tupaia tana. Right ramus of lower jaw. Natural size. A recent insectivorous placental mammal, from Sumatra.)

(Figures 340 and 341. Part of lower jaw of Tupaia tana. Twice natural size.

(FIGURE 340. End view seen from behind, showing the very slight inflection of the angle at c.)

(FIGURE 341. Side view of same.))

(Figures 342 and 343. Part of lower jaw of Didelphys Azarae; recent, Brazil. Natural size.

(FIGURE 342. End view seen from behind, showing the inflection of the angle of the jaw, c, d.)

(FIGURE 343. Side view of same.))

(FIGURE 344. Amphitherium Prevostii, Cuvier sp. Stonesfield Slate. Syn. Thylacotherium Prevostii, Valenc. a. Coronoid process. b. Condyle. c. Angle of jaw. d. Double-fanged molars.)

(FIGURE 345. Amphitheriumm Broderipii, Owen. Natural size. Stonesfield Slate.)

It is now generally admitted that these or really the remains of mammalia (although it was at first suggested that they might be reptiles), and the only question open to controversy is limited to this point, whether the fossil mammalia found in the Lower Oolite of Oxfordshire ought to be referred to the marsupial quadrupeds, or to the ordinary placental series. Cuvier had long ago pointed out a peculiarity in the form of the angular process (c, Figures 342 and 343) of the lower jaw, as a character of the genus Didelphys; and Professor Owen has since confirmed the doctrine of its generality in the entire marsupial series. In all these pouched quadrupeds this process is turned inward, as at c, d, Figure 342, in the Brazilian opossum, whereas in the placental series, as at c, Figures 340 and 341, there is an almost entire absence of such inflection. The Tupaia Tana of Sumatra has been selected by Mr. Waterhouse for this illustration, because the jaws of that small insectivorous quadruped bear a great resemblance to those of the Stonesfield Amphitherium. By clearing away the matrix from the specimen of Amphitherium Prevostii here represented (Figure 344), Professor Owen ascertained that the angular process (c) bent inward in a slighter degree than in any of the known marsupialia; in short, the inflection does not exceed that of the mole or hedgehog. This fact made him doubt whether the Amphitherium might not be an insectivorous placental, although it offered some points of approximation in its osteology to the marsupials, especially to the Myrmecobius, a small insectivorous quadruped of Australia, which has nine molars on each side of the lower jaw, besides a canine and three incisors. (A figure of this recent Myrmecobius will be found in my Principles of Geology chapter 9.) Another species of Amphitherium has been found at Stonesfield (Figure 345), which differs from the former (Figure 344) principally in being larger.

(FIGURE 346. Phascolotherium Bucklandi, Broderip, sp. a. Natural size. b. Molar of same, magnified.)

The second mammiferous genus discovered in the same slates was named originally by Mr. Broderip Didelphys Bucklandi (see Figure 346), and has since been called Phascolotherium by Owen. It manifests a much stronger likeness to the marsupials in the general form of the jaw, and in the extent and position of its inflected angle, while the agreement with the living genus Didelphys in the number of the pre-molar and molar teeth is complete. (Owen's British Fossil Mammals page 62.)

In 1854 the remains of another mammifer, small in size, but larger than any of those previously known, was brought to light. The generic name of Stereognathus was given to it, and, as is usually the case in these old rocks (see above), it consisted of part of a lower jaw, in which were implanted three double-fanged teeth, differing in structure from those of all other known recent or extinct mammals.

PLANTS OF THE OOLITE.

(FIGURE 347. Portion of a fossil fruit of Podocarya Bucklandi, Ung., magnified. (Buckland's Bridgewater Treatise Plate 63.) Inferior Oolite, Charmouth, Dorset.)

(FIGURE 348. Cone of fossil Araucaria sphaerocarpa, Carruthers. Inferior Oolite. Bruton, Somersetshire. One-third diameter of original. In the collection of the British Museum.)

The Araucarian pines, which are now abundant in Australia and its islands, together with marsupial quadrupeds, are found in like manner to have accompanied the marsupials in Europe during the Oolitic period (see Figure 348). In the same rock endogens of the most perfect structure are met with, as, for example, fruits allied to the Pandanus, such as the Kaidacarpum ooliticum of Carruthers in the Great Oolite, and the Podocarya of Buckland (see Figure 347) in the Inferior Oolite.

FULLER'S EARTH.

(FIGURE 349. Ostrea acuminata. Fuller's Earth.)

Between the Great and Inferior Oolite near Bath, an argillaceous deposit, called "the fuller's earth," occurs; but it is wanting in the north of England. It abounds in the small oyster represented in Figure 349. The number of mollusca known in this deposit is about seventy; namely, fifty Lamellibranchiate Bivalves, ten Brachiopods, three Gasteropods, and seven or eight Cephalopods.

INFERIOR OOLITE.

This formation consists of a calcareous freestone, usually of small thickness, but attaining in some places, as in the typical area of Cheltenham and the Western Cotswolds, a thickness of 250 feet. It sometimes rests upon yellow sands, formerly classed as the sands of the Inferior Oolite, but now regarded as a member of the Upper Lias. These sands repose upon the Upper Lias clays in the south and west of England. The Collyweston slate, formerly classed with the Great Oolite, and supposed to represent in Northamptonshire the Stonesfield slate, is now found to belong to the Inferior Oolite, both by community of species and position in the series. The Collyweston beds, on the whole, assume a much more marine character than the Stonesfield slate. Nevertheless, one of the fossil plants Aroides Stutterdi, Carruthers, remarkable, like the Pandanaceous species before mentioned (Figure 347) as a representative of the monocotyledonous class, is common to the Stonesfield beds in Oxfordshire.

(FIGURE 350. Hemitelites Brownii, Goepp. Syn. Phlebopteris contigua, Lind. and Hutt. Lower carbonaceous strata, Inferior Oolite shales. Gristhorpe, Yorkshire.)

The Inferior Oolite of Yorkshire consists largely of shales and sandstones, which assume much the aspect of a true coal-field, thin seams of coal having actually been worked in them for more than a century. A rich harvest of fossil ferns has been obtained from them, as at Gristhorpe, near Scarborough (Figure 350). They contain also Cycadeae, of which family a magnificent specimen has been described by Mr. Williamson under the name Zamia gigas, and a fossil called Equisetum Columnare (see Figure 397), which maintains an upright position in sandstone strata over a wide area. Shells of Estheria and Unio, collected by Mr. Bean from these Yorkshire coal-bearing beds, point to the estuary or fluviatile origin of the deposit.

At Brora, in Sutherlandshire, a coal formation, probably coeval with the above, or at least belonging to some of the lower divisions of the Oolitic period, has been mined extensively for a century or more. It affords the thickest stratum of pure vegetable matter hitherto detected in any secondary rock in England. One seam of coal of good quality has been worked three and a half feet thick, and there are several feet more of pyritous coal resting upon it.

(FIGURE 351. Terebratula fimbria, Sowerby. Inferior Oolite marl. Cotswold Hills.)

(FIGURE 352. Rhynchonella spinosa, Schloth. Inferior Oolite.)

(FIGURE 353. Pholadomya fidicula, Sowerby. One-third natural size. Inferior Oolite.)

(FIGURE 354. Pleurotomaria granulata, Sowerby. Ferruginous Oolite, Normandy. Inferior Oolite, England.)

(FIGURE 355. Pleurotomaria ornata, Sowerby Sp. Inferior Oolite.)

(FIGURE 356. Collyrites (Dysaster) ringens, Agassiz. Inferior Oolite, Somersetshire.)
(FIGURE 357. Ammonites Humphresianus, Sowerby. Inferior Oolite.)
(FIGURE 358. Ammonites Braikenridgii, Sowerby. Oolite, Scarborough. Inferior Oolite, Dundry; Calvados; etc.)
(FIGURE 359. Ostrea Marshii. One-half natural size. Middle and Lower Oolite.)

Among the characteristic shells of the Inferior Oolite, I may instance Terebratula fimbria (Figure 351), Rhynchonella spinosa (Figure 352), and Pholadomya fidicula (Figure 353). The extinct genus Pleurotomaria is also a form very common in this division as well as in the Oolitic system generally. It resembles the Trochus in form, but is marked by a deep cleft (a, Figures 354, 355) on one side of the mouth. The Collyrites (Dysaster) ringens (Figure 356) is an Echinoderm common to the Inferior Oolite of England and France, as are the two Ammonites (Figures 357, 358).

PALAEONTOLOGICAL RELATIONS OF THE OOLITIC STRATA.

Observations have already been made on the distinctness of the organic remains of the Oolitic and Cretaceous strata, and the proportion of species common to the different members of the Oolite. Between the Lower Oolite and the Lias there is a somewhat greater break, for out of 256 mollusca of the Upper Lias, thirty- seven species only pass up into the Inferior Oolite.

In illustration of shells having a great vertical range, it may be stated that in England some few species pass up from the Lower to the Upper Oolite, as, for example, Rhynchonella obsoleta, Lithodomus inclusus, Pholadomya ovalis, and Trigonia costata.

(FIGURE 360. Ammonites macrocephalus, Schloth. One-third natural size. Great Oolite and Oxford Clay.)

Of all the Jurassic Ammonites of Great Britain, A. macrocephalus (Figure 360), which is common to the Great Oolite and Oxford Clay, has the widest range.

We have every reason to conclude that the gaps which occur, both between the larger and smaller sections of the English Oolites, imply intervals of time, elsewhere represented by fossiliferous strata, although no deposit may have taken place in the British area. This conclusion is warranted by the partial extent of many of the minor and some of the larger divisions even in England.

CHAPTER XX.
JURASSIC GROUP— CONTINUED.— LIAS.

Mineral Character of Lias.
Numerous successive Zones in the Lias, marked by distinct Fossils, without Unconformity in the Stratification, or Change in the Mineral Character of the Deposits.
Gryphite Limestone.
Shells of the Lias.
Fish of the Lias.
Reptiles of the Lias.
Ichthyosaur and Plesiosaur.
Marine Reptile of the Galapagos Islands.
Sudden Destruction and Burial of Fossil Animals in Lias.
Fluvio-marine Beds in Gloucestershire, and Insect Limestone.
Fossil Plants.
The origin of the Oolite and Lias, and of alternating Calcareous and Argillaceous Formations.

LIAS.

The English provincial name of Lias has been very generally adopted for a formation of argillaceous limestone, marl, and clay, which forms the base of the Oolite, and is classed by many geologists as part of that group. The peculiar aspect which is most characteristic of the Lias in England, France, and Germany, is an alternation of thin beds of blue or grey limestone, having a surface which becomes light-brown when weathered, these beds being separated by dark-coloured, narrow argillaceous partings, so that the quarries of this rock, at a distance, assume a striped and ribbon-like appearance.

The Lias has been divided in England into three groups, the Upper, Middle, and Lower. The Upper Lias consists first of sands, which were formerly regarded as the base of the Oolite, but which, according to Dr. Wright, are by their fossils more properly referable to the Lias; secondly, of clay shale and thin beds of limestone. The Middle Lias, or marl-stone series, has been divided into three zones; and the Lower Lias, according to the labours of Quenstedt, Oppel, Strickland, Wright, and others, into seven zones, each marked by its own group of fossils. This Lower Lias averages from 600 to 900 feet in thickness.

From Devon and Dorsetshire to Yorkshire all these divisions, observes Professor Ramsay, are constant; and from top to bottom we can not assert that anywhere there is actual unconformity between any two subdivisions, whether of the larger or smaller kind.

In the whole of the English Lias there are at present known about 937 species of mollusca, and of these 267 are Cephalopods, of which class more than two-thirds are Ammonites, the Nautilus and Belemnite also abounding. The whole series has been divided by zones characterised by particular Ammonites; for while other families of shells pass from one division to another in numbers varying from about 20 to 50 per cent, these cephalopods are almost always limited to single zones, as Quenstedt and Oppel have shown for Germany, and Dr. Wright and others for England.

As no actual unconformity is known from the top of the Upper to the bottom of the Lower Lias, and as there is a marked uniformity in the mineral character of almost all the strata, it is somewhat difficult to account even for such partial breaks as have been alluded to in the succession of species, if we reject the hypothesis that the old species were in each case destroyed at the close of the deposition of the rocks containing them, and replaced by the creation of new forms when the succeeding formation began. I agree with Professor Ramsay in not accepting this hypothesis. No doubt some of the old species occasionally died out, and left no representatives in Europe or elsewhere; others were locally exterminated in the struggle for life by species which invaded their ancient domain, or by varieties better fitted for a new state of things. Pauses also of vast duration may have occurred in the deposition of strata, allowing time for the modification of organic life throughout the globe, slowly brought about by variation accompanied by extinction of the original forms.

FOSSILS OF THE LIAS.

(FIGURE 361. Plagiostoma (Lima) giganteum, Sowerby. Inferior Oolite and Lias.)

(FIGURE 362. Gryphaea incurva, Sowerby. (G. arcuata, Lam.) Lias.)

(FIGURE 363. Avicula inaequivalvis, Sowerby. Lower Lias.)

(FIGURE 364. Avicula cygnipes, Phil. Lower Lias, Gloucestershire and Yorkshire. a. Lower valve. b. Upper valve.)

(FIGURE 365. Hippopodium ponderosum, Sowerby. 1/4 diameter. Lias, Cheltenham)

(FIGURE 366. Spiriferina (Spirifera) Walcotti, Sowerby. Lower Lias.)

(FIGURE 367. Leptaena Moorei, Davidson. Upper Lias, Ilminster.)

The name of Gryphite limestone has sometimes been applied to the Lias, in consequence of the great number of shells which it contains of a species of oyster, or Gryphaea (Figure 362). A large heavy shell called Hippopodium (Figure 365), allied to Cypricardia, is also characteristic of the upper part of the Lower Lias. In this formation occur also the Aviculas, Figures 363 and 364. The Lias formation is also remarkable for being the newest of the secondary rocks in which brachiopoda of the genera Spirifer and Leptaena (Figures 366, 367) occur, although the former is slightly modified in structure so as to constitute the subgenus Spiriferina, Davidson, and the Leptaena has dwindled to a shell smaller in size than a pea. No less than eight or nine species of Spiriferina are enumerated by Mr. Davidson as belonging to the Lias. Palliobranchiate mollusca predominate greatly in strata older than the Trias; but, so far as we yet know, they did not survive the Liassic epoch.

(FIGURE 368. Ammonites Bucklandi, Sowerby. Ammonites bisulcatus, Brug. One-eighth diameter of original. a. Side view. b. Front view, showing mouth and bisulcated keel. Characteristic of the lower part of the Lias of England and the Continent.)

(FIGURE 369. Ammonites planorbis, Sowerby. One-half diameter of original. From the base of the Lower Lias of England and the Continent.)

(FIGURE 370. Nautilus truncatus, Sowerby. Lias.)

(FIGURE 371. Ammonites bifrons, Brug. Ammonites Walcotti, Sowerby. Upper Lias shales.)

(FIGURE 372. Ammonites margaritatus, Montf. Syn. Ammonites Stokesi, Sowerby. Middle Lias.)

Allusion has already been made to numerous zones in the Lias having each their peculiar Ammonites. Two of these occur near the base of the Lower Lias, having a united thickness, varying from 40 to 80 feet. The upper of these is characterised by Ammonites Bucklandi, and the lower by Ammonites planorbis (see Figures 368, 369). (Quarterly Journal volume 16 page 376.) Sometimes, however, there is a third intermediate zone, that of Ammonites angulatus, which is the equivalent of the zone called the infra-lias on the Continent, the species of which are for the most part common to the superior group marked by Ammonites Bucklandi.

(FIGURE 373. Extracrinus (Pentacrinus) Briareus. Miller. 1/2 natural size. (Body, arms, and part of stem.) Lower Lias, Lyme Regis.)

(FIGURE 374. Palaeocoma (Ophioderma) tenuibrachiata. E. Forbes. Middle Lias, Seatown, Dorset.)

Among the Crinoids or Stone-lilies of the Lias, the Pentacrinites are conspicuous. (See Figure 373.) Of Palaeocoma (Ophioderma) Egertoni (Figure 374), referable to the Ophiuridae of Muller, perfect specimens have been met with in the Middle Lias beds of Dorset and Yorkshire.

The Extracrinus Briareus (removed by Major Austin from Pentacrinus on account of generic differences) occurs in tangled masses, forming thin beds of considerable extent, in the Lower Lias of Dorset, Gloucestershire, and Yorkshire. The remains are often highly charged with pyrites. This Crinoid, with its innumerable tentacular arms, appears to have been frequently attached to the driftwood of the liassic sea, in the same manner as Barnacles float about on wood at the present day. There is another species of Extracrinus and several of Pentacrinus in the Lias; and the latter genus is found in nearly all the formations from the Lias to the London Clay inclusive. It is represented in the present seas by the delicate and rare Pentacrinus caput-medusae of the Antilles, which, with Comatula, is one of the few surviving members of the ancient family of the Crinoids, represented by so many extinct genera in the older formations.

FISHES OF THE LIAS.

(FIGURE 375. Scales of Lepidotus gigas. Agass. a. Two of the scales detached.)

(FIGURE 376. Aechmodus Leachii and Dapedius monilifer. a. Aechmodus. Restored outline. b. Scales of Aechmodus Leachii. c. Scales of Dapedius monilifer.)

(FIGURE 377. Acrodus nobilis, Agassiz (tooth); commonly called "fossil leech." Lias, Lyme Regis, and Germany.)

The fossil fish, of which there are no less than 117 species known as British, resemble generically those of the Oolite, but differ, according to M. Agassiz, from those of the Cretaceous period. Among them is a species of Lepidotus (L. gigas, Agassiz), Figure 375, which is found in the Lias of England, France, and Germany. (Agassiz Poissons Fossiles volume 2 tab. 28, 29.) This genus was before mentioned (Chapter 18) as occurring in the Wealden, and is supposed to have frequented both rivers and sea-coasts. Another genus of Ganoids (or fish with hard, shining, and enamelled scales), called Aechmodus (Figure 376), is almost exclusively Liassic. The teeth of a species of Acrodus, also, are very abundant in the Lias (Figure 377).

(FIGURE 378. Hybodus reticulatus, Agassiz. Lias, Lyme Regis. a. Part of fin, commonly called Ichthyodorulite. b. Tooth.)

(FIGURE 379. Chimaera monstrosa. (Agassiz Poissons Fossiles volume 3 tab. C Figure 1.) a. Spine forming anterior part of the dorsal fin.)

SBut the remains of fish which have excited more attention than any others are those large bony spines called ichthyodorulites (a, Figure 378), which were once supposed by some naturalists to be jaws, and by others weapons, resembling those of the living Balistes and Silurus; but which M. Agassiz has shown to be neither the one nor the other. The spines, in the genera last mentioned, articulate with the backbone, whereas there are no signs of any such articulation in the ichthyodorulites. These last appear to have been bony spines which formed the anterior part of the dorsal fin, like that of the living genera Cestracion and Chimaera (see a, Figure 379). In both of these genera, the posterior concave face is armed with small spines, as in that of the fossil Hybodus (Figure 378), a placoid fish of the shark family found fossil at Lyme Regis. Such spines are simply imbedded in the flesh, and attached to strong muscles. "They serve," says Dr. Buckland, "as in the Chimaera (Figure 379), to raise and depress the fin, their action resembling that of a movable mast, raising and lowering backward the sail of a barge." (Bridgewater Treatise page 290.)

REPTILES OF THE LIAS.

(FIGURE 380. Skeleton of Ichthyosaurus communis, restored by Conybeare and Cuvier. a. Costal vertebrae.)

(FIGURE 381. Skeleton of Plesiosaurus dolichodeirus, restored by Reverend W.D. Conybeare. a. Cervical vertebra.)

It is not, however, the fossil fish which form the most striking feature in the organic remains of the Lias; but the Enaliosaurian reptiles, which are extraordinary for their number, size, and structure. Among the most singular of these are several species of Ichthyosaurus and Plesiosaurus (Figures 380, 381). The genus Ichthyosaurus, or fish-lizard, is not confined to this formation, but has been found in strata as high as the White Chalk of England, and as low as the Trias of Germany, a formation which immediately succeeds the Lias in the descending order. It is evident from their fish-like vertebrae, their paddles, resembling those of a porpoise or whale, the length of their tail, and other parts of their structure, that the Ichthyosaurs were aquatic. Their jaws and teeth show that they were carnivorous; and the half-digested remains of fishes and reptiles, found within their skeletons, indicate the precise nature of their food.

Mr. Conybeare was enabled, in 1824, after examining many skeletons nearly perfect, to give an ideal restoration of the osteology of this genus, and of that of the Plesiosaurus (Geological Society Transactions Second Series volume 1 page 49.). (See Figures 380, 381.) The latter animal had an extremely long neck and small head, with teeth like those of the crocodile, and paddles analogous to those of the Ichthyosaurus, but larger. It is supposed to have lived in shallow seas and estuaries, and to have breathed air like the Ichthyosaur and our modern cetacea. (Conybeare and De la Beche, Geological Transactions First Series volume 5 page 559; and Buckland Bridgewater Treatise page 203.) Some of the reptiles above mentioned were of formidable dimensions. One specimen of Ichthyosaurus platyodon, from the Lias at Lyme, now in the British Museum, must have belonged to an animal more than 24 feet in length; and there are species of Plesiosaurus which measure from 18 to 20 feet in length. The form of the Ichthyosaurus may have fitted it to cut through the waves like the porpoise; as it was furnished besides its paddles with a tail-fin so constructed as to be a powerful organ of motion; but it is supposed that the Plesiosaurus, at least the long-necked species (Figure 381), was better suited to fish in shallow creeks and bays defended from heavy breakers.

It is now very generally agreed that these extinct saurians must have inhabited the sea; and it was urged that as there are now chelonians, like the tortoise, living in fresh water, and others, as the turtle, frequenting the ocean, so there may have been formerly some saurians proper to salt, others to fresh water. The common crocodile of the Ganges is well-known to frequent equally that river and the brackish and salt water near its mouth; and crocodiles are said in like manner to be abundant both in the rivers of the Isla de Pinos (Isle of Pines), south of Cuba, and in the open sea round the coast. In 1835 a curious lizard (Amblyrhynchus cristatus) was discovered by Mr. Darwin in the Galapagos Islands. (See Darwin Naturalist's Voyage page 385 Murray.) It was found to be exclusively marine, swimming easily by means of its flattened tail, and subsisting chiefly on seaweed. One of them was sunk from the ship by a heavy weight, and on being drawn up after an hour was quite unharmed.

The families of Dinosauria, crocodiles, and Pterosauria or winged reptiles, are also represented in the Lias.

SUDDEN DESTRUCTION OF SAURIANS.

It has been remarked, and truly, that many of the fish and saurians, found fossil in the Lias, must have met with sudden death and immediate burial; and that the destructive operation, whatever may have been its nature, was often repeated.

"Sometimes," says Dr. Buckland, "scarcely a single bone or scale has been removed from the place it occupied during life; which could not have happened had the uncovered bodies of these saurians been left, even for a few hours, exposed to putrefaction, and to the attacks of fishes and other smaller animals at the bottom of the sea." (Bridgewater Treatise page 115.) Not only are the skeletons of the Ichthyosaurs entire, but sometimes the contents of their stomachs still remain between their ribs, as before remarked, so that we can discover the particular species of fish on which they lived, and the form of their excrements. Not unfrequently there are layers of these coprolites, at different depths in the Lias, at a distance from any entire skeletons of the marine lizards from which they were derived; "as if," says Sir H. De la Beche, "the muddy bottom of the sea received small sudden accessions of matter from time to time, covering up the coprolites and other exuviae which had accumulated during the intervals." (Geological Researches page 334.) It is further stated that, at Lyme Regis, those surfaces only of the coprolites which lay uppermost at the bottom of the sea have suffered partial decay, from the action of water before they were covered and protected by the muddy sediment that has afterwards permanently enveloped them.

Numerous specimens of the Calamary or pen-and-ink fish, (Geoteuthis bollensis) have also been met with in the Lias at Lyme, with the ink-bags still distended, containing the ink in a dried state, chiefly composed of carbon, and but slightly impregnated with carbonate of lime. These Cephalopoda, therefore, must, like the saurians, have been soon buried in sediment; for, if long exposed after death, the membrane containing the ink would have decayed. (Buckland Bridgewater Treatise page 307.)

As we know that river-fish are sometimes stifled, even in their own element, by muddy water during floods, it can not be doubted that the periodical discharge of large bodies of turbid fresh water in the sea may be still more fatal to marine tribes. In the "Principles of Geology" I have shown that large quantities of mud and drowned animals have been swept down into the sea by rivers during earthquakes, as in Java in 1699; and that indescribable multitudes of dead fishes have been seen floating on the sea after a discharge of noxious vapours during similar convulsions. But in the intervals between such catastrophes, strata may have accumulated slowly in the sea of the Lias, some being formed chiefly of one description of shell, such as ammonites, others of gryphites.

FRESH-WATER DEPOSITS.— INSECT BEDS.

(FIGURE 382. Wing of a neuropterous insect, from the Lower Lias, Gloucestershire. (Reverend P.B. Brodie.))

From the above remarks the reader will infer that the Lias is for the most part a marine deposit. Some members, however, of the series have an estuarine character, and must have been formed within the influence of rivers. At the base of the Upper and Lower Lias respectively, insect-beds appear to be almost everywhere present throughout the Midland and South-western districts of England. These beds are crowded with the remains of insects, small fish, and crustaceans, with occasional marine shells. One band in Gloucestershire, rarely exceeding a foot in thickness, has been named the "insect limestone." It passes upward, says the Reverend P.B. Brodie, into a shale containing Cypris and Estheria, and is full of the wing-cases of several genera of Coleoptera, with some nearly entire beetles, of which the eyes are preserved. (A History of Fossil Insects etc 1846 London.) The nervures of the wings of neuropterous insects (Figure 382) are beautifully perfect in this bed. Ferns, with Cycads and leaves of monocotyledonous plants, and some apparently brackish and fresh-water shells, accompany the insects in several places, while in others marine shells predominate, the fossils varying apparently as we examine the bed nearer or farther from the ancient land, or the source whence the fresh water was derived. After studying 300 specimens of these insects from the Lias, Mr. Westwood declares that they comprise both wood-eating and herb-devouring beetles, of the Linnean genera Elater, Carabus, etc., besides grasshoppers (Gryllus), and detached wings of dragon-flies and may-flies, or insects referable to the Linnean genera Libellula, Ephemera, Hemerobius, and Panorpa, in all belonging to no less than twenty-four families. The size of the species is usually small, and such as taken alone would imply a temperate climate; but many of the associated organic remains of other classes must lead to a different conclusion.

FOSSIL PLANTS.

Among the vegetable remains of the Lias, several species of Zamia have been found at Lyme Regis, and the remains of coniferous plants at Whitby. M. Ad. Brongniart enumerates forty-seven liassic acrogens, most of them ferns; and fifty gymnosperms, of which thirty-nine are cycads, and eleven conifers. Among the cycads the predominance of Zamites, and among the ferns the numerous genera with leaves having reticulated veins (as in Figure 349), are mentioned as botanical characteristics of this era. (Tableau des Veg. Foss. 1849 page 105.) The absence as yet from the Lias and Oolite of all signs of dicotyledonous angiosperms is worthy of notice. The leaves of such plants are frequent in tertiary strata, and occur in the Cretaceous, though less plentifully (see Chapter 17). The angiosperms seem, therefore, to have been at the least comparatively rare in these older secondary periods, when more space was occupied by the Cycads and Conifers.

ORIGIN OF THE OOLITE AND LIAS.

The entire group of Oolite and Lias consists of repeated alternations of clay, sandstone, and limestone, following each other in the same order. Thus the clays of the Lias are followed by the sands now considered (see Chapter 20) as belonging to the same formation, though formerly referred to the Inferior Oolite, and these sands again by the shelly and coralline limestone called the Great or Bath Oolite. So, in the Middle Oolite, the Oxford Clay is followed by calcareous grit and coral rag; lastly, in the Upper Oolite, the Kimmeridge Clay is followed by the Portland Sand and limestone (see Figure 298). (Conybeare and Philips's Outlines etc. page 166.) The clay beds, however, as Sir H. de la Beche remarks, can be followed over larger areas than the sand or sandstones. (Geological Researches page 337.) It should also be remembered that while the Oolite system becomes arenaceous and resembles a coal-field in Yorkshire, it assumes in the Alps an almost purely calcareous form, the sands and clays being omitted; and even in the intervening tracts it is more complicated and variable than appears in ordinary descriptions. Nevertheless, some of the clays and intervening limestones do retain, in reality, a pretty uniform character for distances of from 400 to 600 miles from east to west and north to south.

In order to account for such a succession of events, we may imagine, first, the bed of the ocean to be the receptacle for ages of fine argillaceous sediment, brought by oceanic currents, which may have communicated with rivers, or with part of the sea near a wasting coast. This mud ceases, at length, to be conveyed to the same region, either because the land which had previously suffered denudation is depressed and submerged, or because the current is deflected in another direction by the altered shape of the bed of the ocean and neighbouring dry land. By such changes the water becomes once more clear and fit for the growth of stony zoophytes. Calcareous sand is then formed from comminuted shell and coral, or, in some cases, arenaceous matter replaces the clay; because it commonly happens that the finer sediment, being first drifted farthest from coasts, is subsequently overspread by coarse sand, after the sea has grown shallower, or when the land, increasing in extent, whether by upheaval or by sediment filling up parts of the sea, has approached nearer to the spots first occupied by fine mud.

The increased thickness of the limestones in those regions, as in the Alps and Jura, where the clays are comparatively thin, arises from the calcareous matter having been derived from species of corals and other organic beings which live in clear water, far from land, to the growth of which the influx of mud would be unfavourable. Portions therefore of these clays and limestones have probably been formed contemporaneously to a greater extent than we can generally prove, for the distinctness of the species of organic beings would be caused by the difference of conditions between the more littoral and the more pelagic areas and the different depths and nature of the sea-bottom. Independently of those ascending and descending movements which have given rise to the superposition of the limestones and clays, and by which the position of land and sea are made in the course of ages to vary, the geologist has the difficult task of allowing for the contemporaneous thinning out in one direction and thickening in another, of the successive organic and inorganic deposits of the same era.

CHAPTER XXI.
TRIAS, OR NEW RED SANDSTONE GROUP.

Beds of Passage between the Lias and Trias, Rhaetic Beds.
Triassic Mammifer.
Triple Division of the Trias.
Keuper, or Upper Trias of England.
Reptiles of the Upper Trias.
Foot-prints in the Bunter formation in England.
Dolomitic Conglomerate of Bristol.
Origin of Red Sandstone and Rock-salt.
Precipitation of Salt from inland Lakes and Lagoons.
Trias of Germany.
Keuper.
St. Cassian and Hallstadt Beds.
Peculiarity of their Fauna.
Muschelkalk and its Fossils.
Trias of the United States.
Fossil Foot-prints of Birds and Reptiles in the Valley of the Connecticut.
Triassic Mammifer of North Carolina.
Triassic Coal-field of Richmond, Virginia.
Low Grade of early Mammals favourable to the Theory of Progressive Development.

BEDS OF PASSAGE BETWEEN THE LIAS AND TRIAS— RHAETIC BEDS.

We have mentioned in the last chapter that the base of the Lower Lias is characterised, both in England and Germany, by beds containing distinct species of Ammonites, the lowest subdivision having been called the zone of Ammonites planorbis. Below this zone, on the boundary line between the Lias and the strata of which we are about to treat, called "Trias," certain cream-coloured limestones devoid of fossils are usually found. These white beds were called by William Smith the White Lias, and they have been shown by Mr. Charles Moore to belong to a formation similar to one in the Rhaetian Alps of Bavaria, to which Mr. Gumbel has applied the name of Rhaetic. They have also long been known as the Koessen beds in Germany, and may be regarded as beds of passage between the Lias and Trias. They are named the Penarth beds by the Government surveyors of Great Britain, from Penarth, near Cardiff, in Glamorganshire, where they sometimes attain a thickness of fifty feet.

(FIGURE 383. Cardium rhaeticum, Merrian. Natural size. Rhaetic Beds.)

(FIGURE 384. Pecten Valoniensis. Dfr. 1/2 natural size. Portrush, Ireland, etc. Rhaetic Beds.)

(FIGURE 385. Avicula contorta. Portlock. Portrush, Ireland, etc. Natural size. Rhaetic Beds.)

The principal member of this group has been called by Dr. Wright the Avicula contorta bed, as this shell is very abundant, and has a wide range in Europe.
(Dr. Wright, on Lias and Bone Bed, Quarterly Geological Journal 1860 volume 16.)
General Portlock first described the formation as it occurs at Portrush, in
Antrim, where the Avicula contorta is accompanied by Pecten Valoniensis, as in Germany.

The best known member of the group, a thin band or bone-breccia, is conspicuous among the black shales in the neighbourhood of Axmouth in Devonshire, and in the cliffs of Westbury-on-Severn, as well as at Aust and other places on the borders of the Bristol Channel. It abounds in the remains of saurians and fish, and was formerly classed as the lowest bed of the Lias; but Sir P. Egerton first pointed out, in 1841, that it should be referred to the Upper New Red Sandstone, because it contained an assemblage of fossil fish which are either peculiar to this stratum, or belong to species well-known in the Muschelkalk of Germany. These fish belong to the genera Acrodus, Hybodus, Gyrolepis, and Saurichthys.

(FIGURE 386. Hybodus plicatilis, Agassiz. Teeth. Bone-bed, Aust and Axmouth.)

(FIGURE 387. Saurichthys apicalis, Agassiz. Tooth; natural size and magnified. Axmouth.)

(FIGURE 388. Gyrolepis tenuistriatus, Agassiz. Scale; natural size and magnified. Axmouth.)

Among those common to the English bone-bed and the Muschelkalk of Germany are Hybodus plicatilis (Figure 386), Saurychthys apicalis (Figure 387), Gyrolepis tenuistriatus (Figure 388), and G. Albertii. Remains of saurians, Plesiosaurus among others, have also been found in the bone-bed, and plates of an Encrinus. It may be questioned whether some of those fossils which have the most Triassic character may not have been derived from the destruction of older strata, since in bone-beds, in general, many of the organic remains are undoubtedly derivative.

TRIASSIC MAMMIFER.

(FIGURE 389. Microlestes antiquus, Plieninger. Molar tooth, magnified. Upper Trias. Diegerloch, near Stuttgart, Wurtemberg. a. View of inner side? b. Same, outer side? c. Same in profile. d. Crown of same.)

In North-western Germany, as in England, there occurs beneath the Lias a remarkable bone breccia. It is filled with shells and with the remains of fishes and reptiles, almost all the genera of which, and some even of the species, agree with those of the subjacent Trias. This breccia has accordingly been considered by Professor Quenstedt, and other German geologists of high authority, as the newest or uppermost part of the Trias. Professor Plieninger found in it, in 1847, the molar tooth of a small Triassic mammifer, called by him Microlestes antiquus. He inferred its true nature from its double fangs, and from the form and number of the protuberances or cusps on the flat crown; and considering it as predaceous, probably insectivorous, he called it Microlestes from micros, little, and lestes, a beast of prey. Soon afterwards he found a second tooth, also at the same locality, Diegerloch, about two miles to the south-east of Stuttgart.

No anatomist had been able to give any feasible conjecture as to the affinities of this minute quadruped until Dr. Falconer, in 1857, recognised an unmistakable resemblance between its teeth and the two back molars of his new genus Plagiaulax (Figure 306), from the Purbeck strata. This would lead us to the conclusion that Microlestes was marsupial and plant-eating.

In Wurtemberg there are two bone-beds, namely, that containing the Microlestes, which has just been described, which constitutes, as we have seen, the uppermost member of the Trias, and another of still greater extent, and still more rich in the remains of fish and reptiles, which is of older date, intervening between the Keuper and Muschelkalk.

The genera Saurichthys, Hybodus, and Gyrolepis are found in both these breccias, and one of the species, Saurichthys Mongeoti, is common to both bone-beds, as is also a remarkable reptile called Nothosaurus mirabilis. The saurian called Belodon by H. von Meyer, of the Thecodont family, is another Triassic form, associated at Diegerloch with Microlestes.

TRIAS OF ENGLAND.

Between the Lias and the Coal (or Carboniferous group) there is interposed, in the midland and western counties of England, a great series of red loams, shales, and sandstones, to which the name of the "New Red Sandstone formation" was first given, to distinguish it from other shales and sandstones called the "Old Red," often identical in mineral character, which lie immediately beneath the coal. The name of "Red Marl" has been incorrectly applied to the red clays of this formation, as before explained (Chapter 2), for they are remarkably free from calcareous matter. The absence, indeed, of carbonate of lime, as well as the scarcity of organic remains, together with the bright red colour of most of the rocks of this group, causes a strong contrast between it and the Jurassic formations before described.

The group in question is more fully developed in Germany than in England or France. It has been called the Trias by German writers, or the Triple Group, because it is separable into three distinct formations, called the "Keuper," the "Muschelkalk," and the "Bunter-sandstein." Of these the middle division, or the Muschelkalk, is wholly wanting in England, and the uppermost (Keuper) and lowest (Bunter) members of the series are not rich in fossils.

UPPER TRIAS OR KEUPER.

In certain grey indurated marls below the bone-bed Mr. Boyd Dawkins has found at Watchet, on the coast of Somersetshire, a molar tooth of Microlestes, enabling him to refer to the Trias strata formerly supposed to be Liassic. Mr. Charles Moore had previously discovered many teeth of mammalia of the same family near Frome, in Somersetshire, in the contents of a vertical fissure traversing a mass of carboniferous limestone. The top of this fissure must have communicated with the bed of the Triassic sea, and probably at a point not far from the ancient shore on which the small marsupials of that era abounded.

This upper division of the Trias called the Keuper is of great thickness in the central counties of England, attaining, according to Mr. Hull's estimate, no less than 3450 feet in Cheshire, and it covers a large extent of country between Lancashire and Devonshire.

(FIGURE 390. Estheria minuta, Bronn.)

In Worcestershire and Warwickshire in sandstone belonging to the uppermost part of the Keuper the bivalve crustacean Estheria minuta occurs. The member of the English "New Red" containing this shell, in those parts of England, is, according to Sir Roderick Murchison and Mr. Strickland, 600 feet thick, and consists chiefly of red marl or slate, with a band of sandstone. Ichthyodorulites, or spines of Hybodus, teeth of fishes, and footprints of reptiles were observed by the same geologists in these strata.

(FIGURE 391. Hyperodapedon Gordoni. Left palate, maxillary. (Showing the two rows of palatal teeth on opposite sides of the jaw.) a. Under surface. b. Exterior right side.)

In the Upper Trias or Keuper the remains of two saurians of the order Lacertilia have been found. The one called Rhynchosaurus occurred at Grinsell near Shrewsbury, and is characterised by having a small bird-like skull and jaws without teeth. The other Hyperodapedon (Figure 391) was first noticed in 1858, near Elgin, in strata now recognised as Upper Triassic, and afterwards in beds of about the same age in the neighbourhood of Warwick. Remains of the same genus have been found both in Central India and Southern Africa in rocks believed to be of Triassic age. The Hyperodapedon has been shown by Professor Huxley to be a terrestrial reptile having numerous palatal teeth, and closely allied to the living Sphenodon of New Zealand.

The recent discoveries of a living saurian in New Zealand so closely allied to this supposed extinct division of the Lacertilia seems to afford an illustration of a principle pointed out by Mr. Darwin of the survival in insulated tracts, after many changes in physical geography, of orders of which the congeners have become extinct on continents where they have been exposed to the severer competition of a larger progressive fauna.

(FIGURE 392. Tooth of Labyrinthodon; natural size. Warwick sandstone.)

(FIGURE 393. Transverse section of upper part of tooth of Labyrinthodon Jaegeri, Owen (Mastodonsaurus Jaegeri, Meyer); natural size, and a segment magnified. a. Pulp cavity, from which the processes of pulp and dentine radiate.)

Teeth of Labyrinthodon (Figure 392) found in the Keuper in Warwickshire were examined microscopically by Professor Owen, and compared with other teeth from the German Keuper. He found after careful investigation that neither of them could be referred to true saurians, although they had been named Mastodonsaurus and Phytosaurus by Jager. It appeared that they were of the Batrachian order, and of gigantic dimensions in comparison with any representatives of that order now living. Both the Continental and English fossil teeth exhibited a most complicated texture, differing from that previously observed in any reptile, whether recent or extinct, but most nearly analogous to the Ichthyosaurus. A section of one of these teeth exhibits a series of irregular folds, resembling the labyrinthic windings of the surface of the brain; and from this character Professor Owen has proposed the name Labyrinthodon for the new genus. Figure 393 of part of one is given from his "Odontography," plate 64, a. The entire length of this tooth is supposed to have been about three inches and a half, and the breadth at the base one inch and a half.

ROCK-SALT.

In Cheshire and Lancashire there are red clays containing gypsum and salt of the age of the Trias which are between 1000 and 1500 feet thick. In some places lenticular masses of pure rock-salt nearly 100 feet thick are interpolated between the argillaceous beds. At the base of the formation beneath the rock- salt occur the Lower Sandstones and Marl, called provincially in Cheshire "water-stones," which are largely quarried for building. They are often ripple-marked, and are impressed with numerous footprints of reptiles.

The basement beds of the Keuper rest with a slight unconformability upon an eroded surface of the "Bunter" next to be described.

LOWER TRIAS OR BUNTER.

(FIGURE 394. Single footstep of Cheirotherium. Bunter-sandstein, Saxony, one-eighth of natural size.)

(FIGURE 395. Line of footsteps on slab of sandstone. Hildburghausen, in Saxony.)

The lower division or English representative of the "Bunter" attains a thickness of 1500 feet in the counties last mentioned, according to Professor Ramsay. Besides red and green shales and red sandstones, it comprises much soft white quartzose sandstone, in which the trunks of silicified trees have been met with at Allesley Hill, near Coventry. Several of them were a foot and a half in diameter, and some yards in length, decidedly of coniferous wood, and showing rings of annual growth. (Buckland Proceedings of the Geological Society volume 2 page 439 and Murchison and Strickland Geological Transactions Second Series volume 5 page 347.) Impressions, also, of the footsteps of animals have been detected in Lancashire and Cheshire in this formation. Some of the most remarkable occur a few miles from Liverpool, in the whitish quartzose sandstone of Storton Hill, on the west side of the Mersey. They bear a close resemblance to tracks first observed in this member of the Upper New Red Sandstone, at the village of Hesseberg, near Hildburghausen, in Saxony. For many years these footprints have been referred to a large unknown quadruped, provisionally named Cheirotherium by Professor Kaup, because the marks both of the fore and hind feet resembled impressions made by a human hand. (See Figure 394.) The foot-marks at Hesseberg are partly concave, and partly in relief, the former, or the depressions, are seen upon the upper surface of the sandstone slabs, but those in relief are only upon the lower surfaces, being, in fact, natural casts, formed in the subjacent footprints as in moulds. The larger impressions, which seem to be those of the hind foot, are generally eight inches in length, and five in width, and one was twelve inches long. Near each large footstep, and at a regular distance (about an inch and a half) before it, a smaller print of a fore foot, four inches long and three inches wide, occurs. The footsteps follow each other in pairs, each pair in the same line, at intervals of fourteen inches from pair to pair. The large as well as the small steps show the great toes alternately on the right and left side; each step makes the print of five toes, the first, or great toe, being bent inward like a thumb. Though the fore and hind foot differ so much in size, they are nearly similar in form.

As neither in Germany nor in England had any bones or teeth been met with in the same identical strata as the footsteps, anatomists indulged, for several years, in various conjectures respecting the mysterious animals from which they might have been derived. Professor Kaup suggested that the unknown quadruped might have been allied to the Marsupialia; for in the kangaroo the first toe of the fore foot is in a similar manner set obliquely to the others, like a thumb, and the disproportion between the fore and hind feet is also very great. But M. Link conceived that some of the four species of animals of which the tracks had been found in Saxony might have been gigantic Batrachians, and when it was afterwards inferred that the Labyrinthodon was an air-breathing reptile, it was conjectured by Professor Owen that it might be one and the same as the Cheirotherium.

DOLOMITIC CONGLOMERATE OF BRISTOL.

(FIGURE 396. Tooth of Thecodontosaurus; three times magnified. Riley and Stutchbury. Dolomitic conglomerate. Redland, near Bristol.)

Near Bristol, in Somersetshire, and in other counties bordering the Severn, the lowest strata belonging to the Triassic series consist of a conglomerate or breccia resting unconformably upon the Old Red Sandstone, and on different members of the Carboniferous rocks, such as the Coal Measures, Millstone Grit, and Mountain Limestone. This mode of superposition will be understood by reference to the section below Dundry Hill (Figure 85), where No. 4 is the dolomitic conglomerate. Such breccias may have been partly the result of the subaerial waste of an old land-surface which gradually sank down and suffered littoral denudation in proportion as it became submerged. The pebbles and fragments of older rocks which constitute the conglomerate are cemented together by a red or yellow base of dolomite, and in some places the encrinites and other fossils derived from the Mountain Limestone are so detached from the parent rocks that they have the deceptive appearance of belonging to a fauna contemporaneous with the dolomitic beds in which they occur. The imbedded fragments are both rounded and angular, some consisting of sandstone from the coal-measures, being of vast size, and weighing nearly a ton. Fractured bones and teeth of saurians which are truly of contemporaneous origin are dispersed through some parts of the breccia, and two of these reptiles called Thecodont saurians, named from the manner in which the teeth were implanted in the jawbone, obtained great celebrity because the patches of red conglomerate in which they were found, near Bristol, were originally supposed to be of Permian or Palaeozoic age, and therefore the only representatives in England of vertebrate animals of so high a grade in rocks of such antiquity. The teeth of these saurians are conical, compressed, and with finely serrated edges (see Figure 396); they are referred by Professor Huxley to the Dinosaurian order.

ORIGIN OF RED SANDSTONE AND ROCK-SALT.

In various parts of the world, red and mottled clays and sandstones, of several distinct geological epochs, are found associated with salt, gypsum, and magnesian limestone, or with one or all of these substances. There is, therefore, in all likelihood, a general cause for such a coincidence. Nevertheless, we must not forget that there are dense masses of red and variegated sandstones and clays, thousands of feet in thickness, and of vast horizontal extent, wholly devoid of saliferous or gypseous matter. There are also deposits of gypsum and of common salt, as in the blue-clay formation of Sicily, without any accompanying red sandstone or red clay.

These red deposits may be accounted for by the decomposition of gneiss and mica schist, which in the eastern Grampians of Scotland has produced a mass of detritus of precisely the same colour as the Old Red Sandstone.

It is a general fact, and one not yet accounted for, that scarcely any fossil remains are ever preserved in stratified rocks in which this oxide of iron abounds; and when we find fossils in the New or Old Red Sandstone in England, it is in the grey, and usually calcareous beds, that they occur. The saline or gypseous interstratified beds may have been produced by submarine gaseous emanations, or hot mineral springs, which often continue to flow in the same spots for ages. Beds of rock-salt are, however, more generally attributed to the evaporation of lakes or lagoons communicating at intervals with the ocean. In Cheshire two beds of salt occur of the extraordinary thickness of 90 or even 100 feet, and extending over an area supposed to be 150 miles in diameter. The adjacent beds present ripple-marked sandstones and footprints of animals at so many levels as to imply that the whole area underwent a slow and gradual depression during the formation of the red sandstone.

Major Harris, in his "Highlands of Ethiopia," describes a salt lake, called the Bahr Assal, near the Abyssinian frontier, which once formed the prolongation of the Gulf of Tadjara, but was afterwards cut off from the gulf by a broad bar of lava or of land upraised by an earthquake. "Fed by no rivers, and exposed in a burning climate to the unmitigated rays of the sun, it has shrunk into an elliptical basin, seven miles in its transverse axis, half filled with smooth water of the deepest caerulean hue, and half with a solid sheet of glittering snow-white salt, the offspring of evaporation." "If," says Mr. Hugh Miller, "we suppose, instead of a barrier of lava, that sand-bars were raised by the surf on a flat arenaceous coast during a slow and equable sinking of the surface, the waters of the outer gulf might occasionally topple over the bar, and supply fresh brine when the first stock had been exhausted by evaporation."

The Runn of Cutch, as I have shown elsewhere (Principles of Geology chapter 27.), is a low region near the delta of the Indus, equal in extent to about a quarter of Ireland, which is neither land nor sea, being dry during part of every year, and covered by salt water during the monsoons. Here and there its surface is incrusted over with a layer of salt caused by the evaporation of sea-water. A subsiding movement has been witnessed in this country during earthquakes, so that a great thickness of pure salt might result from a continuation of such sinking.

TRIAS OF GERMANY.

In Germany, as before hinted, chapter 21, the Trias first received its name as a Triple Group, consisting of two sandstones with an intermediate marine calcareous formation, which last is wanting in England.

NOMENCLATURE OF TRIAS.
COLUMN 1: GERMAN.
COLUMN 2: FRENCH.
COLUMN 3: ENGLISH.

Keuper: Marnes irisees: Saliferous and gypseous shales and sandstone.
Muschelkalk: Muschelkalk, ou calcaire coquillier: Wanting in England.
Bunter-sandstein: Gres bigarre: Sandstone and quartzose conglomerate.

KEUPER.

(FIGURE 397. Equisetites columnaris. (Syn. Equisetum columnare.) Fragment of stem, and a small portion of same magnified. Keuper.)

The first of these, or the Keuper, underlying the beds before described as Rhaetic, attains in Wurtemberg a thickness of about 1000 feet. It is divided by Alberti into sandstone, gypsum, and carbonaceous clay-slate. (Monog. des Bunter- Sandsteins.) Remains of reptiles called Nothosaurus and Phytosaurus, have been found in it with Labyrinthodon; the detached teeth, also, of placoid fish and of Rays, and of the genera Saurichthys and Gyrolepis (Figures 387, 388). The plants of the Keuper are generically very analogous to those of the oolite and lias, consisting of ferns, equisetaceous plants, cycads, and conifers, with a few doubtful monocotyledons. A few species such as Equisetites columnaris, are common to this group and the oolite.

ST. CASSIAN AND HALLSTADT BEDS (SEE MAP, FIGURE 398).

(FIGURE 398. Map of Tyrol and Styria showing St. Cassian and Hallstadt Beds.)

The sandstones and clay of the Keuper resemble the deposits of estuaries and a shallow sea near the land, and afford, in the north-west of Germany, as in France and England, but a scanty representation of the marine life of that period. We might, however, have anticipated, from its rich reptilian fauna, that the contemporaneous inhabitants of the sea of the Keuper period would be very numerous, should we ever have an opportunity of bringing their remains to light. This, it is believed, has at length been accomplished, by the position now assigned to certain Alpine rocks called the "St. Cassian beds," the true place of which in the series was until lately a subject of much doubt and discussion. It has been proved that the Hallstadt beds on the northern flanks of the Austrian Alps correspond in age with the St. Cassian beds on their southern declivity, and the Austrian geologists, M. Suess of Vienna and others, have satisfied themselves that the Hallstadt formation is referable to the period of the Upper Trias. Assuming this conclusion to be correct, we become acquainted suddenly and unexpectedly with a rich marine fauna belonging to a period previously believed to be very barren of organic remains, because in England, France, and Northern Germany the upper Trias is chiefly represented by beds of fresh or brackish water origin.

(FIGURE 399. Scoliostoma, St. Cassian.)
(FIGURE 400. Platystoma Suessii, Hornes. From Hallstadt.)
(FIGURE 401. Koninckia Leonhardi, Wissmann. a. Ventral view. Part of ventral valve removed to show the vascular impressions of dorsal valve. b. Interior of dorsal valve, showing spiral processes restored. c. Vertical section of both valves. Part shaded black showing place occupied by the animal, and the dorsal valve following the curve of the ventral.)

About 600 species of invertebrate fossils occur in the Hallstadt and St. Cassian beds, many of which are still undescribed; some of the Mollusca are of new and peculiar genera, as Scoliostoma, Figure 399, and Platystoma, Figure 400, among the Gasteropoda; and Koninckia, Figure 401, among the Brachiopoda.

TABLE 21.1 GENERA OF FOSSIL MOLLUSCA IN THE ST. CASSIAN AND HALLSTADT BEDS.

COLUMN 1: COMMON TO OLDER ROCKS.
Orthoceras.
Bactrites.
Macrocheilus.

Loxonema.
Holopella.
Murchisonia.
Porcellia.
Athyris.
Retzia.
Cyrtina.
Euomphalus.
    COLUMN 2: CHARACTERISTIC TRIASSIC GENERA.
Ceratites.
Cochloceras.
Choristoceras.
Rhabdoceras.
Aulacoceras.
Scoliostoma (reaches its maximum in the Trias, but passes down to older rocks).
Naticella.
Platystoma.
Ptychostoma.
Euchrysalis.
Halobia.
Hornesia.
Amphiclina.
Koninckia.
Cassianella. (Reach their maximum in the Trias, but pass up to newer rocks.)
Myophoria. (Reach their maximum in the Trias, but pass up to newer rocks.)
    COLUMN 3: COMMON TO NEWER ROCKS.
Ammonites.
Chemnitzia.
Cerithium.
Monodonta.
Opis.
Sphoera.
Cardita.
Myoconcha.
Hinnites.
Monotis.
Plicatula.
Pachyrisma.
Thecidium.

    Table 21.1 of genera of marine shells from the Hallstadt and St. Cassian beds, drawn up first on the joint authority of M. Suess and the late Dr. Woodward, and since corrected by Messrs. Etheridge and Tate, shows how many connecting links between the fauna of primary and secondary Palaeozoic and Mesozoic rocks are supplied by the St. Cassian and Hallstadt beds.

    The first column marks the last appearance of several genera which are characteristic of Palaeozoic strata. The second shows those genera which are characteristic of the Upper Trias, either as peculiar to it, or, as in the three cases marked by asterisks, reaching their maximum of development at this era. The third column marks the first appearance in Triassic rocks of genera destined to become more abundant in later ages.

    It is only, however, when we contemplate the number of species by which each of the above-mentioned genera are represented that we comprehend the peculiarities of what is commonly called the St. Cassian fauna. Thus, for example, the Ammonite, which is not common to older rocks, is represented by no less than seventy-three species; whereas Loxonema, which is only known as common to older rocks, furnishes fifteen Triassic species. Cerithium, so abundant in tertiary strata, and which still lives, is represented by no less than fourteen species. As the Orthoceras had never been met with in the marine

Muschelkalk, much surprise was naturally felt that seven or eight species of the genus should appear in the Hallstadt beds, assuming these last to belong to the Upper Trias. Among these species are some of large dimensions, associated with large Ammonites with foliated lobes, a form never seen before so low in the series, while the Orthoceras had never been seen so high.

On the whole, the rich marine fauna of Hallstadt and St. Cassian, now generally assigned to the lowest members of the Upper Trias or Keuper, leads us to suspect that when the strata of the Triassic age are better known, especially those belonging to the period of the Bunter sandstone, the break between the Palaeozoic and Mesozoic Periods may be almost effaced. Indeed some geologists are not yet satisfied that the true position of the St. Cassian beds (containing so great an admixture of types, having at once both Mesozoic and Palaeozoic affinities) is made out, and doubt whether they have yet been clearly proved to be newer than the Muschelkalk.

MUSCHELKALK.

(FIGURE 402. Ceratites nodosus, Schloth. Muschelkalk, Germany. Side and front view, showing the denticulated outline of the septa dividing the chambers.)

(FIGURE 403. Gervilia (Avicula) socialis, Schloth. Characteristic shell of the Muschelkalk.)

The next member of the Trias in Germany, the Muschelkalk, which underlies the Keuper before described, consists chiefly of a compact greyish limestone, but includes beds of dolomite in many places, together with gypsum and rock-salt. This limestone, a formation wholly unrepresented in England, abounds in fossil shells, as the name implies. Among the Cephalopoda there are no belemnites, and no ammonites with foliated sutures, as in the Lias, and Oolite, and the Hallstadt beds; but we find instead a genus allied to the Ammonite, called Ceratites by de Haan, in which the descending lobes (Figure 402) terminate in a few small denticulations pointing inward. Among the bivalve crustacea, the Estheria minuta, Bronn (see Figure 390), is abundant, ranging through the Keuper, Muschelkalk, and Bunter-sandstein; and Gervillia socialis (Figure 403), having a similar range, is found in great numbers in the Muschelkalk of Germany, France, and Poland.

(FIGURE 404. Encrinus liliiformis, Schlott. Syn. E. moniliformis. Body, arms, and part of stem. a. Section of stem. Muschelkalk.)

(FIGURE 405. Aspidura loricata, Agassiz. a. Upper side. b. Lower side. Muschelkalk.)

(FIGURE 406. Palatal teeth of Placodus gigas. Muschelkalk.)

The abundance of the heads and stems of lily encrinites, Encrinus liliiformis (Figure 404), (or Encrinites moniliformis), shows the slow manner in which some beds of this limestone have been formed in clear sea-water. The star-fish called Aspidura loricata (Figure 405) is as yet peculiar to the Muschelkalk. In the same formation are found the skull and teeth of a reptile of the genus Placodus (see Figure 406), which was referred originally by Munster, and afterwards by Agassiz, to the class of fishes. But more perfect specimens enabled Professor Owen, in 1858, to show that this fossil animal was a Saurian reptile, which probably fed on shell-bearing mollusks, and used its short and flat teeth, so thickly coated with enamel, for pounding and crushing the shells.

BUNTER-SANDSTEIN.

(FIGURE 407. Voltzia heterophylla. (Syn. Voltzia brevifolia.) b. Portion of same magnified to show fructification. Sulzbad. Bunter-sandstein.)

The Bunter-sandstein consists of various-coloured sandstones, dolomites, and red clays, with some beds, especially in the Hartz, of calcareous pisolite or roe- stone, the whole sometimes attaining a thickness of more than 1000 feet. The sandstone of the Vosges is proved, by its fossils, to belong to this lowest member of the Triassic group. At Sulzbad (or Soultz-les-bains), near Strasburg, on the flanks of the Vosges, many plants have been obtained from the "bunter," especially conifers of the extinct genus Voltzia, of which the fructification has been preserved. (See Figure 407.) Out of thirty species of ferns, cycads, conifers, and other plants, enumerated by M. Ad. Brongniart, in 1849, as coming from the "Gres bigarre," or Bunter, not one is common to the Keuper.

The footprints of Labyrinthodon observed in the clays of this formation at Hildburghausen, in Saxony, have already been mentioned. Some idea of the variety and importance of the terrestrial vertebrate fauna of the three members of the Trias in Northern Germany may be derived from the fact that in the great monograph by the late Hermann von Meyer on the reptiles of the Trias, the remains of no less than eighty distinct species are described and figured.

TRIAS OF THE UNITED STATES.
NEW RED SANDSTONE OF THE VALLEY OF THE CONNECTICUT RIVER.

(FIGURE 408. Footprints of a bird, Turner's Falls, Valley of the Connecticut.)

In a depression of the granitic or hypogene rocks in the States of Massachusetts and Connecticut strata of red sandstone, shale, and conglomerate are found, occupying an area more than 150 miles in length from north to south, and about five to ten miles in breadth, the beds dipping to the eastward at angles varying from 5 to 50 degrees. The extreme inclination of 50 degrees is rare, and only observed in the neighbourhood of masses of trap which have been intruded into the red sandstone while it was forming, or before the newer parts of the deposit had been completed. Having examined this series of rocks in many places, I feel satisfied that they were formed in shallow water, and for the most part near the shore, and that some of the beds were from time to time raised above the level of the water, and laid dry, while a newer series, composed of similar sediment, was forming.

According to Professor Hitchcock, the footprints of no less than thirty-two species of bipeds, and twelve of quadrupeds, have been already detected in these rocks. Thirty of these are believed to be those of birds, four of lizards, two of chelonians, and six of batrachians. The tracks have been found in more than twenty places, scattered through an extent of nearly 80 miles from north to south, and they are repeated through a succession of beds attaining at some points a thickness of more than 1000 feet. (Hitchcock Mem. of the American Academy New Series volume 3 page 129 1848.)

The bipedal impressions are, for the most part, trifid, and show the same number of joints as exist in the feet of living tridactylous birds. Now, such birds have three phalangeal bones for the inner toe, four for the middle, and five for the outer one (see Figure 408); but the impression of the terminal joint is that of the nail only. The fossil footprints exhibit regularly, where the joints are seen, the same number; and we see in each continuous line of tracks the three- jointed and five-jointed toes placed alternately outward, first on the one side, and then on the other. In some specimens, besides impressions of the three toes in front, the rudiment is seen of the fourth toe behind. It is not often that the matrix has been fine enough to retain impressions of the integument or skin of the foot; but in one fine specimen found at Turner's Falls, on the Connecticut, by Dr. Deane, these markings are well preserved, and have been recognised by Professor Owen as resembling the skin of the ostrich, and not that of reptiles.

The casts of the footprints show that some of the fossil bipeds of the red sandstone of Connecticut had feet four times as large as the living ostrich, but scarcely, perhaps, larger than the Dinornis of New Zealand, a lost genus of feathered giants related to the Apteryx, of which there were many species which have left their bones and almost entire skeletons in the superficial alluvium of that island. By referring to what was said of the Iguanodon of the Wealden, the reader will perceive that the Dinosaur was somewhat intermediate between reptiles and birds, and left a series of tridactylous impressions on the sand.

To determine the exact age of the red sandstone and shale containing these ancient footprints, in the United States, is not possible at present. No fossil shells have yet been found in the deposit, nor plants in a determinable state. The fossil fish are numerous and very perfect; but they are of a peculiar type, called Ischypterus, by Sir Philip Egerton, from the great size and strength of the fulcral rays of the dorsal fin, from ischus, strength, and pteron, a fin.

The age of the Connecticut beds can not be proved by direct superposition, but may be presumed from the general structure of the country. That structure proves them to be newer than the movements to which the Appalachian or Allegheny chain owes its flexures, and this chain includes the ancient or palaeozoic coal- formation among its contorted rocks.

COAL-FIELD OF RICHMOND, VIRGINIA.

In the State of Virginia, at the distance of about 13 miles eastward of Richmond, the capital of that State, there is a coal-field occurring in a depression of the granite rocks, and occupying a geological position analogous to that of the New Red Sandstone, above-mentioned, of the Connecticut valley. It extends 26 miles from north to south, and from four to twelve from east to west.

The plants consist chiefly of zamites, calamites, equiseta, and ferns, and, upon the whole, are considered by Professor Heer to have the nearest affinity to those of the European Keuper.

The equiseta are very commonly met with in a vertical position more or less compressed perpendicularly. It is clear that they grew in the places where they are now buried in strata of hardened sand and mud. I found them maintaining their erect attitude, at points many miles apart, in beds both above and between the seams of coal. In order to explain this fact, we must suppose such shales and sandstones to have been gradually accumulated during the slow and repeated subsidence of the whole region.

(FIGURE 409. Triassic coal-shale, Richmond, Virginia. a. Estheria ovata. b. Young of same. c. Natural size of a. d. Natural size of b.)

The fossil fish are Ganoids, some of them of the genus Catopterus, others belonging to the liassic genus Tetragonolepis (Aechmodus), see Figure 376. Two species of Entomostraca called Estheria are in such profusion in some shaly beds as to divide them like the plates of mica in micaceous shales (see Figure 409).

These Virginian coal-measures are composed of grits, sandstones, and shales, exactly resembling those of older or primary date in America and Europe, and they rival, or even surpass, the latter in the richness and thickness of the coal-seams. One of these, the main seam, is in some places from 30 to 40 feet thick, composed of pure bituminous coal. The coal is like the finest kinds shipped at Newcastle, and when analysed yields the same proportions of carbon and hydrogen— a fact worthy of notice, when we consider that this fuel has been derived from an assemblage of plants very distinct specifically, and in part generically, from those which have contributed to the formation of the ancient or palaeozoic coal.

TRIASSIC MAMMIFER.

In North Carolina, the late Professor Emmons has described the strata of the Chatham coal-field, which correspond in age to those near Richmond, in Virginia. In beds underlying them he has met with three jaws of a small insectivorous mammal which he has called Dromatherium sylvestre, closely allied to Spalacotherium. Its nearest living analogue, says Professor Owen, "is found in Myrmecobius; for each ramus of the lower jaw contained ten small molars in a continuous series, one canine, and three conical incisors— the latter being divided by short intervals."

LOW GRADE OF EARLY MAMMALS FAVOURABLE TO THE THEORY OF PROGRESSIVE DEVELOPMENT.

There is every reason to believe that this fossil quadruped is at least as ancient as the Microlestes of the European Trias described in Chapter 21; and the fact is highly important, as proving that a certain low grade of marsupials had not only a wide range in time, from the Trias to the Purbeck, or uppermost oolitic strata of Europe, but had also a wide range in space, namely, from Europe to North America, in an east and west direction, and, in regard to latitude, from Stonesfield, in 52 degrees N., to that of North Carolina, 35 degrees N.

If the three localities in Europe where the most ancient mammalia have been found— Purbeck, Stonesfield, and Stuttgart— had belonged all of them to formations of the same age, we might well have imagined so limited an area to have been peopled exclusively with pouched quadrupeds, just as Australia now is, while other parts of the globe were inhabited by placentals; for Australia now supports one hundred and sixty species of marsupials, while the rest of the continents and islands are tenanted by about seventeen hundred species of mammalia, of which only forty-six are marsupial, namely, the opossums of North and South America. But the great difference of age of the strata in each of these three localities seems to indicate the predominance throughout a vast lapse of time (from the era of the Upper Trias to that of the Purbeck beds) of a low grade of quadrupeds; and this persistency of similar generic and ordinal types in Europe while the species were changing, and while the fish, reptiles, and mollusca were undergoing great modifications, would naturally lead us to suspect that there must also have been a vast extension in space of the same marsupial forms during that portion of the Secondary or Mesozoic epoch which has been termed "the age of reptiles." Such an inference as to the wide geographical range of the ancient marsupials has been confirmed by the discovery in the Trias of North America of the above-mentioned Dromatherium. The predominance in earlier ages of these mammalia of a low grade, and the absence, so far as our investigations have yet gone, of species of higher organisation, whether aquatic or terrestrial, is certainly in favour of the theory of progressive development.

PRIMARY OR PALAEOZOIC SERIES.
CHAPTER XXII.
PERMIAN OR MAGNESIAN LIMESTONE GROUP.

Line of Separation between Mesozoic and Palaeozoic Rocks.
Distinctness of Triassic and Permian Fossils.
Term Permian.
Thickness of calcareous and sedimentary Rocks in North of England.
Upper, Middle, and Lower Permian.
Marine Shells and Corals of the English Magnesian Limestone.
Reptiles and Fish of Permian Marl-slate.
Foot-prints of Reptiles.
Angular Breccias in Lower Permian.
Permian Rocks of the Continent.
Zechstein and Rothliegendes of Thuringia.
Permian Flora.
Its generic Affinity to the Carboniferous.

In pursuing our examination of the strata in descending order, we have next to pass from the base of the Secondary or Mesozoic to the uppermost or newest of the Primary or Palaeozoic formations. As this point has been selected as a line of demarkation for one of the three great divisions of the fossiliferous series, the student might naturally expect that by aid of lithological and palaeontological characters he would be able to recognise without difficulty a distinct break between the newer and older group. But so far is this from being the case in Great Britain, that nowhere have geologists found more difficulty in drawing the line of separation than between the Secondary and Primary series. The obscurity has arisen from the great resemblance in colour and mineral character of the Triassic and Permian red marls and sandstones, and the scarcity and often total absence in them of organic remains. The thickness of the strata belonging to each group amounts in some places to several thousand feet; and by dint of a careful examination of their geological position, and of those fossil, animal, and vegetable forms which are occasionally met with in some members of each series, it has at length been made clear that the older or Permian rocks are more connected with the Primary or Palaeozoic than with the Secondary or Mesozoic strata already described.

The term Permian has been proposed for this group by Sir R. Murchison, from Perm, a Russian province, where it occupies an area twice the size of France, and contains a great abundance and variety of fossils, both vertebrate and invertebrate. Professor Sedgwick in 1832 described what is now recognised as the central member of this group, the Magnesian limestone, showing that it attained a thickness of 600 feet along the north-east of England, in the counties of Durham, Yorkshire, and Nottinghamshire, its lower part often passing into a fossiliferous marl-slate and resting on an inferior Red Sandstone, the equivalent of the Rothliegendes of Germany. (Transactions of the Geological Society London Second Series volume 3 page 37.) It has since been shown that some of the Red Sandstones of newer date also belong to the Permian group; and it appears from the observations of Mr. Binney, Sir R. Murchison, Mr. Harkness, and others, that it is in the region where the limestone is most largely developed, as, for example, in the county of Durham, that the associated red sandstones or sedimentary rocks are thinnest, whereas in the country where the latter are thickest the calcareous member is reduced to thirty, or even sometimes to ten feet. It is clear, therefore, says Mr. Hull, that the sedimentary region in the north of England area has been to the westward, and the calcareous area to the eastward; and that in this group there has been a development from opposite directions of the two types of strata.

In illustration of this he has given us the following table:

TABLE 22.1. THICKNESS OF PERMIAN STRATA IN NORTH OF ENGLAND.
COLUMN 1: NAME OF STRATA.
COLUMN 2: THICKNESS IN FEET IN N.W. OF ENGLAND.
COLUMN 3: THICKNESS IN FEET IN N.E. OF ENGLAND.

Upper Permian (Sedimentary): 600 : 50-100.
Middle Permian (Calcareous): 10-30 : 600.
Lower Permian (Sedimentary): 3000 : 100-250. (Edward Hull Ternary Classification Quarterly Journal of Science No. 23 1869.)

UPPER PERMIAN.

What is called in this table the Upper Permian will be seen to attain its chief thickness in the north-west, or on the coast of Cumberland, as at St. Bee's Head, where it is described by Sir Roderick Murchison as consisting of massive red sandstones with gypsum resting on a thin course of Magnesian Limestone with fossils, which again is connected with the Lower Red Sandstone, resembling the upper one in such a manner that the whole forms a continuous series. No fossil footprints have been found in this Upper as in the Lower Red Sandstone.

MIDDLE PERMIAN— MAGNESIAN LIMESTONE AND MARL-SLATE.

(FIGURE 410. Schizodus Schlotheimi, Geinitz. Permian crystalline limestone.)

(FIGURE 411. The hinge of Schizodus truncatus, King. Permian.)

(FIGURE 412. Mytilus septifer, King. Syn. Modiola acuminata, Sowerby. Permian crystalline limestone.)

This formation is seen upon the coast of Durham and Yorkshire, between the Wear and the Tees. Among its characteristic fossils are Schizodus Schlotheimi (Figure 410) and Mytilus septifer (Figure 412). These shells occur at Hartlepool and Sunderland, where the rock assumes an oolitic and botryoidal character. Some of the beds in this division are ripple-marked. In some parts of the coast of Durham, where the rock is not crystalline, it contains as much as 44 per cent of carbonate of magnesia, mixed with carbonate of lime. In other places— for it is extremely variable in structure— it consists chiefly of carbonate of lime, and has concreted into globular and hemispherical masses, varying from the size of a marble to that of a cannon-ball, and radiating from the centre. Occasionally earthy and pulverulent beds pass into compact limestone or hard granular dolomite. Sometimes the limestone appears in a brecciated form, the fragments which are united together not consisting of foreign rocks but seemingly composed of the breaking-up of the Permian limestone itself, about the time of its consolidation. Some of the angular masses in Tynemouth cliff are two feet in diameter.

(FIGURE 413. Magnesian Limestone, Humbleton Hill, near Sunderland. (King's Monograph Plate 2.) a. Fenestella retiformis, Schlot, sp. Syn. Gorgonia infundibuliformis, Goldf.; Retepora flustracea, Phillips. b. Part of the same highly magnified.)

The magnesian limestone sometimes becomes very fossiliferous and includes in it delicate bryozoa, one of which, Fenestella retiformis (Figure 413), is a very variable species, and has received many different names. It sometimes attains a large size, single specimens measuring eight inches in width. The same bryozoan, with several other British species, is also found abundantly in the Permian of Germany.

(FIGURE 414. Productus horridus, Sowerby. (P. calvus, Sowerby) Sunderland and Durham, in Magnesian Limestone; Zechstein and Kupferschiefer, Germany.)

(FIGURE 415. Lingula Crednerii. (Geinitz.) Magnesian Limestone, and Carboniferous Marl-slate, Durham; Zechstein, Thuringia.)

The total known fauna of the Permian series of Great Britain at present numbers 147 species, of which 77, or more than half, are mollusca. Not one of these is common to rocks newer than the Palaeozoic, and the brachiopods are the only group which have furnished species common to the more ancient or Carboniferous rocks. Of these Lingula Crednerii

(Figure 415) is an example. There are 25 Gasteropods and only one cephalopod, Nautilus Freieslebeni, which is also found in the German Zechstein.

(FIGURE 416. Spirifera alata, Schloth. Syn. Trigonotreta undulata, Sowerby., King's Monograph. Magnesian Limestone.)

Shells of the genera Productus (Figure 414) and Strophalosia (the latter of allied form with hinge teeth), which do not occur in strata newer than the Permian, are abundant in the ordinary yellow magnesian limestone, as will be seen in the valuable memoirs of Messrs. King and Howse. They are accompanied by certain species of Spirifera (Figure 416), Lingula Crednerii (Figure 415), and other brachiopoda of the true primary or palaeozoic type. Some of this same tribe of shells, such as Camarophoria, allied to Rhynchonella, Spiriferina, and two species of Lingula, are specifically the same as fossils of the carboniferous rocks. Avicula, Arca, and Schizodus (Figure 410), and other lamellibranchiate bivalves, are abundant, but spiral univalves are very rare.

(FIGURE 417. Restored outline of a fish of the genus Palaeoniscus, Agassiz. Palaeothrissum, Blainville.)

Beneath the limestone lies a formation termed the marl-stone, which consists of hard calcareous shales, marl-slate, and thin-bedded limestones. At East Thickley, in Durham, where it is thirty feet thick, this slate has yielded many fine specimens of fossil fish— of the genera Palaeoniscus ten species, Pygopterus two species, Coelacanthus two species, and Platysomus two species, which as genera are common to the older Carboniferous formation, but the Permian species are peculiar, and, for the most part, identical with those found in the marl-slate or copper-slate of Thuringia.

(FIGURE 418. Shark. Heterocercal.)

(FIGURE 419. Shad. (Clupea. Herring tribe. Homocercal.)

The Palaeoniscus above-mentioned belongs to that division of fishes which M. Agassiz has called "Heterocercal," which have their tails unequally bilobate, like the recent shark and sturgeon, and the vertebral column running along the upper caudal lobe. (See Figure 418.) The "Homocercal" fish, which comprise almost all the 9000 species at present known in the living creation, have the tail-fin either single or equally divided; and the vertebral column stops short, and is not prolonged into either lobe. (See Figure 419.) Now it is a singular fact, first pointed out by Agassiz, that the heterocercal form, which is confined to a small number of genera in the existing creation, is universal in the magnesian limestone, and all the more ancient formations. It characterises the earlier periods of the earth's history, whereas in the secondary strata, or those newer than the Permian, the homocercal tail predominates.

A full description has been given by Sir Philip Egerton of the species of fish characteristic of the marl-slate, in Professor King's monograph before referred to, where figures of the ichthyolites, which are very entire and well preserved, will be found. Even a single scale is usually so characteristically marked as to indicate the genus, and sometimes even the particular species. They are often scattered through the beds singly, and may be useful to a geologist in determining the age of the rock.

(FIGURES 420-425. SCALES OF FISH. MAGNESIAN LIMESTONE.)

(FIGURE 420. Palaeoniscus comptus, Agassiz. Scale, magnified. Marl-slate.)

(FIGURE 421. Palaeoniscus elegans, Sedgwick. Under surface of scale, magnified. Marl-slate.)

(FIGURE 422. Palaeoniscus glaphyrus, Agassiz. Under surface of scale, magnified. Marl-slate.)

(FIGURE 423. Coelacanthus granulatus, Agassiz. Granulated surface of scale, magnified. Marl-slate.)

(FIGURE 424. Pygopterus mandibularis. Agassiz. Marl-slate. a. Outside of scale, magnified. b. Under surface of same.)

(FIGURE 425. Acrolepis Sedgwickii, Agassiz. Outside of scale, magnified. Marl-slate.))

We are indebted to Messrs. Hancock and Howse for the discovery in this marl-slate at Midderidge, Durham, of two species of Protosaurus, a genus of reptiles, one representative of which, P. Speneri, has been celebrated ever since the year 1810 as characteristic of the Kupfer-schiefer or Permian of Thuringia. Professor Huxley informs us that the agreement of the Durham fossil with Hermann von Meyer's figure of the German specimen is most striking. Although the head is wanting in all the examples yet found, they clearly belong to the Lacertian order, and are therefore of a higher grade than any other vertebrate animal hitherto found fossil in a Palaeozoic rock. Remains of Labyrinthodont reptiles have also been met with in the same slate near Durham.

LOWER PERMIAN.

The inferior sandstones which lie beneath the marl-slate consist of sandstone and sand, separating the Magnesian Limestone from the coal, in Yorkshire and Durham. In some instances, red marl and gypsum have been found associated with these beds. They have been classed with the Magnesian Limestone by Professor Sedgwick, as being nearly co-extensive with it in geographical range, though their relations are very obscure. But the principal development of Lower Permian is, as we have seen by Mr. Hull's Table 22.1, in the northwest, where the Penrith sandstone, as it has been called, and the associated breccias and purple shales are estimated by Professor Harkness to attain a thickness of 3000 feet. Organic remains are generally wanting, but the leaves and wood of coniferous plants, and in one case a cone, have been found. Also in the purple marls of Corncockle Muir near Dumfries, very distinct footprints of reptiles occur, originally referred to the Trias, but shown by Mr. Binney in 1856 to be Permian. No bones of the animals which they represent have yet been discovered.

ANGULAR BRECCIAS IN LOWER PERMIAN.

A striking feature in these beds is the occasional occurrence, especially at the base of the formation, of angular and sometimes rounded fragments of Carboniferous and older rocks of the adjoining districts being included in a paste of red marl. Some of the angular masses are of huge size.

In the central and southern counties, where the Middle Permian or Magnesian Limestone is wanting, it is difficult to separate the upper and lower sandstones, and Mr. Hull is of opinion that the patches of this formation found here and there in Worcestershire, Shropshire, and other counties may have been deposited in a sea separated from the northern basin by a barrier of Carboniferous rocks running east and west, and now concealed under the Triassic strata of Cheshire. Similar breccias to those before described are found in the more southern counties last mentioned, where their appearance is rendered more striking by the marked contrast they present to the beds of well-rolled and rounded pebbles of the Trias occupying a large area in the same region.

Professor Ramsay refers the angular form and large size of the fragments composing these breccias to the action of floating ice in the sea. These masses of angular rock, some of them weighing more than half a ton, and lying confusedly in a red, unstratified marl, like stones in boulder-drift, are in some cases polished, striated, and furrowed like erratic blocks in the moraine of a glacier. They can be shown in some cases to have travelled from the parent rocks, thirty or more miles distant, and yet not to have lost their angular shape. (Ramsay Quarterly Geological Journal 1855; and Lyell Principles of Geology volume 1 page 223 10th edition.)

PERMIAN ROCKS OF THE CONTINENT.

Germany is the classic ground of the Magnesian Limestone now called Permian. The formation was well studied by the miners of that country a century ago as containing a thin band of dark-coloured cupriferous shale, characterised at Mansfield in Thuringia by numerous fossil fish. Beneath some variegated sandstones (not belonging to the Trias, though often confounded with it) they came down first upon a dolomitic limestone corresponding to the upper part of our Middle Permian, and then upon a marl-slate richly impregnated with copper pyrites, and containing fish and reptiles (Protosaurus) identical in species with those of the corresponding marl-slate of Durham. To the limestone they gave the name of Zechstein, and to the marl-slate that of Mergel-schiefer or Kupfer- schiefer. Beneath the fossiliferous group lies the Rothliegendes or Rothtodt- liegendes, meaning the red-lyer or red-dead-lyer, so-called by the German miners from its colour, and because the copper had DIED OUT when they reached this underlying non-metalliferous member of the series. This red under-lyer is, in fact, a great deposit of red sandstone, breccia, and conglomerate with associated porphyry, basalt, and amygdaloid.

According to Sir R. Murchison, the Permian rocks are composed, in Russia, of white limestone, with gypsum and white salt; and of red and green grits, occasionally with copper ore; also magnesian limestones, marl-stones, and conglomerates.

PERMIAN FLORA.

(FIGURE 426. Walchia piniformis, Schloth. Permian, Saxony. (Gutbier, Die Versteinerungen des Permischen Systemes in Sachsen volume 2 plate 10.)
a. Branch.
b. Twig of the same.
c. Leaf magnified.)

About 18 or 20 species of plants are known in the Permian rocks of England. None of them pass down into the Carboniferous series, but several genera, such as Alethopteris, Neuropteris, Walchia, and Ullmania, are common to the two groups. The Permian flora on the Continent appears, from the researches of MM. Murchison and de Verneuil in Russia, and of MM. Geinitz and von Gutbier in Saxony, to be, with a few exceptions, distinct from that of the coal.

In the Permian rocks of Saxony no less than 60 species of fossil plants have been met with. Two or three of these, as Calamites gigas, Sphenopteris erosa, and S. lobata, are also met with in the government of Perm in Russia. Seven others, and among them Neuropteris Loshii, Pecopteris arborescens, and P. similis, and several species of Walchia (see Figure 426), a genus of Conifers, called Lycopodites by some authors, are said by Geinitz to be common to the coal-measures.

(FIGURE 427. Cardiocarpon Ottonis. Gutbier, Permian, Saxony. 1/2 diameter.)
(FIGURE 428. Neoggerathia cuneifolia. Brongniart. (Murchison's Russia volume 2 Plate A figure 3.)

Among the genera also enumerated by Colonel Gutbier are the fruit called Cardiocarpon (see Figure 427), Asterophyllites, and Annularia, so characteristic of the Carboniferous period; also Lepidodendron, which is common to the Permian of Saxony, Thuringia, and Russia, although not abundant. Neoggerathia (see Figure 428), the leaves of which have parallel veins without a midrib, and to which various generic synonyms, such as Cordaites, Flabellaria, and Poacites, have been given, is another link between the Permian and Carboniferous vegetation. Coniferae, of the Araucarian division, also occur; but these are likewise met with both in older and newer rocks. The plants called Sigillaria and Stigmaria, so marked a feature in the Carboniferous period, are as yet wanting in the true Permian.

Among the remarkable fossils of the Rothliegendes, or lowest part of the Permian in Saxony and Bohemia, are the silicified trunks of tree-ferns called generically Psaronius. Their bark was surrounded by a dense mass of air-roots, which often constituted a great addition to the original stem, so as to double or quadruple its diameter. The same remark holds good in regard to certain living extra-tropical arborescent ferns, particularly those of New Zealand.

Upon the whole, it is evident that the Permian plants approach much nearer to the Carboniferous flora than to the Triassic; and the same may be said of the Permian fauna.

CHAPTER XXIII.
THE COAL OR CARBONIFEROUS GROUP.

Principal Subdivisions of the Carboniferous Group.
Different Thickness of the sedimentary and calcareous Members in Scotland and the South of England.
Coal-measures.
Terrestrial Nature of the Growth of Coal.
Erect fossil Trees.
Uniting of many Coal-seams into one thick Bed.
Purity of the Coal explained.
Conversion of Coal into Anthracite.
Origin of Clay-ironstone.
Marine and brackish-water Strata in Coal.

Fossil Insects.
Batrachian Reptiles.
Labyrinthodont Foot-prints in Coal-measures.
Nova Scotia Coal-measures with successive Growths of erect fossil Trees.
Similarity of American and European Coal.
Air-breathers of the American Coal.
Changes of Condition of Land and Sea indicated by the Carboniferous Strata of Nova Scotia.

PRINCIPAL SUBDIVISIONS OF THE CARBONIFEROUS GROUP.

The next group which we meet with in the descending order is the Carboniferous, commonly called "The Coal," because it contains many beds of that mineral, in a more or less pure state, interstratified with sandstones, shales, and limestones. The coal itself, even in Great Britain and Belgium, where it is most abundant, constitutes but an insignificant portion of the whole mass. In South Wales, for example, the thickness of the coal-bearing strata has been estimated at between 11,000 and 12,000 feet, while the various coal seams, about 80 in number, do not, according to Professor Phillips, exceed in the aggregate 120 feet.

The Carboniferous formation assumes various characters in different parts even of the British Islands. It usually comprises two very distinct members: first, the sedimentary beds, usually called the Coal-measures, of mixed fresh-water, terrestrial, and marine origin, often including seams of coal; secondly, that named in England the Mountain or Carboniferous Limestone, of purely marine origin, and made up chiefly of corals, shells, and encrinites, and resting on shales called the shales of the Mountain Limestone.

TABLE 23.1.

In the south-western part of our island, in Somersetshire and South Wales, the three divisions usually spoken of are:

1. Coal-measures: strata of shale, sandstone, and grit, from 600 to 12,000 feet thick, with occasional seams of coal.

2. Millstone grit: a coarse quartzose sandstone passing into a conglomerate, sometimes used for millstones, with beds of shale; usually devoid of coal; occasionally above 600 feet thick.

3. Mountain or Carboniferous Limestone: a calcareous rock containing marine shells, corals, and encrinites; devoid of coal; thickness variable, sometimes more than 1500 feet.

If the reader will refer to the section in Figure 85, he will see that the Upper and Lower Coal-measures of the coal-field near Bristol are divided by a micaceous flaggy sandstone called the Pennant Rock. The Lower Coal-measures of the same section rest sometimes, especially in the north part of the basin, on a base of coarse grit called the Millstone Grit (No. 2 of the above Table 23.1.)

In the South Welsh coal-field Millstone Grit occurs in like manner at the base of the productive coal. It is called by the miners the "Farewell Rock," as when they reach it they have no longer any hopes of obtaining coal at a greater depth in the same district. In the central and northern coal-fields of England this same grit, including quartz pebbles, with some accompanying sandstones and shales containing coal plants, acquires a thickness of several thousand feet, lying beneath the productive coal-measures, which are nearly 10,000 feet thick.

Below the Millstone Grit is a continuation of similar sandstones and shales called by Professor Phillips the Yoredale series, from Yoredale, in Yorkshire, where they attain a thickness of from 800 to 1000 feet. At several intervals bands of limestone divide this part of the series, one of which, called the Main Limestone or Upper Scar Limestone, composed in great part of encrinites, is 70 feet thick. Thin seams of coal also occur in these lower Yoredale beds in Yorkshire, showing that in the same region there were great alternations in the state of the surface. For at successive periods in the same area there prevailed first terrestrial conditions favourable to the growth of pure coal, secondly, a sea of some depth suited to the formation of Carboniferous Limestone, and, thirdly, a supply of muddy sediment and sand, furnishing the materials for sandstone and shale. There is no clear line of demarkation between the Coal- measures and the Millstone Grit, nor between the Millstone Grit and underlying Yoredale rocks.

On comparing a series of vertical sections in a north-westerly direction from Leicestershire and Warwickshire into North Lancashire, we find, says Mr. Hull, within a distance of 120 miles an augmentation of the sedimentary materials to the extent of 16,000 feet.

Leicestershire and Warwickshire: 2,600 feet.
North Staffordshire: 9,000 feet.
South Lancashire: 12,130 feet.
North Lancashire: 18,700 feet.

In central England, where the sedimentary beds are reduced to about 3000 feet in all, the Carboniferous Limestone attains an enormous thickness, as much as 4000 feet at Ashbourne, near Derby, according to Mr. Hull's estimate. To a certain extent, therefore, we may consider the calcareous member of the formation as having originated simultaneously with the accumulation of the materials of grit, sandstone, and shale, with seams of coal; just as strata of mud, sand, and pebbles, several thousand feet thick, with layers of vegetable matter, are now in the process of formation in the cypress swamps and delta of the Mississippi, while coral reefs are forming on the coast of Florida and in the sea of the Bermuda islands. For we may safely conclude that in the ancient Carboniferous ocean those marine animals which were limestone builders were never freely developed in areas where the rivers poured in fresh water charged with sand or clay; and the limestone could only become several thousand feet thick in parts of the ocean which remained perfectly clear for ages.

The calcareous strata of the Scotch coal-fields, those of Lanarkshire, the Lothians, and Fife, for example, are very insignificant in thickness when compared to those of England. They consist of a few beds intercalated between the sandstones and shales containing coal and ironstone, the combined thickness of all the limestones amounting to no more than 150 feet. The vegetation of some of these northern sedimentary beds containing coal may be older than any of the coal-measures of central and southern England, as being coeval with the Mountain Limestone of the south. In Ireland the limestone predominates over the coal- bearing sands and shales. We may infer the former continuity of several of the coal-fields in northern and central England, not only from the abrupt manner in which they are cut off at their outcrop, but from their remarkable correspondence in the succession and character of particular beds. But the limited extent to which these strata are exposed at the surface is not merely owing to their former denudation, but even in a still greater degree to their having been largely covered by the New Red Sandstone, as in Cheshire, and here and there by the Permian strata, as in Durham.

It has long been the opinion of the most eminent geologists that the coal-fields of Yorkshire and Lancashire were once united, the upper Coal-measures and the overlying Millstone Grit and Yoredale rocks having been subsequently removed; but what is remarkable, is the ancient date now assigned to this denudation, for it seems that a thickness of no less than 10,000 feet of the coal-measures had been carried away before the deposition even of the lower Permian rocks which were thrown down upon the already disturbed truncated edges of the coal-strata. (Edward Hull Quarterly Geological Journal volume 24 page 327.) The carboniferous strata most productive of workable coal have so often a basin-shaped arrangement that these troughs have sometimes been supposed to be connected with the original conformation of the surface upon which the beds were deposited. But it is now admitted that this structure has been owing to movements of the earth's crust of upheaval and subsidence, and that the flexure and inclination of the beds has no connection with the original geographical configuration of the district.

COAL-MEASURES.

I shall now treat more particularly of the productive coal-measures, and their mode of origin and organic remains.

COAL-FORMED ON LAND.

In South Wales, already alluded to, where the coal-measures attain a thickness of 12,000 feet, the beds throughout appear to have been formed in water of moderate depth, during a slow, but perhaps intermittent, depression of the ground, in a region to which rivers were bringing a never-failing supply of muddy sediment and sand. The same area was sometimes covered with vast forests, such as we see in the deltas of great rivers in warm climates, which are liable to be submerged beneath fresh or salt water should the ground sink vertically a few feet.

In one section near Swansea, in South Wales, where the total thickness of strata is 3246 feet, we learn from Sir H. De la Beche that there are ten principal masses of sandstone. One of these is 500 feet thick, and the whole of them make together a thickness of 2125 feet. They are separated by masses of shale, varying in thickness from 10 to 50 feet. The intercalated coal-beds, sixteen in number, are generally from one to five feet thick, one of them, which has two or three layers of clay interposed, attaining nine feet. At other points in the same coal-field the shales predominate over the sandstones. Great as is the diversity in the horizontal extent of individual coal-seams, they all present one characteristic feature, in having, each of them, what is called its UNDERCLAY. These underclays, co-extensive with every layer of coal, consist of arenaceous shale, sometimes called fire-stone, because it can be made into bricks which stand the fire of a furnace. They vary in thickness from six inches to more than ten feet; and Sir William Logan first announced to the scientific world in 1841 that they were regarded by the colliers in South Wales as an essential accompaniment of each of the eighty or more seams of coal met with in their coal-field. They are said to form the FLOOR on which the coal rests; and some of them have a slight admixture of carbonaceous matter, while others are quite blackened by it.

All of them, as Sir William Logan pointed out, are characterised by inclosing a peculiar species of fossil vegetable called Stigmaria, to the exclusion of other plants. It was also observed that, while in the overlying shales, or "roof" of the coal, ferns and trunks of trees abound without any Stigmariae, and are flattened and compressed, those singular plants of the underclay most commonly retain their natural forms, unflattened and branching freely, and sending out their slender rootlets, formerly thought to be leaves, through the mud in all directions. Several species of Stigmaria had long been known to botanists, and described by them, before their position under each seam of coal was pointed out, and before their true nature as the roots of trees (some having been actually found attached to the base of Sigillaria stumps) was recognised. It was conjectured that they might be aquatic, perhaps floating plants, which sometimes extended their branches and leaves freely in fluid mud, in which they were finally enveloped.

Now that all agree that these underclays are ancient soils, it follows that in every instance where we find them they attest the terrestrial nature of the plants which formed the overlying coal, which consists of the trunks, branches, and leaves of the same plants. The trunks have generally fallen prostrate in the coal, but some of them still remain at right angles to the ancient soils (see Figure 440). Professor Goppert, after examining the fossil vegetables of the coal-fields of Germany, has detected, in beds of pure coal, remains of plants of every family hitherto known to occur fossil in the carboniferous rocks. Many seams, he remarks, are rich in Sigillariae, Lepidodendra, and Stigmariae, the latter in such abundance as to appear to form the bulk of the coal. In some places, almost all the plants were calamites, in others ferns. (Quarterly Geological Journal volume 5 Mem. page 17.)

Between the years 1837 and 1840, six fossil trees were discovered in the coal-fields of Lancashire, where it is intersected by the Bolton railway. They were all at right angles to the plane of the bed, which dips about 15 degrees to the south. The distance between the first and the last was more than 100 feet, and the roots of all were imbedded in a soft argillaceous shale. In the same plane with the roots is a bed of coal, eight or ten inches thick, which has been found to extend across the railway, or to the distance of at least ten yards. Just above the covering of the roots, yet beneath the coal-seam, so large a quantity of the Lepidostrobus variabilis was discovered inclosed in nodules of hard clay, that more than a bushel was collected from the small openings around the base of some of the trees (see Figure 457 of this genus). The exterior trunk of each was marked by a coating of friable coal, varying from one-quarter to three-quarters of an inch in thickness; but it crumbled away on removing the matrix. The dimensions of one of the trees is 15 1/2 feet in circumference at the base, 7 1/2 feet at the top, its height being eleven feet. All the trees have large spreading roots, solid and strong, sometimes branching, and traced to a distance of several feet, and presumed to extend much farther.

In a colliery near Newcastle a great number of Sigillariae occur in the rock as if they had retained the position in which they grew. No less than thirty, some of them four or five feet in diameter, were visible within an area of 50 yards square, the interior being sandstone, and the bark having been converted into coal. Such vertical stems are familiar to our coal-miners, under the name of coal-pipes. They are much dreaded, for almost every year in the Bristol, Newcastle, and other coal-fields, they are the cause of fatal accidents. Each cylindrical cast of a tree, formed of solid sandstone, and increasing gradually in size towards the base, and being without branches, has its whole weight thrown downward, and receives no support from the coating of friable coal which has replaced the bark. As soon, therefore, as the cohesion of this external layer is overcome, the heavy column falls suddenly in a perpendicular or oblique direction from the roof of the gallery whence coal has been extracted, wounding or killing the workman who stands below. It is strange to reflect how many thousands of these trees fell originally in their native forests in obedience to the law of gravity; and how the few which continued to stand erect, obeying, after myriads of ages, the same force, are cast down to immolate their human victims.

(FIGURE 429. Ground-plan of a fossil forest, Parkfield Colliery, near Wolverhampton, showing the position of 73 trees in a quarter of an acre.)

It has been remarked that if, instead of working in the dark, the miner was accustomed to remove the upper covering of rock from each seam of coal, and to expose to the day the soils on which ancient forests grew, the evidence of their former growth would be obvious. Thus in South Staffordshire a seam of coal was laid bare in the year 1844, in what is called an open work at Parkfield colliery, near Wolverhampton. In the space of about a quarter of an acre the stumps of no less than 73 trees with their roots attached appeared, as shown in Figure 429, some of them more than eight feet in circumference. The trunks, broken off close to the root, were lying prostrate in every direction, often crossing each other. One of them measured 15, another 30 feet in length, and others less. They were invariably flattened to the thickness of one or two inches, and converted into coal. Their roots formed part of a stratum of coal ten inches thick, which rested on a layer of clay two inches thick, below which was a second forest resting on a two-foot seam of coal. Five feet below this, again, was a third forest with large stumps of Lepidodendra, Calamites, and other trees.

BLENDING OF COAL-SEAMS.

Both in England and North America seams of coal are occasionally observed to be parted from each other by layers of clay and sand, and, after they have been persistent for miles, to come together and blend in one single bed, which is then found to be equal in the aggregate to the thickness of the several seams. I was shown by Mr. H.D. Rogers a remarkable example of this in Pennsylvania. In the Shark Mountain, near Pottsville, in that State, there are thirteen seams of anthracite coal, some of them more than six feet thick, separated by beds of white quartzose grit and a conglomerate of quartz pebbles, often of the size of a hen's egg. Between Pottsville and the Lehigh Summit Mine, seven of these seams of coal, at first widely separated, are, in the course of several miles, brought nearer and nearer together by the gradual thinning out of the intervening coarse-grained strata and their accompanying shales, until at length they successively unite and form one mass of coal between forty and fifty feet thick, very pure on the whole, though with a few thin partings of clay. This mass of coal I saw quarried in the open air at Mauch Chunk, on the Bear Mountain. The origin of such a vast thickness of vegetable remains, so unmixed, on the whole, with earthy ingredients, can be accounted for in no other way than by the growth, during thousands of years, of trees and ferns in the manner of peat— a theory which the presence of the Stigmaria in situ under each of the seven layers of anthracite fully bears out. The rival hypothesis, of the drifting of plants into a sea or estuary, leaves the non-intermixture of sediment, or of clay, sand, and pebbles, with the pure coal wholly unexplained.

(FIGURE 430. Uniting of distinct coal-seams.)

The late Mr. Bowman was the first who gave a satisfactory explanation of the manner in which distinct coal-seams, after maintaining their independence for miles, may at length unite, and then persist throughout another wide area with a thickness equal to that which the separate seams had previously maintained.

Let A-C (Figure 430) be a three-foot seam of coal originally laid down as a mass of vegetable matter on the level area of an extensive swamp, having an under- clay, f-g, through which the Stigmariae or roots of the trees penetrate as usual. One portion, B-C, of this seam of coal is now inclined; the area of the swamp having subsided as much as 25 feet at E-C, and become for a time submerged under salt, fresh, or brackish water. Some of the trees of the original forest A-B-C fell down, others continued to stand erect in the new lagoon, their stumps and part of their trunks becoming gradually enveloped in layers of sand and mud, which at length filled up the new piece of water C-E.

When this lagoon has been entirely silted up and converted into land, the forest-covered surface A-B will extend once more over the whole area A-B-E, and a second mass of vegetable matter, D-E, forming three feet more of coal, will accumulate. We then find in the region E-C two seams of coals, each three feet thick, with their respective under-clays, with erect buried trees based upon the surface of the lower coal, the two seams being separated by 25 feet of intervening shale and sandstone. Whereas in the region A-B, where the growth of the forest has never been interrupted by submergence, there will simply be one seam, two yards thick, corresponding to the united thickness of the beds B-E and B-C. It may be objected that the uninterrupted growth of plants during the interval of time required for the filling up of the lagoon will have caused the vegetable matter in the region D-A-B to be thicker than the two distinct seams E and C, and no doubt there would actually be a slight excess representing one or more generation of trees and plants forming the undergrowth; but this excess of vegetable matter, when compressed into coal, would be so insignificant in thickness that the miner might still affirm that the seam D-A throughout the area D-A-B was equal to the two seams C and E.

CAUSE OF THE PURITY OF COAL.

The purity of the coal itself, or the absence in it of earthy particles and sand, throughout areas of vast extent, is a fact which appears very difficult to explain when we attribute each coal-seam to a vegetation growing in swamps. It has been asked how, during river inundations capable of sweeping away the leaves of ferns and the stems and roots of Sigillariae and other trees, could the waters fail to transport some fine mud into the swamps? One generation after another of tall trees grew with their roots in mud, and their leaves and prostrate trunks formed layers of vegetable matter, which was afterwards covered with mud since turned to shale. Yet the coal itself, or altered vegetable matter, remained all the while unsoiled by earthy particles. This enigma, however perplexing at first sight, may, I think, be solved by attending to what is now taking place in deltas. The dense growth of reeds and herbage which encompasses the margins of forest-covered swamps in the valley and delta of the Mississippi is such that the fluviatile waters, in passing through them, are filtered and made to clear themselves entirely before they reach the areas in which vegetable matter may accumulate for centuries, forming coal if the climate be favourable. There is no possibility of the least intermixture of earthy matter in such cases. Thus in the large submerged tract called the "Sunk Country," near New Madrid, forming part of the western side of the valley of the Mississippi, erect trees have been standing ever since the year 1811-12, killed by the great earthquake of that date; lacustrine and swamp plants have been growing there in the shallows, and several rivers have annually inundated the whole space, and yet have been unable to carry in any sediment within the outer boundaries of the morass, so dense is the marginal belt of reeds and brush-wood. It may be affirmed that generally, in the "cypress swamps" of the Mississippi, no sediment mingles with the vegetable matter accumulated there from the decay of trees and semi-aquatic plants. As a singular proof of this fact, I may mention that whenever any part of a swamp in Louisiana is dried up, during an unusually hot season, and the wood set on fire, pits are burnt into the ground many feet deep, or as far down as the fire can descend without meeting with water, and it is then found that scarcely any residuum or earthy matter is left. At the bottom of all these "cypress swamps" a bed of clay is found, with roots of the tall cypress (Taxodium distichum), just as the under-clays of the coal are filled with Stigmaria.

CONVERSION OF COAL INTO ANTHRACITE.

It appears from the researches of Liebig and other eminent chemists, that when wood and vegetable matter are buried in the earth exposed to moisture, and partially or entirely excluded from the air, they decompose slowly and evolve carbonic acid gas, thus parting with a portion of their original oxygen. By this means they become gradually converted into lignite or wood-coal, which contains a larger proportion of hydrogen than wood does. A continuance of decomposition changes this lignite into common or bituminous coal, chiefly by the discharge of carbureted hydrogen, or the gas by which we illuminate our streets and houses. According to Bischoff, the inflammable gases which are always escaping from mineral coal, and are so often the cause of fatal accidents in mines, always contain carbonic acid, carbureted hydrogen, nitrogen, and olefiant gas. The disengagement of all these gradually transforms ordinary or bituminous coal into anthracite, to which the various names of glance-coal, coke, hard-coal, culm, and many others, have been given.

There is an intimate connection between the extent to which the coal has in different regions parted with its gaseous contents, and the amount of disturbance which the strata have undergone. The coincidence of these phenomena may be attributed partly to the greater facility afforded for the escape of volatile matter, when the fracturing of the rocks has produced an infinite number of cracks and crevices. The gases and water which are made to penetrate these cracks are probably rendered the more effective as metamorphic agents by increased temperature derived from the interior. It is well known that, at the present period, thermal waters and hot vapours burst out from the earth during earthquakes, and these would not fail to promote the disengagement of volatile matter from the Carboniferous rocks.

In Pennsylvania the strata of coal are horizontal to the westward of the Alleghany Mountains, where the late Professor H.D. Rogers pointed out that they were most bituminous; but as we travel south-eastward, where they no longer remain level and unbroken, the same seams become progressively debituminized in proportion as the rocks become more bent and distorted. At first, on the Ohio River, the proportion of hydrogen, oxygen, and other volatile matters ranges from forty to fifty per cent. Eastward of this line, on the Monongahela, it still approaches forty per cent, where the strata begin to experience some gentle flexures. On entering the Alleghany Mountains, where the distinct anticlinal axes begin to show themselves, but before the dislocations are considerable, the volatile matter is generally in the proportion of eighteen or twenty per cent. At length, when we arrive at some insulated coal-fields associated with the boldest flexures of the Appalachian chain, where the strata have been actually turned over, as near Pottsville, we find the coal to contain only from six per cent of volatile matter, thus becoming a genuine anthracite.

CLAY-IRONSTONE.

Bands and nodules of clay-ironstone are common in coal-measures, and are formed, says Sir H. De la Beche, of carbonate of iron mingled mechanically with earthy matter, like that constituting the shales. Mr. Hunt, of the Museum of Practical Geology, instituted a series of experiments to illustrate the production of this substance, and found that decomposing vegetable matter, such as would be distributed through all coal strata, prevented the further oxidation of the proto-salts of iron, and converted the peroxide into protoxide by taking a portion of its oxygen to form carbonic acid. Such carbonic acid, meeting with the protoxide of iron in solution, would unite with it and form a carbonate of iron; and this mingling with fine mud, when the excess of carbonic acid was removed, might form beds or nodules of argillaceous ironstone. (Memoirs of the Geological Survey pages 51, 255, etc.)

INTERCALATED MARINE BEDS IN COAL.

(FIGURE 431. Microconchus (Spirorbis) carbonarius, Murchison. Natural size and magnified. b. Variety of same.)

(FIGURE 432. Cythere (Leperditia) inflata. Natural size and magnified. Murchison.)

(FIGURE 433. Goniatites Listeri, Martin. Coal-measures, Yorkshire and Lancashire.)

(FIGURE 434. Aviculopecten papyraceus, Goldf. (Pecten papyraceus, Sowerby.))

Both in the coal-fields of Europe and America the association of fresh, brackish-water, and marine strata with coal-seams of terrestrial origin is frequently recognised. Thus, for example, a deposit near Shrewsbury, probably formed in brackish water, has been described by Sir R. Murchison as the youngest member of the coal-measures of that district, at the point where they are in contact with the overlying Permian group. It consists of shales and sandstones about 150 feet thick, with coal and traces of plants; including a bed of limestone varying from two to nine feet in thickness, which is cellular, and resembles some lacustrine limestones of France and Germany. It has been traced for 30 miles in a straight line, and can be recognised at still more distant points. The characteristic fossils are a small bivalve, having the form of a Cyclas or Cyrena, also a small entomostracan, Cythere inflata (Figure 432), and the microscopic shell of an annelid of an extinct genus called Microconchus (Figure 431), allied to Spirorbis. In the coal-field of Yorkshire there are fresh-water strata, some of which contain shells referred to the family Unionidae; but in the midst of the series there is one thin but very widely- spread stratum, abounding in fishes and marine shells, such as Goniatites Listeri (Figure 433), Orthoceras, and Aviculopecten papyraceus, Goldf. (Figure 434).

INSECTS IN EUROPEAN COAL.

Articulate animals of the genus Scorpion were found by Count Sternberg in 1835 in the coal-measures of Bohemia, and about the same time in those of Coalbrook Dale by Mr. Prestwich, were also true insects, such as beetles of the family Curculionidae, a neuropterous insect of the genus Corydalis, and another related to the Phasmidae, have been found.

(FIGURE 435. Wing of a Grasshopper. Gryllacris lithanthraca, Goldenberg. Coal, Saarbruck, near Treves.)

From the coal of Wetting, in Westphalia, several specimens of the cockroach or Blatta family, and the wing of a cricket (Acridites) have been described by Germar. Professor Goldenberg published, in 1854, descriptions of no less than twelve species of insects from the nodular clay-ironstone of Saarbruck, near Treves. (Dunker and V. Meyer Palaeontology volume 4 page 17.) Among them are several Blattinae, three species of Neuroptera, one beetle of the Scarabaeus family, a grasshopper or locust, Gryllacris (see Figure 435), and several white ants or Termites. Professor Goldenberg showed me, in 1864, the wing of a white ant, found low down in the productive coal-measures of Saarbruck, in the interior of a flattened Lepidodendron. It is much larger than that of any known living species of the same genus.

BATRACHIAN REPTILES IN COAL.

(FIGURE 436. Archegosaurus minor, Goldfuss. Fossil reptile from the coal-measures, Saarbruck.)

(FIGURE 437. Imbricated covering of skin of Archegosaurus medius, Goldf. Magnified.)

No vertebrated animals more highly organised than fish were known in rocks of higher antiquity than the Permian until the year 1844, when the Apateon pedestris, Meyer, was discovered in the coal-measures of Munster-Appel in Rhenish Bavaria, and three years later, in 1847, Professor von Dechen found three other distinct species of the same family of Amphibia in the Saarbruck coal-field above alluded to. These were described by the late Professor Goldfuss under the generic name of Archegosaurus. The skulls, teeth, and the greater portions of the skeleton, nay, even a large part of the skin, of two of these reptiles have been faithfully preserved in the centre of spheroidal concretions of clay-ironstone. The largest of these, Archegosaurus Decheni, must have been three feet six inches long. Figure 436 represents the skull and neck bones of the smallest of the three, of the natural size. They were considered by Goldfuss as saurians, but by Herman von Meyer as most nearly allied to the Labyrinthodon before mentioned (Chapter 21), and the remains of the extremities leave no doubt they were quadrupeds, "provided," says Von Meyer, "with hands and feet terminating in distinct toes; but these limbs were weak, serving only for swimming or creeping." The same anatomist has pointed out certain points of analogy between their bones and those of the Proteus anguinus; and Professor Owen has observed that they make an approach to the Proteus in the shortness of their ribs. Two specimens of these ancient reptiles retain a large part of the outer skin, which consisted of long, narrow, wedge-shaped, tile-like, and horny scales, arranged in rows (see Figure 437).

In 1865, several species belonging to three different genera of the same family of perennibranchiate Batrachians were found in the coal-field of Kilkenny in bituminous shale at the junction of the coal with the underlying Stigmaria- bearing clay. They were, probably, inhabitants of a marsh, and the large processes projecting from the vertebrae of their tail imply, according to Professor Huxley, great powers of swimming. They were of the Labyrinthodont family, and their association with the fish of the coal, of which so large a proportion are ganoids, reminds us that the living perennibranchiate amphibia of America frequent the same rivers as the ganoid Lepidostei or bony pikes.

LABYRINTHODONT FOOTPRINTS IN COAL-MEASURES.

(FIGURE 438. Slab of sandstone from the coal-measures of Pennsylvania, with footprints of air-breathing reptile and casts of cracks. Scale one-sixth the original.)

In 1844, the very year when the Apateon, before mentioned, of the coal was first met with in the country between the Moselle and the Rhine, Dr. King published an account of the footprints of a large reptile discovered by him in North America. These occur in the coal-strata of Greensburg, in Westmoreland County, Pennsylvania; and I had an opportunity of examining them when in that country in 1846. The footmarks were first observed standing out in relief from the lower surface of slabs of sandstone, resting on thin layers of fine unctuous clay. I brought away one of these masses, which is represented in Figure 438. It displays, together with footprints, the casts of cracks (a, a') of various sizes. The origin of such cracks in clay, and casts of the same, has before been explained, and referred to the drying and shrinking of mud, and the subsequent pouring of sand into open crevices. It will be seen that some of the cracks, as at b, c, traverse the footprints, and produce distortion in them, as might have been expected, for the mud must have been soft when the animal walked over it and left the impressions; whereas, when it afterwards dried up and shrank, it would be too hard to receive such indentations.

We may assume that the reptile which left these prints on the ancient sands of the coal-measures was an air-breather, because its weight would not have been sufficient under water to have made impressions so deep and distinct. The same conclusion is also borne out by the casts of the cracks above described, for they show that the clay had been exposed to the air and sun, so as to have dried and shrunk.

NOVA SCOTIA COAL-MEASURES.

The sedimentary strata in which thin seams of coal occur attain a thickness, as we have seen, of 18,000 feet in the north of England exclusive of the Mountain Limestone, and are estimated by Von Dechen at over 20,000 feet in Rhenish Prussia. But the finest example in the world of a natural exposure in a continuous section ten miles long, occurs in the sea-cliffs bordering a branch of the Bay of Fundy, in Nova Scotia. These cliffs, called the "South Joggins," which I first examined in 1842, and afterwards with Dr. Dawson in 1845, have lately been admirably described by the last-mentioned geologist in detail, and his evidence is most valuable as showing how large a portion of this dense mass was formed on land, or in swamps where terrestrial vegetation flourished, or in fresh-water lagoons. (Acadian Geology second edition 1868.) His computation of the thickness of the whole series of carboniferous strata as exceeding three miles, agrees with the measurement made independently by Sir William Logan in his survey of this coast.

There is no reason to believe that in this vast succession of strata, comprising some marine as well as many fresh-water and terrestrial formations, there is any repetition of the same beds. There are no faults to mislead the geologist, and cause him to count the same beds over more than once, while some of the same plants have been traced from the top to the bottom of the whole series, and are distinct from the flora of the antecedent Devonian formation of Canada. Eighty- one seams of coal, varying in thickness from an inch to about five feet, have been discovered, and no less than seventy-one of these have been actually exposed in the sea-cliffs.

(FIGURE 439. Section of the cliffs of the South Joggins, near Minudie, Nova Scotia (from north to south through coal with upright trees and sandstone and shale). c. Grindstone. d, g. Alternations of sandstone, shale, and coal containing upright trees. e, f. Portion of cliff, given on a larger scale in Figure 440. f. Four-foot coal, main seam. h, i. Shale with fresh-water mussels, see below.)

In the section in Figure 439, which I examined in 1842, the beds from c to i are seen all dipping the same way, their average inclination being at an angle of 24 degrees S.S.W. The vertical height of the cliffs is from 150 to 200 feet; and between d and g— in which space I observed seventeen trees in an upright position, or, to speak more correctly, at right angles to the planes of stratification— I counted nineteen seams of coal, varying in thickness from two inches to four feet. At low tide a fine horizontal section of the same beds is exposed to view on the beach, which at low tide extends sometimes 200 yards from the base of the cliff. The thickness of the beds alluded to, between d and g, is about 2500 feet, the erect trees consisting chiefly of large Sigillariae, occurring at ten distinct levels, one above the other. The usual height of the buried trees seen by me was from six to eight feet; but one trunk was about 25 feet high and four feet in diameter, with a considerable bulge at the base. In no instance could I detect any trunk intersecting a layer of coal, however thin; and most of the trees terminated downward in seams of coal. Some few only were based on clay and shale; none of them, except Calamites, on sandstone. The erect trees, therefore, appeared in general to have grown on beds of vegetable matter. In the underclays Stigmaria abounds.

These root-bearing beds have been found under all the coal-seams, and such old soils are at present the most destructible masses in the whole cliff, the sandstones and laminated shales being harder and more capable of resisting the action of the waves and the weather. Originally the reverse was doubtless true, for in the existing delta of the Mississippi those clays in which the innumerable roots of the deciduous cypress and other swamp trees ramify in all directions are seen to withstand far more effectually the undermining power of the river, or of the sea at the base of the delta, than do beds of loose sand or layers of mud not supporting trees. It is obvious that if this sand or mud be afterwards consolidated and turned to sandstone and hard shale, it would be the least destructible.

(FIGURE 440. Erect fossil trees. Coal-measures, Nova Scotia.)

In regard to the plants, they belonged to the same genera, and most of them to the same species, as those met with in the distant coal-fields of Europe. Dr. Dawson has enumerated more than 150 species, two-thirds of which are European, a greater agreement than can be said to exist between the same Nova Scotia flora and that of the coal-fields of the United States. By referring to the section in Figure 439, the position of the four-foot coal will be perceived, and in Figure 440 (a section made by me in 1842 of a small portion) that from e to f of the same cliff is exhibited, in order to show the manner of occurrence of erect fossil trees at right angles to the planes of the inclined strata.

In the sandstone which filled their interiors, I frequently observed fern- leaves, and sometimes fragments of Stigmaria, which had evidently entered together with sediment after the trunk had decayed and become hollow, and while it was still standing under water. Thus the tree, a, Figure 440, represented in the bed e in the section, Figure 439, is a hollow trunk five feet eight inches in length, traversing various strata, and cut off at the top by a layer of clay two feet thick, on which rests a seam of coal (b, Figure 440) one foot thick. On this coal again stood two large trees (c and d), while at a greater height the trees f and g rest upon a thin seam of coal (e), and above them is an underclay, supporting the four-foot coal.

Occasionally the layers of matter in the inside of the tree are more numerous than those without; but it is more common in the coal-measures of all countries to find a cylinder of pure sandstone— the cast of the interior of a tree— intersecting a great many alternating beds of shale and sandstone, which originally enveloped the trunk as it stood erect in the water. Such a want of correspondence in the materials outside and inside, is just what we might expect if we reflect on the difference of time at which the deposition of sediment will take place in the two cases; the imbedding of the tree having gone on for many years before its decay had made much progress. In many places distinct proof is seen that the enveloping strata took years to accumulate, for some of the sandstones surrounding erect sigillarian trunks support at different levels roots and stems of Calamites; the Calamites having begun to grow after the older Sigillariae had been partially buried.

The general absence of structure in the interior of the large fossil trees of the Coal implies the very durable nature of their bark, as compared with their woody portion. The same difference of durability of bark and wood exists in modern trees, and was first pointed out to me by Dr. Dawson, in the forests of Nova Scotia, where the Canoe Birch (Betula papyracea) has such tough bark that it may sometimes be seen in the swamps looking externally sound and fresh, although consisting simply of a hollow cylinder with all the wood decayed and gone. When portions of such trunks have become submerged in the swamps they are sometimes found filled with mud. One of the erect fossil trees of the South Joggins fifteen feet in height, occurring at a higher level than the main coal, has been shown by Dr. Dawson to have a coniferous structure, so that some Coniferae of the Coal period grew in the same swamps as Sigillariae, just as now the deciduous Cypress (Taxodium distichum) abounds in the marshes of Louisiana even to the edge of the sea.

When the carboniferous forests sank below high-water mark, a species of Spirorbis or Serpula (Figure 431), attached itself to the outside of the stumps and stems of the erect trees, adhering occasionally even to the interior of the bark— another proof that the process of envelopment was very gradual. These hollow upright trees, covered with innumerable marine annelids, reminded me of a "cane-brake," as it is commonly called, consisting of tall reeds, Arundinaria macrosperma, which I saw in 1846, at the Balize, or extremity of the delta of the Mississippi. Although these reeds are fresh-water plants, they were covered with barnacles, having been killed by an incursion of salt-water over an extent of many acres, where the sea had for a season usurped a space previously gained from it by the river. Yet the dead reeds, in spite of this change, remained standing in the soft mud, enabling us to conceive how easily the larger Sigillariae, hollow as they were but supported by strong roots, may have resisted an incursion of the sea.

The high tides of the Bay of Fundy, rising more than 60 feet, are so destructive as to undermine and sweep away continually the whole face of the cliffs, and thus a new crop of erect fossil trees is brought into view every three or four years. They are known to extend over a space between two and three miles from north to south, and more than twice that distance from east to west, being seen in the banks of streams intersecting the coal-field.

STRUCTURE OF COAL.

The bituminous coal of Nova Scotia is similar in composition and structure to that of Great Britain, being chiefly derived from sigillarioid trees mixed with leaves of ferns and of a Lycopodiaceous tree called Cordaites (Noeggerathia, etc., for genus, see Figure 428), supposed by Dawson to have been deciduous, and which had broad parallel veined leaves without a mid-rib. On the surface of the seams of coal are large quantities of mineral charcoal, which doubtless consist, as Dr. Dawson suggests, of fragments of wood which decayed in the open air, as would naturally be expected in swamps where so many erect trees were preserved. Beds of cannel-coal display, says Dr. Dawson, such a microscopical structure and chemical composition as shows them to have been of the nature of fine vegetable mud such as accumulates in the shallow ponds of modern swamps. The underclays are loamy soils, which must have been sufficiently above water to admit of drainage, and the absence of sulphurets, and the occurrence of carbonate of iron in them, prove that when they existed as soils, rain-water, and not sea-water, percolated them. With the exception, perhaps, of Asterophyllites (see Figure 461), there is a remarkable absence from the coal-measures of any form of vegetation properly aquatic, the true coal being a sub-aerial accumulation in soil that was wet and swampy but not permanently submerged.

AIR-BREATHERS OF THE COAL.

If we have rightly interpreted the evidence of the former existence at more than eighty different levels of forests of trees, some of them of vast extent, and which lasted for ages, giving rise to a great accumulation of vegetable matter, it is natural to ask whether there were not many air-breathing inhabitants of these same regions. As yet no remains of mammalia or birds have been found, a negative character common at present to all the Palaeozoic formations; but in 1852 the osseous remains of a reptile, the first ever met with in the carboniferous strata of the American continent, were found by Dr. Dawson and myself.

We detected them in the interior of one of the erect Sigillariae before alluded to as of such frequent occurrence in Nova Scotia. The tree was about two feet in diameter, and consisted of an external cylinder of bark, converted into coal, and an internal stony axis of black sandstone, or rather mud and sand stained black by carbonaceous matter, and cemented together with fragments of wood into a rock. These fragments were in the state of charcoal, and seem to have fallen to the bottom of the hollow tree while it was rotting away. The skull, jaws, and vertebrae of a reptile, probably about 2 1/2 feet in length (Dendrerpeton Acadianum, Owen), were scattered through this stony matrix. The shell, also, of a Pupa (see Figure 442), the first land-shell ever met with in the coal or in beds older than the tertiary, was observed in the same stony mass. Dr. Wyman of Boston pronounced the reptile to be allied in structure to Menobranchus and Menopoma, species of batrachians, now inhabiting the North American rivers. The same view was afterwards confirmed by Professor Owen, who also pointed out the resemblance of the cranial plates to those seen in the skull of Archegosaurus and Labyrinthodon. (Quarterly Geological Journal volume 9 page 58.) Whether the creature had crept into the hollow tree while its top was still open to the air, or whether it was washed in with mud during a flood, or in whatever other manner it entered, must be matter of conjecture.

Footprints of two reptiles of different sizes had previously been observed by Dr. Harding and Dr. Gesner on ripple-marked flags of the lower coal-measures in Nova Scotia (No. 2, Figure 447), evidently made by quadrupeds walking on the ancient beach, or out of the water, just as the recent Menopoma is sometimes observed to do.

The remains of a second and smaller species of Dendrerpeton, D. Oweni, were also found accompanying the larger one, and still retaining some of its dermal appendages; and in the same tree were the bones of a third small lizard-like reptile, Hylonomus Lyelli, seven inches long, with stout hind limbs, and fore limbs comparatively slender, supposed by Dr. Dawson to be capable of walking and running on land. (Dawson Air-Breathers of the Coal in Nova Scotia Montreal 1863.)

(FIGURE 441. Xylobius Sigillariae, Dawson. Coal, Nova Scotia. a. Natural size. b. Anterior part, magnified. c. Caudal extremity, magnified.)

(FIGURE 442. Pupa vetusta, Dawson. a. Natural size. b. Magnified.)

In a second specimen of an erect stump of a hollow tree 15 inches in diameter, the ribbed bark of which showed that it was a Sigillaria, and which belonged to the same forest as the specimen examined by us in 1852, Dr. Dawson obtained not only fifty specimens of Pupa vetusta (Figure 442), and nine skeletons of reptiles belonging to four species, but also several examples of an articulated animal resembling the recent centipede or gally-worm, a creature which feeds on decayed vegetable matter (see Figure 441). Under the microscope, the head, with the eyes, mandible, and labrum, are well seen. It is interesting, as being the earliest known representative of the myriapods, none of which had previously been met with in rocks older than the oolite or lithographic slate of Germany.

Some years after the discovery of the first Pupa, Dr. Dawson, carefully examining the same great section containing so many buried forests in the cliffs of Nova Scotia, discovered another bed, separated from the tree containing Dendrerpeton by a mass of strata more than 1200 feet thick. As there were 21 seams of coal in this intervening mass, the length of time comprised in the interval is not to be measured by the mere thickness of the sandstones and shales. This lower bed is an underclay seven feet thick, with stigmarian rootlets, and the small land-shells occurring in it are in all stages of growth. They are chiefly confined to a layer about two inches thick, and are unmixed with any aquatic shells. They were all originally entire when imbedded, but are most of them now crushed, flattened, and distorted by pressure; they must have been accumulated, says Dr. Dawson, in mud deposited in a pond or creek.

(FIGURE 443. Zonites (Conulus) priscus, Carpenter. a. Natural size. b. Magnified.)

The surface striae of Pupa vetusta, when magnified 50 diameters, present exactly the same appearance as a portion corresponding in size of the common English Pupa juniperi, and the internal hexagonal cells, magnified 500 diameters, show the internal structure of the fossil and recent Pupa to be identical. In 1866 Dr. Dawson discovered in this lower bed, so full of the Pupa, another land-shell of the genus Helix (sub-genus Zonites), see Figure 443. (Dawson Acadian Geology 1868 page 385.)

None of the reptiles obtained from the coal-measures of the South Joggins are of a higher grade than the Labyrinthodonts, but some of these were of very great size, two caudal vertebrae found by Mr. Marsh in 1862 measuring two and a half inches in diameter, and implying a gigantic aquatic reptile with a powerful swimming tail.

Except some obscure traces of an insect found by Dr. Dawson in a coprolite of a terrestrial reptile occurring in a fossil tree, no specimen of this class has been brought to light in the Joggins. But Mr. James Barnes found in a bed of shale at Little Grace Bay, Cape Breton, the wing of an Ephemera, which must have measured seven inches from tip to tip of the expanded wings— larger than any known living insect of the Neuropterous family.

That we should have made so little progress in obtaining a knowledge of the terrestrial fauna of the Coal is certainly a mystery, but we have no reason to wonder at the extreme rarity of insects, seeing how few are known in the carboniferous rocks of Europe, worked for centuries before America was discovered, and now quarried on so enormous a scale. These European rocks have not yet produced a single land-shell, in spite of the millions of tons of coal annually extracted, and the many hundreds of soils replete with the fossil roots of trees, and the erect trunks and stumps preserved in the position in which they grew. In many large coal-fields we continue as much in the dark respecting the invertebrate air-breathers then living, as if the coal had been thrown down in mid-ocean. The early date of the carboniferous strata can not explain the enigma, because we know that while the land supported a luxuriant vegetation, the contemporaneous seas swarmed with life— with Articulata, Mollusca, Radiata, and Fishes. The perplexity in which we are involved when we attempt to solve this problem may be owing partly to our want of diligence as collectors, but still more perhaps to ignorance of the laws which govern the fossilisation of land-animals, whether of high or low degree.

CARBONIFEROUS RAIN-PRINTS.

(FIGURES 444 and 445. On green shale, from Cape Breton, Nova Scotia.

(FIGURE 444. Carboniferous rain-prints with worm-tracks (a, b) on green shale, from Cape Breton, Nova Scotia. Natural size.)

(FIGURE 445. Casts of rain-prints on a portion of the same slab (Figure 444), seen to project on the under side of an incumbent layer of arenaceous shale. Natural size. The arrow represents the supposed direction of the shower.))

At various levels in the coal measures of Nova Scotia, ripple-marked sandstones, and shales with rain-prints, were seen by Dr. Dawson and myself, but still more perfect impressions of rain were discovered by Mr. Brown, near Sydney, in the adjoining island of cape Breton. They consist of very delicate markings on greenish slates, accompanied by worm-tracks (a, b, Figure 444), such as are often seen between high and low water mark on the recent mud of the Bay of Fundy.

The great humidity of the climate of the Coal period had been previously inferred from the number of its ferns and the continuity of its forests for hundreds of miles; but it is satisfactory to have at length obtained such positive proofs of showers of rain, the drops of which resembled in their average size those which now fall from the clouds. From such data we may presume that the atmosphere of the Carboniferous period corresponded in density with that now investing the globe, and that different currents of air varied then as now in temperature, so as to give rise, by their mixture, to the condensation of aqueous vapour.

FOLDING AND DENUDATION OF THE BEDS INDICATED BY THE NOVA SCOTIA COAL-STRATA.

(FIGURE 446. Cone and branch of Lepidodendron corrugatum. Lower Carboniferous,
New Brunswick.)

The series of events which are indicated by the great section of the coal-strata in Nova Scotia consist of a gradual and long-continued subsidence of a tract which throughout most of the period was in the state of a delta, though occasionally submerged beneath a sea of moderate depth. Deposits of mud and sand were first carried down into a shallow sea on the low shores of which the footprints of reptiles were sometimes impressed (see above).

Though no regular seams of coal were formed, the characteristic imbedded coal-plants are of the genera Cyclopteris and Alethopteris, agreeing with species occurring at much higher levels, and distinct from those of the antecedent Devonian group. The Lepidodendron corrugatum (see Figure 446), a plant predominating in the Lower Carboniferous group of Europe, is also conspicuous in these shallow-water beds, together with many fishes and entomostracans. A more rapid rate of subsidence sometimes converted part of the sea into deep clear water, in which there was a growth of coral which was afterwards turned into crystalline limestone, and parts of it, apparently by the action of sulphuric acid, into gypsum. In spite of continued sinking, amounting to several thousand feet, the sea might in time have been rendered shallow by the growth of coral, had not its conversion into land or swampy ground been accelerated by the pouring in of sand and the advance of the delta accompanied with such fluviatile and brackish-water formations as are common in lagoons.

(FIGURE 447. Diagram section from north, through Minudie, S. Joggins, Shoulie R. and Cobequid Mountains, south, showing the curvature and supposed denudation of the Carboniferous strata in Nova Scotia. A. Anticlinal axis of Minudie. B. Synclinal of Shoulie River. 1. Coal-measures. 2. Lower Carboniferous.)

The amount to which the bed of the sea sank down in order to allow of the formation of so vast a thickness of rock of sedimentary and organic origin is expressed by the total thickness of the Carboniferous strata, including the coal-measures, No. 1, and the rocks which underlie them, No. 2, Figure 447.

After the strata No. 2 had been elaborated, the conditions proper to a great delta exclusively prevailed, the subsidence still continuing so that one forest after another grew and was submerged until their under-clays with roots, and usually seams of coal, were left at more than eighty distinct levels. Here and there, also, deposits bearing testimony to the existence of fresh or brackish- water lagoons, filled with calcareo-bituminous mud, were formed. In these beds (h and i, Figure 439) are found fresh-water bivalves or mussels allied to Anodon, though not identical with that or any living genus, and called Naiadites carbonarius by Dawson. They are associated with small entomostracous crustaceans of the genus Cythere, and scales of small fishes. Occasionally some of the calamite brakes and forests of Sigillariae and Coniferae were exposed in the flood season, or sometimes, perhaps, by slight elevatory movements to the denuding action of the river or the sea.

In order to interpret the great coast section exposed to view on the shores of the Bay of Fundy, the student must, in the first place, understand that the newest or last-mentioned coal formations would have been the only ones known to us (for they would have covered all the others), had there not been two great movements in opposite directions, the first consisting of a general sinking of three miles, which took place during the Carboniferous Period, and the second an upheaval of more limited horizontal extent, by which the anticlinal axis A was formed. That the first great change of level was one of subsidence is proved by the fact that there are shallow-water deposits at the base of the Carboniferous series, or in the lowest beds of No. 2.

Subsequent movements produced in the Nova Scotia and the adjoining New Brunswick coal-fields the usual anticlinal and synclinal flexures. In order to follow these, we must survey the country for about thirty miles round the South Joggins, or the region where the erect trees described in the foregoing pages are seen. As we pass along the cliffs for miles in a southerly direction, the beds containing these fossil trees, which were mentioned as dipping about 18 degrees south, are less and less inclined, until they become nearly horizontal in the valley of a small river called the Shoulie, as ascertained by Dr. Dawson. After passing this synclinal line the beds begin to dip in an opposite or north- easterly direction, acquiring a steep dip where they rest unconformably on the edges of the Upper Silurian strata of the Cobequid Hills, as shown in Figure 447. But if we travel northward towards Minudie from the region of the coal- seams and buried forests, we find the dip of the coal-strata increasing from an angle of 18 degrees to one of more than 40 degrees, lower beds being continually exposed to view until we reach the anticlinal axis A and see the lower Carboniferous formation, No. 2, at the surface. The missing rocks removed by denudation are expressed by the faint lines at A, and thus the student will see that, according to the principles laid down in the seventh chapter, we are enabled, by the joint operations of upheaval and denudation, to look, as it were, about three miles into the interior of the earth without passing beyond the limits of a single formation.

CHAPTER XXIV.
FLORA AND FAUNA OF THE CARBONIFEROUS PERIOD.

Vegetation of the Coal Period.
Ferns, Lycopodiaceae, Equisetaceae, Sigillariae, Stigmariae, Coniferae.
Angiosperms.
Climate of the Coal Period.
Mountain Limestone.
Marine Fauna of the Carboniferous Period.
Corals.
Bryozoa, Crinoidea.
Mollusca.
Great Number of fossil Fish.
Foraminifera.

VEGETATION OF THE COAL PERIOD.

In the last chapter we have seen that the seams of coal, whether bituminous or anthracitic, are derived from the same species of plants, and Goppert has ascertained that the remains of every family of plants scattered through the shales and sandstones of the coal-measures are sometimes met with in the pure coal itself— a fact which adds greatly to the geological interest of this flora.

The coal-period was called by Adolphe Brongniart the age of Acrogens, so great appears to have been the numerical preponderance of flowerless or cryptogamic plants of the families of ferns, club-mosses, and horse-tails. (For botanical nomenclature see Chapter 17.) He reckoned the known species in 1849 at 500, and the number has been largely increased by recent research in spite of reductions owing to the discovery that different parts of even the same plants had been taken for distinct species. Notwithstanding these changes, Brongniart's generalisation concerning this flora still holds true, namely, that the state of the vegetable world was then extremely different from that now prevailing, not only because the cryptogamous plants constituted nearly the whole flora, but also because they were, on the whole, more highly developed than any belonging to the same class now existing, and united some forms of structure now only found separately and in distinct orders. The only phaenogamous plants were constitute any feature in the coal are the coniferae; monocotyledonous angiosperms appear to have been very rare, and the dicotyledonous, with one or two doubtful exceptions, were wanting. For this we are in some measure prepared by what we have seen of the Secondary or Mesozoic floras if, consistently with the belief in the theory of evolution, we expect to find the prevalence of simpler and less specialised organisms in older rocks.

FERNS.

(FIGURE 448. Pecopteris elliptica, Bunbury. (Sir C. Bunbury Quarterly Geological Journal volume 2 1845.) Frostburg.)

We are struck at the first glance with the similarity of the ferns to those now living. In the fossil genus Pecopteris, for example (Figure 448), it is not easy to decide whether the fossils might not be referred to the same genera as those established for living ferns; whereas, in regard to some of the other contemporary families of plants, with the exception of the fir tribe, it is not easy to guess even the class to which they belong. The ferns of the Carboniferous period are generally without organs of fructification, but in the few instances in which these do occur in a fit state for microscopical investigations they agree with those of the living ferns.

(FIGURE 449. Caulopteris primaeva, Lindley.)

When collecting fossil specimens from the coal-measures of Frostburg, in Maryland, I found in the iron-shales several species with well-preserved rounded spots or marks of the sori (see Figure 448). In the general absence of such characters they have been divided into genera distinguished chiefly by the branching of the fronds and the way in which the veins of the leaves are disposed. The larger portion are supposed to have been of the size of ordinary European ferns, but some were decidedly arborescent, especially the group called

Caulopteris (see Figure 449) by Lindley, and the Psaronius of the upper or newest coal-measures, before alluded to (Chapter 22). All the recent tree-ferns belong to one tribe (Polypodiaceae), and to a small number only of genera in that tribe, in which the surface of the trunk is marked with scars, or cicatrices, left after the fall of the fronds. These scars resemble those of Caulopteris.

(FIGURES 450, 451 and 452. Living tree-ferns of different genera. (Ad. Brong.)
(FIGURE 450. Tree-fern from Isle of Bourbon.)
(FIGURE 451. Cyathea glauca, Mauritius.)
(FIGURE 452. Tree-fern from Brazil.))

No less than 130 species of ferns are enumerated as having been obtained from the British coal-strata, and this number is more than doubled if we include the Continental and American species. Even if we make some reduction on the ground of varieties which have been mistaken, in the absence of their fructification, for species, still the result is singular, because the whole of Europe affords at present no more than sixty-seven indigenous species.

LYCOPODIACEAE— LEPIDODENDRON.

(FIGURES 453, 454 and 455. Lepidodendron Sternbergii. Coal-measures, near Newcastle.
(FIGURE 453. Branching trunk, 49 feet long, supposed to have belonged to L. Sternbergii. (Foss. Flo. 203.))
(FIGURE 454. Branching stem with bark and leaves of L. Sternbergii. (Foss. Flo. 4.)
(FIGURE 455. Portion of same nearer the root. Natural size. (Ibid.)))
(FIGURE 456. Lycopodium densum. a. Living species. New Zealand. b. Branch; natural size. c. Part of same, magnified.)

About forty species of fossil plants of the Coal have been referred to this genus, more than half of which are found in the British coal-measures. They consist of cylindrical stems or trunks, covered with leaf-scars. In their mode of branching, they are always dichotomous (see Figure 454). They belong to the Lycopodiaceae, bearing sporangia and spores similar to those of the living representatives of this family (Figure 457); and although most of the Carboniferous species grew to the size of large trees, Mr. Carruthers has found by careful measurement that the volume of the fossil spores did not exceed that of the recent club-moss, a fact of some geological importance, as it may help to explain the facility with which these seeds may have been transported by the wind, causing the same wide distribution of the species of the fossil forests in Europe and America which we now observe in the geographical distribution of so many living families of cryptogamous plants. The Figures 453-455 represent a fossil Lepidodendron, 49 feet long, found in Jarrow Colliery, near Newcastle, lying in shale parallel to the planes of stratification. Fragments of others, found in the same shale, indicate, by the size of the rhomboidal scars which cover them, a still greater magnitude. The living club-mosses, of which there are about 200 species, are most abundant in tropical climates. They usually creep on the ground, but some stand erect, as the Lycopodium densum from New Zealand (see Figure 456), which attains a height of three feet.

(FIGURE 457. Lepidostrobus ornatus, Brong. Shropshire. a. (Body) half natural size. b. Portion of a section, showing the large sporangia in their natural position, and each supported by its bract or scale. c. Spores in these sporangia, highly magnified. (Hooker Mem. Geological Survey volume 2 part 2 page 440.)

In the Carboniferous strata of Coalbrook Dale, and in many other coal-fields, elongated cylindrical bodies, called fossil cones, named Lepidostrobus by M. Adolphe Brongniart, are met with. (See Figure 457.) They often form the nucleus of concretionary balls of clay-ironstone, and are well preserved, exhibiting a conical axis, around which a great quantity of scales were compactly imbricated. The opinion of M. Brongniart that the Lepidostrobus is the fruit of Lepidodendron has been confirmed, for these strobili or fruits have been found terminating the tip of a branch of a well-characterised Lepidodendron in Coalbrook Dale and elsewhere.

EQUISETACEAE.

(FIGURE 458. Calamites Sucowii, Brong.; natural size. Common in coal throughout Europe.)

(FIGURE 459. Stem of Figure 458, as restored by Dr. Dawson.)

(FIGURE 460. Radical termination of a Calamite. Nova Scotia.)

To this family belong two fossil genera of the coal, Equisetites and Calamites. The Calamites were evidently closely related to the modern horse-tails (Equiseta) differing principally in their great size, the want of sheaths at the joints, and some details of fructification. They grew in dense brakes on sandy and muddy flats in the manner of modern Equisetaceae, and their remains are frequent in the coal. Seven species of this plant occur in the great Nova Scotia section before described, where the stems of some of them five inches in diameter, and sometimes eight feet high, may be seen terminating downward in a tapering root (see Figure 460).

(FIGURE 461. Asterophillites foliosus. (Foss. Flo. 25.) Coal-measures, Newcastle.)

(FIGURE 462. Annularia sphenophylloides, Dawson.)

(FIGURE 463. Sphenophyllum erosum, Dawson.)

Botanists are not yet agreed whether the Asterophyllites, a species of which is represented in Figure 461, can form a separate genus from the Calamite, from which, however, according to Dr. Dawson, its foliage is distinguished by a true mid-rib, which is wanting in the leaves known to belong to some Calamites. Figures 462 and 463 represent leaves of Annularia and Sphenophyllum, common in the coal, and believed by Mr. Carruthers to be leaves of Calamites. Dr. Williamson, who has carefully studied the Calamites, thinks that they had a fistular pith, exogenous woody stem, and thick smooth bark, which last having always disappeared, leaves a fluted stem, as represented in Figure 459.

SIGILLARIA.

(FIGURE 464. Sigillaria laevigata, Brong.)

A large portion of the trees of the Carboniferous period belonged to this genus, of which as many as 28 species are enumerated as British. The structure, both internal and external, was very peculiar, and, with reference to existing types, very anomalous. They were formerly referred, by M. Ad. Brongniart, to ferns, which they resemble in the scalariform texture of their vessels and, in some degree, in the form of the cicatrices left by the base of the leaf-stalks which have fallen off (see Figure 464). But some of them are ascertained to have had long linear leaves, quite unlike those of ferns. They grew to a great height, from 30 to 60, or even 70 feet, with regular cylindrical stems, and without branches, although some species were dichotomous towards the top. Their fluted trunks, from one to five feet in diameter, appear to have decayed more rapidly in the interior than externally, so that they became hollow when standing; and when thrown prostrate, they were squeezed down and flattened. Hence, we find the bark of the two opposite sides (now converted into bright shining coal) constitute two horizontal layers, one upon the other, half an inch, or an inch, in their united thickness. These same trunks, when they are placed obliquely or vertically to the planes of stratification, retain their original rounded form, and are uncompressed, the cylinder of bark having been filled with sand, which now affords a cast of the interior.

Dr. Hooker inclined to the belief that the Sigillariae may have been cryptogamous, though more highly developed than any flowerless plants now living. Dr. Dawson having found in some species what he regards as medullary rays, thinks with Brongniart that they have some relation to gymnogens, while Mr. Carruthers leans to the opinion that they belong to the Lycopodiaceae.

STIGMARIA.

(FIGURE 465. Stigmaria attached to a trunk of Sigillaria.)

This fossil, the importance of which has already been pointed out in Chapter 23, was originally conjectured to be an aquatic plant. It is now ascertained to be the root of Sigillaria. The connection of the roots with the stem, previously suspected, on botanical grounds, by Brongniart, was first proved, by actual contact, in the Lancashire coal-field, by Mr. Binney. The fact has lately been shown, even more distinctly, by Mr. Richard Brown, in his description of the Stigmariae occurring in the under-clays of the coal-seams of the Island of

Cape Breton, in Nova Scotia. In a specimen of one of these, represented in Figure 465, the spread of the roots was sixteen feet, and some of them sent out rootlets, in all directions, into the surrounding clay.

(FIGURE 466. Stigmaria ficoides, Brong. 1/4 natural size. (Foss. Flo. 32.))

(FIGURE 467. Stigmaria ficoides, Brong. Surface of another individual of same species, showing form of tubercles. (Foss. Flo. 34.))

In the sea-cliffs of the South Joggins in Nova Scotia, I examined several erect Sigillariae, in company with Dr. Dawson, and we found that from the lower extremities of the trunk they sent out Stigmariae as roots. All the stools of the fossil trees dug out by us divided into four parts, and these again bifurcated, forming eight roots, which were also dichotomous when traceable far enough. The cylindrical rootlets formerly regarded as leaves are now shown by more perfect specimens to have been attached to the root by fitting into deep cylindrical pits. In the fossil there is rarely any trace of the form of these cavities, in consequence of the shrinkage of the surrounding tissues. Where the rootlets are removed, nothing remains on the surface of the Stigmaria but rows of mammillated tubercles (see Figures 466, 467), which have formed the base of each rootlet. These protuberances may possibly indicate the place of a joint at the lower extremity of the rootlet. Rows of these tubercles are arranged spirally round each root, which have always a medullary axis and woody system much resembling that of Sigillaria, the structure of the vessels being, like it, scalariform.

CONIFERAE.

(FIGURE 468. Fragment of coniferous wood, Dadoxylon, of Endlicher, fractured longitudinally; from Coalbrook Dale. W.C. Williamson. (Manchester Philosophical Mem. volume 9 1851.) a. Bark. b. Woody zone or fibre (pleurenchyma). c. Medulla or pith. d. Cast of hollow pith or "Sternbergia.")

(FIGURE 469. Fragment of coniferous wood, Dadoxylon, of Endlicher. Magnified portion of Figure 468; transverse section. b-b. Woody fibre. c. Pith. e, e, e. Medullary rays.)

The coniferous trees of this period are referred to five genera; the woody structure of some of them showing that they were allied to the Araucarian division of pines, more than to any of our common European firs. Some of their trunks exceeded forty-four feet in height. Many, if not all of them, seem to have differed from living Coniferae in having large piths; for Professor Williamson has demonstrated the fossil of the coal-measures called Sternbergia to be the pith of these trees, or rather the cast of cavities formed by the shrinking or partial absorption of the original medullary axis (see Figures 468, 469). This peculiar type of pith is observed in living plants of very different families, such as the common Walnut and the White Jasmine, in which the pith becomes so reduced as simply to form a thin lining of the medullary cavity, across which transverse plates of pith extend horizontally, so as to divide the cylindrical hollow into discoid interspaces. When these interspaces have been filled up with inorganic matter, they constitute an axis to which, before their true nature was known, the provisional name of Sternbergia (d, d, Figure 468) was given. In the above specimen the structure of the wood (b, Figures 468 and 469) is coniferous, and the fossil is referable to Endlicher's fossil genus Dadoxylon.

(FIGURE 470. Trigonocarpum ovatum, Lindley and Hutton. Peel Quarry, Lancashire.)

(FIGURE 471. Trigonocarpum olivaeforme, Lindley, with its fleshy envelope. Felling Colliery, Newcastle.)

The fossil named Trigonocarpon (Figures 470 and 471), formerly supposed to be the fruit of a palm, may now, according to Dr. Hooker, be referred, like the Sternbergia, to the Coniferae. Its geological importance is great, for so abundant is it in the coal-measures, that in certain localities the fruit of some species may be procured by the bushel; nor is there any part of the formation where they do not occur, except the under-clays and limestone. The sandstone, ironstone, shales, and coal itself, all contain them. Mr. Binney has at length found in the clay-ironstone of Lancashire several specimens displaying structure, and from these, says Dr. Hooker, we learn that the Trigonocarpon belonged to that large section of existing coniferous plants which bear fleshy solitary fruits, and not cones. It resembled very closely the fruit of the Chinese genus Salisburia, one of the Yew tribe, or Taxoid conifers.

ANGIOSPERMS.

(FIGURE 472. Antholithes. Felling Colliery, Newcastle.)

The curious fossils called Antholithes by Lindley have usually been considered to be flower spikes, having what seems a calyx and linear petals (see Figure 472). Dr. Hooker, after seeing very perfect specimens, also thought that they resembled the spike of a highly-organised plant in full flower, such as one of the Bromeliaceae, to which Professor Lindley had at first compared them. Mr. Carruthers, who has lately examined a large series in different museums, considers it to be a dicotyledonous angiosperm allied to Orobanche (broom-rape), which grew, not on the soil, but parasitically on the trees of the coal forests.

(FIGURE 473. Pothocites Grantonii, Pat. Coal-measures, Edinburgh. c. Stem and spike; 1/2 natural size. b. Remains of the spathe magnified. c. Portion of spike magnified. d. One of the calyces magnified.)

In the coal-measures of Granton, near Edinburgh, a remarkable fossil (Figure 473) was found and described in 1840, by Dr. Robert Paterson. (Transactions of the Botanical Society of Edinburgh volume 1 1844.) It was compressed between layers of bituminous shale, and consists of a stem bearing a cylindrical spike, a, which in the portion preserved in the slate exhibits two subdivisions and part of a third. The spike is covered on the exposed surface with the four-cleft calyces of the flowers arranged in parallel rows. The stem shows, at b, a little below the spike, remains of a lateral appendage, which is supposed to indicate the beginning of the spathe. The fossil has been referred to the Aroidiae, and there is every probability that it is a true member of this order. There can at least be no doubt as to the high grade of its organisation, and that it belongs to the monocotyledonous angiosperms. Mr. Carruthers has carefully examined the original specimen in the Botanical Museum, Edinburgh, and thinks it may have been an epiphyte.

CLIMATE OF THE COAL PERIOD.

As to the climate of the Coal, the Ferns and the Coniferae are perhaps the two classes of plants which may be most relied upon as leading us to safe conclusions, as the genera are nearly allied to living types. All botanists admit that the abundance of ferns implies a moist atmosphere. But the coniferae, says Hooker, are of more doubtful import, as they are found in hot and dry, and in cold and dry climates; in hot and moist, and in cold and moist regions. In New Zealand the coniferae attain their maximum in numbers, constituting 1/62 part of all the flowering plants; whereas in a wide district around the Cape of Good Hope they do not form 1/1600 of the phenogamic flora. Besides the conifers, many species of ferns flourish in New Zealand, some of them arborescent, together with many lycopodiums; so that a forest in that country may make a nearer approach to the carboniferous vegetation than any other now existing on the globe.

MARINE FAUNA OF THE CARBONIFEROUS PERIOD.

It has already been stated that the Carboniferous or Mountain Limestone underlies the coal-measures in the South of England and Wales, whereas in the North, and in Scotland, marine calcareous rocks partly of the age of the Mountain Limestone alternate with shales and sandstones, containing seams of coal. In its most calcareous form the Mountain Limestone is destitute of land- plants, and is loaded with marine remains— the greater part, indeed, of the rock being made up bodily of crinoids, corals, and bryozoa with interspersed mollusca.

CORALS.

(FIGURE 474. Palaeozoic type of lamelliferous cup-shaped Coral. Order ZOANTHARIA RUGOSA, Milne Edwards and Jules Haime. a. Vertical section of Campophyllum flexuosum, (Cyathophyllum, Goldfuss); 1/2 natural size: from the Devonian of the Eifel. The lamellae are seen around the inside of the cup; the walls consist of cellular tissue; and large transverse plates, called tubulae, divide the interior into chambers. b. Arrangement of the lamellae in Polycoelia profunda, Germar, sp.; natural size: from the Magnesian Limestone, Durham. This diagram shows the quadripartite arrangement of the primary septa, characteristic of palaeozoic corals, there being four principal and eight intermediate lamellae, the whole number in this type being always a multiple of four. c. Stauria astraeiformis, Milne Edwards. Young group, natural size. Upper Silurian, Gothland.

The lamellae or septal system in each cup are divided by four prominent ridges into four groups.)

(FIGURE 475. Neozoic type of lamelliferous cup-shaped Coral. Order ZOANTHARIA APOROSA, M. Edwards and J. Haime. a. Parasmilia centralis, Mantell, sp. Vertical section; natural size. Upper Chalk, Gravesend. In this type the lamellae are massive, and extend to the axis or columella composed of loose cellular tissue, without any transverse plates like those in Figure 474, a. b. Cyathina Bowerbankii, Ed. and H. Transverse section, enlarged. Gault, Folkestone. In this coral the primary septa are a multiple of six. The twelve principal plates reach the columella, and between each pair there are three secondaries, in all forty-eight. The short intermediate plates which proceed from the columella are not counted. They are called pali. c. Fungia patellaris, Lamarck. Recent; very young state. Diagram of its six primary and six secondary septa, magnified. The sextuple arrangement is always more manifest in the young than in the adult state.)

The corals deserve especial notice, as the cup-and-star corals, which have the most massive and stony skeletons, display peculiarities of structure by which they may be distinguished generally, as MM. Milne Edwards and Haime first pointed out, from all species found in strata newer than the Permian. There is, in short, an ancient or PALAEOZOIC, and a modern or NEOZOIC type, if, by the latter term, we designate (as proposed by Professor E. Forbes) all strata from the triassic to the most modern, inclusive. The accompanying diagrams (Figures 474, 475) may illustrate these types.

It will be seen that the more ancient corals have what is called a quadripartite arrangement of the chief plates or LAMELLAE— parts of the skeleton which support the organs of reproduction. The number of these lamellae in the Palaeozoic type is 4, 8, 16, etc.; while in the Neozoic type the number is 6, 12, 24, or some other multiple of six; and this holds good, whether they be simple forms, as in Figures 474, a, and 475, a, or aggregate clusters of corallites, as in 474, c. But further investigations have shown in this, as in all similar grand generalisations in natural history, that there are exceptions to the rule. Thus in the Lower Greensand Holocystis elegans (Ed. and H.) and other forms have the Palaeozoic type, and Dr. Duncan has shown to what extent the Neozoic forms penetrate downward into the Carboniferous and Devonian rocks.

(FIGURE 476. Lithostrotion basaltiforme, Phil. sp. (Lithostrotion striatum, Fleming; Astraea basaltiformis, Conyb. and Phill.). England, Ireland, Russia, Iowa, and westward of the Mississippi, United States. (D.D. Owen.)

(FIGURE 477. Lonsdaleia floriformis, Martin, sp., M. Edwards. (Lithostrotion floriforme, Fleming. Strombodes.) a. Young specimen, with buds or corallites on the disk, illustrating calicular gemmation. b. Part of a full-grown compound mass. Bristol, etc.; Russia.)

From a great number of lamelliferous corals met with in the Mountain Limestone, two species (Figures 476, 477) have been selected, as having a very wide range, extending from the eastern borders of Russia to the British Isles, and being found almost everywhere in each country. These fossils, together with numerous species of Zaphrentis, Amplexus, Cyathophyllum, Clisiophyllum, Syringopora, and Michelinia, form a group of rugose corals widely different from any that followed them. (For figures of these corals, see Palaeontographical Society's Monographs 1852.)

BRYOZOA AND CRINOIDEA.

(FIGURE 478. Cyathocrinus planus, Miller. Body and arms. Mountain Limestone.)

(FIGURE 479. Cyathocrinus caryocrinoides, M'Coy. a. Surface of one of the joints of the stem. b. Pelvis or body; called also calyx or cup. c. One of the pelvic plates.)

Of the Bryozoa, the prevailing forms are Fenestella, Hemitrypa, and Polypora, and these often form considerable beds. Their net-like fronds are easily recognised. Crinoidea are also numerous in the Mountain Limestone (see Figures 478, 479), two genera, Pentremites and Codonaster, being peculiar to this formation in Europe and North America.

(FIGURE 480. Palaechinus gigas, M'Coy. Reduced one-third. Mountain Limestone. Ireland.)

In the greater part of them, the cup or pelvis, Figure 479, b, is greatly developed in size in proportion to the arms, although this is not the case in Figure 478. The genera

Poteriocrinus, Cyathocrinus, Pentremites, Actinocrinus, and Platycrinus, are all of them characteristic of this formation. Other Echinoderms are rare, a few Sea-Urchins only being known: these have a complex structure, with many more plates on their surface than are seen in the modern genera of the same group. One genus, the Palaechinus (Figure 480), is the analogue of the modern Echinus, but has four, five, or six rows of plates in the interambulacral region or area, whereas the modern genera have only two. The other, Archaeocidaris, represents, in like manner, the Cidaris of the present seas.

MOLLUSCA.

(FIGURE 481. Productus semireticulatus, Martin, sp. (P. antiquatus, Sowerby.) Mountain Limestone. England, Russia, the Andes, etc.)

(FIGURE 482. Spirifera trigonalis, Martin, sp. Mountain Limestone. Derbyshire, etc.)

(FIGURE 483. Spirifera glabra, Martin, sp. Mountain Limestone.)

The British Carboniferous mollusca enumerated by Mr. Etheridge comprise 653 species referable to 86 genera, occurring chiefly in the Mountain Limestone. (Quarterly Geological Journal volume 23 page 674 1867.) Of this large number only 40 species are common to the underlying Devonian rocks, 9 of them being Cephalopods, 7 Gasteropods, and the rest bivalves, chiefly Brachiopoda (or Palliobranchiates). This latter group constitutes the larger part of the Carboniferous Mollusca, 157 species being known in Great Britain alone, and it will be found to increase in importance in the fauna of the primary rocks the lower we descend in the series. Perhaps the most characteristic shells of the formation are large species of Productus, such as P. giganteus, p. hemisphericus, P. semireticulatus (Figure 481), and P. scabriculus. Large plaited spirifers, as Spirifera striata, S. rotundata, and S. trigonalis (Figure 482), also abound; and smooth species, such as Spirifera glabra (Figure 483), with its numerous varieties.

(FIGURE 484. Terebratula hastata, Sowerby, with radiating bands of colour. Mountain Limestone. Derbyshire, Ireland, Russia, etc.)

(FIGURE 485. Aviculopecten sublobatus, Phill. Mountain Limestone. Derbyshire, Yorkshire.)

(FIGURE 486. Pleurotomaria carinata, Sowerby. (P. flammigera, Phillips). Mountain Limestone. Derbyshire, etc.)

Among the brachiopoda, Terebratula hastata (Figure 484) deserves mention, not only for its wide range, but because it often retains the pattern of the original coloured stripes which ornamented the living shell. These coloured bands are also preserved in several lamellibranchiate bivalves, as in Aviculopecten (Figure 485), in which dark stripes alternate with a light ground. In some also of the spiral univalves the pattern of the original painting is distinctly retained, as in Pleurotomaria (Figure 486), which displays wavy blotches, resembling the colouring in many recent trochidae.

(FIGURE 487. Euomphalus pentangulatus, Sowerby. Mountain Limestone. a. Upper side. b. Lower or umbilical side. c. View showing mouth, which is less pentagonal in older individuals. d. View of polished section, showing internal chambers.)

Some few of the carboniferous mollusca, such as Avicula, Nucula (sub-genus Ctenodonta), Solemya, and Lithodomus, belong no doubt to existing genera; but the majority, though often referred to as living types, such as Isocardia, Turritella, and Buccinum, belong really to forms which appear to have become extinct at the close of the Palaeozoic epoch. Euomphalus is a characteristic univalve shell of this period. In the interior it is divided into chambers (Figure 487, d), the septa or partitions not being perforated as in foraminiferous shells, or in those having siphuncles, like the Nautilus. The animal appears to have retreated at different periods of its growth from the internal cavity previously formed, and to have closed all communication with it by a septum. The number of chambers is irregular, and they are generally wanting in the innermost whorl. The animal of the recent Turritella communis partitions off in like manner as it advances in age a part of its spire, forming a shelly septum.

(FIGURE 488. Bellerophon costatus, Sowerby. Mountain Limestone.)

More than twenty species of the genus Bellerophon (see Figure 488), a shell like the living Argonaut without chambers, occur in the Mountain Limestone. The genus is not met with in strata of later date. It is most generally regarded as belonging to the pelagic

Nucleobranchiata and the family Atlantidae, partly allied to the Glass-Shell, Carinaria; but by some few it is thought to be a simple form of Cephalopod.

(FIGURE 489. Portion of Orthoceras laterale. Phill. Mountain Limestone.)

(FIGURE 490. Goniatites crenistra, Phillips. Mountain Limestone. North America, Britain, Germany, etc.

a. Lateral view.

b. Front view, showing the mouth.)

The carboniferous Cephalopoda do not depart so widely from the living type (the Nautilus) as do the more ancient Silurian representatives of the same order; yet they offer some remarkable forms. Among these is Orthoceras, a siphuncled and chambered shell, like a Nautilus uncoiled and straightened (Figure 489). Some species of this genus are several feet long. The Goniatite is another genus, nearly allied to the Ammonite, from which it differs in having the lobes of the septa free from lateral denticulations, or crenatures; so that the outline of these is angular, continuous, and uninterrupted. The species represented in Figure 490 is found in most localities, and presents the zigzag character of the septal lobes in perfection. The dorsal position of the siphuncle, however, clearly distinguishes the Goniatite from the Nautilus, and proves it to have belonged to the family of the Ammonites, from which, indeed, some authors do not believe it to be generically distinct.

FOSSIL FISH.

(FIGURE 491. Psammodus porosus, Agassiz. Bone-bed, Mountain Limestone. Bristol, Armagh.)

(FIGURE 492. Cochliodus contortus, Agassiz. Bone-bed, Mountain Limestone. Bristol, Armagh.)

The distribution of these is singularly partial; so much so, that M. De Koninck of Liege, the eminent palaeontologist, once stated to me that, in making his extensive collection of the fossils of the Mountain Limestone of Belgium, he had found no more than four or five examples of the bones or teeth of fishes. Judging from Belgian data, he might have concluded that this class of vertebrata was of extreme rarity in the Carboniferous seas; whereas the investigation of other countries has led to quite a different result. Thus, near Clifton, on the Avon, as well as at numerous places around the Bristol basin from the Mendip Hills to Tortworth, there is a celebrated "bone-bed," almost entirely made up of ichthyolites. It occurs at the base of the Lower Limestone shales immediately resting upon the passage beds of the Old Red Sandstone. Similar bone-beds occur in the Carboniferous Limestone of Armagh, in Ireland, where they are made up chiefly of the teeth of fishes of the Placoid order, nearly all of them rolled as if drifted from a distance. Some teeth are sharp and pointed, as in ordinary sharks, of which the genus Cladodus afford an illustration; but the majority, as in Psammodus and Cochliodus, are, like the teeth of the Cestracion of Port Jackson (see Figure 261), massive palatal teeth fitted for grinding. (See Figures 491, 492.)

There are upward of seventy other species of fossil fish known in the Mountain Limestone of the British Islands. The defensive fin-bones of these creatures are not infrequent at Armagh and Bristol; those known as Oracanthus, Ctenocanthus, and Onchus are often of a very large size. Ganoid fish, such as Holoptychius, also occur; but these are far less numerous. The great Megalichthys Hibberti appears to range from the Upper Coal-measures to the lowest Carboniferous strata.

FORAMINIFERA.

(FIGURE 493. Fusulina cylindrica, d'Orbigny. Magnified 3 diameters. Mountain Limestone.)

In the upper part of the Mountain Limestone group in the south-west of England, near Bristol, limestones having a distinct oolitic structure alternate with shales. In these rocks the nucleus of every minute spherule is seen, under the microscope, to consist of a small rhizopod or foraminifer. This division of the lower animals, which is represented so fully at later epochs by the Nummulites and their numerous minute allies, appears in the Mountain Limestone to be restricted to a very few species, among which Textularia, Nodosaria, Endothyra, and Fusulina (Figure 493), have been recognised. The first two genera are common to this and all the after periods; the third has been found in the Upper Silurian, but is not known above the Carboniferous strata; the fourth (Figure 493) is characteristic of the Mountain Limestone in the United States, Arctic America, Russia, and Asia Minor, but is also known in the Permian.

CHAPTER XXV.
DEVONIAN OR OLD RED SANDSTONE GROUP.

Classification of the Old Red Sandstone in Scotland and in Devonshire.
Upper Old Red Sandstone in Scotland, with Fish and Plants.
Middle Old Red Sandstone.
Classification of the Ichthyolites of the Old Red, and their Relation to Living Types.
Lower Old Red Sandstone, with Cephalaspis and Pterygotus.
Marine or Devonian Type of Old Red Sandstone.
Table of Devonian Series.
Upper Devonian Rocks and Fossils.
Middle.
Lower.
Eifel Limestone of Germany.
Devonian of Russia.
Devonian Strata of the United States and Canada.
Devonian Plants and Insects of Canada.

CLASSIFICATION OF THE TWO TYPES OF OLD RED SANDSTONE.

We have seen that the Carboniferous strata are surmounted by the Permian and Trias, both originally included in England under the name "New Red Sandstone," from the prevailing red colour of the strata. Under the coal came other red sandstones and shales which were distinguished by the title of "Old Red Sandstone." Afterwards the name of "Devonian" was given by Sir R. Murchison and Professor Sedgwick to marine fossiliferous strata which, in the south of England, occupy a similar position between the overlying coal and the underlying Silurian formations.

It may be truly said that in the British Isles the rocks of this age present themselves in their mineral aspect, and even to some extent in their fossil contents, under two very different forms; the one as distinct from the other as are often lacustrine or fluviatile from marine strata. It has indeed been suggested that by far the greater part of the deposits belonging to what may be termed the Old Red Sandstone type are of fresh-water origin. The number of land- plants, the character of the fishes, and the fact that the only shell yet discovered belongs to the genus Anodonta, must be allowed to lend no small countenance to this opinion. In this case the difficulty of classification when the strata of this type are compared in different regions, even where they are contiguous, may arise partly from their having been formed in distinct hydrographical basins, or in the neighbourhood of the land in shallow parts of the sea into which large bodies of fresh-water entered, and where no marine mollusca or corals could flourish. Under such geographical conditions the limited extent of some kinds of sediment, as well as the absence of those marine forms by which we are able to identify or contrast marine formations, may be explained, while the great thickness of the rocks, which might seem at first sight to require a corresponding depth of water, can often be shown to have been due to the gradual sinking down of the bottom of the estuary or sea where the sediment was accumulated.

Another active cause of local variation in Scotland was the frequency of contemporaneous volcanic eruptions; some of the rocks derived from this source, as between the Grampians and the Tay, having formed islands in the sea, and having been converted into shingle and conglomerate, before the upper portions of the red shales and sandstones were superimposed.

The dearth of calcareous matter over wide areas is characteristic of the Old Red Sandstone. This is, no doubt, in great part due to the absence of shells and corals; but why should these be so generally wanting in all sedimentary rocks the colour of which is determined by the red oxide of iron? Some geologists are of opinion that the waters impregnated with this oxide were prejudicial to living beings, others that strata permeated with this oxide would not preserve such fossil remains.

In regard to the two types, the Old Red Sandstone and the Devonian, I shall first treat of them separately, and then allude to the proofs of their having been to a great extent contemporaneous. That they constitute a series of rocks intermediate in date between the lowest Carboniferous and the uppermost Silurian is not disputed by the ablest geologists; and it can no longer be contended that the Upper, Middle, and Lower Old Red Sandstone preceded in date the three divisions to which, by aid of the marine shells, the Devonian rocks have been referred, while, on the other hand, we have not yet data for enabling us to affirm to what extent the subdivisions of the one series may be the equivalents in time of those of the other.

UPPER OLD RED SANDSTONE.

(FIGURE 494. Anodonta Jukesii, Forbes. Upper Devonian, Kiltorkan, Ireland.)

(FIGURE 495. Bifurcating branch of Lepidodendron Griffithsii, Brongn. Upper Devonian, Kilkenny.)

(FIGURE 496. Palaeopteris Hibernica, Schimp. (Cyclopteris Hibernica), Edward Forbes (Adiantites, Gop.). Upper Devonian, Kilkenny.)

The highest beds of the series in Scotland, lying immediately below the coal in Fife, are composed of yellow sandstone well seen at Dura Den, near Coupar, in Fife, where, although the strata contain no mollusca, fish have been found abundantly, and have been referred to the genera Holoptychius, Pamphractus, Glyptopomus, and many others. In the county of Cork, in Ireland, a similar yellow sandstone occurs containing fish of genera characteristic of the Scotch Old Red Sandstone, as for example Coccosteus (a form represented by many species in the Old Red Sandstone and by one only in the Carboniferous group), and Glytolepis and Asterolepis, both exclusively confined to the "Old Red." In the same Irish sandstone at Kiltorkan has been found an Anodonta or fresh-water mussel, the only shell hitherto discovered in the Old Red Sandstone of the British Isles (see Figure 494). In the same formation are found the fern (Figure 496) and the Lepidodendron (Figure 495), and other species of plants, some of which, Professor Heer remarks, agree specifically with species from the lower carboniferous beds. This induces him to lean to the opinion long ago advocated by Sir Richard Griffiths, that the yellow sandstone, in spite of its fish remains, should be classed as Lower Carboniferous, an opinion which I am not yet prepared to adopt. Between the Mountain Limestone and the yellow sandstone in the south-west of Ireland there intervenes a formation no less than 5000 feet thick, called the "Carboniferous slate," and at the base of this, in some places, are local deposits, such as the Glengariff Grits, which appear to be beds of passage between the Carboniferous and Old Red Sandstone groups.

It is a remarkable result of the recent examination of the fossil flora of Bear Island, latitude 74 degrees 30' N., that Professor Heer has described as occurring in that part of the Arctic region (nearly twenty-six degrees to the north of the Irish locality) a flora agreeing in several of its species with that of the yellow sandstones of Ireland. This Bear Island flora is believed by Professor Heer to comprise species of plants some of which ascend even to the higher stages of the European Carboniferous formation, or as high as the Mountain Limestone and Millstone Grit. Palaeontologists have long maintained that the same species which have a wide range in space are also the most persistent in time, which may prepare us to find that some plants having a vast geographical range may also have endured from the period of the Upper Devonian to that of the Millstone Grit.

(FIGURE 497. Scale of Holoptychius nobilissimus, Agassiz. Clashbinnie. 1/2 natural size.)

(FIGURE 498. Holoptychius, as restored by Professor Huxley. a. The fringed pectoral fins. b. The fringed ventral fins. c. Anal fin. d, e. Dorsal fins.)

Outliers of the Upper "Old Red" occur unconformably on older members of the group, and the formation represented at Whiteness, near Arbroath, a, Figure 55, may probably be one of these outliers, though the want of organic remains renders this uncertain. It is not improbable that the beds given in this section as Nos. 1, 2, and 3, may all belong to the early part of the period of the Upper Old Red, as some scales of Holoptychius nobilissimus have been found scattered through these beds, No. 2, in Strathmore. Another nearly allied Holoptychius occurs in Dura Den, see Figure 498 of this fish and also Figure 497 of one of its scales, as these last are often the only parts met with; being scattered in Forfarshire through red-coloured shales and sandstones, as are scales of a large species of the same genus in a corresponding matrix in Herefordshire. (Siluria 4th edition page 265.) The number of fish obtained from the British Upper Old Red Sandstone amounts to fifteen species referred to eleven genera.

Sir R. Murchison groups with this upper division of the Old Red of Scotland certain light-red and yellow sandstones and grits which occur in the northernmost part of the mainland, and extend also into the Orkney and Shetland Islands. They contain Calamites and other plants which agree generically with Carboniferous forms.

MIDDLE OLD RED SANDSTONE.

In the northern part of Scotland there occur a great series of bituminous schists and flagstones, to the fossil fish of which attention was first called by the late Hugh Miller. They were afterwards described by Agassiz, and the rocks containing them were examined by Sir R. Murchison and Professor Sedgwick, in Caithness, Cromarty, Moray, Nairn, Gamrie in Banff, and the Orkneys and Shetlands, in which great numbers of fossil fish have been found. These were at first supposed to be the oldest known vertebrate animals, as in Cromarty the beds in which they occur seem to form the base of the Old Red system resting almost immediately on the crystalline or metamorphic rocks. But in fact these fish-bearing beds, when they are traced from north to south, or to the central parts of Scotland, thin out, so that their relative age to the Lower Old Red Sandstone, presently to be mentioned, was not at first detected, the two formations not appearing in superposition in the same district. In Caithness, however, many hundred feet below the fish-zone of the middle division, remains of Pteraspis were found by Mr. Peach in 1861. This genus has never yet been found in either of the two higher divisions of the Old Red Sandstone, and confirms Sir R. Murchison's previous suspicion that the rocks in which it occurs belong to the Lower "Old Red," or agree in age with the Arbroath paving-stone. (Siluria 4th edition page 258.)

FOSSIL FISH OF THE MIDDLE OLD RED SANDSTONE.

The Devonian fish were referred by Agassiz to two of his great orders, namely, the Placoids and Ganoids. Of the first of these, which in the Recent period comprise the shark, the dog-fish, and the ray, no entire skeletons are preserved, but fin-spines, called ichthyodorulites, and teeth occur. On such remains the genera Onchus, Odontacanthus, and Ctenodus, a supposed cestracient, and some others, have been established.

(FIGURE 499. Polypterus. See Agassiz, "Recherces sur les Poissons Fossiles." Living in the Nile and other African rivers. a. One of the fringed pectoral fins. b. One of the ventral fins. c. Anal fin. d. Dorsal fin, or row of finlets.)

(FIGURE 500. Restoration of Osteolepis. Pander. Old Red Sandstone, or Devonian. a. One of the fringed pectoral fins. b. One of the ventral fins. c. Anal fin. d, e. Dorsal fins.)

By far the greater number of the Old Red Sandstone fishes belong to a sub-order of Ganoids instituted by Huxley in 1861, and for which he has proposed the name of Crossopterygidae (Abridged from crossotos, a fringe, and pteryx, a fin.), or the fringe-finned, in consideration of the peculiar manner in which the fin-rays of the paired fins are arranged so as to form a fringe round a central lobe, as in the Polypterus (see a, Figure 499), a genus of which there are several species now inhabiting the Nile and other African rivers. The reader will at once recognise in Osteolepis (Figure 500), one of the common fishes of the Old Red Sandstone, many points of analogy with Polypterus. They not only agree in the structure of the fin, at first pointed out by Huxley, but also in the position of the pectoral, ventral, and anal fins, and in having an elongated body and rhomboidal scales. On the other hand, the tail is more symmetrical in the recent fish, which has also an apparatus of dorsal finlets of a very abnormal character, both as to number and structure. As to the dorsals of Osteolepis, they are regular in structure and position, having nothing remarkable about them, except that there are two of them, which is comparatively unusual in living fish.

Among the "fringe-finned" Ganoids we find some with rhomboidal scales, such as Osteolepis, Figure 500; others with cycloidal scales, as Holoptychius, before mentioned (see Figure 498). In the genera Dipterus and Diplopterus, as Hugh Miller pointed out, and in several other of the fringe-finned genera, as in Gyroptychius and Glyptolepis, the two dorsals are placed far backward, or directly over the ventral and anal fins. The Asterolepis was a ganoid fish of gigantic dimensions. A. Asmusii, Eichwald, a species characteristic of the Old Red Sandstone of Russia, as well as that of Scotland, attained the length of between twenty and thirty feet. It was clothed with strong bony armour, embossed with star-like tubercles, but it had only a cartilaginous skeleton. The mouth was furnished with two rows of teeth, the outer ones small and fish-like, the inner larger and with a reptilian character. The Asterolepis occurs also in the Devonian rocks of North America.

If we except the Placoids already alluded to, and a few other families of doubtful affinities, all the Old Red Sandstone fishes are Ganoids, an order so named by Agassiz from the shining outer surface of their scales; but Professor Huxley has also called our attention to the fact that, while a few of the primary and the great majority of the secondary Ganoids resemble the living bony pike, Lepidosteus, or the Amia, genera now found in North American rivers, and one of them, Lepidosteus, extending as far south as Guatemala, the Crossopterygii, or fringe-finned Ichthyolites, of the Old Red are closely related to the African Polypterus, which is represented by five or six species now inhabiting the Nile and the rivers of Senegal. These North American and African Ganoids are quite exceptional in the living creation; they are entirely confined to the northern hemisphere, unless some species of Polypterus range to the south of the line in Africa; and, out of about 9000 living species of fish known to M. Gunther, and of which more than 6000 are now preserved in the British Museum, they probably constitute no more than nine.

If many circumstances favour the theory of the fresh-water origin of the Old Red Sandstone, this view of its nature is not a little confirmed by our finding that it is in Llake Superior and the other inland Canadian seas of fresh water, and in the Mississippi and African rivers, that we at present find those fish which have the nearest affinity to the fossil forms of this ancient formation.

(FIGURE 501. Pterichthys, Agassiz; Upper side, showing mouth; as restored by H. Miller.)

Among the anomalous forms of Old Red fishes not referable to Huxley's Crossopterygii is the Pterichthys, of which five species have been found in the middle division of the Old Red of Scotland. Some writers have compared their shelly covering to that of Crustaceans, with which, however, they have no real affinity. The wing-like appendages, whence the genus is named, were first supposed by Hugh Miller to be paddles, like those of the turtle; and there can now be no doubt that they do really correspond with the pectoral fins.

The number of species of fish already obtained from the middle division of the Old Red Sandstone in Great Britain is about 70, and the principal genera, besides Osteolepis and Pterichthys, already mentioned, are Glyptolepis, Diplacanthus, Dendrodus, Coccosteus, Cheirancanthus, and Acanthoides.

LOWER OLD RED SANDSTONE.

(FIGURE 502. Cephalaspis Lyellii, Agassiz. Length 6 3/4 inches. From a specimen in my collection found at Glammiss, in Forfarshire. (See other figures, Agassiz, volume 2 table 1 a and 1 b. a. One of the peculiar scales with which the head is covered when perfect. These scales are generally removed, as in the specimen above figured. b, c. Scales from different parts of the body and tail.)

The third or lowest division south of the Grampians consists of grey paving- stone and roofing-slate, with associated red and grey shales; these strata underlie a dense mass of conglomerate. In these grey beds several remarkable fish have been found of the genus named by Agassiz Cephalaspis, or "buckler- headed," from the extraordinary shield which covers the head (see Figure 502), and which has often been mistaken for that of a trilobite, such as Asaphus. A species of Pteraspis, of the same family, has also been found by the Reverend Hugh Mitchell in beds of corresponding age in Perthshire; and Mr. Powrie enumerates no less than five genera of the family Acanthodidae, the spines, scales, and other remains of which have been detected in the grey flaggy sandstones. (Powrie Geological Quarterly Journal volume 20 page 417.)

(FIGURE 503. Pterygotus anglicus, Agassiz. Middle portion of the back of the head called the seraphim.)

(FIGURE 504. Pterygotus anglicus, Agassiz. Forfarshire. Ventral aspect. Restored by H. Wodward, F.G.S. a. Carapace, showing the large sessile eyes at the anterior angles. b. The metastoma or post-oral plate (serving the office of a lower lip). c, c. Chelate appendages (antennules). d. First pair of simple palpi (antennae). e. Second pair of simple palpi (mandibles). f. Third pair of simple palpi (first maxillae). g. Pair of swimming feet with their broad basal joints, whose serrated edges serve the office of maxillae. h. Thoracic plate covering the first two thoracic segments, which are indicated by the figures 1, 2, and a dotted line. 1-6. Thoracic segments. 7-12. Abdominal segments. 13. Telson, or tail-plate.)

In the same formation at Carmylie, in Forfarshire, commonly known as the Arbroath paving-stone, fragments of a huge crustacean have been met with from time to time. They are called by the Scotch quarrymen the "Seraphim," from the wing-like form and feather-like ornament of the thoracic appendage, the part most usually met with. Agassiz, having previously referred some of these fragments to the class of fishes, was the first to recognise their crustacean character, and, although at the time unable correctly to determine the true relation of the several parts, he figured the portions on which he founded his opinion, in the first plate of his "Poissons Fossiles du Vieux Gres Rouge."

A restoration in correct proportion to the size of the fragments of P. anglicus (Figure 504), from the Lower Old Red Sandstone of Perthshire and Forfarshire, would give us a creature measuring from five to six feet in length, and more than one foot across.

The largest crustaceans living at the present day are the Inachus Kaempferi, of De Haan, from Japan (a brachyurous or short-tailed crab), chiefly remarkable for the extraordinary length of its limbs; the fore-arm measuring four feet in length, and the others in proportion, so that it covers about 25 square feet of ground; and the Limulus Moluccanus, the great King Crab of China and the Eastern seas, which, when adult, measures 1 1/2 foot across its carapace, and is three feet in length.

(FIGURE 505. Parka decipiens, Fleming. In sandstone of lower beds of Old Red, Ley's Mill, Forfarshire.)

(FIGURE 506. Parka decipiens, Fleming. In shale of Lower Old Red, Park Hill, Fife.)

(FIGURE 507. Shale of Old Red Sandstone. Forfarshire. With impression of plants and eggs of Crustaceans. a. Two pair of ova? resembling those of large Salamanders or Tritons— on the same leaf. b, b. Detached ova.)

Besides some species of Pterygotus, several of the allied genus Eurypterus occur in the Lower Old Red Sandstone, and with them the remains of grass-like plants so abundant in Forfarshire and Kincardineshire as to be useful to the geologist by enabling him to identify the inferior strata at distant points. Some botanists have suggested that these plants may be of the family Fluviales, and of fresh-water genera. They are accompanied by fossils, called "berries" by the quarrymen, which they compared to a compressed blackberry (see Figures 505, 506), and which were called "Parka" by Dr. Fleming. They are now considered by Mr. Powrie to be the eggs of crustaceans, which is highly probable, for they have not only been found with Pterygotus anglicus in Forfarshire and Perthshire, but also in the Upper Silurian strata of England, in which species of the same genus, Pterygotus, occur.

The grandest exhibitions, says Sir R. Murchison, of the Old Red Sandstone in England and Wales appear in the escarpments of the Black Mountains and in the Fans of Brecon and Carmarthen, the one 2862, and the other 2590 feet above the sea. The mass of red and brown sandstone in these mountains is estimated at not less than 10,000 feet, clearly intercalated between the Carboniferous and Silurian strata. No shells or corals have ever been found in the whole series, not even where the beds are calcareous, forming irregular courses of concretionary lumps called "corn-stones," which may be described as mottled red and green earthy limestones. The fishes of this lowest English Old Red are Cephalaspis and Pteraspis, specifically different from species of the same genera which occur in the uppermost Ludlow or Silurian tilestones. Crustaceans also of the genus Eurypterus are met with.

MARINE OR DEVONIAN TYPE.

We may now speak of the marine type of the British strata intermediate between the Carboniferous and Silurian, in treating of which we shall find it much more easy to identify the Upper, Middle, and Lower divisions with strata of the same age in other countries. It was not until the year 1836 that Sir R. Murchison and Professor Sedgwick discovered that the culmiferous or anthracitic shales and sandstones of North Devon, several thousand feet thick, belonged to the coal, and that the beds below them, which are of still greater thickness, and which, like the carboniferous strata, had been confounded under the general name "graywacke," occupied a geological position corresponding to that of the Old Red Sandstone already described. In this reform they were aided by a suggestion of Mr. Lonsdale, who, after studying the Devonshire fossils, perceived that they belonged to a peculiar palaeontological type of intermediate character between the Carboniferous and Silurian.

It is in the north of Devon that these formations may best be studied, where they have been divided into an Upper, Middle, and Lower Group, and where, although much contorted and folded, they have for the most part escaped being altered by intrusive trap-rocks and by granite, which in Dartmoor and the more southern parts of the same county have often reduced them to a crystalline or metamorphic state.

TABLE 25.1 DEVONIAN SERIES IN NORTH DEVON.
UPPER DEVONIAN OR PILTON GROUP.

a. Sandy slates and schists with fossils, 36 species out of 110 common to the Carboniferous group (Pilton, Barnstaple, etc.), resting on soft schists in which fossils are very abundant (Croyde, etc.), and which pass down into

b. Yellow, brown, and red sandstone, with land plants (Cyclopteris, etc.) and marine shells. One zone, characterised by the abundance of cucullaea (Baggy Point, Marwood, Sloly, etc.) resting on hard grey and reddish sandstone and micaceous flags, no fossils yet found (Dulverton, Pickwell, Down, etc.)

MIDDLE DEVONIAN OR ILFRACOMBE GROUP.

a. Green glossy slates of considerable thickness, no fossils yet recorded from these beds (Mortenoe, Lee Bay, etc.).

b. Slates and schists, with several irregular courses of limestone containing shells and corals like those of the Plymouth Limestone (Combe Martin, Ilfracombe, etc.).

LOWER DEVONIAN OR LYNTON GROUP.

a. Hard, greenish, red, and purple sandstone— no fossils yet found (Hangman Hill, etc.).

b. Soft slates with subordinate sandstones— fossils numerous at various horizons— Orthis, Corals, Encrinites, etc. (Valley of Rocks, Lynmouth, etc.).

Table 25.1 exhibits the sequence of the strata or subdivisions as seen both on the sea-coast of the British Channel and in the interior of Devon. It will be seen that in all main points it agrees with the table drawn up in 1864 for the sixth edition of my "Elements." Mr. Etheridge has since published an excellent account of the different subdivisions of the rocks and their fossils, and has also pointed out their relation to the corresponding marine strata of the Continent. (Quarterly Geological Journal volume 23 1867.) The slight modifications introduced in my table since 1864 are the result of a tour made in 1870 in company with Mr. T. Mck. Hughes, when we had the advantage of Mr. Etheridge's memoir as our guide.

The place of the sandstones of the Foreland is not yet clearly made out, as they are cut off by a great fault and disturbance.

UPPER DEVONIAN ROCKS.

(FIGURE 508. Spirifera disjuncta, Sowerby. Syn. Sp. Verneuilii, Murch. Upper Devonian, Boulogne.)

(FIGURE 509. Phacops latifrons, Bronn. Characteristic of the Devonian in Europe, Asia, and N. and S. America.)

(FIGURE 510. Clymenia linearis, Munster. Petherwyn, Cornwall; Elbersreuth, Bavaria.)

(FIGURE 511. Cypridina serrato-striata, Sandberger, Weilburg, etc.; Cornwall, Nassau, Saxony, Belgium.)

The slates and sandstones of Barnstaple (a and b of the preceding section) contain the shell Spirifera disjuncta, Sowerby (S. Verneuilii, Murch.), (see Figure 508), which has a very wide range in Europe, Asia Minor, and even China; also Strophalosia caperata, together with the large trilobite Phacops latifrons, Bronn. (See Figure 509), which is all but world-wide in its distribution. The fossils are numerous, and comprise about 150 species of mollusca, a fifth of which pass up into the overlying Carboniferous rocks. To this Upper Devonian belong a series of limestones and slates well developed at Petherwyn, in Cornwall, where they have yielded 75 species of fossils. The genus of Cephalopoda called Clymenia (Figure 510) is represented by no less than eleven species, and strata occupying the same position in Germany are called Clymenien- Kalk, or sometimes Cypridinen-Schiefer, on account of the number of minute bivalve shells of the crustacean called Cypridina serrato-striata (Figure 511), which is found in these beds, in the Rhenish provinces, the Harz, Saxony, and Silesia, as well as in Cornwall and Belgium.

MIDDLE DEVONIAN ROCKS.

(FIGURE 512. Heliolites porosa, Goldf. sp. (Porites pyriformis, Lonsd.) a. Portion of the same magnified. Middle Devonian, Torquay, Plymouth; Eifel.)

(FIGURE 513. Favosites cervicornis, Blainv. S. Devon, from a polished specimen. a. Portion of the same magnified, to show the pores.)

(FIGURE 514. Cyathophyllum caespitosum, Goldf.; Plymouth and Ilfracombe. b. A terminal star. c. Vertical section, exhibiting transverse plates, and part of another branch.)

We come next to the most typical portion of the Devonian system, including the great limestones of Plymouth and Torbay, replete with shells, trilobites, and corals. Of the corals 51 species are enumerated by Mr. Etheridge, none of which pass into the Carboniferous formation. Among the genera we find Favosites, Heliolites, and Cyathophyllum. The two former genera are very frequent in Silurian rocks: some few even of the species are said to be common to the Devonian and Silurian groups, as, for example, Favosites cervicornis (Figure 513), one of the commonest of all the Devonshire fossils. The Cyathophyllum caespitosum (Figure 514) and Heliolites pyriformis (Figure 512) are species peculiar to this formation.

(FIGURE 515. Stringocephalus Burtini, Def. a. Valves united. b. Interior of ventral or large valve, showing thick partition and portion of a large process which projects from the dorsal valve across the shell.)

(FIGURE 516. Uncites Gryphus, Def. Middle Devonian. S. Devon and the Continent.)

With the above are found no less than eleven genera of stone-lilies or crinoids, some of them, such as Cupressocrinites, distinct from any Carboniferous forms. The mollusks, also, are no less characteristic; of 68 species of Brachiopoda, ten only are common to the Carboniferous Limestone. The Stringocephalus Burtini (Figure 515) and Uncites Gryphus (Figure 516) may be mentioned as exclusively Middle Devonian genera, and extremely characteristic of the same division in Belgium. The Stringocephalus is also so abundant in the Middle Devonian of the banks of the Rhine as to have suggested the name of Stringocephalus Limestone. The only two species of Brachiopoda common to the Silurian and Devonian formations are Atrypa reticularis (Figure 532), which seems to have been a cosmopolite species, and Strophomena rhomboidalis.

(FIGURE 517. Megalodon cucullatus, Sowerby. Eifel; also Bradley, S. Devon. a. The valves united. b. Interior of valve, showing the large cardinal tooth.)

(FIGURE 518. Conularia ornata, D'Arch. and De Vern. (Geological Transactions Sec. Ser. volume 6. Plate 29.) Refrath, near Cologne.)

(FIGURE 519. Bronteus flabellifer, Goldf. Mid. Devon; S. Devon; and the Eifel.)

Among the peculiar lamellibranchiate bivalves common to the Plymouth limestone of Devonshire and the Continent, we find the Megalodon (Figure 517). There are also twelve genera of Gasteropods which have yielded 36 species, four of which pass to the Carboniferous group, namely Macrocheilus, Acroculia, Euomphalus, and Murchisonia. Pteropods occur, such as Conularia (Figure 518), and Cephalopods, such as Cyrtoceras, Gyroceras, Orthoceras, and others, nearly all of genera distinct from those prevailing in the Upper Devonian Limestone, or Clymenien- kalk of the Germans already mentioned. Although but few species of Trilobites occur, the characteristic Bronteus flabellifer (Figure 519) is far from rare, and all collectors are familiar with its fan-like tail. In this same group, called, as before stated, the Stringocephalus, or Eifel Limestone, in Germany, several fish remains have been detected, and among others the remarkable genus Coccosteus, covered with its tuberculated bony armour; and these ichthyolites serve, as Sir R. Murchison observes (Siluria page 362), to identify this middle marine Devonian with the Old Red Sandstone of Britain and Russia.

(FIGURE 520. Calceola sandalina, Lam. Eifel; also South Devon. a. Ventral valve. b. Inner side of dorsal valve.)

Beneath the Eifel Limestone (the great central and typical member of "the Devonian" on the Continent) lie certain schists called by German writers "Calceola-schiefer," because they contain in abundance a fossil body of very curious structure, Calceola sandalina (Figure 520), which has been usually considered a brachiopod, but which some naturalists have lately referred to a Goniophyllum, supposing it to be an abnormal form of the order Zoantharia rugosa (see Figure 474), differing from all other corals in being furnished with a strong operculum. This is by no means a rare fossil in the slaty limestone of South Devon, and, like the Eifel form, is confined to the middle group of this country.

LOWER DEVONIAN ROCKS.

(FIGURE 521. Spirifera mucronata, Hall. Devonian of Pennsylvania.)

A great series of sandstones and glossy slates, with Crinoids, Brachiopods, and some corals, occurring on the coast at Lynmouth and the neighbourhood, and called the Lynton Group (see Table 25.1), form the lowest member of the Devonian in North Devon. Among the 18 species of all classes enumerated by Mr. Etheridge, two-thirds are common to the Middle Devonian, but only one, the ubiquitous Atrypa reticularis, can with certainty be identified with Silurian species. Among the characteristic forms are Alveolites suborbicularis, also common to this formation in the Rhine, and Orthis arcuata, very widely spread in the North Devon localities. But we may expect a large addition to the number of fossils whenever these strata shall have been carefully searched. The Spirifer Sandstone of Sandberger, as exhibited in the rocks bordering the Rhine between Coblentz and Caub, belong to this Lower division, and the same broad-winged Spirifers distinguish the Devonian strata of North America.

(FIGURE 522. Homalonotus armatus, Burmeister. Lower Devonian; Daun, in the Eifel; and S. Devon.

Obs. The two rows of spines down the body give an appearance of more distinct trilobation than really occurs in this or most other species of the genus.)

Among the Trilobites of this era several large species of Homalonotus (Figure 522) are conspicuous. The genus is still better known as a Silurian form, but the spinose species appear to belong exclusively to the "Lower Devonian," and are found in Britain, Europe, and the Cape of Good Hope.

DEVONIAN OF RUSSIA.

The Devonian strata of Russia extend, according to Sir R. Murchison, over a region more spacious than the British Isles; and it is remarkable that, where they consist of sandstone like the "Old Red" of Scotland and Central England, they are tenanted by fossil fishes often of the same species and still oftener of the same genera as the British, whereas when they consist of limestone they contain shells similar to those of Devonshire, thus confirming, as Sir Roderick has pointed out, the contemporaneous origin which had been previously assigned to formations exhibiting two very distinct mineral types in different parts of Britain. (Murchison's Siluria page 329.) The calcareous and the arenaceous rocks of Russia above alluded to alternate in such a manner as to leave no doubt of their having been deposited in different parts of the same great period.

DEVONIAN STRATA IN THE UNITED STATES AND CANADA.

(FIGURE 523. Psilophyton princeps, Dawson, Quarterly Geological Journal volume 15 1863; and Canada Survey 1863. Species characteristic of the whole Devonian series in North America. a. Fruit; natural size. b. Stem; natural size. c. Scalariform tissue of the axis highly magnified.)

Between the Carboniferous and Silurian strata there intervenes, in the United States and Canada, a great series of formations referable to the Devonian group, comprising some strata of marine origin abounding in shells and corals, and others of shallow-water and littoral origin in which terrestrial plants abound. The fossils, both of the deep and shallow water strata, are very analogous to those of Europe, the species being in some cases the same. In Eastern Canada Sir W. Logan has pointed out that in the peninsula of Gaspe, south of the estuary of St. Lawrence, a mass of sandstone, conglomerate, and shale referable to this period occurs, rich in vegetable remains, together with some fish-spines. Far down in the sandstones of Gaspe, Dr. Dawson found, in 1869, an entire specimen of the genus Cephalaspis, a form so characteristic, as we have already seen, of the Scotch Lower Old Red Sandstone. Some of the sandstones are ripple-marked, and towards the upper part of the whole series a thin seam of coal has been observed, measuring, together with some associated carbonaceous shale, about three inches in thickness. It rests on an under-clay in which are the roots of Psilophyton (see Figure 523). At many other levels rootlets of this same plant have been shown by Principal Dawson to penetrate the clays, and to play the same part as do the rootlets of Stigmaria in the coal formation.

We had already learnt from the works of Goppert, Unger, and Bronn that the European plants of the Devonian epoch resemble generically, with few exceptions, those already known as Carboniferous; and Dr. Dawson, in 1859, enumerated 32 genera and 69 species which he had then obtained from the State of New York and Canada. A perusal of his catalogue (Quarterly Geological Journal volume 15 page 477 1859; also volume 18 page 296 1862.), comprising Coniferae, Sigillariae, Calamites, Asterophyllites, Lepidodendra, and ferns of the genera Cyclopteris, Neuropteris, Sphenopteris, and others, together with fruits, such as Cardiocarpum and Trigonocarpum, might dispose geologists to believe that they were presented with a list of Carboniferous fossils, the difference of the species from those of the coal-measures, and even a slight admixture of genera unknown in Europe, being naturally ascribed to geographical distribution and the distance of the New from the Old World. But fortunately the coal formation is fully developed on the other side of the Atlantic, and is singularly like that of Europe, both lithologically and in the species of its fossil plants. There is also the most unequivocal evidence of relative age afforded by superposition, for the Devonian strata in the United States are seen to crop out from beneath the Carboniferous on the borders of Pennsylvania and New York, where both formations are of great thickness.

The number of American Devonian plants has now been raised by Dr. Dawson to 120, to which we may add about 80 from the European flora of the same age, so that already the vegetation of this period is beginning to be nearly half as rich as that of the coal-measures which have been studied for so much longer a time and over so much wider an area. The Psilophyton above alluded to is believed by Dr. Dawson to be a lycopodiaceous plant, branching dichotomously (see P. princeps, Figure 523), with stems springing from a rhizome, which last has circular areoles, much resembling those of Stigmaria, and like it sending forth cylindrical rootlets. The extreme points of some of the branchlets are rolled up so as to resemble the croziers or circinate vernation of ferns; the leaves or bracts, a, supposed to belong to the same plant, are described by Dawson as having inclosed the fructification. The remains of Psilophyton princeps have been traced through all the members of the Devonian series in America, and Dr. Dawson has lately recognised it in specimens of Old Red Sandstone from the north of Scotland.

The monotonous character of the Carboniferous flora might be explained by imagining that we have only the vegetation handed down to us of one set of stations, consisting of wide swampy flats. But Dr. Dawson supposes that the geographical conditions under which the Devonian plants grew were more varied, and had more of an upland character. If so, the limitation of this more ancient flora, represented by so many genera and species, to the gymnospermous and cryptogamous orders, and the absence or extreme rarity of plants of higher grade, lead us naturally to speculate on the theory of progressive development, however difficult it may be to avail ourselves of this explanation, so long as we meet with even a few exceptional cases of what may seem to be monocotyledonous or dicotyledonous exogens.

DEVONIAN INSECTS OF CANADA.

The earliest known insects were brought to light in 1865 in the Devonian strata of St. John's, New Brunswick, and are referred by Mr. Scudder to four species of Neuroptera. One of them is a gigantic Ephemera, and measured five inches in expanse of wing.

Like many other ancient animals, says Dr. Dawson, they show a remarkable union of characters now found in distinct orders of insects, or constitute what have been named "synthetic types." Of this kind is a stridulating or musical apparatus like that of the cricket in an insect otherwise allied to the Neuroptera. This structure, as Dr. Dawson observes, if rightly interpreted by Mr. Scudder, introduces us to the sounds of the Devonian woods, bringing before our imagination the trill and hum of insect life that enlivened the solitudes of these strange old forests.

CHAPTER XXVI.
SILURIAN GROUP.

Classification of the Silurian Rocks.
Ludlow Formation and Fossils.
Bone-bed of the Upper Ludlow.
Lower Ludlow Shales with Pentamerus.
Oldest known Remains of fossil Fish.
Table of the progressive Discovery of Vertebrata in older Rocks.
Wenlock Formation, Corals, Cystideans and Trilobites.
Llandovery Group or Beds of Passage.
Lower Silurian Rocks.
Caradoc and Bala Beds.
Brachiopoda.
Trilobites.
Cystideae.
Graptolites.
Llandeilo Flags.
Arenig or Stiper-stones Group.
Foreign Silurian Equivalents in Europe.
Silurian Strata of the United States.
Canadian Equivalents.
Amount of specific Agreement of Fossils with those of Europe.

CLASSIFICATION OF THE SILURIAN ROCKS.

We come next in descending order to that division of Primary or Palaeozoic rocks which immediately underlie the Devonian group or Old Red Sandstone. For these strata Sir Roderick Murchison first proposed the name of Silurian when he had studied and classified them in that part of Wales and some of the contiguous counties of England which once constituted the kingdom of the Silures, a tribe of ancient Britons. Table 26.1 will explain the two principal divisions, Upper and Lower, of the Silurian rocks, and the minor subdivisions usually adopted, comprehending all the strata originally embraced in the Silurian system by Sir Roderick Murchison. The formations below the Arenig or Stiper-stones group are treated of in the next chapter, when the "Primordial" or Cambrian group is described.

TABLE 26.1. SILURIAN ROCKS (THICKNESS GIVEN IN FEET).
UPPER SILURIAN ROCKS.
1. LUDLOW FORMATION:
a. Upper Ludlow beds: 780. b. Lower Ludlow beds: 1,050.
2. WENLOCK FORMATION:
a. Wenlock limestone and shale and b. Woolhope limestone and shale, and Denbighshire grits: above 4,000.
3. LLANDOVERY FORMATION (Beds of passage between Upper and Lower Silurian):
a. Upper Llandovery (May-Hill beds): 800. b. Lower Llandovery: 600-1,000.
LOWER SILURIAN ROCKS.
1. BALA AND CARADOC BEDS, including volcanic rocks: 12,000.
2. LLANDEILO FLAGS, including volcanic rocks: 4,500.
3. ARENIG OR STIPER-STONES GROUP, including volcanic rocks: above 10,000.

UPPER SILURIAN ROCKS.
1. LUDLOW FORMATION.

This member of the Upper Silurian group, as will be seen by Table 26.1, is of great thickness, and subdivided into two parts— the Upper Ludlow and the Lower Ludlow. Each of these may be distinguished near the town of Ludlow, and at other places in Shropshire and Herefordshire, by peculiar organic remains; but out of more than 500 species found in the Ludlow formation as a whole, not more than five species per hundred are common to the overlying Devonian. The student may refer to the excellent tables given in the last edition of Sir R. Murchison's Siluria for a list of the organic remains of all classes distributed through the different subdivisions of the Upper and Lower Silurian.

A. UPPER LUDLOW: DOWNTON SANDSTONE.

At the top of this subdivision there occur beds of fine-grained yellowish sandstone and hard reddish grits which were formerly referred by Sir R. Murchison to the Old Red Sandstone, under the name of "Tilestones." In mineral character this group forms a transition from the Silurian to the Old Red Sandstone, the strata of both being conformable; but it is now ascertained that the fossils agree in great part specifically, and in general character entirely, with those of the underlying Upper Ludlow rocks. Among these are Orthoceras bullatum, Platyschisma helicites, Bellerophon trilobatus, Chonetes lata, etc., with numerous defenses of fishes.

These beds, therefore, now generally called the "Downton Sandstone," are classed as the newest member of the Upper Silurian. They are well seen at Downton Castle, near Ludlow, where they are quarried for building, and at Kington, in Herefordshire. In the latter place, as well as at Ludlow, crustaceans of the genera Pterygotus (for genus see Figure 504) and Eurypterus are met with.

BONE-BED OF THE UPPER LUDLOW.

At the base of the Downton sandstones there occurs a bone-bed which deserves especial notice as affording the most ancient example of fossil fish occurring in any considerable quantity. It usually consists of one or two thin layers of brown bony fragments near the junction of the Old Red Sandstone and the Ludlow rocks, and was first observed by Sir R. Murchison near the town of Ludlow, where it is three or four inches thick. It has since been traced to a distance of 45 miles from that point into Gloucestershire and other counties, and is commonly not more than an inch thick, but varies to nearly a foot. Near Ludlow two bone- beds are observable, with 14 feet of intervening strata full of Upper Ludlow fossils. (Murchison's Siluria page 140.) At that point immediately above the upper fish-bed numerous small globular bodies have been found, which were determined by Dr. Hooker to be the sporangia of a cryptogamic land-plant, probably lycopodiaceous.

(FIGURE 524. Onchus tenuistriatus, Agassiz. Bone-bed. Upper Silurian. Ludlow.)

(FIGURE 525. Shagreen-scales of a placoid fish, Thelodus parvidens, Agassiz. Bone-bed, Upper Ludlow.)

(FIGURE 526. Plectrodus mirabilis, Agassiz. Bone-bed, Upper Ludlow.)

Most of the fish have been referred by Agassiz to his placoid order, some of them to the genus Onchus, to which the spine (Figure 524) and the minute scales (Figure 525) are supposed to belong. It has been suggested, however, that Onchus may be one of those Acanthodian fish referred by Agassiz to his Ganoid order, which are so characteristic of the base of the Old Red Sandstone in Forfarshire, although the species of the Old Red are all different from these of the Silurian beds now under consideration. The jaw and teeth of another predaceous genus (Figure 526) have also been detected, together with some specimens of Pteraspis Ludensis. As usual in bone-beds, the teeth and bones are, for the most part, fragmentary and rolled.

GREY SANDSTONE AND MUDSTONE, ETC.

(FIGURE 527. Orthis elegantula, Dalm. Var. Orbicularis, Sowerby. Upper Ludlow.)

(FIGURE 528. Rhynchonella navicula, Sowerby. Ludlow Beds.)

The next subdivision of the Upper Ludlow consists of grey calcareous sandstone, or very commonly a micaceous stone, decomposing into soft mud, and contains, besides the shells mentioned above, Lingula cornea, Orthis orbicularis, a round variety of O. elegantula, Modiolopsis platyphylla, Grammysia cingulata, all characteristic of the Upper Ludlow. The lowest or mud-stone beds contain Rhynchonella navicula (Figure 528), which is common to this bed and the Lower Ludlow. As usual in Palaeozoic strata older than the coal, the brachiopodous mollusca greatly outnumber the lamellibranchiate (see below); but the latter are by no means unrepresented. Among other genera, for example, we observe Avicula and Pterinea, Cardiola, Ctenodonta (sub-genus of Nucula), Orthonota, Modiolopsis, and Palaearca.

Some of the Upper Ludlow sandstones are ripple-marked, thus affording evidence of gradual deposition; and the same may be said of the accompanying fine argillaceous shales, which are of great thickness, and have been provincially named "mud-stones." In some of these shales stems of crinoidea are found in an erect position, having evidently become fossil on the spots where they grew at the bottom of the sea. The facility with which these rocks, when exposed to the weather, are resolved into mud, proves that, notwithstanding their antiquity, they are nearly in the state in which they were first thrown down.

b. LOWER LUDLOW BEDS.

(FIGURE 529. Pentamerus Knightii, Sowerby. Aymestry. One-half natural size. a. View of both valves united. b. Longitudinal section through both valves, showing the central plates or septa.)

The chief mass of this formation consists of a dark grey argillaceous shale with calcareous concretions, having a maximum thickness of 1000 feet. In some places, and especially at Aymestry, in Herefordshire, a subcrystalline and argillaceous limestone, sometimes 50 feet thick, overlies the shale. Sir R. Murchison therefore classes this Aymestry limestone as holding an intermediate position between the Upper and Lower Ludlow, but Mr. Lightbody remarks that at Mocktrie, near Leintwardine, the Lower Ludlow shales, with their characteristic fossils, occur both above and below a similar limestone. This limestone around Aymestry and Sedgeley is distinguished by the abundance of Pentamerus Knightii, Sowerby (Figure 529), also found in the Lower Ludlow and Wenlock shale. This genus of brachiopoda was first found in Silurian strata, and is exclusively a palaeozoic form. The name was derived from pente, five, and meros, a part, because both valves are divided by a central septum, making four chambers, and in one valve the septum itself contains a small chamber, making five. The size of these septa is enormous compared with those of any other brachiopod shell; and they must nearly have divided the animal into two equal halves; but they are, nevertheless, of the same nature as the septa or plates which are found in the interior of Spirifera, Terebratula, and many other shells of this order. Messrs. Murchison and De Verneuil discovered this species dispersed in myriads through a white limestone of Upper Silurian age, on the banks of the Is, on the eastern flank of the Urals in Russia, and a similar species is frequent in Sweden.

(FIGURE 530. Lingula Lewisii, J. Sowb. Abberley Hills.)

(FIGURE 531. Rhynchonella (Terebratula) Wilsoni, Sowerby. Aymestry.)

(FIGURE 532. Atrypa reticularis, Linn. (Terebratula affinis, Min. Con.) Aymestry. a. Upper valve. b. Lower valve. c. Anterior margin of the valves.)

Three other abundant shells in the Aymestry limestone are, 1st, Lingula Lewisii (Figure 530); second, Rhynchonella Wilsoni, Sowerby. (Figure 531), which is also common to the Lower Ludlow and Wenlock limestone; third, Atrypa reticularis, Linn. (Figure 532), which has a very wide range, being found in every part of the Upper Silurian system, and even ranging up into the Middle Devonian series.

The Aymestry Limestone contains many shells, especially brachiopoda, corals, trilobites, and other fossils, amounting on the whole to 74 species, all except three or four being common to the beds either above or below.

(FIGURE 533. Phragmoceras ventricosum, J. Sowerby. (Orthoceras ventricosum, Stein.) Aymestry; one-quarter natural size.)

(FIGURE 534. Lituites (Trochoceras) giganteus, J. Sowerby. Near Ludlow; also in the Aymestry and Wenlock Limestones; 1/4 natural size.)

(FIGURE 535. Fragment of Orthoceras Ludense, J. Sowerby. Leintwardine, Shropshire.)

The Lower Ludlow Shale contains, among other fossils, many large cephalopoda not known in newer rocks, as the Phragmoceras of Broderip, and the Lituites of Breynius (see Figures 533, 534). The latter is partly straight and partly convoluted in a very flat spire. The

Orthoceras Ludense (Figure 535), as well as the cephalopod last mentioned, occurs in this member of the species.

A species of Graptolite, G. priodon, Bronn (Figure 545), occurs plentifully in the Lower Ludlow. This fossil, referred, though somewhat doubtfully, to a form of hydrozoid or sertularian polyp, has not yet been met with in strata above the Silurian.

Star-fish, as Sir R. Murchison points out, are by no means rare in the Lower Ludlow rock. These fossils, of which six extinct genera are now known in the Ludlow series, represented by 18 species, remind us of various living forms now found in our British seas, both of the families Asteriadae and Ophiuridae.

OLDEST KNOWN FOSSIL FISH.

Until 1859 there was no example of a fossil fish older than the bone-bed of the Upper Ludlow, but in that year a specimen of Pteraspis was found at Church Hill, near Leintwardine, in Shropshire, by Mr. J.E. Lee of Caerleon, F.G.S., in shale below the Aymestry limestone, associated with fossil shells of the Lower Ludlow formation— shells which differ considerably from those characterising the Upper Ludlow already described. This discovery is of no small interest as bearing on the theory of progressive development, because, according to Professor Huxley, the genus Pteraspis is allied to the sturgeon, and therefore by no means of low grade in the piscine class.

It is a fact well worthy of notice that no remains of vertebrata have yet been met with in any strata older than the Lower Ludlow.

When we reflect on the hundreds of Mollusks, Echinoderms, Trilobites, Corals, and other fossils already obtained from more ancient Silurian formations, Upper, Middle, and Lower, we may well ask whether any set of fossiliferous rocks newer in the series were ever studied with equal diligence, and over so vast an area, without yielding a single ichthyolite. Yet we must hesitate before we accept, even on such evidence, so sweeping a conclusion, as that the globe, for ages after it was inhabited by all the great classes of invertebrata, remained wholly untenanted by vertebrate animals.

TABLE 26.2. DATES OF THE DISCOVERY OF DIFFERENT CLASSES OF FOSSIL VERTEBRATA; SHOWING THE GRADUAL PROGRESS MADE IN TRACING THEM TO ROCKS OF HIGHER ANTIQUITY.

COLUMN 1: YEAR.
COLUMN 2: FORMATIONS.
COLUMN 3: GEOGRAPHICAL LOCALITIES.

MAMMALIA:

1798: Upper Eocene: Paris (Gypsum of Montmartre). (George Cuvier, Bulletin Soc. Philom. 20.)

1818: Lower Oolite: Stonesfield. (In 1818, Cuvier, visiting the Museum of Oxford, decided on the mammalian character of a jaw from Stonesfield. See also above Chapter 19.)

1847: Upper Trias: Stuttgart. (Professor Plieninger. See above Chapter 21.)

AVES:

1782: Upper Eocene: Paris (Gypsum of Montmartre). (Cuvier, Ossemens Foss. Art. "Oiseaux.")

1839: Lower Eocene: Isle of Sheppey (London Clay). (Professor Owen Geological Transactions second series volume 6 page 203 1839.)

1854: Lower Eocene: Woolwich Beds. (Upper part of the Woolwich beds. Prestwich Quarterly Geological Journal volume 10 page 157.)

1855: Lower Eocene: Meudon (Plastic Clay). (Gastornis Parisiensis. Owen Quarterly Geological Journal volume 12 page 204 1856.)

1858: Chloritic Series, or Upper Greensand: Cambridge. (Coprolitic bed, in the Upper Greensand. See above Chapter 17.)

1863: Upper Oolite: Solenhofen. (The Archaeopteryx macrura, Owen. See above Chapter 19.)

REPTILIA:

1710!: Permian (or Zechstein): Thuringia. (The fossil monitor of Thuringia (Protosaurus Speneri, V. Meyer) was figured by Spener of Berlin in 1710. (Miscel. Berlin.))

1844: Carboniferous: Saarbruck, near Treves. (See Chapter 23.)

PISCES:
1709: Permian (or Kupferschiefer): Thuringia. (Memorabilia Saxoniae Subterr. Leipsic 1709.)
1793: Carboniferous (Mountain Limestone): Glasgow. (History of Rutherglen by David Ure, 1793.)
1828: Devonian: Caithness. (Sedgwick and Murchison Geological Transactions second series volume 3 page 141 1828.)
1840: Upper Ludlow: Ludlow. (Sir R. Murchison. See Chapter 26.)
1859: Lower Ludlow: Leintwardine. (See Chapter 26.)
Obs.— The evidence derived from foot-prints, though often to be relied on, is omitted in the above table, as being less exact than that founded on bones and teeth.

In Table 26.2 a few dates are set before the reader of the discovery of different classes of animals in ancient rocks, to enable him to perceive at a glance how gradual has been our progress in tracing back the signs of vertebrata to formations of high antiquity. Such facts may be useful in warning us not to assume too hastily that the point which our retrospect may have reached at the present moment can be regarded as fixing the date of the first introduction of any one class of beings upon the earth.

2. WENLOCK FORMATION.
We next come to the Wenlock formation, which has been divided into Wenlock limestone, Wenlock shale, and Woolhope limestone and Denbighshire grits.

a. WENLOCK LIMESTONE.

This limestone, otherwise well known to collectors by the name of the Dudley Limestone, forms a continuous ridge in Shropshire, ranging for about 20 miles from S.W. to N.E., about a mile distant from the nearly parallel escarpment of the Aymestry limestone. This ridgy prominence is due to the solidity of the rock, and to the softness of the shales above and below it. Near Wenlock it consists of thick masses of grey subcrystalline limestone, replete with corals, encrinites, and trilobites. It is essentially of a concretionary nature; and the concretions, termed "ball-stones" in Shropshire, are often enormous, even 80 feet in diameter. They are of pure carbonate of lime, the surrounding rock being more or less argillaceous (Murchison's Siluria chapter 6.) Sometimes in the Malvern Hills this limestone, according to Professor Phillips, is oolitic.

(FIGURE 536. Halysites catenularius, Linn. sp. Upper and Lower Silurian.)

(FIGURE 537. Favosites Gothlandica, Lam. Dudley. a. Portion of a large mass; less than the natural size. b. Magnified portion, to show the pores and the partitions in the tubes.)

(FIGURE 538. Omphyma turbinatum, Linn. Sp. (Cyathophyllum, Goldfuss) Wenlock Limestone, Shropshire.)

Among the corals, in which this formation is so rich, 53 species being known, the "chain-coral," Halysites catenularius (Figure 536), may be pointed out as one very easily recognised, and widely spread in Europe, ranging through all parts of the Silurian group, from the Aymestry limestone to near the bottom of the Llandeilo rocks. Another coral, the Favosites Gothlandica (Figure 537), is also met with in profusion in large hemispherical masses, which break up into columnar and prismatic fragments, like that here figured (Figure 537, b). Another common form in the Wenlock limestone is the Omphyma turbinatum (Figure 538), which, like many of its modern companions, reminds us of some cup-corals; but all the Silurian genera belong to the palaeozoic type before mentioned (Chapter 24), exhibiting the quadripartite arrangement of the septalamellae within the cup.

(FIGURE 539. Pseudocrintes bifasciatus, Pearce. Wenlock Limestone, Dudley.)

Among the numerous Crinoids, several peculiar species of Cyathocrinus (for genus see Figures 478, 479) contribute their calcareous stems, arms, and cups towards the composition of the Wenlock limestone. Of Cystideans there are a few very remarkable forms, most of them peculiar to the Upper Silurian formation, as, for example, the Pseudocrinites, which was furnished with pinnated fixed arms, as represented in Figure 539. (E. Forbes Mem. Geological Survey volume 2 page 496.)

(FIGURE 540. Strophomena (Leptaena) depressa, Sowerby. Wenlock and Ludlow Rocks.)

The Brachiopoda are, many of them, of the same species as those of the Aymestry limestone; as, for example, Atrypa reticularis (Figure 532), and Strophomena depressa (Figure 540); but the latter species ranges also from the Ludlow rocks, through the Wenlock shale, to the Caradoc Sandstone.

(FIGURE 541. Calymene Blumenbachii, Brong. Ludlow, Wenlock, and Bala beds.)

(FIGURE 542. Phacops (Asaphus) caudatus, Brong. Wenlock and Ludlow Rocks.)

(FIGURE 543. Sphaerexochus mirus, Beyrich; coiled up. Wenlock Limestone, Dudley; also found in Ohio, North America.)

(FIGURE 544. Homalonotus delphinocephalus, Konig. Wenlock Limestone, Dudley Castle.)

The crustaceans are represented almost exclusively by Trilobites, which are very conspicuous, 22 being peculiar. The Calymene Blumenbachii (Figure 541), called the "Dudley Trilobite," was known to collectors long before its true place in the animal kingdom was ascertained. It is often found coiled up like the common Oniscus or wood-louse, and this is so usual a circumstance among certain genera of trilobites as to lead us to conclude that they must have habitually resorted to this mode of protecting themselves when alarmed. The other common species is the Phacops caudatus (Asaphus caudatus), Brong. (see Figure 542), and this is conspicuous for its large size and flattened form. Sphaerexochus mirus (Figure 543) is almost a globe when rolled up, the forehead or glabellum of this species being extremely inflated. The Homalonotus, a form of Trilobite in which the tripartite division of the dorsal crust is almost lost (see Figure 544), is very characteristic of this division of the Silurian series.

WENLOCK SHALE.

(FIGURE 545. Graptolithus priodon, Bronn. Ludlow and Wenlock shales.)

The Wenlock Shale, observes Sir R. Murchison, is infinitely the largest and most persistent member of the Wenlock formation, for the limestone often thins out and disappears. The shale, like the Lower Ludlow, often contains elliptical concretions of impure earthy limestone. In the Malvern district it is a mass of finely levigated argillaceous matter, attaining, according to Professor Phillips, a thickness of 640 feet, but it is sometimes more than 1000 feet thick in Wales, and is worked for flag-stones and slates. The prevailing fossils, besides corals and trilobites, and some crinoids, are several small species of Orthis, Cardiola, and numerous thin-shelled species of Orthoceratites.

About six species of Graptolite, a peculiar group of sertularian fossils before alluded to as being confined to Silurian rocks, occur in this shale. Of fossils of this genus, which is very characteristic of the Lower Silurian, I shall again speak in the sequel.

b. WOOLHOPE BEDS.

Though not always recognised as a separate subdivision of the Wenlock, the Woolhope beds, which underlie the Wenlock shale, are of great importance. Usually they occur as massive or nodular limestones, underlaid by a fine shale or flag-stone; and in other cases, as in the noted Denbighshire sandstones, as a coarse grit of very great thickness. This grit forms mountain ranges through North and South Wales, and is generally marked by the great sterility of the soil where it occurs. It contains the usual Wenlock fossils, but with the addition of some common in the uppermost Ludlow rock, such as Chonetes lata and Bellerophon trilobatus. The chief fossils of the Woolhope limestone are Illaenus Barriensis, Homalonotus delphinocephalus (Figure 544), Strophomena imbrex, and Rhynchonella Wilsoni (Figure 531). The latter attains in the Woolhope beds an unusual size for the species, the specimens being sometimes twice as large as those found in the Wenlock limestone.

In some places below the Wenlock formation there are shales of a pale or purple colour, which near Tarannon attain a thickness of about 1000 feet; they can be traced through Radnor and Montgomery to North Wales, according to Messrs. Jukes and Aveline. By the latter geologist they have been identified with certain shales above the May-Hill Sandstone, near Llandovery, but, owing to the extreme scarcity of fossils, their exact position remains doubtful.

3. LLANDOVERY GROUP— BEDS OF PASSAGE.

We now come to beds respecting the classification of which there has been much difference of opinion, and which in fact must be considered as beds of passage between

Upper and Lower Silurian. I formerly adopted the plan of those who class them as Middle Silurian, but they are scarcely entitled to this distinction, since after about 1400 Silurian species have been compared the number peculiar to the group in question only gives them an importance equal to such minor subdivisions as the Ludlow or Bala groups. I therefore prefer to regard them as the base of the Upper Silurian, to which group they are linked by more than twice as many species as to the Lower Silurian. By this arrangement the line of demarkation between the two great divisions, though confessedly arbitrary, is less so than by any other. They are called Llandovery Rocks, from a town in South Wales, in the neighbourhood of which they are well developed, and where, especially at a hill called Noeth Grug, in spite of several faults, their relations to one another can be clearly seen.

a. UPPER LLANDOVERY OR MAY-HILL SANDSTONE.

(FIGURE 546. Pentamerus oblongus, Sowerby. Upper and Lower Llandovery beds. a, b. Views of the shell itself, from figures in Murchison's Sil. Syst. c. Cast with portion of shell remaining, and with the hollow of the central septum filled with spar. d. Internal cast of a valve, the space once occupied by the septum being represented by a hollow in which is seen a cast of the chamber within the septum.)

(FIGURE 547. Stricklandinia (Pentamerus) lirata, Sowerby.)

The May-Hill group, which has also been named "Upper Llandovery," by Sir R. Murchison, ranges from the west of the Longmynd to Builth, Llandovery, and Llandeilo, and to the sea in Marlow's Bay, where it is seen in the cliffs. It consists of brownish and yellow sandstones with calcareous nodules, having sometimes a conglomerate at the base derived from the waste of the Lower Silurian rocks. These May-Hill beds were formerly supposed to be part of the Caradoc formation, but their true position was determined by Professor Sedgwick to be at the base of the Upper Silurian proper. (Quarterly Geological Journal volume 4 page 215 1853.) The more calcareous portions of the rock have been called the Pentamerus limestone, because Pentamerus oblongus (Figure 546) is very abundant in them. It is usually accompanied by P. (Stricklandinia) lirata (Figure 547); both forms have a wide geographical range, being also met with in the same part of the Silurian series in Russia and the United States.

About 228 species of fossils are known in the May-Hill division, more than half of which are Wenlock species. They consist of trilobites of the genera Illaenus and Calymene; Brachiopods of the genera Orthis, Atrypa, Leptaena, Pentamerus, Strophomena, and others; Gasteropods of the genera Turbo, Murchisonia (for genus, see Figure 567), and Bellerophon; and Pteropods of the genus Conularia. The Brachiopods, of which there are 66 species, are almost all Upper Silurian.

(FIGURE 548. Tentaculites annulatus, Schlot. Interior casts in sandstone. Upper Llandovery, Eastnor Park, near Malvern. Natural size and magnified.)

Among the fossils of the May-Hill shelly sandstone at Malvern, Tentaculites annulatus (Figure 548), an annelid, probably allied to Serpula, is found.

LOWER LLANDOVERY ROCKS.

Below the May-Hill Group are the Lower Llandovery Rocks, which consist chiefly of hard slaty rocks, and beds of conglomerate from 600 to 1000 feet in thickness. The fossils, which are somewhat rare in the lower beds, consist of 128 known species, only eleven of which are peculiar, 83 being common to the May-Hill group above, and 93 common to the rocks below. Stricklandinia (Pentamerus) levis, which is common in the Lower Llandovery, becomes rare in the Upper, while Pentamerus oblongus (Figure 546), which is the characteristic shell of the Upper Llandovery, occurs but seldom in the Lower.

LOWER SILURIAN ROCKS.

The Lower Silurian has been divided into, first, the Bala Group; secondly, the Llandeilo flags; and, thirdly, the Arenig or Lower Llandeilo formation.

BALA AND CARADOC BEDS.

(FIGURE 549. Orthis tricenaria, Conrad. New York; Canada. 1/2 natural size.)

(FIGURE 550. Orthis vespertilio, Sowerby. Shropshire, N. and S. Wales. One-half natural size.)

(FIGURE 551. Orthis (Strophomena) grandis, Sowerby. Two-thirds natural size. Caradoc Beds, Horderley, Shropshire, and Coniston, Lancashire.)

The Caradoc sandstone was originally so named by Sir R.I. Murchison from the mountain called Caer Caradoc, in Shropshire; it consists of shelly sandstones of great thickness, and sometimes containing much calcareous matter. The rock is frequently laden with the beautiful trilobite called by Murchison Trinucleus Caractaci (see Figure 553), which ranges from the base to the summit of the formation, usually accompanied by Strophomena grandis (see Figure 551), and Orthis vespertilio (Figure 550), with many other fossils.

BRACHIOPODA.

Nothing is more remarkable in these beds, and in the Silurian strata generally of all countries, than the preponderance of brachiopoda over other forms of mollusca. Their proportional numbers can by no means be explained by supposing them to have inhabited seas of great depth, for the contrast between the palaeozoic and the present state of things has not been essentially altered by the late discoveries made in our deep-sea dredgings. We find the living brachiopoda so rare as to form about one forty-fourth of the whole bivalve fauna, whereas in the Lower Silurian rocks of which we are now about to treat, and where the brachiopoda reach their maximum, they are represented by more than twice as many species as the Lamellibranchiate bivalves.

There may, indeed, be said to be a continued decrease of the proportional number of this lower tribe of mollusca as we proceed from older to newer rocks. In the British Devonian, for example, the Brachiopoda number 99, the Lamellibranchiata 58; while in the Carboniferous their proportions are more than reversed, the Lamellibranchiata numbering 334 species, and the Brachiopoda only 157. In the Secondary or Cainozoic formations the preponderance of the higher grade of bivalves becomes more and more marked, till in the tertiary strata it approaches that observed in the living creation.

While on this subject, it may be useful to the student to know that a Brachiopod differs from ordinary bivalves, mussels, cockles, etc., in being always equal- sided and never quite equi-valved; the form of each valve being symmetrical, it may be divided into two equal parts by a line drawn from the apex to the centre of the margin.

TRILOBITES.

In the Bala and Caradoc beds the trilobites reach their maximum, being represented by 111 species referred to 23 genera.

(FIGURE 552. Young individuals of Trinucleus concentricus (T. ornatus, Barr.). a. Youngest state. Natural size and magnified; the body rings not at all developed. b. A little older. One thorax joint. c. Still more advanced. Three thorax joints. The fourth, fifth, and sixth segments are successively produced, probably each time the animal moulted its crust.)

(FIGURE 553. Trinucleus concentricus, Eaton. Syn. T. Caractaci, Murch. Ireland; Wales; Shropshire; North America; Bohemia.)

Burmeister, in his work on the organisation of trilobites, supposes that they swam at the surface of the water in the open sea and near coasts, feeding on smaller marine animals, and to have had the power of rolling themselves into a ball as a defence against injury. He was also of opinion that they underwent various transformations analogous to those of living crustaceans. M. Barrande, author of an admirable work on the Silurian rocks of Bohemia, confirms the doctrine of their metamorphosis, having traced more than twenty species through different stages of growth from the young state just after its escape from the egg to the adult form. He has followed some of them from a point in which they show no eyes, no joints, or body rings, and no distinct tail, up to the complete form with the full number of segments. This change is brought about before the animal has attained a tenth part of its full dimensions, and hence such minute and delicate specimens are rarely met with. Some of his figures of the metamorphoses of the common Trinucleus are copied in Figures 552 and 553. It was not till 1870 that Mr. Billings was enabled, by means of a specimen found in Canada, to prove that the trilobite was provided with eight legs.

(FIGURE 554. Palaeaster asperimus, Salt. Caradoc, Welshpool.)

(FIGURE 555. Echinosphaeronites balticus, Eichwald. (Of the family Cystideae.) a. Mouth. b. Point of attachment of stem. Lower Silurian S. and N. Wales.)

It has been ascertained that a great thickness of slaty and crystalline rocks of South Wales, as well as those of Snowdon and Bala, in North Wales, which were first supposed to be of older date than the Silurian sandstones and mudstones of Shropshire, are in fact identical in age, and contain the same organic remains. At Bala, in Merionethshire, a limestone rich in fossils occurs, in which two genera of star-fish, Protaster and Palaeaster, are found; the fossil specimen of the latter (Figure 554) being almost as uncompressed as if found just washed up on the sea-beach. Besides the star-fish there occur abundance of those peculiar bodies called Cystideae. They are the Sphaeronites of old authors, and were considered by Professor E. Forbes as intermediate between the crinoids and echinoderms. The Echinosphaeronite here represented (Figure 555) is characteristic of the Caradoc beds in Wales, and of their equivalents in Sweden and Russia.

With it have been found several other genera of the same family, such as Sphaeronites, Hemicosmites, etc. Among the mollusca are Pteropods of the genus Conularia of large size (for genus, see Figure 518). About eleven species of Graptolite are reckoned as belonging to this formation; they are chiefly found in peculiar localities where black mud abounded. The formation, when traced into South Wales and Ireland, assumes a greatly altered mineral aspect, but still retains its characteristic fossils. The known fauna of the Bala group comprises 565 species, 352 of which are peculiar, and 93, as before stated, are common to the overlying Llandovery rocks. It is worthy of remark that, when it occurs under the form of trappean tuff (volcanic ashes of De la Beche), as in the crest of Snowdon, the peculiar species which distinguish it from the Llandeilo beds are still observable. The formation generally appears to be of shallow-water origin, and in that respect is contrasted with the group next to be described. Professor Ramsay estimates the thickness of the Bala Beds, including the contemporaneous volcanic rocks, stratified and unstratified, as being from 10,000 to 12,000 feet.

LLANDEILO FLAGS.

(FIGURE 556. Didymograpsus (Graptolites) Murchisonii, Beck. Llandeilo flags, Wales.)

The Lower Silurian strata were originally divided by Sir R. Murchison into the upper group already described, under the name of Caradoc Sandstone, and a lower one, called, from a town in Carmarthenshire, the Llandeilo flags. The last mentioned strata consist of dark-coloured micaceous flags, frequently calcareous, with a great thickness of shales, generally black, below them. The same beds are also seen at Builth, in Radnorshire, where they are interstratified with volcanic matter.

(FIGURE 557. Diplograpsus pristis, Hisinger. Llandeilo beds, Waterford.)

(FIGURE 558. Rastrites peregrinus, Barrande. Scotland; Bohemia; Saxony. Llandeilo flags.)

(FIGURE 559. Diplograpsus folium, Hisinger. Dumfriesshire; Sweden. Llandeilo flags.)

A still lower part of the Llandeilo rocks consists of a black carbonaceous slate of great thickness, frequently containing sulphate of alumina, and sometimes, as in Dumfriesshire, beds of anthracite. It has been conjectured that this carbonaceous matter may be due in great measure to large quantities of imbedded animal remains, for the number of Graptolites included in these slates was certainly very great. In Great Britain eleven genera and about 40 species of Graptolites occur in the Llandeilo flags and underlying Arenig beds. The double Graptolites, or those with two rows of cells, such as Diplograpsus (Figure 557), are conspicuous.

The brachiopoda of the Llandeilo flags, which number 47 species, are in the main the same as those of the Caradoc Sandstone, but the other mollusca are in great part of different species.

(FIGURE 560. Orthoceras duplex, Wahlenberg. Russia and Sweden. (From Murchison's Siluria.))

(FIGURE 561. Asaphus tyrannus, Murchison. Llandeilo; Bishop's Castle; etc.)

(FIGURE 562. Ogygia Buchii, Burm. Syn. Asaphus Buchii, Brongn. Builth, Radnorshire; Llandeilo, Carmarthenshire.)

In Europe generally, as, for example, in Sweden and Russia, no shells are so characteristic of this formation as Orthoceratites, usually of great size, and with a wide siphuncle placed on one side instead of being central (see Figure 560). Among other

Cephalopods in the Llandeilo flags is Cyrtoceras; in the same beds also are found Bellerophon (see Figure 488) and some Pteropod shells (Conularia, Theca, etc.), also in spots where sand abounded, lamellibranchiate bivalves of large size. The Crustaceans were plentifully represented by the Trilobites, which appear to have swarmed in the Silurian seas just as crabs and shrimps do in our own; no less than 263 species have been found in the British Silurian fauna. The genera Asaphus (Figure 561), Ogygia (Figure 562), and Trinucleus (Figures 552 and 553) form a marked feature of the rich and varied Trilobitic fauna of this age.

Beneath the black slates above described of the Llandeilo formation, Graptolites are still found in great variety and abundance, and the characteristic genera of shells and trilobites of the Lower Silurian rocks are still traceable downward, in Shropshire, Cumberland, and North and South Wales, through a vast depth of shaly beds, in some districts interstratified with trappean formations of contemporaneous origin; these consist of tuffs and lavas, the tuffs being formed of such materials as are ejected from craters and deposited immediately on the bed of the ocean, or washed into it from the land. According to Professor Ramsay, their thickness is about 3300 feet in North Wales, including those of the Lower Llandeilo. The lavas are feldspathic, and of porphyritic structure, and, according to the same authority, of an aggregate thickness of 2500 feet.

ARENIG OR STIPER-STONES GROUP (LOWER LLANDEILO OF MURCHISON).

(FIGURE 563. Arenicolites linearis, Hall. Arenig beds, Stiper-Stones. a. Parting between the beds, or planes of bedding.)

(FIGURE 564. Didymograpsus geminus, Hisinger, sp. Sweden.)

Next in the descending order are the shales and sandstones in which the quartzose rocks called Stiper-Stones in Shropshire occur. Originally these Stiper-Stones were only known as arenaceous quartzose strata in which no organic remains were conspicuous, except the tubular burrows of annelids (see Figure 563, Arenicolites linearis), which are remarkably common in the Lowest Silurian in Shropshire, and in the State of New York, in America. They have already been alluded to as occurring by thousands in the Silurian strata unconformably overlying the Cambrian, in the mountain of Queenaig, in Sutherlandshire (Figure 82). I have seen similar burrows now made on the retiring of the tides in the sands of the Bristol Channel, near Minehead, by lob-worms which are dug out by fishermen and used as bait. When the term Silurian was given by Sir R. Murchison, in 1835, to the whole series, he considered the Stiper-Stones as the base of the Silurian system, but no fossil fauna had then been obtained, such as could alone enable the geologist to draw a line between this member of the series and the Llandeilo flags above, or a vast thickness of rock below, which was seen to form the Longmynd hills, and was called "unfossiliferous graywacke." Professor Sedgwick had described, in 1843, strata now ascertained to be of the same age as largely developed in the Arenig mountain, in Merionethshire; and the Skiddaw slates in the Lake-District of Cumberland, studied by the same author, were of corresponding date, though the number of fossils was, in both cases, too few for the determination of their true chronological relations. The subsequent researches of Messrs. Sedgwick and Harkness, in Cumberland, and of Sir R.I. Murchison and the Government surveyors in Shropshire, have increased the species to more than sixty. These were examined by Mr. Salter, and shown in the third edition of "Siluria" (page 52, 1859) to be quite distinct from the fossils of the overlying Llandeilo flags. Among these the Obolella plumbea, Aeglina binodosa, Ogygia Selwynii, and Didymograpsus geminus (Figure 564), and D. Hirundo, are characteristic.

But, although the species are distinct, the genera are the same as those which characterise the Silurian rocks above, and none of the characteristic primordial or Cambrian forms, presently to be mentioned, are intermixed. The same may be said of a set of beds underlying the Arenig rocks at Ramsay Island and other places in the neighbourhood of St. David's. These beds, which have only lately become known to us through the labours of Dr. Hicks (Transactions of the British Association 1866. Proceedings of the Liverpool Geological Society 1869.), present already twenty new species, the greater part of them allied generically to the Arenig rocks. This Arenig group may therefore be conveniently regarded as the base of the great Silurian system, a system which, by the thickness of its strata and the changes in animal life of which it contains the record, is more than equal in value to the Devonian, or Carboniferous, or other principal divisions, whether of primary or secondary date.

It would be unsafe to rely on the mere thickness of the strata, considered apart from the great fluctuations in organic life which took place between the era of the Llandeilo and that of the Ludlow formation, especially as the enormous pile of Silurian rocks observed in Great Britain (in Wales more particularly) is derived in great part from igneous action, and is not confined to the ordinary deposition of sediment from rivers or the waste of cliffs.

In volcanic archipelagoes, such as the Canaries, we see the most active of all known causes, aqueous and igneous, simultaneously at work to produce great results in a comparatively moderate lapse of time. The outpouring of repeated streams of lava— the showering down upon land and sea of volcanic ashes— the sweeping seaward of loose sand and cinders, or of rocks ground down to pebbles and sand, by rivers and torrents descending steeply inclined channels— the undermining and eating away of long lines of sea-cliff exposed to the swell of a deep and open ocean— these operations combine to produce a considerable volume of superimposed matter, without there being time for any extensive change of species. Nevertheless, there would seem to be a limit to the thickness of stony masses formed even under such favourable circumstances, for the analogy of tertiary volcanic regions lends no countenance to the notion that sedimentary and igneous rocks 25,000, much less 45,000 feet thick, like those of Wales, could originate while one and the same fauna should continue to people the earth. If, then, we allow that about 25,000 feet of matter may be ascribed to one system, such as the Silurian, as above described, we may be prepared to discover in the next series of subjacent rocks a distinct assemblage of species, or even in great part of genera, of organic remains. Such appears to be the fact, and I shall therefore conclude with the Arenig beds my enumeration of the Silurian formations in Great Britain, and proceed to say something of their foreign equivalents, before treating of rocks older than the Silurian.

SILURIAN STRATA OF THE CONTINENT OF EUROPE.

When we turn to the continent of Europe, we discover the same ancient series occupying a wide area, but in no region as yet has it been observed to attain great thickness. Thus, in Norway and Sweden, the total thickness of strata of Silurian age is considerably less than 1000 feet, although the representatives both of the Upper and Lower Silurian of England are not wanting there. In Russia the Silurian strata, so far as they are yet known, seem to be even of smaller vertical dimensions than in Scandinavia, and they appear to consist chiefly of the Llandovery group, or of a limestone containing Pentamerus oblongus, below which are strata with fossils corresponding to those of the Llandeilo beds of England. The lowest rock with organic remains yet discovered is "the Ungulite or Obolus grit" of St. Petersburg, probably coeval with the Llandeilo flags of Wales.

(Figures 565 and 566. Shells of the lowest known Fossiliferous Beds in Russia.

(FIGURE 565. Siphonotreta unguiculata, Eichwald. From the Lowest Silurian Sandstone, "Obolus grits," of St. Petersburg. a. Outside of perforated valve. b. Interior of same, showing the termination of the foramen within. (Davidson.))

(FIGURE 566. Obolus Apollinis, Eichwald. From the same locality. a. Interior of the larger or ventral valve. b. Exterior of the upper (dorsal) valve. (Davidson, "Palaeontographic Monograph.")))

The shales and grits near St. Petersburg, above alluded to, contain green grains in their sandy layers, and are in a singularly unaltered state, taking into account their high antiquity. The prevailing Brachiopods consist of the Obolus or Ungulite of Pander, and a Siphonotreta (Figures 565, 566). Notwithstanding the antiquity of this Russian formation, it should be stated that both of these genera of brachiopods have been also found in the Upper Silurian of England, i.e. In the Wenlock limestone.

Among the green grains of the sandy strata above-mentioned, Professor Ehrenberg announced in 1854 his discovery of remains of foraminifera. These are casts of the cells; and among five or six forms three are considered by him as referable to existing genera (e.g., Textularia, Rotalia, and Guttulina).

SILURIAN STRATA OF THE UNITED STATES.

Table 26.3. SUBDIVISIONS OF THE SILURIAN STRATA OF NEW YORK. (Strata below the Oriskany sandstone or base of the Devonian.)
COLUMN 1: NEW YORK NAMES.
COLUMN 2: BRITISH EQUIVALENTS.

1. Upper Pentamerus Limestone: Upper Silurian (or Ludlow and Wenlock formations).
2. Encrinal Limestone: Upper Silurian (or Ludlow and Wenlock formations).
3. Delthyris Shaly Limestone: Upper Silurian (or Ludlow and Wenlock formations).
4. Pentamerus and Tentaculite Limestones: Upper Silurian (or Ludlow and Wenlock formations).
5. Water Lime Group: Upper Silurian (or Ludlow and Wenlock formations).
6. Onondaga Salt Group: Upper Silurian (or Ludlow and Wenlock formations).
7. Niagara Group: Upper Silurian (or Ludlow and Wenlock formations).
8. Clinton Group: Beds of Passage, Llandovery Group.
9. Medina Sandstone: Beds of Passage, Llandovery Group.
10. Oneida Conglomerate: Beds of Passage, Llandovery Group.
11. Gray Sandstone: Beds of Passage, Llandovery Group.
12. Hudson River Group: Lower Silurian (or Caradoc and Bala, Llandeilo and Arenig Formations).
13. Trenton Limestone: Lower Silurian (or Caradoc and Bala, Llandeilo and Arenig Formations).
14. Black-River Limestone: Lower Silurian (or Caradoc and Bala, Llandeilo and Arenig Formations).
15. Bird's-eye Limestone: Lower Silurian (or Caradoc and Bala, Llandeilo and Arenig Formations).
16. Chazy Limestone: Lower Silurian (or Caradoc and Bala, Llandeilo and Arenig Formations).
17. Calciferous Sandstone: Lower Silurian (or Caradoc and Bala, Llandeilo and Arenig Formations).

The Silurian formations can be advantageously studied in the States of New York, Ohio, and other regions north and south of the great Canadian lakes. Here they are often found, as in Russia, nearly in horizontal position, and are more rich in well-preserved fossils than in almost any spot in Europe. In the State of New York, where the succession of the beds and their fossils have been most carefully worked out by the Government surveyors, the subdivisions given in the first column of Table 26.3 have been adopted.

In the second column of the same table I have added the supposed British equivalents. All Palaeontologists, European and American, such as MM. De Verneuil, D. Sharpe, Professor Hall, E. Billings, and others, who have entered upon this comparison, admit that there is a marked general correspondence in the succession of fossil forms, and even species, as we trace the organic remains downward from the highest to the lowest beds; but it is impossible to parallel each minor subdivision.

That the Niagara Limestone, over which the river of that name is precipitated at the great cataract, together with its underlying shales, corresponds to the Wenlock limestone and shale of England there can be no doubt. Among the species common to this formation in America and Europe are Calymene Blumenbachii, Homalonotus delphinocephalus (Figure 544), with several other trilobites; Rhynchonella Wilsoni, Figure 531, and Retzia cuneata; Orthis elegantula, Pentamerus galeatus, with many more brachiopods; Orthoceras annulatum, among the cephalopodous shells; and Favosites gothlandica, with other large corals.

The Clinton Group, containing Pentamerus oblongus and Stricklandinia, and related more nearly by its fossil species with the beds above than with those below, is the equivalent of the Llandovery Group or beds of passage.

(FIGURE 567. Murchisonia gracilis, Hall. A fossil characteristic of the Trenton Limestone. The genus is common in Lower Silurian rocks.)

The Hudson River Group, and the Trenton Limestone, agree palaeontologically with the Caradoc or Bala group, containing in common with them several species of trilobites, such as Asaphus (Isotelus) gigas, Trinucleus concentricus (Figure 553); and various shells, such as Orthis striatula, Orthis biforata (or O. lynx), O. porcata (O. occidentalis of Hall), and Bellerophon bilobatus. In the Trenton limestone occurs Murchisonia gracilis, Figure 567, a fossil also common to the Llandeilo beds in England.

Mr. D. Sharpe, in his report on the mollusca collected by me from these strata in North America (Quarterly Geological Journal volume 4.), has concluded that the number of species common to the Silurian rocks on both sides of the Atlantic is between 30 and 40 per cent; a result which, although no doubt liable to future modification, when a larger comparison shall have been made, proves, nevertheless, that many of the species had a wide geographical range. It seems that comparatively few of the gasteropods and lamellibranchiate bivalves of North America can be identified specifically with European fossils, while no less than two-fifths of the brachiopoda, of which my collection chiefly consisted, are the same. In explanation of these facts, it is suggested that most of the recent brachiopoda (especially the orthidiform ones) are inhabitants of deep water, and that they may have had a wider geographical range than shells living near shore. The predominance of bivalve mollusca of this peculiar class has caused the Silurian period to be sometimes styled "the age of brachiopods."

In Canada, as in the State of New York, the Potsdam Sandstone underlies the above-mentioned calcareous rocks, but contains a different suite of fossils, as will be hereafter explained. In parts of the globe still more remote from Europe the Silurian strata have also been recognised, as in South America, Australia, and India. In all these regions the facies of the fauna, or the types of organic life, enable us to recognise the contemporaneous origin of the rocks; but the fossil species are distinct, showing that the old notion of a universal diffusion throughout the "primaeval seas" of one uniform specific fauna was quite unfounded, geographical provinces having evidently existed in the oldest as in the most modern times.

CHAPTER XXVII.
CAMBRIAN AND LAURENTIAN GROUPS.

Classification of the Cambrian Group, and its Equivalent in Bohemia.
Upper Cambrian Rocks.
Tremadoc Slates and their Fossils.
Lingula Flags.
Lower Cambrian Rocks.
Menevian Beds.
Longmynd Group.
Harlech Grits with large Trilobites.
Llanberis Slates.
Cambrian Rocks of Bohemia.
Primordial Zone of Barrande.
Metamorphosis of Trilobites.
Cambrian Rocks of Sweden and Norway.
Cambrian Rocks of the United States and Canada.
Potsdam Sandstone.
Huronian Series.
Laurentian Group, upper and lower.
Eozoon Canadense, oldest known Fossil.
Fundamental Gneiss of Scotland.

CAMBRIAN GROUP.

The characters of the Upper and Lower Silurian rocks were established so fully, both on stratigraphical and palaeontological data, by Sir Roderick Murchison after five years' labour, in 1839, when his "Silurian System" was published, that these formations could from that period be recognised and identified in all other parts of Europe and in North America, even in countries where most of the fossils differed specifically from those of the classical region in Britain, where they were first studied.

TABLE 27.1. SHOWING THE SUCCESSION OF THE STRATA IN ENGLAND AND WALES WHICH BELONG TO THE CAMBRIAN GROUP OR THE FOSSILIFEROUS ROCKS OLDER THAN THE ARENIG OR LOWER LLANDEILO ROCKS:

UPPER CAMBRIAN.
TREMADOC SLATES. (Primordial of Barrande in part.)
LINGULA FLAGS. (Primordial of Barrande.)
LOWER CAMBRIAN.
MENEVIAN BEDS. (Primordial of Barrande.)
LONGMYND GROUP. a. Harlech Grits. b. Llanberis slates.

While Sir R.I. Murchison was exploring in 1833, in Shropshire and the borders of Wales, the strata which in 1835 he first called Silurian, Professor Sedgwick was surveying the rocks of North Wales, which both these geologists considered at that period as of older date, and for which in 1836 Sedgwick proposed the name of Cambrian. It was afterwards found that a large portion of the slaty rocks of North Wales, which had been considered as more ancient than the Llandeilo beds and Stiper-Stones before alluded to, were, in reality, not inferior in position to those Lower Silurian beds of Murchison, but merely extensive undulations of the same, bearing fossils identical in species, though these were generally rarer and less perfectly preserved, owing to the changes which the rocks had undergone from metamorphic action. To such rocks the term "Cambrian" was no longer applicable, although it continued to be appropriate to strata inferior to the Stiper-Stones, and which were older than those of the Lower Silurian group as originally defined. It was not till 1846 that fossils were found in Wales in the Lingula flags, the place of which will be seen in Table 27.1. By this time Barrande had already published an account of a rich collection of fossils which he had discovered in Bohemia, portions of which he recognised as of corresponding age with Murchison's Upper and Lower Silurian, while others were more ancient, to which he gave the name of "Primordial," for the fossils were sufficiently distinct to entitle the rocks to be referred to a new period. They consisted chiefly of trilobites of genera distinct from those occurring in the overlying Silurian formations. These peculiar genera were afterwards found in rocks holding a corresponding position in Wales, and I shall retain for them the term Cambrian, as recent discoveries in our own country seem to carry the first fauna of Barrande, or his primordial type, even into older strata than any which he found to be fossiliferous in Bohemia.

The term primordial was intended to express M. Barrande's own belief that the fossils of the rocks so-called afforded evidence of the first appearance of vital phenomena on this planet, and that consequently no fossiliferous strata of older date would or could ever be discovered. The acceptance of such a nomenclature would seem to imply that we despaired of extending our discoveries of new and more ancient fossil groups at some future day when vast portions of the globe, hitherto unexplored, should have been thoroughly surveyed. Already the discovery of the Laurentian Eozoon in Canada, presently to be mentioned, discountenances such views.

UPPER CAMBRIAN.
TREMADOC SLATES.
(FIGURE 568. Theca (Cleidotheca) operculata. Lower Tremadoc beds. Tremadoc.)

The Tremadoc slates of Sedgwick are more than 1000 feet in thickness, and consist of dark earthy slates occurring near the little town of Tremadoc, situated on the north side of Cardigan Bay, in Carnarvonshire. These slates were first examined by Sedgwick in 1831, and were re-examined by him and described in 1846 (Quarterly Geological Journal volume 3 page 156.), after some fossils had been found in the underlying Lingula flags by Mr. Davis. The inferiority in position of these Lingula flags to the Tremadoc beds was at the same time established. The overlying Tremadoc beds were traced by their pisolitic ore from Tremadoc to Dolgelly. No fossils proper to the Tremadoc slates were then observed, but subsequently, thirty-six species of all classes have been found in them, thanks to the researches of Messrs. Salter, Homfray, and Ash. We have already seen that in the Arenig or Stiper-Stones group, where the species are distinct, the genera agree with Silurian types; but in these Tremadoc slates, where the species are also peculiar, there is about an equal admixture of Silurian types with those which Barrande has termed "primordial." Here, therefore, it may truly be said that we are entering upon a new domain of life in our retrospective survey of the past. The trilobites of new species, but of Lower Silurian genera, belong to Ogygia, Asaphus, and Cheirurus; whereas those belonging to primordial types, or Barrande's first fauna as well as to the Lingula flags of Wales, comprise Dikelocephalus, Conocoryphe (for genera see Figures 577 and 581 (This genus has been substituted for Barrande's Conocephalus, as the latter term had been preoccupied by the entomologists.)), Olenus, and Angelina. In the Tremadoc slates are found Bellerophon, Orthoceras, and Cyrtoceras, all specifically distinct from Lower Silurian fossils of the same genera: the Pteropods Theca (Figure 568) and Conularia range throughout these slates; there are no Graptolites. The Lingula (Lingulella) Davisii ranges from the top to the bottom of the formation, and links it with the zone next to be described. The Tremadoc slates are very local, and seem to be confined to a small part of North Wales; and Professor Ramsay supposes them to lie unconformably on the Lingula flags, and that a long interval of time elapsed between these formations. Cephalopoda have not yet been found lower than this group, but it will be observed that they occur here associated with genera of Trilobites considered by Barrande as characteristically Primordial, some of which belong to all the divisions of the British Cambrian about to be mentioned. This renders the absence of cephalopoda of less importance as bearing on the theory of development.

LINGULA FLAGS.

(FIGURES 569 to 571. "Lingula flags" of Dolgelly, and Ffestiniog; N. Wales.

(FIGURE 569. Hymenocaris vermicauda, Salter. A phyllopod crustacean. One-half natural size.)

(FIGURE 570. Lingulella Davisii, M'Coy. a. One-half natural size. b. Distorted by cleavage.)

(FIGURE 571. Olenus micrurus, Salter. One-half natural size.))

Next below the Tremadoc slates in North Wales lie micaceous flagstones and slates, in which, in 1846, Mr. E. Davis discovered the Lingula (Lingulella), Figure 570, named after him, and from which was derived the name of Lingula flags. These beds, which are palaeontologically the equivalents of Barrande's primordial zone, are represented by more than 5000 feet of strata, and have been studied chiefly in the neighbourhood of Dolgelly, Ffestiniog, and Portmadoc in North Wales, and at St. David's in South Wales. They have yielded about forty species of fossils, of which six only are common to the overlying Tremadoc rocks, but the two formations are closely allied by having several characteristic "primordial" genera in common. Dikelocephalus, Olenus (Figure 571), and Conocoryphe are prominent forms, as is also Hymenocaris (Figure 569), a genus of phyllopod crustacean entirely confined to the Lingula Flags. According to Mr. Belt, who has devoted much attention to these beds, there are already palaeontological data for subdividing the Lingula Flags into three sections. (Geological Magazine volume 4.)

In Merionethshire, according to Professor Ramsay, the Lingula Flags attain their greatest development; in Carnarvonshire they thin out so as to have lost two- thirds of their thickness in eleven miles, while in Anglesea and on the Menai Straits both they and the Tremadoc beds are entirely absent, and the Lower Silurian rests directly on Lower Cambrian strata.

LOWER CAMBRIAN.
MENEVIAN BEDS.

(FIGURE 572. Paradoxides Davidis, Salter. One-tenth natural size. Menevian beds. St. David's and Dolgelly.)

Immediately beneath the Lingula Flags there occurs a series of dark grey and black flags and slates alternating at the upper part with some beds of sandstone, the whole reaching a thickness of from 500 to 600 feet. These beds were formerly classed, on purely lithological grounds, as the base of the Lingula Flags, but Messrs. Hicks and Salter, to whose exertions we owe almost all our knowledge of the fossils, have pointed out that the most characteristic genera found in them are quite unknown in the Lingula Flags, while they possess many of the strictly Lower Cambrian genera, such as Microdiscus and Paradoxides. (British Association Report 1865, 1866, 1868 and Quarterly Geological Journal volumes 21,

25.) They therefore proposed to place them, and it seems to me with good reason, at the top of the Lower Cambrian under the term "Menevian," Menevia being the classical name of St. David's. The beds are well exhibited in the neighbourhood of St. David's in South Wales, and near Dolgelly and Maentwrog in North Wales. They are the equivalents of the lowest part of Barrande's Primordial Zone (Etage C). More than forty species have been found in them, and the group is altogether very rich in fossils for so early a period. The trilobites are of large size; Paradoxides Davidis (see Figure 572), the largest trilobite known in England, 22 inches or nearly two feet long, is peculiar to the Menevian Beds. By referring to the Bohemian trilobite of the same genus (Figure 576), the reader will at once see how these fossils (though of such different dimensions) resemble each other in Bohemia and Wales, and other closely allied species from the two regions might be added, besides some which are common to both countries. The Swedish fauna, presently to be mentioned, will be found to be still more nearly connected with the Welsh Menevian. In all these countries there is an equally marked difference between the Cambrian fossils and those of the Upper and Lower Silurian rocks. The trilobite with the largest number of rings, Erinnys venulosa, occurs here in conjunction with Agnostus and Microdiscus, the genera with the smallest number. Blind trilobites are also found as well as those which have the largest eyes, such as Microdiscus on the one hand, and Anoplenus on the other.

LONGMYND GROUP.

Older than the Menevian Beds are a thick series of olive green, purple, red and grey grits and conglomerates found in North and South Wales, Shropshire, and parts of Ireland and Scotland. They have been called by Professor Sedgwick the Longmynd or Bangor Group, comprising, first, the Harlech and Barmouth sandstones; and secondly, the Llanberis slates.

HARLECH GRITS.

(FIGURE 573. Histioderma Hibernica, Kinahan. Oldhamia beds. Bray Head, Ireland. 1. Showing opening of burrow, and tube with wrinklings or crossing ridges, probably produced by a tentacled sea worm or annelid. 2. Lower and curved extremity of tube with five transverse lines.)

The sandstones of this period attain in the Longmynd hills a thickness of no less than 6000 feet without any interposition of volcanic matter; in some places in Merionethshire they are still thicker. Until recently these rocks possessed but a very scanty fauna.

With the exception of five species of annelids (see Figure 460) brought to light by Mr. Salter in Shropshire, and Dr. Kinahan in Wicklow, and an obscure crustacean form, Palaeopyge Ramsayi, they were supposed to be barren of organic remains. Now, however, through the labours of Mr. Hicks, they have yielded at St. David's a rich fauna of trilobites, brachiopods, phyllopods, and pteropods, showing, together with other fossils, a by no means low state of organisation at this early period. (British Association Report 1868.) Already the fauna amounts to 20 species referred to 17 genera.

A new genus of trilobite called Plutonia Sedgwickii, not yet figured and described, has been met with in the Harlech grits. It is comparable in size to the large Paradoxides Davidis before mentioned, has well-developed eyes, and is covered all over with tubercles. In the same strata occur other genera of trilobites, namely, Conocoryphe, Paradoxides, Microdiscus, and the Pteropod Theca (Figure 568), all represented by species peculiar to the Harlech grits. The sands of this formation are often rippled, and were evidently left dry at low tides, so that the surface was dried by the sun and made to shrink and present sun-cracks. There are also distinct impressions of rain-drops on many surfaces, like those in Figures 444 and 445.

LANBERIS SLATES.

(FIGURE 574. Oldhamia radiata, Forbes. Wicklow, Ireland.)

(FIGURE 575. Oldhamia antiqua, Forbes. Wicklow, Ireland.)

The slates of Llanberis and Penrhyn in Carnarvonshire, with their associated sandy strata, attain a great thickness, sometimes about 3000 feet. They are perhaps not more ancient than the Harlech and Barmouth beds last mentioned, for they may represent the deposits of fine mud thrown down in the same sea, on the borders of which the sands above-mentioned were accumulating. In some of these slaty rocks in Ireland, immediately opposite Anglesea and Carnarvon, two species of fossils have been found, to which the late Professor E. Forbes gave the name of Oldhamia. The nature of these organisms is still a matter of discussion among naturalists.

CAMBRIAN ROCKS OF BOHEMIA (PRIMORDIAL ZONE OF BARRANDE).

In the year 1846, as before stated, M. Joachim Barrande, after ten years' exploration of Bohemia, and after collecting more than a thousand species of fossils, had ascertained the existence in that country of three distinct faunas below the Devonian. To his first fauna, which was older than any then known in this country, he gave the name of Etage C; his two first stages A and B consisting of crystalline and metamorphic rocks and unfossiliferous schists. This Etage C or primordial zone proved afterwards to be the equivalent of those subdivisions of the Cambrian groups which have been above described under the names of Menevian and Lingula Flags. The second fauna tallies with Murchison's Lower Silurian, as originally defined by him when no fossils had been discovered below the Stiper-Stones. The third fauna agrees with the Upper Silurian of the same author. Barrande, without government assistance, had undertaken single-handed the geological survey of Bohemia, the fossils previously obtained from that country having scarcely exceeded 20 in number, whereas he had already acquired, in 1850, no less than 1100 species, namely, 250 crustaceans (chiefly Trilobites), 250 Cephalopods, 160 gasteropods and pteropods, 130 acephalous mollusks, 210 brachiopods, and 110 corals and other fossils. These numbers have since been almost doubled by subsequent investigations in the same country.

(Figures 576 to 580. Fossils of the lowest Fossiliferous Beds in Bohemia, or "Primordial Zone" of Barrande.

(FIGURE 576. Paradoxides Bohemicus, Barr. About one-half natural size.)

(FIGURE 577. Conocoryphe striata. Syn. Conocephalus striatus, Emmrich. One-half natural size. Ginetz and Skrey.)

(FIGURE 578. Agnostus integer, Beyrich. Natural size and magnified.)

(FIGURE 579. Agnostus Rex, Barr. Natural size, Skrey.)

(FIGURE 580. Sao hirsuta, Barrande, in its various stages of growth. The small lines beneath indicate the true size. In the youngest state, a, no segments are visible; as the metamorphosis progresses, b, c, the body segments begin to be developed: in the stage d the eyes are introduced, but the facial sutures are not completed; at e the full-grown animal, half its true size, is shown.))

In the primordial zone C, he discovered trilobites of the genera Paradoxides, Conocoryphe, Ellipsocephalus, Sao, Arionellus, Hydrocephalus, and Agnostus. M. Barrande pointed out that these primordial trilobites have a peculiar facies of their own dependent on the multiplication of their thoracic segments and the diminution of their caudal shield or pygidium.

One of the "primordial" or Upper Cambrian Trilobites of the genus Sao, a form not found as yet elsewhere in the world, afforded M. Barrande a fine illustration of the metamorphosis of these creatures, for he traced them through no less than twenty stages of their development. A few of these changes have been selected for representation in Figure 580, that the reader may learn the gradual manner in which different segments of the body and the eyes make their appearance.

In Bohemia the primordial fauna of Barrande derived its importance exclusively from its numerous and peculiar trilobites. Besides these, however, the same ancient schists have yielded two genera of brachiopods, Orthis and Orbicula, a Pteropod of the genus Theca, and four echinoderms of the cystidean family.

CAMBRIAN OF SWEDEN AND NORWAY.

The Cambrian beds of Wales are represented in Sweden by strata the fossils of which have been described by a most able naturalist, M. Angelin, in his "Palaeontologica Suecica" (1852-4). The "alum-schists," as they are called in Sweden, are horizontal argillaceous rocks which underlie conformably certain Lower Silurian strata in the mountain called Kinnekulle, south of the great Wener Lake in Sweden. These schists contain trilobites belonging to the genera Paradoxides, Olenus, Agnostus, and others, some of which present rudimentary forms, like the genus last mentioned, without eyes, and with the body segments scarcely developed, and others, again, have the number of segments excessively multiplied, as in Paradoxides. Such peculiarities agree with the characters of the crustaceans met with in the Cambrian strata of Wales; and Dr. Torell has recently found in Sweden the Paradoxides Hicksii, a well-known Lower Cambrian fossil.

At the base of the Cambrian strata in Sweden, which in the neighbourhood of Lake Wener are perfectly horizontal, lie ripple-marked quartzose sandstones with worm-tracks and annelid borings, like some of those found in the Harlech grits of the Longmynd. Among these are some which have been referred doubtfully to plants. These sandstones have been called in Sweden "fucoid sandstones." The whole thickness of the Cambrian rocks of Sweden does not exceed 300 feet from the equivalents of the Tremadoc beds to these sandstones, which last seem to correspond with the Longmynd, and are regarded by Torell as older than any fossiliferous primordial rocks in Bohemia.

CAMBRIAN OF THE UNITED STATES AND CANADA (POTSDAM SANDSTONE).

(FIGURE 581. Dikelocephalus Minnesotensis. Dale Owen. One-third diameter. A large crustacean of the Olenoid group. Potsdam sandstone. Falls of St. Croix, on the Upper Mississippi.)

This formation, as we learn from Sir W. Logan, is 700 feet thick in Canada; the upper part consists of sandstone containing fucoids, and perforated by small vertical holes, which are very characteristic of the rock, and appear to have been made by annelids (Scolithus linearis). The lower portion is a conglomerate with quartz pebbles. I have seen the Potsdam sandstone on the banks of the St. Lawrence, and on the borders of Lake Champlain, where, as at Keesville, it is a white quartzose fine-grained grit, almost passing into quartzite. It is divided into horizontal ripple-marked beds, very like those of the Lingula Flags of Britain, and replete with a small round-shaped Obolella, in such numbers as to divide the rock into parallel planes, in the same manner as do the scales of mica in some micaceous sandstones. Among the shells of this formation in Wisconsin are species of Lingula and Orthis, and several trilobites of the primordial genus Dikelocephalus (Figure 581). On the banks of the St. Lawrence, near Beauharnois and elsewhere, many fossil footprints have been observed on the surface of the rippled layers. They are supposed by Professor Owen to be the trails of more than one species of articulate animal, probably allied to the King Crab, or Limulus.

Recent investigations by the naturalists of the Canadian survey have rendered it certain that below the level of the Potsdam Sandstone there are slates and schists extending from New York to Newfoundland, occupied by a series of trilobitic forms similar in genera, though not in species, to those found in the European Upper Cambrian strata.

HURONIAN SERIES.

Next below the Upper Cambrian occur strata called the Huronian by Sir W. Logan, which are of vast thickness, consisting chiefly of quartzite, with great masses of greenish chloritic slate, which sometimes include pebbles of crystalline rocks derived from the Laurentian formation, next to be described. Limestones are rare in this series, but one band of 300 feet in thickness has been traced for considerable distances to the north of Lake Huron. Beds of greenstone are intercalated conformably with the quartzose and argillaceous members of this series. No organic remains have yet been found in any of the beds, which are about 18,000 feet thick, and rest unconformably on the Laurentian rocks.

LAURENTIAN GROUP.

In the course of the geological survey carried on under the direction of Sir W.E. Logan, it has been shown that, northward of the river St. Lawrence, there is a vast series of crystalline rocks of gneiss, mica-schist, quartzite, and limestone, more than 30,000 feet in thickness, which have been called Laurentian, and which are already known to occupy an area of about 200,000 square miles. They are not only more ancient than the fossiliferous Cambrian formations above described, but are older than the Huronian last mentioned, and had undergone great disturbing movements before the Potsdam sandstone and the other "primordial" or Cambrian rocks were formed. The older half of this Laurentian series is unconformable to the newer portion of the same.

UPPER LAURENTIAN OR LABRADOR SERIES.

The Upper Group, more than 10,000 feet thick, consists of stratified crystalline rocks in which no organic remains have yet been found. They consist in great part of feldspars, which vary in composition from anorthite to andesine, or from those kinds in which there is less than one per cent of potash and soda to those in which there is more than seven per cent of these alkalies, the soda preponderating greatly. These feldsparites sometimes form mountain masses almost without any admixture of other minerals; but at other times they include augite, which passes into hypersthene. They are often granitoid in structure. One of the varieties is the same as the apolescent labradorite rock of Labrador. The Adirondack Mountains in the State of New York are referred to the same series, and it is conjectured that the hypersthene rocks of Skye, which resemble this formation in mineral character, may be of the same geological age.

LOWER LAURENTIAN.

This series, about 20,000 feet in thickness, is, as before stated, unconformable to that last mentioned; it consists in great part of gneiss of a reddish tint with orthoclase feldspar. Beds of nearly pure quartz, from 400 to 600 feet thick, occur in some places. Hornblendic and micaceous schists are often interstratified, and beds of limestone, usually crystalline. Beds of plumbago also occur. That this pure carbon may have been of organic origin before metamorphism has naturally been conjectured.

(FIGURES 582 and 583. Eozoon Canadense, Daw. (after Carpenter). Oldest known organic body.

(FIGURE 582. Eozoon Canadense, Daw. (after Carpenter). Oldest known organic body. a. Chambers of lower tier communicating at +, and separated from adjoining chambers at o by an intervening septum, traversed by passages. b. Chambers of an upper tier. c. Walls of the chambers traversed by fine tubules. (These tubules pass with uniform parallelism from the inner to the outer surface, opening at regular distances from each other.) d. Intermediate skeleton, composed of homogeneous shell substance, traversed by f. Stoloniferous passages connecting the chambers of the two tiers. e. Canal system in intermediate skeleton, showing the arborescent saceodic prolongations. (Figure 583 shows these bodies in a decalcified state.))

(FIGURE 583. Eozoon Canadense, Daw. (after Carpenter). Oldest known organic body. Decalcified portion of natural rock, showing CANAL SYSTEM and the several layers; the acuteness of the planes prevents more than one or two parallel tiers being observed. Natural size.))

There are several of these limestones which have been traced to great distances, and one of them is from 700 to 1500 feet thick. In the most massive of them Sir W. Logan observed, in 1859, what he considered to be an organic body much resembling the Silurian fossil called Stromatopora rugosa. It had been obtained the year before by Mr. J. MacMullen at the Grand Calumet, on the river Ottawa. This fossil was examined in 1864 by Dr. Dawson of Montreal, who detected in it, by aid of the microscope, the distinct structure of a Rhizopod or Foraminifer. Dr. Carpenter and Professor T. Rupert Jones have since confirmed this opinion, comparing the structure to that of the well-known nummulite. It appears to have grown one layer over another, and to have formed reefs of limestone as do the living coral-building polyp animals. Parts of the original skeleton, consisting of carbonate of lime, are still preserved; while certain inter-spaces in the calcareous fossil have been filled up with serpentine and white augite. On this oldest of known organic remains Dr. Dawson has conferred the name of Eozoon Canadense (see Figures 582, 583); its antiquity is such that the distance of time which separated it from the Upper Cambrian period, or that of the Potsdam sandstone, may, says Sir W. Logan, be equal to the time which elapsed between the Potsdam sandstone and the nummulitic limestones of the Tertiary period. The Laurentian and Huronian rocks united are about 50,000 feet in thickness, and the Lower Laurentian was disturbed before the newer series was deposited. We may naturally expect the other proofs of unconformability will hereafter be detected at more than one point in so vast a succession of strata.

The mineral character of the Upper Laurentian differs, as we have seen, from that of the Lower, and the pebbles of gneiss in the Huronian conglomerates are thought to prove that the Laurentian strata were already in a metamorphic state before they were broken up to supply materials for the Huronian. Even if we had not discovered the Eozoon, we might fairly have inferred from analogy that as the quartzites were once beds of sand, and the gneiss and mica-schist derived from shales and argillaceous sandstones, so the calcareous masses, from 400 to 1000 feet and more in thickness, were originally of organic origin. This is now generally believed to have been the case with the Silurian, Devonian, Carboniferous, Oolitic, and Cretaceous limestones and those nummulitic rocks of tertiary date which bear the closest affinity to the Eozoon reefs of the Lower Laurentian. The oldest stratified rock in Scotland is that called by Sir R. Murchison "the fundamental gneiss," which is found in the north-west of Ross- shire, and in Sutherlandshire (see Figure 82), and forms the whole of the adjoining island of Lewis, in the Hebrides. It has a strike from north-west to south-east, nearly at right angles to the metamorphic strata of the Grampians. On this Laurentian gneiss, in parts of the western Highlands, the Lower Cambrian and various metamorphic rocks rest unconformably. It seems highly probable that this ancient gneiss of Scotland may correspond in date with part of the great Laurentian group of North America.

CHAPTER XXVIII.
VOLCANIC ROCKS.

External Form, Structure, and Origin of Volcanic Mountains.
Cones and Craters.
Hypothesis of "Elevation Craters" considered.
Trap Rocks.
Name whence derived.
Minerals most abundant in Volcanic Rocks.
Table of the Analysis of Minerals in the Volcanic and Hypogene Rocks.
Similar Minerals in Meteorites.
Theory of Isomorphism.
Basaltic Rocks.
Trachytic Rocks.
Special Forms of Structure.
The columnar and globular Forms.
Trap Dikes and Veins.
Alteration of Rocks by volcanic Dikes.
Conversion of Chalk into Marble.
Intrusion of Trap between Strata.
Relation of trappean Rocks to the Products of active Volcanoes.

(FIGURE 584. Section through formations from a, low, to c, high. a. Hypogene formations, stratified and unstratified. b. Aqueous formations. c. Volcanic rocks.)

The aqueous or fossiliferous rocks having now been described, we have next to examine those which may be called volcanic, in the most extended sense of that term. In the diagram (Figure 584) suppose a, a to represent the crystalline formations, such as the granitic and metamorphic; b, b the fossiliferous strata; and c, c the volcanic rocks. These last are sometimes found, as was explained in the first chapter, breaking through a and b, sometimes overlying both, and occasionally alternating with the strata b, b.

EXTERNAL FORM, STRUCTURE, AND ORIGIN OF VOLCANIC MOUNTAINS.

The origin of volcanic cones with crater-shaped summits has been explained in the "Principles of Geology" (Chapters 23 to 27), where Vesuvius, Etna, Santorin, and Barren Island are described. The more ancient portions of those mountains or islands, formed long before the times of history, exhibit the same external features and internal structure which belong to most of the extinct volcanoes of still higher antiquity; and these last have evidently been due to a complicated series of operations, varied in kind according to circumstances; as, for example, whether the accumulation took place above or below the level of the sea, whether the lava issued from one or several contiguous vents, and, lastly, whether the rocks reduced to fusion in the subterranean regions happened to have contained more or less silica, potash, soda, lime, iron, and other ingredients. We are best acquainted with the effects of eruptions above water, or those called subaerial or supramarine; yet the products even of these are arranged in so many ways that their interpretation has given rise to a variety of contradictory opinions, some of which will have to be considered in this chapter.

CONES AND CRATERS.

(FIGURE 585. Part of the chain of extinct volcanoes called the Monts Dome, Auvergne. (Scrope.))

In regions where the eruption of volcanic matter has taken place in the open air, and where the surface has never since been subjected to great aqueous denudation, cones and craters constitute the most striking peculiarity of this class of formations. Many hundreds of these cones are seen in central France, in the ancient provinces of Auvergne, Velay, and Vivarais, where they observe, for the most part, a linear arrangement, and form chains of hills. Although none of the eruptions have happened within the historical era, the streams of lava may still be traced distinctly descending from many of the craters, and following the lowest levels of the existing valleys. The origin of the cone and crater- shaped hill is well understood, the growth of many having been watched during volcanic eruptions. A chasm or fissure first opens in the earth, from which great volumes of steam are evolved. The explosions are so violent as to hurl up into the air fragments of broken stone, parts of which are shivered into minute atoms. At the same time melted stone or LAVA usually ascends through the chimney or vent by which the gases make their escape. Although extremely heavy, this lava is forced up by the expansive power of entangled gaseous fluids, chiefly steam or aqueous vapour, exactly in the same manner as water is made to boil over the edge of a vessel when steam has been generated at the bottom by heat. Large quantities of the lava are also shot up into the air, where it separates into fragments, and acquires a spongy texture by the sudden enlargement of the included gases, and thus forms SCORIAE, other portions being reduced to an impalpable powder or dust. The showering down of the various ejected materials round the orifice of eruption gives rise to a conical mound, in which the successive envelopes of sand and scoriae form layers, dipping on all sides from a central axis. In the mean time a hollow, called a CRATER, has been kept open in the middle of the mound by the continued passage upward of steam and other gaseous fluids. The lava sometimes flows over the edge of the crater, and thus thickens and strengthens the sides of the cone; but sometimes it breaks down the cone on one side (see Figure 585), and often it flows out from a fissure at the base of the hill, or at some distance from its base.

Some geologists had erroneously supposed, from observations made on recent cones of eruption, that lava which consolidates on steep slopes is always of a scoriaceous or vesicular structure, and never of that compact texture which we find in those rocks which are usually termed "trappean." Misled by this theory, they have gone so far as to believe that if melted matter has originally descended a slope at an angle exceeding four or five degrees, it never, on cooling, acquires a stony compact texture. Consequently, whenever they found in a volcanic mountain sheets of stony materials inclined at angles of from 5 degrees to 20 degrees or even more than 30 degrees, they thought themselves warranted in assuming that such rocks had been originally horizontal, or very slightly inclined, and had acquired their high inclination by subsequent upheaval. To such dome-shaped mountains with a cavity in the middle, and with the inclined beds having what was called a quaquaversal dip or a slope outward on all sides, they gave the name of "Elevation craters."

As the late Leopold Von Buch, the author of this theory, had selected the Isle of Palma, one of the Canaries, as a typical illustration of this form of volcanic mountain, I visited that island in 1854, in company with my friend Mr. Hartung, and I satisfied myself that it owes its origin to a series of eruptions of the same nature as those which formed the minor cones, already alluded to. In some of the more ancient or Miocene volcanic mountains, such as Mont Dor and Cantal in central France, the mode of origin by upheaval as above described is attributed to those dome-shaped masses, whether they possess or not a great central cavity, as in Palma. Where this cavity is present, it has probably been due to one or more great explosions similar to that which destroyed a great part of ancient Vesuvius in the time of Pliny. Similar paroxysmal catastrophes have caused in historical times the truncation on a grand scale of some large cones in Java and elsewhere. (Principles volume 2 pages 56 and 145.)

Among the objections which may be considered as fatal to Von Buch's doctrine of upheaval in these cases, I may state that a series of volcanic formations extending over an area six or seven miles in its shortest diameter, as in Palma, could not be accumulated in the form of lavas, tuffs, and volcanic breccias or agglomerates without producing a mountain as lofty as that which they now constitute. But assuming that they were first horizontal, and then lifted up by a force acting most powerfully in the centre and tilting the beds on all sides, a central crater having been formed by explosion or by a chasm opening in the middle, where the continuity of the rocks was interrupted, we should have a right to expect that the chief ravines or valleys would open towards the central cavity, instead of which the rim of the great crater in Palma and other similar ancient volcanoes is entire for more than three parts of the whole circumference.

If dikes are seen in the precipices surrounding such craters or central cavities, they certainly imply rents which were filled up with liquid matter. But none of the dislocations producing such rents can have belonged to the supposed period of terminal and paroxysmal upheaval, for had a great central crater been already formed before they originated, or at the time when they took place, the melted matter, instead of filling the narrow vents, would have flowed down into the bottom of the cavity, and would have obliterated it to a certain extent. Making due allowance for the quantity of matter removed by subaerial denudation in volcanic mountains of high antiquity, and for the grand explosions which are known to have caused truncation in active volcanoes, there is no reason for calling in the violent hypothesis of elevation craters to explain the structure of such mountains as Teneriffe, the Grand Canary, Palma, or those of central France, Etna, or Vesuvius, all of which I have examined. With regard to Etna, I have shown, from observations made by me in 1857, that modern lavas, several of them of known date, have formed continuous beds of compact stone even on slopes of 15, 36, and 38 degrees, and, in the case of the lava of 1852, more than 40 degrees. The thickness of these tabular layers varies from 1 1/2 foot to 26 feet. And their planes of stratification are parallel to those of the overlying and underlying scoriae which form part of the same currents. (Memoir on Mount Etna Philosophical Transactions 1858.)

NOMENCLATURE OF TRAPPEAN ROCKS.

When geologists first began to examine attentively the structure of the northern and western parts of Europe, they were almost entirely ignorant of the phenomena of existing volcanoes. They found certain rocks, for the most part without stratification, and of a peculiar mineral composition, to which they gave different names, such as basalt, greenstone, porphyry, trap tuff, and amygdaloid. All these, which were recognised as belonging to one family, were called "trap" by Bergmann, from trappa, Swedish for a flight of steps— a name since adopted very generally into the nomenclature of the science; for it was observed that many rocks of this class occurred in great tabular masses of unequal extent, so as to form a succession of terraces or steps. It was also felt that some general term was indispensable, because these rocks, although very diversified in form and composition, evidently belonged to one group, distinguishable from the Plutonic as well as from the non-volcanic fossiliferous rocks.

By degrees familiarity with the products of active volcanoes convinced geologists more and more that they were identical with the trappean rocks. In every stream of modern lava there is some variation in character and composition, and even where no important difference can be recognised in the proportions of silica, alumina, lime, potash, iron, and other elementary materials, the resulting materials are often not the same, for reasons which we are as yet unable to explain. The difference also of the lavas poured out from the same mountain at two distinct periods, especially in the quantity of silica which they contain, is often so great as to give rise to rocks which are regarded as forming distinct families, although there may be every intermediate gradation between the two extremes, and although some rocks, forming a transition from the one class to the other, may often be so abundant as to demand special names. These species might be multiplied indefinitely, and I can only afford space to name a few of the principal ones, about the composition and aspect of which there is the least discordance of opinion.

MINERALS MOST ABUNDANT IN VOLCANIC ROCKS.

TABLE 28.1. ANALYSIS OF MINERALS MOST ABUNDANT IN THE VOLCANIC AND HYPOGENE ROCKS.
COLUMN 1: SILICA.
COLUMN 2: ALUMINA.
COLUMN 3: SESQUIOXIDE OF IRON.
COLUMN 4: PROTOXIDES OF IRON AND MANGANESE.
COLUMN 5: LIME.
COLUMN 6: MAGNESIA.
COLUMN 7: POTASH.
COLUMN 8: SODA.
COLUMN 9: OTHER CONSTITUENTS.
In this column the following signs are used:
F. Fluorine;
Li. Lithia;
W. Loss on igniting the mineral, in most instances only Water.
COLUMN 10: SPECIFIC GRAVITY.

THE QUARTZ GROUP:
1 2 3 4 5 6 7 8 9 10.
Quartz: 100.0 2.6.
Tridymite: 100.0 2.3.

THE FELDSPAR GROUP:
1 2 3 4 5 6 7 8 9 10.
Orthoclase. Carlsbad, in granite (Bulk):
65.23 18.26 0.27 .... trace .... 14.66 1.45 .... 2.55.
Orthoclase. Sanadine, Drachenfels in trachyte (Rammelsberg).
65.87 18.53 .... 0.95 0.30 10.32 3.42 W 0.44 2.55.
Albite. Arendal, in granite (G. Rose).
68.46 19.30 .... 0.28 0.68 .... 11.27 .... 2.61.
Oligoclase. Ytterby, in granite (Berzelius).
61.55 23.80 .... 3.18 0.80 0.38 9.67 .... 2.65.
Oligoclase. Teneriffe, in trachyte (Deville).
61.55 22.03 .... 2.81 0.47 3.44 7.74 .... 2.59.
Labradorite. Hitteroe, in Labrador-Rock (Waage).
51.39 29.42 2.90 .... 9.44 0.37 1.10 5.03 W 0.71 2.72.
Labradorite. Iceland, in volcanic (Damour).
52.17 29.22 1.90 .... 13.11 .... 3.40 .... 2.71.
Anorthite. Harzburg, in diorite (Streng).
45.37 34.81 0.59 .... 16.52 0.83 0.40 1.45 W 0.87 2.74.
Anorthite. Hecla, in volcanic (Waltershausen).
45.14 32.10 2.03 0.78 18.32 .... 0.22 1.06 .... 2.74.

Leucite. Vesuvius, 1811, in lava (Rammelsberg).
56.10 23.22 .... 20.59 0.57 .... 2.48.
Nepheline. Miask, in Miascite (Scheerer).
44.30 33.25 0.82 .... 0.32 0.07 5.82 16.02 .... 2.59.
Nepheline. Vesuvius, in volcanic (Arfvedson).
44.11 33.73 .... 20.46 W 0.62 2.60.

THE MICA GROUP:
1 2 3 4 5 6 7 8 9 10.

Muscovite. Finland, in granite (Rose).
46.36 36.80 4.53 .... 9.22 .... F 0.67 2.90. W 1.84.
Lepidolite. Cornwall, in granite (Regnault).
52.40 26.80 .... 1.50 .... 9.14 .... F 4.18 2.90. Li 4.85.
Biotite. Bodennais (V. Kobell).
40.86 15.13 13.00 .... 22.00 8.83 .... W 0.44 2.70.
Biotite. Vesuvius, in volcanic (Chodnef).
40.91 17.71 11.02 .... 0.30 19.04 9.96 .... 2.75.
Phlogopite. New York, in metamorphic limestone (Rammelsberg).
41.96 13.47 .... 2.67 0.34 27.12 9.37 .... F 2.93 2.81. W 0.60.

Margarite. Nexos (Smith).
30.02 49.52 1.65 .... 10.82 0.48 1.25 W 5.55 2.99.
Chlorite. Dauphiny (Marignac).
26.88 17.52 29.76 .... 13.84 .... W 11.33 2.87.
Rapidolite. Pyrenees (Delesse).
32.10 18.50 .... 0.06 .... 36.70 .... W 12.10 2.61.
Talc. Zillerthal (Delesse).
63.00 .... trace .... 33.60 .... W 3.10 2.78.

THE AMPHIBOLE AND PYROXENE GROUP.
1 2 3 4 5 6 7 8 9 10.
Tremolite. St. Gothard (Rammelsberg)
58.55 .... 13.90 26.63 .... FW 0.34 2.93.
Actinolite. Arendal, in granite (Rammelsberg).
56.77 0.97 .... 5.88 13.56 21.48 .... W 2.20 3.02.
Hornblende. Faymont, in diorite (Deville).
41.99 11.66 .... 22.22 9.55 12.59 .... 1.02 W 1.47 3.20.
Hornblende Etna, in volcanic (Waltershausen).
40.91 13.68 .... 17.49 13.44 13.19 .... W 0.85 3.01.

Uralite. Ural (Rammelsberg)
50.75 5.65 .... 17.27 11.59 12.28 .... W 1.80 3.14.

Augite. Bohemia, in dolerite (Rammelsberg).
51.12 3.38 0.95 8.08 23.54 12.82 .... 3.35.
Augite. Vesuvius, in lava of 1858 (Rammelsberg).
49.61 4.42 .... 9.08 22.83 14.22 .... 3.25.
Diallage. Harz, in Gabbro (Rammelsberg).
52.00 3.10 .... 9.36 16.29 18.51 .... W 1.10 3.23.
Hypersthene. Labrador, in Labrador-Rock (Damour).
51.36 0.37 .... 22.59 3.09 21.31 .... 3.39.

THE OLIVINE GROUP.
1 2 3 4 5 6 7 8 9 10
Bronzite. Greenland (V. Kobell).
58.00 1.33 11.14 .... 29.66 .... 3.20.
Olivine. Carlsbad, in basalt (Rammelsberg).
39.34 .... 14.85 .... 45.81 .... 3.40.
Olivine. Mount Somma, in volcanic (Walmstedt).
40.08 0.18 .... 15.74 .... 44.22 .... 3.33.

The minerals which form the chief constituents of these igneous rocks are few in number. Next to quartz, which is nearly pure silica or silicic acid, the most important are those silicates commonly classed under the several heads of feldspar, mica, hornblende or augite, and olivine. In Table 28.1, in drawing up which I have received the able assistance of Mr. David Forbes, the chemical analysis of these minerals and their varieties is shown, and he has added the specific gravity of the different mineral species, the geological application of which in determining the rocks formed by these minerals will be explained in the sequel.

From Table 28.1 it will be observed that many minerals are omitted which, even if they are of common occurrence, are more to be regarded as accessory than as essential components of the rocks in which they are found. (For analyses of these minerals see the Mineralogies of Dana and Bristow.) Such are, for example, Garnet, Epidote, Tourmaline, Idocrase, Andalusite, Scapolite, the various Zeolites, and several other silicates of somewhat rarer occurrence. Magnetite, Titanoferrite, and Iron-pyrites also occur as normal constituents of various igneous rocks, although in very small amount, as also Apatite, or phosphate of lime. The other salts of lime, including its carbonate or calcite, although often met with, are invariably products of secondary chemical action.

The Zeolites, above mentioned, so named from the manner in which they froth up under the blow-pipe and melt into a glass, differ in their chemical composition from all the other mineral constituents of volcanic rocks, since they are hydrated silicates containing from 10 to 25 per cent of water. They abound in some trappean rocks and ancient lavas, where they fill up vesicular cavities and interstices in the substance of the rocks, but are rarely found in any quantity in recent lavas; in most cases they are to be regarded as secondary products formed by the action of water on the other constituents of the rocks. Among them the species Analcime, Stilbite, Natrolite, and Chabazite may be mentioned as of most common occurrence.

QUARTZ GROUP.

The microscope has shown that pure quartz is oftener present in lavas than was formerly supposed. It had been argued that the quartz in granite having a specific gravity of 2.6, was not of purely igneous origin, because the silica resulting from fusion in the laboratory has only a specific gravity of 2.3. But Mr. David Forbes has ascertained that the free quartz in trachytes, which are known to have flowed as lava, has the same specific gravity as the ordinary quartz of granite; and the recent researches of Von Rath and others prove that the mineral Tridymite, which is crystallised silica of specific gravity 2.3 (see Table 28.1), is of common occurrence in the volcanic rocks of Mexico, Auvergne, the Rhine, and elsewhere, although hitherto entirely overlooked.

FELDSPAR GROUP.

In the Feldspar group (Table 28.1) the five mineral species most commonly met with as rock constituents are: 1. Orthoclase, often called common or potash- feldspar. 2. Albite, or soda-feldspar, a mineral which plays a more subordinate part than was formerly supposed, this name having been given to much which has since been proved to be Oligoclase. 3. Oligoclase, or soda-lime feldspar, in which soda is present in much larger proportion than lime, and of which mineral andesite are andesine, is considered to be a variety. 4. Labradorite, or lime- soda-feldspar, in which the proportions of lime and soda are the reverse to what they are in Oligoclase. 5. Anorthite or lime-feldspar. The two latter feldspars are rarely if ever found to enter into the composition of rocks containing quartz.

In employing such terms as potash-feldspar, etc., it must, however, always be borne in mind that it is only intended to direct attention to the predominant alkali or alkaline earth in the mineral, not to assert the absence of the others, which in most cases will be found to be present in minor quantity. Thus potash-feldspar (orthoclase) almost always contains a little soda, and often traces of lime or magnesia; and in like manner with the others. The terms "glassy" and "compact" feldspars only refer to structure, and not to species or composition; the student should be prepared to meet with any of the above feldspars in either of these conditions: the glassy state being apparently due to quick cooling, and the compact to conditions unfavourable to crystallisation; the so-called "compact feldspar" is also very commonly found to be an admixture of more than one feldspar species, and frequently also contains quartz and other extraneous mineral matter only to be detected by the microscope.

Feldspars when arranged according to their system of crystallisation are MONOCLINIC, having one axis obliquely inclined; or TRICLINIC, having the three axes all obliquely inclined to each other. If arranged with reference to their cleavage they are ORTHOCLASTIC, the fracture taking place always at a right angle; or PLAGIOCLASTIC, in which the cleavages are oblique to one another. Orthoclase is orthoclastic and monoclinic; all the other feldspars are plagioclastic and triclinic.

MINERALS IN METEORITES.

That variety of the Feldspar Group which is called Anorthite has been shown by Rammelsberg to occur in a meteoric stone, and his analysis proves it to be almost identical in its chemical proportions to the same mineral in the lavas of modern volcanoes. So also Bronzite (Enstatite) and Olivine have been met with in meteorites shown by analysis to come remarkably near to these minerals in ordinary rocks.

MICA GROUP.

With regard to the micas, the four principal species (Table 28.1) all contain potash in nearly the same proportion, but differ greatly in the proportion and nature of their other ingredients. Muscovite is often called common or potash mica; Lepidolite is characterised by containing lithia in addition; Biotite contains a large amount of magnesia and oxide of iron; whilst Phlogopite contains still more of the former substance in quantity. In rocks containing quartz, muscovite or lepidolite are most common. The mica in recent volcanic rocks, gabbros, and diorites is usually Biotite, while that so common in metamorphic limestones is usually, if not always, Phlogopite.

AMPHIBOLE AND PYROXENE GROUP.

The minerals included in Table 28.1 under the Amphibole and Pyroxene Group differ somewhat in their crystallisation form, though they all belong to the monoclinic system. Amphibole is a general name for all the different varieties of Hornblende, Actinolite, Tremolite, etc., while Pyroxene includes Augite, Diallage, Malacolite, Sahlite, etc. The two divisions are so much allied in chemical composition and crystallographic characters, and blend so completely one into the other in Uralite, that it is perhaps best to unite them in one group.

THEORY OF ISOMORPHISM.

The history of the changes of opinion on this point is curious and instructive. Werner first distinguished augite from hornblende; and his proposal to separate them obtained afterwards the sanction of Hauy, Mohs, and other celebrated mineralogists. It was agreed that the form of the crystals of the two species was different, and also their structure, as shown by CLEAVAGE— that is to say, by breaking or cleaving the mineral with a chisel, or a blow of the hammer, in the direction in which it yields most readily. It was also found by analysis that augite usually contained more lime, less alumina, and no fluoric acid; which last, though not always found in hornblende, often enters into its composition in minute quantity. In addition to these characters, it was remarked as a geological fact, that augite and hornblende are very rarely associated together in the same rock. It was also remarked that in the crystalline slags of furnaces augitic forms were frequent, the hornblendic entirely absent; hence it was conjectured that hornblende might be the result of slow, and augite of rapid cooling. This view was confirmed by the fact that Mitscherlich and Berthier were able to make augite artificially, but could never succeed in forming hornblende. Lastly, Gustavus Rose fused a mass of hornblende in a porcelain furnace, and found that it did not, on cooling, assume its previous shape, but invariably took that of augite. The same mineralogist observed certain crystals called Uralite (see Table 28.1) in rocks from Siberia, which possessed the cleavage and chemical composition of hornblende, while they had the external form of augite.

If, from these data, it is inferred that the same substance may assume the crystalline forms of hornblende or augite indifferently, according to the more or less rapid cooling of the melted mass, it is nevertheless certain that the variety commonly called augite, and recognised by a peculiar crystalline form, has usually more lime in it, and less alumina, than that called hornblende, although the quantities of these elements do not seem to be always the same. Unquestionably the facts and experiments above mentioned show the very near affinity of hornblende and augite; but even the convertibility of one into the other, by melting and recrystallising, does not perhaps demonstrate their absolute identity. For there is often some portion of the materials in a crystal which are not in perfect chemical combination with the rest. Carbonate of lime, for example, sometimes carries with it a considerable quantity of silex into its own form of crystal, the silex being mechanically mixed as sand, and yet not preventing the carbonate of lime from assuming the form proper to it. This is an extreme case, but in many others some one or more of the ingredients in a crystal may be excluded from perfect chemical union; and after fusion, when the mass recrystallises, the same elements may combine perfectly or in new proportions, and thus a new mineral may be produced. Or some one of the gaseous elements of the atmosphere, the oxygen for example, may, when the melted matter reconsolidates, combine with some one of the component elements.

The different quantity of the impurities or the refuse above alluded to, which may occur in all but the most transparent and perfect crystals, may partly explain the discordant results at which experienced chemists have arrived in their analysis of the same mineral. For the reader will often find that crystals of a mineral determined to be the same by physical characters, crystalline form, and optical properties, have been declared by skilful analysers to be composed of distinct elements. This disagreement seemed at first subversive of the atomic theory, or the doctrine that there is a fixed and constant relation between the crystalline form and structure of a mineral and its chemical composition. The apparent anomaly, however, which threatened to throw the whole science of mineralogy into confusion, was reconciled to fixed principles by the discoveries of Professor Mitscherlich at Berlin, who ascertained that the composition of the minerals which had appeared so variable was governed by a general law, to which he gave the name of ISOMORPHISM (from isos, equal, and morphe, form). According to this law, the ingredients of a given species of mineral are not absolutely fixed as to their kind and quality; but one ingredient may be replaced by an equivalent portion of some analogous ingredient. Thus, in augite, the lime may be in part replaced by portions of protoxide of iron, or of manganese, while the form of the crystal, and the angle of its cleavage planes, remain the same. These vicarious substitutions, however, of particular elements can not exceed certain defined limits.

BASALTIC ROCKS.

The two principal families of trappean or volcanic rocks are the basalts and the trachytes, which differ chiefly from each other in the quantity of silica which they contain. The basaltic rocks are comparatively poor in silica, containing less than 50 per cent of that mineral, and none in a pure state or as free quartz, apart from the rest of the matrix. They contain a larger proportion of lime and magnesia than the trachytes, so that they are heavier, independently of the frequent presence of the oxides of iron which in some cases forms more than a fourth part of the whole mass. Abich has, therefore, proposed that we should weigh these rocks, in order to appreciate their composition in cases where it is impossible to separate their component minerals. Thus, basalt from Staffa, containing 47.80 per cent of silica, has a specific gravity of 2.95; whereas trachyte, which has 66 per cent of silica, has a specific gravity of only 2.68; trachytic porphyry, containing 69 per cent of silica, a specific gravity of only 2.58. If we then take a rock of intermediate composition, such as that prevailing in the Peak of Teneriffe, which Abich calls Trachyte-dolerite, its proportion of silica being intermediate, or 58 per cent, it weighs 2.78, or more than trachyte, and less than basalt. (Dr. Daubeny on Volcanoes second edition pages 14, 15.)

BASALT.

The different varieties of this rock are distinguished by the names of basalts, anamezites, and dolerites, names which, however, only denote differences in texture without implying any difference in mineral or chemical composition: the term BASALT being used only when the rock is compact, amorphous, and often semi- vitreous in texture, and when it breaks with a perfect conchoidal fracture; when, however, it is uniformly crystalline in appearance, yet very close- grained, the name ANAMESITE (from anamesos, intermediate) is employed, but if the rock be so coarsely crystallised that its different mineral constituents can be easily recognised by the eye, it is called DOLERITE (from doleros, deceitful), in allusion to the difficulty of distinguishing it from some of the rocks known as Plutonic.

MELAPHYRE is often quite undistinguishable in external appearance from basalt, for although rarely so heavy, dark-coloured, or compact, it may present at times all these varieties of texture. Both these rocks are composed of triclinic feldspar and augite with more or less olivine, magnetic or titaniferous oxide of iron, and usually a little nepheline, leucite, and apatite; basalt usually contains considerably more olivine than melaphyre, but chemically they are closely allied, although the melaphyres usually contain more silica and alumina, with less oxides of iron, lime, and magnesia, than the basalts. The Rowley Hills in Staffordshire, commonly known as Rowley Ragstone, are melaphyre.

GREENSTONE.

This name has usually been extended to all granular mixtures, whether of hornblende and feldspar, or of augite and feldspar. The term DIORITE has been applied exclusively to compounds of hornblende and triclinic feldspar. LABRADOR- ROCK is a term used for a compound of labradorite or labrador-feldspar and hypersthene; when the hypersthene predominates it is sometimes known under the name of HYPERSTHENE-ROCK. GABBRO and DIABASE are rocks mainly composed of triclinic feldspars and diallage. All these rocks become sometimes very crystalline, and help to connect the volcanic with the Plutonic formations, which will be treated of in Chapter 31.

The name trachyte (from trachus, rough) was originally given to a coarse granular feldspathic rock which was rough and gritty to the touch. The term was subsequently made to include other rocks, such as clinkstone and obsidian, which have the same mineral composition, but to which, owing to their different texture, the word in its original meaning would not apply. The feldspars which occur in Trachytic rocks are invariably those which contain the largest proportion of silica, or from 60 to 70 per cent of that mineral. Through the base are usually disseminated crystals of glassy feldspar, mica, and sometimes hornblende. Although quartz is not a necessary ingredient in the composition of this rock, it is very frequently present, and the quartz trachytes are very largely developed in many volcanic districts. In this respect the trachytes differ entirely from the members of the Basaltic family, and are more nearly allied to the granites.

OBSIDIAN.

Obsidian, Pitchstone, and Pearlstone are only different forms of a volcanic glass produced by the fusion of trachytic rocks. The distinction between them is caused by different rates of cooling from the melted state, as has been proved by experiment. Obsidian is of a black or ash-grey colour, and though opaque in mass is transparent in thin edges.

CLINKSTONE OR PHONOLITE.

Among the rocks of the trachytic family, or those in which the feldspars are rich in silica, that termed Clinkstone or Phonolite is conspicuous by its fissile structure, and its tendency to lamination, which is such as sometimes to render it useful as roofing-slate. It rings when struck with the hammer, whence its name; is compact, and usually of a greyish blue or brownish colour; is variable in composition, but almost entirely composed of feldspar. When it contains disseminated crystals of feldspar, it is called CLINKSTONE PORPHYRY.

VOLCANIC ROCKS DISTINGUISHED BY SPECIAL FORMS OF STRUCTURE.

Many volcanic rocks are commonly spoken of under names denoting structure alone, which must not be taken to imply that they are distinct rocks, i.e., that they differ from one another either in mineral or chemical composition. Thus the terms Trachytic porphyry, Trachytic tuff, etc., merely refer to the same rock under different conditions of mechanical aggregation or crystalline development which would be more correctly expressed by the use of the adjective, as porphyritic trachyte, etc., but as these terms are so commonly employed it is considered advisable to direct the student's attention to them.

PORPHYRY.

(FIGURE 586. Porphyry. White crystals of feldspar in a dark base of hornblende and feldspar.)

PORPHYRY is one of this class, and very characteristic of the volcanic formations. When distinct crystals of one or more minerals are scattered through an earthy or compact base, the rock is termed a porphyry (see Figure 586). Thus trachyte is usually porphyritic; for in it, as in many modern lavas, there are crystals of feldspar; but in some porphyries the crystals are of augite, olivine, or other minerals. If the base be greenstone, basalt, or pitchstone, the rock may be denominated greenstone-porphyry, pitchstone-porphyry, and so forth. The old classical type of this form of rock is the red porphyry of Egypt, or the well-known "Rosso antico." It consists, according to Delesse, of a red feldspathic base in which are disseminated rose-coloured crystals of the feldspar called oligoclase, with some plates of blackish hornblende and grains of oxide of iron (iron-glance). RED QUARTZIFEROUS PORPHYRY is a much more siliceous rock, containing about 70 or 80 per cent of silex, while that of Egypt has only 62 per cent.

AMYGDALOID.

This is also another form of igneous rock, admitting of every variety of composition. It comprehends any rock in which round or almond-shaped nodules of some mineral, such as agate, chalcedony, calcareous spar, or zeolite, are scattered through a base of wacke, basalt, greenstone, or other kind of trap. It derives its name from the Greek word amygdalon, an almond. The origin of this structure can not be doubted, for we may trace the process of its formation in modern lavas. Small pores or cells are caused by bubbles of steam and gas confined in the melted matter. After or during consolidation, these empty spaces are gradually filled up by matter separating from the mass, or infiltered by water permeating the rock. As these bubbles have been sometimes lengthened by the flow of the lava before it finally cooled, the contents of such cavities have the form of almonds. In some of the amygdaloidal traps of Scotland, where the nodules have decomposed, the empty cells are seen to have a glazed or vitreous coating, and in this respect exactly resemble scoriaceous lavas, or the slags of furnaces.

(FIGURE 587. Scoriaceous lava in part converted into an amygdaloid. Montagne de la Veille, Department of Puy de Dome, France.)

Figure 587 represents a fragment of stone taken from the upper part of a sheet of basaltic lava in Auvergne. One-half is scoriaceous, the pores being perfectly empty; the other part is amygdaloidal, the pores or cells being mostly filled up with carbonate of lime, forming white kernels.

LAVA.

This term has a somewhat vague signification, having been applied to all melted matter observed to flow in streams from volcanic vents. When this matter consolidates in the open air, the upper part is usually scoriaceous, and the mass becomes more and more stony as we descend, or in proportion as it has consolidated more slowly and under greater pressure. At the bottom, however, of a stream of lava, a small portion of scoriaceous rock very frequently occurs, formed by the first thin sheet of liquid matter, which often precedes the main current, and solidifies under slight pressure.

The more compact lavas are often porphyritic, but even the scoriaceous part sometimes contains imperfect crystals, which have been derived from some older rocks, in which the crystals pre-existed, but were not melted, as being more infusible in their nature. Although melted matter rising in a crater, and even that which enters a rent on the side of a crater, is called lava, yet this term belongs more properly to that which has flowed either in the open air or on the bed of a lake or sea. If the same fluid has not reached the surface, but has been merely injected into fissures below ground, it is called trap. There is every variety of composition in lavas; some are trachytic, as in the Peak of Teneriffe; a great number are basaltic, as in Vesuvius and Auvergne; others are andesitic, as those of Chili; some of the most modern in Vesuvius consist of green augite, and many of those of Etna of augite and labrador-feldspar. (G. Hose, Ann. des Mines tome 8 page 32.)

SCORIAE and PUMICE may next be mentioned, as porous rocks produced by the action of gases on materials melted by volcanic heat. SCORIAE are usually of a reddish-brown and black colour, and are the cinders and slags of basaltic or augitic lavas. PUMICE is a light, spongy, fibrous substance, produced by the action of gases on trachytic and other lavas; the relation, however, of its origin to the composition of lava is not yet well understood. Von Buch says that it never occurs where only labrador-feldspar is present.

VOLCANIC ASH OR TUFF, TRAP TUFF.

Small angular fragments of the scoriae and pumice, above-mentioned, and the dust of the same, produced by volcanic explosions, form the tuffs which abound in all regions of active volcanoes, where showers of these materials, together with small pieces of other rocks ejected from the crater, and more or less burnt, fall down upon the land or into the sea. Here they often become mingled with shells, and are stratified. Such tuffs are sometimes bound together by a calcareous cement, and form a stone susceptible of a beautiful polish. But even when little or no lime is present, there is a great tendency in the materials of ordinary tuffs to cohere together. The term VOLCANIC ASH has been much used for rocks of all ages supposed to have been derived from matter ejected in a melted state from volcanic orifices. We meet occasionally with extremely compact beds of volcanic materials, interstratified with fossiliferous rocks. These may sometimes be tuffs, although their density or compactness is such as the cause them to resemble many of those kinds of trap which are found in ordinary dikes.

WACKE is a name given to a decomposed state of various trap rocks of the basaltic family, or those which are poor in silica. It resembles clay of a yellowish or brown colour, and passes gradually from the soft state to the hard dolerite, greenstone, or other trap rock from which it has been derived.

AGGLOMERATE.

In the neighbourhood of volcanic vents, we frequently observe accumulations of angular fragments of rocks formed during eruptions by the explosive action of steam, which shatters the subjacent stony formations, and hurls them up into the air. They then fall in showers around the cone or crater, or may be spread for some distance over the surrounding country. The fragments consist usually of different varieties of scoriaceous and compact lavas; but other kinds of rock, such as granite or even fossiliferous limestones, may be intermixed; in short, any substance through which the expansive gases have forced their way. The dispersion of such materials may be aided by the wind, as it varies in direction or intensity, and by the slope of the cone down which they roll, or by floods of rain, which often accompany eruptions. But if the power of running water, or of the waves and currents of the sea, be sufficient to carry the fragments to a distance, it can scarcely fail to wear off their angles, and the formation then becomes a CONGLOMERATE. If occasionally globular pieces of scoriae abound in an agglomerate, they may not owe their round form to attrition. When all the angular fragments are of volcanic rocks the mass is usually termed a volcanic breccia.

Laterite is a red or brick-like rock composed of silicate of alumina and oxide of iron. The red layers called "ochre beds," dividing the lavas of the Giant's Causeway, are laterites. These were found by Delesse to be trap impregnated with the red oxide of iron, and in part reduced to kaolin. When still more decomposed, they were found to be clay coloured by red ochre. As two of the lavas of the Giant's Causeway are parted by a bed of lignite, it is not improbable that the layers of laterite seen in the Antrim cliffs resulted from atmospheric decomposition. In Madeira and the Canary Islands streams of lava of subaerial origin are often divided by red bands of laterite, probably ancient soils formed by the decomposition of the surfaces of lava-currents, many of these soils having been coloured red in the atmosphere by oxide of iron, others burnt into a red brick by the overflowing of heated lavas. These red bands are sometimes prismatic, the small prisms being at right angles to the sheets of lava. Red clay or red marl, formed as above stated by the disintegration of lava, scoriae, or tuff, has often accumulated to a great thickness in the valleys of Madeira, being washed into them by alluvial action; and some of the thick beds of laterite in India may have had a similar origin. In India, however, especially in the Deccan, the term "laterite" seems to have been used too vaguely to answer the above definition. The vegetable soil in the gardens of the suburbs of Catania which was overflowed by the lava of 1669 was turned or burnt into a layer of red brick-coloured stone, or in other words, into laterite, which may now be seen supporting the old lava-current.

COLUMNAR AND GLOBULAR STRUCTURE.

One of the characteristic forms of volcanic rocks, especially of basalt, is the columnar, where large masses are divided into regular prisms, sometimes easily separable, but in other cases adhering firmly together. The columns vary, in the number of angles, from three to twelve; but they have most commonly from five to seven sides. They are often divided transversely, at nearly equal distances, like the joints in a vertebral column, as in the Giant's Causeway, in Ireland. They vary exceedingly in respect to length and diameter. Dr. MacCulloch mentions some in Skye which are about 400 feet long; others, in Morven, not exceeding an inch. In regard to diameter, those of Ailsa measure nine feet, and those of Morven an inch or less. (MacCulloch System of Geology volume 2 page 137.) They are usually straight, but sometimes curved; and examples of both these occur in the island of Staffa. In a horizontal bed or sheet of trap the columns are vertical; in a vertical dike they are horizontal.

(FIGURE 588. Lava of La Coupe d'Ayzac, near Antraigue, in the Department of Ardeche.)

It being assumed that columnar trap has consolidated from a fluid state, the prisms are said to be always at right angles to the COOLING SURFACES. If these surfaces, therefore, instead of being either perpendicular or horizontal, are curved, the columns ought to be inclined at every angle to the horizon; and there is a beautiful exemplification of this phenomenon in one of the valleys of the Vivarais, a mountainous district in the South of France, where, in the midst of a region of gneiss, a geologist encounters unexpectedly several volcanic cones of loose sand and scoriae. From the crater of one of these cones, called La Coupe d'Ayzac, a stream of lava has descended and occupied the bottom of a narrow valley, except at those points where the river Volant, or the torrents which join it, have cut away portions of the solid lava. Figure 588 represents the remnant of the lava at one of these points. It is clear that the lava once filled the whole valley up to the dotted line d-a; but the river has gradually swept away all below that line, while the tributary torrent has laid open a transverse section; by which we perceive, in the first place, that the lava is composed, as usual in this country, of three parts: the uppermost, at a, being scoriaceous, the second b, presenting irregular prisms; and the third, c, with regular columns, which are vertical on the banks of the Volant, where they rest on a horizontal base of gneiss, but which are inclined at an angle of 45 degrees, at g, and are nearly horizontal at f, their position having been everywhere determined, according to the law before mentioned, by the form of the original valley.

(FIGURE 589. Columnar basalt in the Vincentin. (Fortis.)

In Figure 589, a view is given of some of the inclined and curved columns which present themselves on the sides of the valleys in the hilly region north of Vicenza, in Italy, and at the foot of the higher Alps. (Fortis Mem. sur l'Hist. Nat. de l'Italie tome 1 page 233 plate 7.) Unlike those of the Vivarais, last mentioned, the basalt of this country was evidently submarine, and the present valleys have since been hollowed out by denudation.

(FIGURE 590. Basaltic pillars of the Kasegrotte, Bertrich-Baden, half-way between Treves and Coblentz. Height of grotto, from 7 to 8 feet.)

The columnar structure is by no means peculiar to the trap rocks in which augite abounds; it is also observed in trachyte, and other feldspathic rocks of the igneous class, although in these it is rarely exhibited in such regular polygonal forms. It has been already stated that basaltic columns are often divided by cross-joints. Sometimes each segment, instead of an angular, assumes a spheroidal form, so that a pillar is made up of a pile of balls, usually flattened, as in the Cheese-grotto at Bertrich-Baden, in the Eifel, near the Moselle (Figure 590). The basalt there is part of a small stream of lava, from 30 to 40 feet thick, which has proceeded from one of several volcanic craters, still extant, on the neighbouring heights.

In some masses of decomposing greenstone, basalt, and other trap rocks, the globular structure is so conspicuous that the rock has the appearance of a heap of large cannon balls. According to M. Delesse, the centre of each spheroid has been a centre of crystallisation, around which the different minerals of the rock arranged themselves symmetrically during the process of cooling. But it was also, he says, a centre of contraction, produced by the same cooling, the globular form, therefore, of such spheroids being the combined result of crystallisation and contraction. (Delesse sur les Roches Globuleuses Mem. de la Soc. Geol. de France 2 ser. tome 4.)

(FIGURE 591. Globiform pitchstone. Chiaja di Luna, Isle of Ponza. (Scrope.))

Mr. Scrope gives as an illustration of this structure a resinous trachyte or pitchstone-porphyry in one of the Ponza islands, which rise from the Mediterranean, off the coast of Terracina and Gaeta. The globes vary from a few inches to three feet in diameter, and are of an ellipsoidal form (see Figure 591). The whole rock is in a state of decomposition, "and when the balls," says Mr. Scrope, "have been exposed a short time to the weather, they scale off at a touch into numerous concentric coats, like those of a bulbous root, inclosing a compact nucleus. The laminae of this nucleus have not been so much loosened by decomposition; but the application of a ruder blow will produce a still further exfoliation." (Scrope Geological Transactions second series volume 2 page 205.)

VOLCANIC OR TRAP DIKES.

(FIGURE 592. Dike in valley, near Brazen Head, Madeira. (From a drawing of Captain Basil Hall, R.N.))

The leading varieties of the trappean rocks— basalt, greenstone, trachyte, and the rest— are found sometimes in dikes penetrating stratified and unstratified formations, sometimes in shapeless masses protruding through or overlying them, or in horizontal sheets intercalated between strata. Fissures have already been spoken of as occurring in all kinds of rocks, some a few feet, others many yards in width, and often filled up with earth or angular pieces of stone, or with sand and pebbles. Instead of such materials, suppose a quantity of melted stone to be driven or injected into an open rent, and there consolidated, we have then a tabular mass resembling a wall, and called a trap dike. It is not uncommon to find such dikes passing through strata of soft materials, such as tuff, scoriae, or shale, which, being more perishable than the trap, are often washed away by the sea, rivers, or rain, in which case the dike stands prominently out in the face of precipices, or on the level surface of a country (see Figure 592).

(FIGURE 593. Ground-plan of greenstone dikes traversing sandstone. Arran.)

In the islands of Arran and Skye, and in other parts of Scotland, where sandstone, conglomerate, and other hard rocks are traversed by dikes of trap, the converse of the above phenomenon is seen. The dike, having decomposed more rapidly than the containing rock, has once more left open the original fissure, often for a distance of many yards inland from the sea-coast. There is yet another case, by no means uncommon in Arran and other parts of Scotland, where the strata in contact with the dike, and for a certain distance from it, have been hardened, so as to resist the action of the weather more than the dike itself, or the surrounding rocks. When this happens, two parallel walls of indurated strata are seen protruding above the general level of the country and following the course of the dike. In Figure 593 a ground plan is given of a ramifying dike of greenstone, which I observed cutting through sandstone on the beach near Kildonan Castle, in Arran. The larger branch varies from five to seven feet in width, which will afford a scale of measurement for the whole.

(FIGURE 594. Trap dividing and covering sandstone near Suishnish, in Skye. (MacCulloch.))

In the Hebrides and other countries, the same masses of trap which occupy the surface of the country far and wide, concealing the subjacent stratified rocks, are seen also in the sea-cliffs, prolonged downward in veins or dikes, which probably unite with other masses of igneous rock at a greater depth. The largest of the dikes represented in Figure 594, and which are seen in part of the coast of Skye, is no less than 100 feet in width.

Every variety of trap-rock is sometimes found in dikes, as basalt, greenstone, feldspar-porphyry, and trachyte. The amygdaloidal traps also occur, though more rarely, and even tuff and breccia, for the materials of these last may be washed down into open fissures at the bottom of the sea, or during eruption on the land may be showered into them from the air. Some dikes of trap may be followed for leagues uninterruptedly in nearly a straight direction, as in the north of England, showing that the fissures which they fill must have been of extraordinary length.

ROCKS ALTERED BY VOLCANIC DIKES.

After these remarks on the form and composition of dikes themselves, I shall describe the alterations which they sometimes produce in the rocks in contact with them. The changes are usually such as the heat of melted matter and of the entangled steam and gases might be expected to cause.

PLAS-NEWYDD: DIKE CUTTING THROUGH SHALE.

A striking example, near Plas-Newydd, in Anglesea, has been described by Professor Henslow. (Cambridge Transactions volume 1 page 402.) The dike is 134 feet wide, and consists of a rock which is a compound of feldspar and augite (dolerite of some authors). Strata of shale and argillaceous limestone, through which it cuts perpendicularly, are altered to a distance of 30, or even, in some places, of 35 feet from the edge of the dike. The shale, as it approaches the trap, becomes gradually more compact, and is most indurated where nearest the junction. Here it loses part of its schistose structure, but the separation into parallel layers is still discernible. In several places the shale is converted into hard porcelanous jasper. In the most hardened part of the mass the fossil shells, principally Producti, are nearly obliterated; yet even here their impressions may frequently be traced. The argillaceous limestone undergoes analogous mutations, losing its earthy texture as it approaches the dike, and becoming granular and crystalline. But the most extraordinary phenomenon is the appearance in the shale of numerous crystals of analcime and garnet, which are distinctly confined to those portions of the rock affected by the dike. (Ibid. volume 1 page 410.) Some garnets contain as much as 20 per cent of lime, which they may have derived from the decomposition of the fossil shells or Producti. The same mineral has been observed, under very analogous circumstances, in High Teesdale, by Professor Sedgwick, where it also occurs in shale and limestone, altered by basalt. (Ibid. volume 2 page 175.)

ANTRIM: DIKE CUTTING THROUGH CHALK.

In several parts of the county of Antrim, in the north of Ireland, chalk with flints is traversed by basaltic dikes. The chalk is there converted into granular marble near the basalt, the change sometimes extending eight or ten feet from the wall of the dike, being greatest near the point of contact, and thence gradually decreasing till it becomes evanescent. "The extreme effect," says Dr. Berger, "presents a dark brown crystalline limestone, the crystals running in flakes as large as those of coarse primitive (METAMORPHIC) limestone; the next state is saccharine, then fine grained and arenaceous; a compact variety, having a porcelanous aspect and a bluish-grey colour, succeeds: this, towards the outer edge, becomes yellowish-white, and insensibly graduates into the unaltered chalk. The flints in the altered chalk usually assume a grey yellowish colour." (Dr. Berger Geological Transactions 1st series volume 3 page 172.) All traces of organic remains are effaced in that part of the limestone which is most crystalline.

(FIGURE 595. Basaltic dikes in chalk in Island of Rathlin, Antrim. Ground-plan as seen on the beach. (Conybeare and Buckland. (Geological Transactions 1st series volume 3 page 210 and plate 10. From left to right: chalk; dike 35 ft.: dike 1 ft.: dike 20 ft.: chalk.)

Figure 595 represents three basaltic dikes traversing the chalk, all within the distance of 90 feet. The chalk contiguous to the two outer dikes is converted into a finely granular marble, m, m, as are the whole of the masses between the outer dikes and the central one. The entire contrast in the composition and colour of the intrusive and invaded rocks, in these cases, renders the phenomena peculiarly clear and interesting. Another of the dikes of the north-east of Ireland has converted a mass of red sandstone into hornstone. By another, the shale of the coal-measures has been indurated, assuming the character of flinty slate; and in another place the slate-clay of the lias has been changed into flinty slate, which still retains numerous impressions of ammonites. (Ibid. volume 3 page 213; and Playfair Illustration of Huttonian Theory s. 253.)

It might have been anticipated that beds of coal would, from their combustible nature, be affected in an extraordinary degree by the contact of melted rock. Accordingly, one of the greenstone dikes of Antrim, on passing through a bed of coal, reduces it to a cinder for the space of nine feet on each side. At Cockfield Fell, in the north of England, a similar change is observed. Specimens taken at the distance of about thirty yards from the trap are not distinguishable from ordinary pit-coal; those nearer the dike are like cinders, and have all the character of coke; while those close to it are converted into a substance resembling soot. (Sedgwick Cambridge Transactions volume 2 page 37.)

It is by no means uncommon to meet with the same rocks, even in the same districts, absolutely unchanged in the proximity of volcanic dikes. This great inequality in the effects of the igneous rocks may often arise from an original difference in their temperature, and in that of the entangled gases, such as is ascertained to prevail in different lavas, or in the same lava near its source and at a distance from it. The power also of the invaded rocks to conduct heat may vary, according to their composition, structure, and the fractures which they may have experienced, and perhaps, also, according to the quantity of water (so capable of being heated) which they contain. It must happen in some cases that the component materials are mixed in such proportions as to prepare them readily to enter into chemical union, and form new minerals; while in other cases the mass may be more homogeneous, or the proportions less adapted for such union.

We must also take into consideration, that one fissure may be simply filled with lava, which may begin to cool from the first; whereas in other cases the fissure may give passage to a current of melted matter, which may ascend for days or months, feeding streams which are overflowing the country above, or being ejected in the shape of scoriae from some crater. If the walls of a rent, moreover, are heated by hot vapour before the lava rises, as we know may happen on the flanks of a volcano, the additional heat supplied by the dike and its gases will act more powerfully.

INTRUSION OF TRAP BETWEEN STRATA.

Masses of trap are not unfrequently met with intercalated between strata, and maintaining their parallelism to the planes of stratification throughout large areas. They must in some places have forced their way laterally between the divisions of the strata, a direction in which there would be the least resistance to an advancing fluid, if no vertical rents communicated with the surface, and a powerful hydrostatic pressure were caused by gases propelling the lava upward.

RELATION OF TRAPPEAN ROCKS TO THE PRODUCTS OF ACTIVE VOLCANOES.

When we reflect on the changes above described in the strata near their contact with trap dikes, and consider how complete is the analogy or often identity in composition and structure of the rocks called trappean and the lavas of active volcanoes, it seems difficult at first to understand how so much doubt could have prevailed for half a century as to whether trap was of igneous or aqueous origin. To a certain extent, however, there was a real distinction between the trappean formations and those to which the term volcanic was almost exclusively confined. A large portion of the trappean rocks first studied in the north of Germany, and in Norway, France, Scotland, and other countries, were such as had been formed entirely under water, or had been injected into fissures and intruded between strata, and which had never flowed out in the air, or over the bottom of a shallow sea. When these products, therefore, of submarine or subterranean igneous action were contrasted with loose cones of scoriae, tuff, and lava, or with narrow streams of lava in great part scoriaceous and porous, such as were observed to have proceeded from Vesuvius and Etna, the resemblance seemed remote and equivocal. It was, in truth, like comparing the roots of a tree with its leaves and branches, which, although the belong to the same plant, differ in form, texture, colour, mode of growth, and position. The external cone, with its loose ashes and porous lava, may be likened to the light foliage and branches, and the rocks concealed far below, to the roots. But it is not enough to say of the volcano,

"Quantum vertice in auras Aetherias, tantum radice in Tartara tendit,"

for its roots do literally reach downward to Tartarus, or to the regions of subterranean fire; and what is concealed far below is probably always more important in volume and extent than what is visible above ground.

(FIGURE 596. Strata intercepted by a trap dike, and covered with alluvium.)

We have already stated how frequently dense masses of strata have been removed by denudation from wide areas (see Chapter 6); and this fact prepares us to expect a similar destruction of whatever may once have formed the uppermost part of ancient submarine or subaerial volcanoes, more especially as those superficial parts are always of the lightest and most perishable materials. The abrupt manner in which dikes of trap usually terminate at the surface (see Figure 596), and the water-worn pebbles of trap in the alluvium which covers the dike, prove incontestably that whatever was uppermost in these formations has been swept away. It is easy, therefore, to conceive that what is gone in regions of trap may have corresponded to what is now visible in active volcanoes.

As to the absence of porosity in the trappean formations, the appearances are in a great degree deceptive, for all amygdaloids are, as already explained, porous rocks, into the cells of which mineral matter such as silex, carbonate of lime, and other ingredients, have been subsequently introduced (see above); sometimes, perhaps, by secretion during the cooling and consolidation of lavas. In the Little Cumbray, one of the Western Islands, near Arran, the amygdaloid sometimes contains elongated cavities filled with brown spar; and when the nodules have been washed out, the interior of the cavities is glazed with the vitreous varnish so characteristic of the pores of slaggy lavas. Even in some parts of this rock which are excluded from air and water, the cells are empty, and seem to have always remained in this state, and are therefore undistinguishable from some modern lavas. (MacCulloch Western Islands volume 2 page 487.)

Dr. MacCulloch, after examining with great attention these and the other igneous rocks of Scotland, observes, "that it is a mere dispute about terms, to refuse to the ancient eruptions of trap the name of submarine volcanoes; for they are such in every essential point, although they no longer eject fire and smoke." The same author also considers it not improbable that some of the volcanic rocks of the same country may have been poured out in the open air. (System of Geology volume 2 page 114.)

It will be seen in the following chapters that in the earth's crust there are volcanic tuffs of all ages, containing marine shells, which bear witness to eruptions at many successive geological periods. These tuffs, and the associated trappean rocks, must not be compared to lava and scoriae which had cooled in the open air. Their counterparts must be sought in the products of modern submarine volcanic eruptions. If it be objected that we have no opportunity of studying these last, it may be answered, that subterranean movements have caused, almost everywhere in regions of active volcanoes, great changes in the relative level of land and sea, in times comparatively modern, so as to expose to view the effects of volcanic operations at the bottom of the sea.

CHAPTER XXIX.
ON THE AGES OF VOLCANIC ROCKS.

Tests of relative Age of Volcanic Rocks.
Why ancient and modern Rocks can not be identical.
Tests by Superposition and intrusion.
Test by Alteration of Rocks in Contact.
Test by Organic Remains.
Test of Age by Mineral Character.
Test by Included Fragments.
Recent and Post-pliocene volcanic Rocks.
Vesuvius, Auvergne, Puy de Come, and Puy de Pariou.
Newer Pliocene volcanic Rocks.
Cyclopean Isles, Etna, Dikes of Palagonia, Madeira.
Older Pliocene volcanic Rocks.
Italy.
Pliocene Volcanoes of the Eifel.
Trass.

Having in the former part of this work referred the sedimentary strata to a long succession of geological periods, we have now to consider how far the volcanic formations can be classed in a similar chronological order. The tests of relative age in this class of rocks are four: first, superposition and intrusion, with or without alteration of the rocks in contact; second, organic remains; third, mineral characters; fourth, included fragments of older rocks.

Besides these four tests it may be said, in a general way, that volcanic rocks of Primary or Palaeozoic antiquity differ from those of the Secondary or Mesozoic age, and these again from the Tertiary and Recent. Not, perhaps, that they differed originally in a greater degree than the modern volcanic rocks of one region, such as that of the Andes, differ from those of another, such as Iceland, but because all rocks permeated by water, especially if its temperature be high, are liable to undergo a slow transmutation, even when they do not assume a new crystalline form like that of the hypogene rocks.

Although subaerial and submarine denudation, as before stated, remove, in the course of ages, large portions of the upper or more superficial products of volcanoes, yet these are sometimes preserved by subsidence, becoming covered by the sea or by superimposed marine deposits. In this way they may be protected for ages from the waves of the sea, or the destroying action of rivers, while, at the same time, they may not sink so deep as to be exposed to that Plutonic action (to be spoken of in Chapter 31) which would convert them into crystalline rocks. But even in this case they will not remain unaltered, because they will be percolated by water often of high temperature, and charged with carbonate of lime, silex, iron, and other mineral ingredients, whereby gradual changes in the constitution of the rocks may be superinduced. Every geologist is aware how often silicified trees occur in volcanic tuffs, the perfect preservation of their internal structure showing that they have not decayed before the petrifying material was supplied.

The porous and vesicular nature of a large part, both of the basaltic and trachytic lavas, affords cavities in which silex and carbonate of lime are readily deposited. Minerals of the zeolite family, the composition of which has already been alluded to in Chapter 28, occur in amygdaloids and other trap-rocks in great abundance, and Daubree's observations have proved that they are not always simple deposits of substances held in solution by the percolating waters, being occasionally products of the chemical action of that water on the rock through which they are filtered, and portions of which are decomposed. From these considerations it follows that the perfect identity of very ancient and very modern volcanic formations is scarcely possible.

TESTS BY SUPERPOSITION.

(FIGURE 597. Section through sedimentary mass with melted matter.)

If a volcanic rock rest upon an aqueous deposit, the volcanic must be the newest of the two; but the like rule does not hold good where the aqueous formation rests upon the volcanic, for melted matter, rising from below, may penetrate a sedimentary mass without reaching the surface, or may be forced in conformably between two strata, as b below D in Figure 597, after which it may cool down and consolidate. Superposition, therefore, is not of the same value as a test of age in the unstratified volcanic rocks as in fossiliferous formations. We can only rely implicitly on this test where the volcanic rocks are contemporaneous, not where they are intrusive. Now, they are said to be contemporaneous if produced by volcanic action which was going on simultaneously with the deposition of the strata with which they are associated. Thus in the section at D (Figure 597), we may perhaps ascertain that the trap b flowed over the fossiliferous bed c, and that, after its consolidation, a was deposited upon it, a and c both belonging to the same geological period. But, on the other hand, we must conclude the trap to be intrusive, if the stratum a be altered by b at the point of contact, or if, in pursuing b for some distance, we find at length that it cuts through the stratum a, and then overlies it as at E.

(FIGURE 598. Section through sedimentary mass with melted matter.)

We may, however, be easily deceived in supposing the volcanic rock to be intrusive, when in reality it is contemporaneous; for a sheet of lava, as it spreads over the bottom of the sea, can not rest everywhere upon the same stratum, either because these have been denuded, or because, if newly thrown down, they thin out in certain places, thus allowing the lava to cross their edges. Besides, the heavy igneous fluid will often, as it moves along, cut a channel into beds of soft mud and sand. Suppose the submarine lava F (Figure 598) to have come in contact in this manner with the strata a, b, c, and that after its consolidation the strata d, e are thrown down in a nearly horizontal position, yet so as to lie unconformably to F, the appearance of subsequent intrusion will here be complete, although the trap is in fact contemporaneous. We must not, therefore, hastily infer that the rock F is intrusive, unless we find the overlying strata, d, e, to have been altered at their junction, as if by heat.

The test of age by superposition is strictly applicable to all stratified volcanic tuffs, according to the rules already explained in the case of sedimentary deposits (see Chapter 8).

TEST OF AGE BY ORGANIC REMAINS.

We have seen how, in the vicinity of active volcanoes, scoriae, pumice, fine sand, and fragments of rock are thrown up into the air, and then showered down upon the land, or into neighbouring lakes or seas. In the tuffs so formed shells, corals, or any other durable organic bodies which may happen to be strewed over the bottom of a lake or sea will be imbedded, and thus continue as permanent memorials of the geological period when the volcanic eruption occurred. Tufaceous strata thus formed in the neighbourhood of Vesuvius, Etna, Stromboli, and other volcanoes now in islands or near the sea, may give information of the relative age of these tuffs at some remote future period when the fires of these mountains are extinguished. By evidence of this kind we can establish a coincidence in age between volcanic rocks and the different primary, secondary, and tertiary fossiliferous strata.

The tuffs alluded to may not always be marine, but may include, in some places, fresh-water shells; in others, the bones of terrestrial quadrupeds. The diversity of organic remains in formations of this nature is perfectly intelligible, if we reflect on the wide dispersion of ejected matter during late eruptions, such as that of the volcano of Coseguina, in the province of Nicaragua, January 19, 1835. Hot cinders and fine scoriae were then cast up to a vast height, and covered the ground as they fell to the depth of more than ten feet, for a distance of eight leagues from the crater, in a southerly direction. Birds, cattle, and wild animals were scorched to death in great numbers, and buried in ashes. Some volcanic dust fell at Chiapa, upward of 1200 miles, not to leeward of the volcano, as might have been anticipated, but to windward, a striking proof of a counter-current in the upper region of the atmosphere; and some on Jamaica, about 700 miles distant to the north-east. In the sea, also, at the distance of 1100 miles from the point of eruption, Captain Eden of the "Conway" sailed 40 miles through floating pumice, among which were some pieces of considerable size. (Caldcleugh Philosophical Transactions 1836 page 27.)

TEST OF AGE BY MINERAL COMPOSITION.

As sediment of homogeneous composition, when discharged from the mouth of a large river, is often deposited simultaneously over a wide space, so a particular kind of lava flowing from a crater during one eruption may spread over an extensive area; thus in Iceland, in 1783, the melted matter, pouring from Skaptar Jokul, flowed in streams in opposite directions, and caused a continuous mass the extreme points of which were 90 miles distant from each other. This enormous current of lava varied in thickness from 100 feet to 600 feet, and in breadth from that of a narrow river gorge to 15 miles. (See Principles Index "Skaptar Jokul.") Now, if such a mass should afterwards be divided into separate fragments by denudation, we might still, perhaps, identify the detached portions by their similarity in mineral composition. Nevertheless, this test will not always avail the geologist; for, although there is usually a prevailing character in lava emitted during the same eruption, and even in the successive currents flowing from the same volcano, still, in many cases, the different parts even of one lava-stream, or, as before stated, of one continuous mass of trap, vary much in mineral composition and texture.

In Auvergne, the Eifel, and other countries where trachyte and basalt are both present, the trachytic rocks are for the most part older than the basaltic. These rocks do, indeed, sometimes alternate partially, as in the volcano of Mont Dor, in Auvergne; and in Madeira trachytic rocks overlie an older basaltic series; but the trachyte occupies more generally an inferior position, and is cut through and overflowed by basalt. It can by no means be inferred that trachyte predominated at one period of the earth's history and basalt at another, for we know that trachytic lavas have been formed at many successive periods, and are still emitted from many active craters; but it seems that in each region, where a long series of eruptions have occurred, the lavas containing feldspar more rich in silica have been first emitted, and the escape of the more augitic kinds has followed. The hypothesis suggested by Mr. Scrope may, perhaps, afford a solution of this problem. The minerals, he observes, which abound in basalt are of greater specific gravity than those composing the feldspathic lavas; thus, for example, hornblende, augite, and olivine are each more than three times the weight of water; whereas common feldspar and albite have each scarcely more than 2 1/2 times the specific gravity of water; and the difference is increased in consequence of there being much more iron in a metallic state in basalt and greenstone than in trachyte and other allied feldspathic lavas. If, therefore, a large quantity of rock be melted up in the bowels of the earth by volcanic heat, the denser ingredients of the boiling fluid may sink to the bottom, and the lighter remaining above would in that case be first propelled upward to the surface by the expansive power of gases. Those materials, therefore, which occupy the lowest place in the subterranean reservoir will always be emitted last, and take the uppermost place on the exterior of the earth's crust.

TEST BY INCLUDED FRAGMENTS.

We may sometimes discover the relative age of two trap-rocks, or of an aqueous deposit and the trap on which it rests, by finding fragments of one included in the other in cases such as those before alluded to, where the evidence of superposition alone would be insufficient. It is also not uncommon to find a conglomerate almost exclusively composed of rolled pebbles of trap, associated with some fossiliferous stratified formation in the neighbourhood of massive trap. If the pebbles agree generally in mineral character with the latter, we are then enabled to determine its relative age by knowing that of the fossiliferous strata associated with the conglomerate. The origin of such conglomerates is explained by observing the shingle beaches composed of trap- pebbles in modern volcanoes, as at the base of Etna.

RECENT AND POST-PLIOCENE VOLCANIC ROCKS.

I shall now select examples of contemporaneous volcanic rocks of successive geological periods, to show that igneous causes have been in activity in all past ages of the world. They have been perpetually shifting the places where they have broken out at the earth's surface, and we can sometimes prove that those areas which are now the great theatres of volcanic action were in a state of perfect tranquillity at remote geological epochs, and that, on the other hand, in places where at former periods the most violent eruptions took place at the surface and continued for a great length of time, there has been an entire suspension of igneous action in historical times, and even, as in the British Isles, throughout a large part of the antecedent Tertiary Period.

In the absence of British examples of volcanic rocks newer than the Upper Miocene, I may state that in other parts of the world, especially in those where volcanic eruptions are now taking place from time to time, there are tuffs and lavas belonging to that part of the Tertiary era the antiquity of which is proved by the presence of the bones of extinct quadrupeds which co-existed with terrestrial, fresh-water, and marine mollusca of species still living. One portion of the lavas, tuffs, and trap-dikes of Etna, Vesuvius, and the island of Ischia has been produced within the historical era; another and a far more considerable part originated at times immediately antecedent, when the waters of the Mediterranean were already inhabited by the existing testacea, but when certain species of elephant, rhinoceros, and other quadrupeds now extinct, inhabited Europe.

VESUVIUS.

I have traced in the "Principles of Geology" the history of the changes which the volcanic region of Campania is known to have undergone during the last 2000 years. The aggregate effect of igneous operations during that period is far from insignificant, comprising as it does the formation of the modern cone of Vesuvius since the year 79, and the production of several minor cones in Ischia, together with that of Monte Nuovo in the year 1538. Lava-currents have also flowed upon the land and along the bottom of the sea— volcanic sand, pumice, and scoriae have been showered down so abundantly that whole cities were buried- - tracts of the sea have been filled up or converted into shoals— and tufaceous sediment has been transported by rivers and land-floods to the sea. There are also proofs, during the same recent period, of a permanent alteration of the relative levels of the land and sea in several places, and of the same tract having, near Puzzuoli, been alternately upheaved and depressed to the amount of more than twenty feet. In connection with these convulsions, there are found, on the shores of the Bay of Baiae, recent tufaceous strata, filled with articles fabricated by the hands of man, and mingled with marine shells.

It has also been stated (Chapter 13), that when we examine this same region, it is found to consist largely of tufaceous strata, of a date anterior to human history or tradition, which are of such thickness as to constitute hills from 500 to more than 2000 feet in height. Some of these strata contain marine shells which are exclusively of living species, others contain a slight mixture, one or two per cent of species not known as living.

The ancient part of Vesuvius is called Somma, and consists of the remains of an older cone which appears to have been partly destroyed by explosion. In the great escarpment which this remnant of the ancient mountain presents towards the modern cone of Vesuvius, there are many dikes which are for the most part vertical, and traverse the inclined beds of lava and scoriae which were successively superimposed during those eruptions by which the old cone was formed. They project in relief several inches, or sometimes feet, from the face of the cliff, being extremely compact, and less destructible than the intersected tuffs and porous lavas. In vertical extent they vary from a few yards to 500 feet, and in breadth from one to twelve feet. Many of them cut all the inclined beds in the escarpment of Somma from top to bottom, others stop short before they ascend above halfway. In mineral composition they scarcely differ from the lavas of Somma, the rock consisting of a base of leucite and augite, through which large crystals of augite and some of leucite are scattered.

Nothing is more remarkable than the usual parallelism of the opposite sides of the dikes, which correspond almost as regularly as the two opposite faces of a wall of masonry. This character appears at first the more inexplicable, when we consider how jagged and uneven are the rents caused by earthquakes in masses of heterogeneous composition, like those composing the cone of Somma. In explanation of this phenomenon, M. Necker refers us to Sir W. Hamilton's account of an eruption of Vesuvius in the year 1779, who records the following fact: "The lavas, when they either boiled over the crater, or broke out from the conical parts of the volcano, constantly formed channels as regular as if they had been cut by art down the steep part of the mountain; and whilst in a state of perfect fusion, continued their course in those channels, which were sometimes full to the brim, and at other times more or less so, according to the quantity of matter in motion.

"These channels (says the same observer), I have found, upon examination after an eruption, to be in general from two to five or six feet wide, and seven or eight feet deep. They were often hid from the sight by a quantity of scoriae that had formed a crust over them; and the lava, having been conveyed in a covered way for some yards, came out fresh again into an open channel. After an eruption, I have walked in some of those subterraneous or covered galleries, which were exceedingly curious, the sides, top, and bottom BEING WORN PERFECTLY SMOOTH AND EVEN in most parts by the violence of the currents of the red-hot lavas which they had conveyed for many weeks successively." I was able to verify this phenomenon in 1858, when a stream of lava issued from a lateral cone. (Principles of Geology volume 1 page 626.) Now, the walls of a vertical fissure, through which lava has ascended in its way to a volcanic vent, must have been exposed to the same erosion as the sides of the channels before adverted to. The prolonged and uniform friction of the heavy fluid, as it is forced and made to flow upward, can not fail to wear and smooth down the surfaces on which it rubs, and the intense heat must melt all such masses as project and obstruct the passage of the incandescent fluid.

The rock composing the dikes both in the modern and ancient part of Vesuvius is far more compact than that of ordinary lava, for the pressure of a column of melted matter in a fissure greatly exceeds that in an ordinary stream of lava; and pressure checks the expansion of those gases which give rise to vesicles in lava. There is a tendency in almost all the Vesuvian dikes to divide into horizontal prisms, a phenomenon in accordance with the formation of vertical columns in horizontal beds of lava; for in both cases the divisions which give rise to the prismatic structure are at right angles to the cooling surfaces. (See Chapter 28.)

AUVERGNE.

Although the latest eruptions in central France seem to have long preceded the historical era, they are so modern as to have a very intimate connection with the present superficial outline of the country and with the existing valleys and river-courses. Among a great number of cones with perfect craters, one called the Puy de Tartaret sent forth a lava-current which can be traced up to its crater, and which flowed for a distance of thirteen miles along the bottom of the present valley to the village of Nechers, covering the alluvium of the old valley in which were preserved the bones of an extinct species of horse, and of a lagomys and other quadrupeds all closely allied to recent animals, while the associated land-shells were of species now living, such as Cyclostoma elegans, Helix hortensis, H. nemoralis, H. lapicida, and Clausilia rugosa. That the current which has issued from the Puy de Tartaret may, nevertheless, be very ancient in reference to the events of human history, we may conclude, not only from the divergence of the mammiferous fauna from that of our day, but from the fact that a Roman bridge of such form and construction as continued in use only down to the fifth century, but which may be older, is now seen at a place about a mile and a half from St. Nectaire. This ancient bridge spans the river Couze with two arches, each about fourteen feet wide. These arches spring from the lava of Tartaret, on both banks, showing that a ravine precisely like that now existing had already been excavated by the river through that lava thirteen or fourteen centuries ago.

While the river Couze has in most cases, as at the site of this ancient bridge, been simply able to cut a deep channel through the lava, the lower portion of which is shown to be columnar, the same torrent has in other places, where the valley was contracted to a narrow gorge, had power to remove the entire mass of basaltic rock, causing for a short space a complete breach of continuity in the volcanic current. The work of erosion has been very slow, as the basalt is tough and hard, and one column after another must have been undermined and reduced to pebbles, and then to sand. During the time required for this operation, the perishable cone of Tartaret, occupying the lowest part of the great valley descending from Mont Dor (see Chapter 30), and damming up the river so as to cause the Lake of Chambon, has stood uninjured, proving that no great flood or deluge can have passed over this region in the interval between the eruption of Tartaret and our own times.

PUY DE COME.

The Puy de Come and its lava-current, near Clermont, may be mentioned as another minor volcano of about the same age. This conical hill rises from the granitic platform, at an angle of between 30 and 40 degrees, to the height of more than 900 feet. Its summit presents two distinct craters, one of them with a vertical depth of 250 feet. A stream of lava takes its rise at the western base of the hill instead of issuing from either crater, and descends the granitic slope towards the present site of the town of Pont Gibaud. Thence it pours in a broad sheet down a steep declivity into the valley of the Sioule, filling the ancient river-channel for the distance of more than a mile. The Sioule, thus dispossessed of its bed, has worked out a fresh one between the lava and the granite of its western bank; and the excavation has disclosed, in one spot, a wall of columnar basalt about fifty feet high. (Scrope's Central France page 60 and plate.)

The excavation of the ravine is still in progress, every winter some columns of basalt being undermined and carried down the channel of the river, and in the course of a few miles rolled to sand and pebbles. Meanwhile the cone of Come remains unimpaired, its loose materials being protected by a dense vegetation, and the hill standing on a ridge not commanded by any higher ground, so that no floods of rain-water can descend upon it. There is no end to the waste which the hard basalt may undergo in future, if the physical geography of the country continue unchanged— no limit to the number of years during which the heap of incoherent and transportable materials called the Puy de Come may remain in an almost stationary condition.

PUY DE PARIOU.

The brim of the crater of the Puy de Pariou, near Clermont, is so sharp, and has been so little blunted by time, that it scarcely affords room to stand upon. This and other cones in an equally remarkable state of integrity have stood, I conceive, uninjured, not IN SPITE of their loose porous nature, as might at first be naturally supposed, but in consequence of it. No rills can collect where all the rain is instantly absorbed by the sand and scoriae, as is remarkably the case on Etna; and nothing but a water-spout breaking directly upon the Puy de Pariou could carry away a portion of the hill, so long as it is not rent or ingulfed by earthquakes.

NEWER PLIOCENE VOLCANIC ROCKS.

The more ancient portion of Vesuvius and Etna originated at the close of the Newer Pliocene period, when less than ten, sometimes only one, in a hundred of the shells differed from those now living. In the case of Etna, it was before stated (Chapter 13) that Post-pliocene formations occur in the neighbourhood of Catania, while the oldest lavas of the great volcano are Pliocene. These last are seen associated with sedimentary deposits at Trezza and other places on the southern and eastern flanks of the great cone (see Chapter 13).

CYCLOPEAN ISLANDS.

The Cyclopean Islands, called by the Sicilians Dei Faraglioni, in the sea-cliffs of which these beds of clay, tuff, and associated lava are laid open to view, are situated in the Bay of Trezza, and may be regarded as the extremity of a promontory severed from the main land. Here numerous proofs are seen of submarine eruptions, by which the argillaceous and sandy strata were invaded and cut through, and tufaceous breccias formed. Inclosed in these breccias are many angular and hardened fragments of laminated clay in different states of alteration by heat, and intermixed with volcanic sands.

(FIGURE 599. View of the Isle of Cyclops, in the Bay of Trezza. (Drawn by Captain Basil Hall, R.N.))

The loftiest of the Cyclopean islets, or rather rocks, is about 200 feet in height, the summit being formed of a mass of stratified clay, the laminae of which are occasionally subdivided by thin arenaceous layers. These strata dip to the N.W., and rest on a mass of columnar lava (see Figure 599) in which the tops of the pillars are weathered, and so rounded as to be often hemispherical. In some places in the adjoining and largest islet of the group, which lies to the north-eastward of that represented in Figure 599), the overlying clay has been greatly altered and hardened by the igneous rock, and occasionally contorted in the most extraordinary manner; yet the lamination has not been obliterated, but, on the contrary, rendered much more conspicuous, by the indurating process.

(FIGURE 600. Contortions of strata in the largest of the Cyclopean Islands.)

(FIGURE 601. Newer Pliocene strata invaded by lava. Isle of Cyclops (horizontal section). a. Lava. b. Laminated clay and sand. c. The same altered.)

In Figure 600 I have represented a portion of the altered rock, a few feet square, where the alternating thin laminae of sand and clay are contorted in a manner often observed in ancient metamorphic schists. A great fissure, running from east to west, nearly divides this larger island into two parts, and lays open its internal structure. In the section thus exhibited, a dike of lava is seen, first cutting through an older mass of lava, and then penetrating the superincumbent tertiary strata. In one place the lava ramifies and terminates in thin veins, from a few feet to a few inches in thickness (see Figure 601). The arenaceous laminae are much hardened at the point of contact, and the clays are converted into siliceous schist. In this island the altered rocks assume a honey-comb structure on their weathered surface, singularly contrasted with the smooth and even outline which the same beds present in their usual soft and yielding state. The pores of the lava are sometimes coated, or entirely filled with carbonate of lime, and with a zeolite resembling analcime, which has been called cyclopite. The latter mineral has also been found in small fissures traversing the altered marl, showing that the same cause which introduced the minerals into the cavities of the lava, whether we suppose sublimation or aqueous infiltration, conveyed it also into the open rents of the contiguous sedimentary strata.

DIKES OF PALAGONIA.

(FIGURES 602 and 603. Ground-plan of dikes near Palagonia.)

(FIGURE 602. Ground-plan of dikes near Palagonia. a. Lava. b. Peperino, consisting of volcanic sand, mixed with fragments of lava and limestone.)

(FIGURE 603. Ground-plan of dikes near Palagonia. a. Lava. b. Peperino, consisting of volcanic sand, mixed with fragments of lava and limestone.))

Dikes of vesicular and amygdaloidal lava are also seen traversing marine tuff or peperino, west of Palagonia, some of the pores of the lava being empty, while others are filled with carbonate of lime. In such cases we may suppose the tuff to have resulted from showers of volcanic sand and scoriae, together with fragments of limestone, thrown out by a submarine explosion, similar to that which gave rise to Graham Island in 1831. When the mass was, to a certain degree, consolidated, it may have been rent open, so that the lava ascended through fissures, the walls of which were perfectly even and parallel. In one case, after the melted matter that filled the rent (Figure 602) had cooled down, it must have been fractured and shifted horizontally by a lateral movement.

In Figure 603, the lava has more the appearance of a vein, which forced its way through the peperino. It is highly probable that similar appearances would be seen, if we could examine the floor of the sea in that part of the Mediterranean where the waves have recently washed away the new volcanic island; for when a superincumbent mass of ejected fragments has been removed by denudation, we may expect to see sections of dikes traversing tuff, or, in other words, sections of the channels of communication by which the subterranean lavas reached the surface.

MADEIRA.

Although the more ancient portion of the volcanic eruptions by which the island of Madeira and the neighbouring one of Porto Santo were built up occurred, as we shall presently see, in the Upper Miocene Period, a still larger part of the island is of Pliocene date. That the latest outbreaks belonged to the Newer Pliocene Period, I infer from the close affinity to the present flora of Madeira of the fossil plants preserved in a leaf-bed in the north-eastern part of the island. These fossils, associated with some lignite in the ravine of the river San Jorge, can none of them be proved to be of extinct species, but their antiquity may be inferred from the following considerations: Firstly— The leaf- bed, discovered by Mr. Hartung and myself in 1853, at the height of 1000 feet above the level of the sea, crops out at the base of a cliff formed by the erosion of a gorge cut through alternating layers of basalt and scoriae, the product of a vast succession of eruptions of unknown date, piled up to a thickness of 1000 feet, and which were all poured out after the plants, of which about twenty species have been recognised, flourished in Madeira. These lavas are inclined at an angle of about 15 degrees to the north, and came down from the great central region of eruption. Their accumulation implies a long period of intermittent volcanic action, subsequently to which the ravine of San Jorge was hollowed out. Secondly— Some few of the plants, though perhaps all of living species, are supposed to be of genera not now existing in the island. They have been described by Sir Charles Bunbury and Professor Heer, and the former first pointed out that many of the leaves are of the laurel type, and analogous to those now flourishing in the modern forests of Madeira. He also recognised among them the leaves of Woodwardia radicans, and Davallia Canariensis, ferns now abundant in Madeira. Thirdly— the great age of this leaf-bed of San Jorge, which was perhaps originally formed in the crater of some ancient volcanic cone afterwards buried under lava, is proved by its belonging to a part of the eastern extremity of Madeira, which, after the close of the igneous eruptions, became covered in the adjoining district of Canical with blown sand in which a vast number of land-shells were buried. These fossil shells belonged to no less than 36 species, among which are many now extremely rare in the island, and others, about five per cent, extinct or unknown in any part of the world. Several of these of the genus Helix are conspicuous from the peculiarity of their forms, others from their large dimensions. The geographical configuration of the country shows that this shell-bed is considerably more modern than the leaf-bed; it must therefore be referred to the Newer Pliocene, according to the definition of this period given in Chapter 9.

OLDER PLIOCENE PERIOD.— ITALY.

In Tuscany, as at Radicofani, Viterbo, and Aquapendente, and in the Campagna di Roma, submarine volcanic tuffs are interstratified with the Older Pliocene strata of the Sub-apennine hills in such a manner as to leave no doubt that they were the products of eruptions which occurred when the shelly marls and sands of the Sub-appenine hills were in the course of deposition. This opinion I expressed after my visit to Italy in 1828 (See 1st edition of Principles of Geology volume 3 chapters 8 and 14 1833 and former editions of this work chapter 31.), and it has recently (1850) been confirmed by the argument adduced by Sir R. Murchison in favour of the submarine origin of the tertiary volcanic rocks of Italy. (Quarterly Geological Journal volume 6 page 281.) These rocks are well- known to rest conformably on the Sub-apennine marls, even as far south as Monte Mario, in the suburbs of Rome. On the exact age of the deposits of Monte Mario new light has recently been thrown by a careful study of their marine fossil shells, undertaken by MM. Rayneval, Van den Hecke, and Ponzi. They have compared no less than 160 species with the shells of the

Coralline Crag of Suffolk, so well described by Mr. Searles Wood; and the specific agreement between the British and Italian fossils is so great, if we make due allowance for geographical distance and the difference of latitude, that we can have little hesitation in referring both to the same period, or to the Older Pliocene of this work. It is highly probable that, between the oldest trachytes of Tuscany and the newest rocks in the neighbourhood of Naples, a series of volcanic products might be detected of every age from the Older Pliocene to the historical epoch.

PLIOCENE VOLCANOES OF THE EIFEL.

Some of the most perfect cones and craters in Europe, not even excepting those of the district round Vesuvius, may be seen on the left or west bank of the Rhine, near Bonn and Andernach. They exhibit characters distinct from any which I have observed elsewhere, owing to the large part which the escape of aqueous vapour has played in the eruptions and the small quantities of lava emitted. The fundamental rocks of the district are grey and red sandstones and shales, with some associated limestones, replete with fossils of the Devonian or Old Red Sandstone group. The volcanoes broke out in the midst of these inclined strata, and when the present systems of hills and valleys had already been formed. The eruptions occurred sometimes at the bottom of deep valleys, sometimes on the summit of hills, and frequently on intervening platforms. In travelling through this district we often come upon them most unexpectedly, and may find ourselves on the very edge of a crater before we had been led to suspect that we were approaching the site of any igneous outburst. Thus, for example, on arriving at the village of Gemund, immediately south of Daun, we leave the stream, which flows at the bottom of a deep valley in which strata of sandstone and shale crop out. We then climb a steep hill, on the surface of which we see the edges of the same strata dipping inward towards the mountain. When we have ascended to a considerable height, we see fragments of scoriae sparingly scattered over the surface; until at length, on reaching the summit, we find ourselves suddenly on the edge of a tarn, or deep circular lake-basin called the Gemunder Maar. In it we recognise the ordinary form of a crater, for which we have been prepared by the occurrence of scoriae scattered over the surface of the soil. But on examining the walls of the crater we find precipices of sandstone and shale which exhibit no signs of the action of heat; and we look in vain for those beds of lava and scoriae, dipping outward on every side, which we have been accustomed to consider as characteristic of volcanic vents. As we proceed, however, to the opposite side of the lake, we find a considerable quantity of scoriae and some lava, and see the whole surface of the soil sparkling with volcanic sand, and strewed with ejected fragments of half-fused shale, which preserves its laminated texture in the interior, while it has a vitrified or scoriform coating.

Other crater lakes of circular or oval form, and hollowed out of similar ancient strata, occur in the Upper Eifel, where copious aeriform discharges have taken place, throwing out vast heaps of pulverized shale into the air. I know of no other extinct volcanoes where gaseous explosions of such magnitude have been attended by the emission of so small a quantity of lava. Yet I looked in vain in the Eifel for any appearances which could lend support to the hypothesis that the sudden rushing out of such enormous volumes of gas had ever lifted up the stratified rocks immediately around the vent so as to form conical masses, having their strata dipping outward on all sides from a central axis, as is assumed in the theory of elevation craters, alluded to in the last chapter.

I have already given (Figure 590) an example in the Eifel of a small stream of lava which issued from one of the craters of that district at Bertrich-Baden. It shows that when some of these volcanoes were in action the valleys had already been eroded to their present depth.

TRASS.

The tufaceous alluvium called trass, which has covered large areas in the Eifel, and choked up some valleys now partially re-excavated, is unstratified. Its base consists almost entirely of pumice, in which are included fragments of basalt and other lavas, pieces of burnt shale, slate, and sandstone, and numerous trunks and branches of trees. If, as is probable, this trass was formed during the period of volcanic eruptions, it may have originated in the manner of the moya of the Andes.

We may easily conceive that a similar mass might now be produced, if a copious evolution of gases should occur in one of the lake-basins. If a breach should be made in the side of the cone, the flood would sweep away great heaps of ejected fragments of shale and sandstone, which would be borne down into the adjoining valleys. Forests might be torn up by such a flood, and thus the occurrence of the numerous trunks of trees dispersed irregularly through the trass can be explained. The manner in which this trass conforms to the shape of the present valleys implies its comparatively modern origin, probably not dating farther back than the Pliocene Period.

CHAPTER XXX.
AGE OF VOLCANIC ROCKS CONTINUED.

Volcanic Rocks of the Upper Miocene Period.
Madeira.
Grand Canary.
Azores.
Lower Miocene Volcanic Rocks.
Isle of Mull.
Staffa and Antrim.
The Eifel.
Upper and Lower Miocene Volcanic Rocks of Auvergne.
Hill of Gergovia.
Eocene Volcanic Rocks of Monte Bolca.
Trap of Cretaceous Period.
Oolitic Period.
Triassic Period.
Permian Period.
Carboniferous Period.
Erect Trees buried in Volcanic Ash in the Island of Arran.
Old Red Sandstone Period.
Silurian Period.
Cambrian Period.
Laurentian Volcanic Rocks.

VOLCANIC ROCKS OF THE UPPER MIOCENE PERIOD.
MADEIRA.

The greater part of the volcanic eruptions of Madeira, as we have already seen (Chapter 29), belong to the Pliocene Period, but the most ancient of them are of Upper Miocene date, as shown by the fossil shells included in the marine tuffs which have been upraised at San Vicente, in the northern part of the island, to the height of 1300 feet above the level of the sea. A similar marine and volcanic formation constitutes the fundamental portion of the neighbouring island of Porto Santo, forty miles distant from Madeira, and is there elevated to an equal height, and covered, as in Madeira, with lavas of supra-marine origin.

The largest number of fossils have been collected from the tuffs and conglomerates and some beds of limestone in the island of Baixo, off the southern extremity of Porto Santo. They amount in this single locality to more than sixty in number, of which about fifty are mollusca, but many of these are only casts. Some of the shells probably lived on the spot during the intervals between eruptions, and some may have been cast up into the water or air together with muddy ejections, and, falling down again, have been deposited on the bottom of the sea. The hollows in some of the fragments of vesicular lava of which the breccias and conglomerates are composed are partially filled with calc-sinter, being thus half converted into amygdaloids. Among the fossil shells common to Madeira and Porto Santo, large cones, strombs, and cowries are conspicuous among the univalves, and Cardium, Spondylus, and Lithodomus among the lamellibranchiate bivalves, and among the Echinoderms the large Clypeaster called C. altus, an extinct European Miocene fossil.

The largest list of fossils has been published by Mr. Karl Meyer, in Hartung's "Madeira;" but in the collection made by myself, and in a still larger one formed by Mr. J. Yate Johnson, several remarkable forms not in Meyer's list occur, as, for example,

Pholadomya, and a large Terebra. Mr. Johnson also found a fine specimen of Nautilus (Atruria) ziczac (Figure 211), a well-known Falunian fossil of Europe; and in the same volcanic tuff of Baixo, the Echinoderm Brisus Scillae, a living Mediterranean species, found fossil in the Miocene strata of Malta. Mr. Meyer identifies one-third of the Madeira shells with known European Miocene (or Falunian) forms. The huge Strombus of San Vicente and Porto Santo, S. Italicus, is an extinct shell of the Sub-apennine or Older Pliocene formations. The mollusca already obtained from various localities of Madeira and Porto Santo are not less than one hundred in number, and, according to the late Dr. S.P. Woodward, rather more than a third are of species still living, but many of these are not now inhabitants of the neighbouring sea.

It has been remarked (Chapter 16), that in the Older Pliocene and Upper Miocene deposits of Europe many forms occur of a more southern aspect than those now inhabiting the nearest sea. In like manner the fossil corals, or Zoantharia, six in number, which I obtained from Madeira, of the genera Astraea, Sarcinula, Hydnophora, were pronounced by Mr. Lonsdale to be forms foreign to the adjacent coasts, and agreeing with the fauna of a sea warmer than that now separating Madeira from the nearest part of the African coast. We learn, indeed, from the observations made in 1859, by the Reverend R.T. Lowe, that more than one-half, or fifty-three in ninety, of the marine mollusks collected by him from the sandy beach of Mogador are common British species, although Mogador is 18 1/2 degrees south of the nearest shores of England. The living shells of Madeira and Porto Santo are in like manner those of a temperate climate, although in great part differing specifically from those of Mogador. (Linnean Proceedings Zoology 1860.)

GRAND CANARY.

In the Canaries, especially in the Grand Canary, the same marine Upper Miocene formation is found. Stratified tuffs, with intercalated conglomerates and lavas, are there seen in nearly horizontal layers in sea-cliffs about 300 feet high, near Las Palmas. Mr. Hartung and I were unable to find marine shells in these tuffs at a greater elevation than 400 feet above the sea; but as the deposit to which they belong reaches to the height of 1100 feet or more in the interior, we conceive that an upheaval of at least that amount has taken place. The Clypeaster altus, Spondylus gaederopus, Pectunculus pilosus, Cardita calyculata, and several other shells, serve to identify this formation with that of the Madeiras, and Ancillaria glandiformis, which is not rare, and some other fossils, remind us of the faluns of Touraine.

The sixty-two Miocene species which I collected in the Grand Canary were referred by the late Dr. S.P. Woodward to forty-seven genera, ten of which are no longer represented in the neighbouring sea, namely Corbis, an African form, Hinnites, now living in Oregon, Thecidium (T. Mediterranean, identical with the Miocene fossil of St. Juvat, in Brittany), Calyptraea, Hipponyx, Nerita, Erato, Oliva, Ancillaria, and Fasciolaria.

These tuffs of the southern shores of the Grand Canary, containing the Upper Miocene shells, appear to be about the same age as the most ancient volcanic rocks of the island, composed of slaty diabase, phonolite, and trachyte. Over the marine lavas and tuffs trachytic and basaltic products of subaerial volcanic origin, between 4000 and 5000 feet in thickness, have been piled, the central parts of the Grand Canary reaching the height of about 6000 feet above the level of the sea. A large portion of this mass is of Pliocene date, and some of the latest lavas have been poured out since the time when the valleys were already excavated to within a few feet of their present depth.

On the whole, the rocks of the Grand Canary, an island of a nearly circular shape, and 6 1/2 geographical miles diameter, exhibit proofs of a long series of eruptions beginning like those of Madeira, Porto Santo, and the Azores, in the Upper Miocene period, and continued to the Post-Pliocene. The building up of the Grand Canary by subaerial eruptions, several thousand feet thick, went on simultaneously with the gradual upheaval of the earliest products of submarine eruptions, in the same manner as the Pliocene marine strata of the oldest parts of Vesuvius and Etna have been upraised during eruptions of Post-tertiary date.

In proof that movements of elevation have actually continued down to Post- tertiary times, I may remark that I found raised beaches containing shells of the Recent Period in the Grand Canary, Teneriffe, and Porto Santo. The most remarkable raised beach which I observed in the Grand Canary, in the study of which I was assisted by Don Pedro Maffiotte, is situated in the north-eastern part of the island at San Catalina, about a quarter of a mile north of Las Palmas. It intervenes between the base of the high cliff formed of the tuffs with Miocene shells and the sea-shore. From this beach, at an elevation of twenty-five feet above high-water mark, and at a distance of about 150 feet from the present shore, I obtained more than fifty species of living marine shells. Many of them, according to Dr. S.P. Woodward, are no longer inhabitants of the contiguous sea, as, for example, Strombus bubonius, which is still living on the West Coast of Africa, and Cerithium procerum, found at Mozambique; others are Mediterranean species, as Pecten Jacobaeus and P. polymorphus. Some of these testacea, such as Cardita squamosa, are inhabitants of deep water, and the deposit on the whole seems to indicate a depth of water exceeding a hundred feet.

AZORES.

In the island of St. Mary's, one of the Azores, marine fossil shells have long been known. They are found on the north-east coast on a small projecting promontory called Ponta do Papagaio (or Point-Parrot), chiefly in a limestone about twenty feet thick, which rests upon, and is again covered by, basaltic lavas, scoriae, and conglomerates. The pebbles in the conglomerate are cemented together with carbonate of lime.

Mr. Hartung, in his account of the Azores, published in 1860, describes twenty- three shells from St. Mary's (Hartung Die Azoren 1860 also Insel Gran Canaria, Madeira und Porto Santo 1864 Leipsig.), of which eight perhaps are identical with living species, and twelve are with more or less certainty referred to European Tertiary forms, chiefly Upper Miocene. One of the most characteristic and abundant of the new species, Cardium Hartungi, not known as fossil in Europe, is very common in Porto Santo and Baixo, and serves to connect the Miocene fauna of the Azores and the Madeiras. In some of the Azores, as well as in the Canary islands, the volcanic fires are not yet extinct, as the recorded eruptions of Lanzerote, Teneriffe, Palma, St. Michael's, and others, attest.

LOWER MIOCENE VOLCANIC ROCKS.
ISLE OF MULL AND ANTRIM.

I may refer the reader to the account already given (Chapter 15) of leaf-beds at Ardtun, in the Isle of Mull in the Hebrides, which bear a relation to the associated volcanic rocks of Lower Miocene date analogous to that which the Madeira leaf-bed, above described (Chapter 29), bears to the Pliocene lavas of that island. Mr. Geikie has shown that the volcanic rocks in Mull are above 3000 feet in thickness. There seems little doubt that the well-known columnar basalt of Staffa, as well as that of Antrim in Ireland, are of the same age, and not of higher antiquity, as once suspected.

THE EIFEL.

A large portion of the volcanic rocks of the Lower Rhine and the Eifel are coeval with the Lower Miocene deposits to which most of the "Brown-Coal" of Germany belongs. The Tertiary strata of that age are seen on both sides of the Rhine, in the neighbourhood of Bonn, resting unconformably on highly inclined and vertical strata of Silurian and Devonian rocks. The Brown-Coal formation of that region consists of beds of loose sand, sandstone, and conglomerate, clay with nodules of clay-iron-stone, and occasionally silex. Layers of light brown and sometimes black lignite are interstratified with the clays and sands, and often irregularly diffused through them. They contain numerous impressions of leaves and stems of trees, and are extensively worked for fuel, whence the name of the formation. In several places layers of trachytic tuff are interstratified, and in these tuffs are leaves of plants identical with those found in the brown-coal, showing that, during the period of the accumulation of the latter, some volcanic products were ejected. The igneous rocks of the Westerwald, and of the mountains called the Siebengebirge, consist partly of basaltic and partly of trachytic lavas, the latter being in general the more ancient of the two. There are many varieties of trachyte, some of which are highly crystalline, resembling a coarse-grained granite, with large separate crystals of feldspar. Trachytic tuff is also very abundant.

M. Von Dechen, in his work on the Siebengebirge, has given a copious list of the animal and vegetable remains of the fresh-water strata associated with the brown-coal of that part of Germany. (Geognost. Beschreib. des Siebengebirges am Rhein Bonn 1852.) Plants of the genera Flabellaria, Ceanothus, and Daphnogene, including D. cinnamomifolia (Figure 155), occur in these beds, with nearly 150 other plants. The fishes of the brown-coal near

Bonn are found in a bituminous shale, called paper-coal, from being divisible into extremely thin leaves. The individuals are very numerous; but they appear to belong to a small number of species, some of which were referred by Agassiz to the genera Leuciscus, Aspius, and Perca. The remains of frogs also, of extinct species, have been discovered in the paper-coal; and a complete series may be seen in the museum at Bonn, from the most imperfect state of the tadpole to that of the full-grown animal. With these a salamander, scarcely distinguishable from the recent species, has been found, and the remains of many insects.

UPPER AND LOWER MIOCENE VOLCANIC ROCKS OF AUVERGNE.

The extinct volcanoes of Auvergne and Cantal, in central France, seem to have commenced their eruptions in the Lower Miocene period, but to have been most active during the Upper Miocene and Pliocene eras. I have already alluded to the grand succession of events of which there is evidence in Auvergne since the last retreat of the sea (see Chapter 29).

The earliest monuments of the Tertiary Period in that region are lacustrine deposits of great thickness, in the lowest conglomerates of which are rounded pebbles of quartz, mica-schist, granite, and other non-volcanic rocks, without the slightest intermixture of igneous products. To these conglomerates succeed argillaceous and calcareous marls and limestones, containing Lower Miocene shells and bones of mammalia, the higher beds of which sometimes alternate with volcanic tuff of contemporaneous origin. After the filling up or drainage of the ancient lakes, huge piles of trachytic and basaltic rocks, with volcanic breccias, accumulated to a thickness of several thousand feet, and were superimposed upon granite, or the contiguous lacustrine strata. The greater portion of these igneous rocks appear to have originated during the Upper Miocene and Pliocene periods; and extinct quadrupeds of those eras, belonging to the genera Mastodon, Rhinoceros, and others, were buried in ashes and beds of alluvial sand and gravel, which owe their preservation to overspreading sheets of lava.

In Auvergne, the most ancient and conspicuous of the volcanic masses is Mont Dor, which rests immediately on the granitic rocks standing apart from the fresh-water strata. This great mountain rises suddenly to the height of several thousand feet above the surrounding platform, and retains the shape of a flattened and somewhat irregular cone, the slope of which is gradually lost in the high plain around. This cone is composed of layers of scoriae, pumice- stones, and their fine detritus, with interposed beds of trachyte and basalt, which descend often in uninterrupted sheets until they reach and spread themselves round the base of the mountain. (Scrope Central France page 98.) Conglomerates, also, composed of angular and rounded fragments of igneous rocks, are observed to alternate with the above; and the various masses are seen to dip off from the central axis, and to lie parallel to the sloping flanks of the mountain. The summit of Mont Dor terminates in seven or eight rocky peaks, where no regular crater can now be traced, but where we may easily imagine one to have existed, which may have been shattered by earthquakes, and have suffered degradation by aqueous agents. Originally, perhaps, like the highest crater of Etna, it may have formed an insignificant feature in the great pile, and, like it, may frequently have been destroyed and renovated.

Respecting the age of the great mass of Mont Dor, we can not come at present to any positive decision, because no organic remains have yet been found in the tuffs, except impressions of the leaves of trees of species not yet determined. It has already been stated (Chapter 15) that the earliest eruptions must have been posterior in origin to those grits and conglomerates of the fresh-water formation of the Limagne which contain no pebbles of volcanic rocks. But there is evidence at a few points, as in the hill of Gergovia, presently to be mentioned, that some eruptions took place before the great lakes were drained, while others occurred after the desiccation of those lakes, and when deep valleys had already been excavated through fresh-water strata.

The valley in which the cone of Tartaret, above-mentioned (Chapter 29), is situated affords an impressive monument of the very different dates at which the igneous eruptions of Auvergne have happened; for while the cone itself is of Post-Pliocene date, the valley is bounded by lofty precipices composed of sheets of ancient columnar trachyte and basalt, which once flowed from the summit of Mont Dor in some part of the Miocene period.

These Miocene lavas had accumulated to a thickness of nearly 1000 feet before the ravine was cut down to the level of the river Couze, a river which was at length dammed up by the modern cone and the upper part of its course transformed into a lake.

GERGOVIA.
(FIGURE 604. Hill of Gergovia.
Section through (bottom to top) White and green marls: Altered Marl: Dike: Altered Marl: Limestone and peperino: Tuffs: Blue marls: White and yellow marl: Basaltic capping.)

It has been supposed by some observers that there is an alternation of a contemporaneous sheet of lava with fresh-water strata in the hill of Gergovia, near Clermont. But this idea has arisen from the intrusion of the dike represented in Figure 604, which has altered the green and white marls both above and below. Nevertheless, there is a real alternation of volcanic tuff with strata containing Lower Miocene fresh-water shells, among others a Melania allied to M. inquinata (Figure 217), with a Melanopsis and a Unio; there can, therefore, be no doubt that in Auvergne some volcanic explosions took place before the drainage of the lakes, and at a time when the Lower Miocene species of animals and plants still flourished.

EOCENE VOLCANIC ROCKS.
MONTE BOLCA.

The fissile limestone of Monte Bolca, near Verona, has for many centuries been celebrated in Italy for the number of perfect Ichthyolites which it contains. Agassiz has described no less than 133 species of fossil fish from this single deposit, and the multitude of individuals by which many of the species are represented is attested by the variety of specimens treasured up in the principal museums of Europe. They have been all obtained from quarries worked exclusively by lovers of natural history, for the sake of the fossils. Had the lithographic stone of Solenhofen, now regarded as so rich in fossils, been in like manner quarried solely for scientific objects, it would have remained almost a sealed book to palaeontologists, so sparsely are the organic remains scattered through it. When I visited Monte Bolca, in company with Sir Roderick Murchison, in 1828, we ascertained that the fish-bearing beds were of Eocene date, containing well-known species of Nummulites, and that a long series of submarine volcanic eruptions, evidently contemporaneous, had produced beds of tuff, which are cut through by dikes of basalt. There is evidence here of a long series of submarine volcanic eruptions of Eocene date, and during some of them, as Sir R. Murchison has suggested, shoals of fish were probably destroyed by the evolution of heat, noxious gases, and tufaceous mud, just as happened when Graham's Island was thrown up between Sicily and Africa in 1831, at which time the waters of the Mediterranean were seen to be charged with red mud, and covered with dead fish over a wide area. (Principles of Geology chapter 26 9th edition page 432.)

Associated with the marls and limestones of Monte Bolca are beds containing lignite and shale with numerous plants, which have been described by Unger and Massalongo, and referred by them to the Eocene period. I have already cited (Chapter 16) Professor Heer's remark, that several of the species are common to Monte Bolca and the white clay of Alum Bay, a Middle Eocene deposit; and the same botanist dwells on the tropical character of the flora of Monte Bolca and its distinctness from the sub-tropical flora of the Lower Miocene of Switzerland and Italy, in which last there is a far more considerable mixture of forms of a temperate climate, such as the willow, poplar, birch, elm, and others. That scarcely any one of the Monte Bolca fish should have been found in any other locality in Europe, is a striking illustration of the extreme imperfection of the palaeontological record. We are in the habit of imagining that our insight into the geology of the Eocene period is more than usually perfect, and we are certainly acquainted with an almost unbroken succession of assemblages of shells passing one into the other from the era of the Thanet sands to that of the Bembridge beds or Paris gypsum. The general dearth, therefore, of fish in the different members of the Eocene series, Upper, Middle, and Lower, might induce a hasty reasoner to conclude that there was a poverty of ichthyic forms during this period; but when a local accident, like the volcanic eruptions of Monte Bolca, occurs, proofs are suddenly revealed to us of the richness and variety of this great class of vertebrata in the Eocene sea. The number of genera of Monte Bolca fish is, according to Agassiz, no less than seventy-five, twenty of them peculiar to that locality, and only eight common to the antecedent Cretaceous period. No less than forty-seven out of the seventy-five genera make their appearance for the first time in the Monte Bolca rocks, none of them having been met with as yet in the antecedent formations. They form a great contrast to the fish of the secondary strata, as, with the exception of the Placoids, they are all Teleosteans, only one genus, Pycnodus, belonging to the order of Ganoids, which form, as before stated, the vast majority of the ichthyolites entombed in the secondary are Mesozoic rocks.

CRETACEOUS PERIOD.

M. Virlet, in his account of the geology of the Morea, page 205, has clearly shown that certain traps in Greece are of Cretaceous date; as those, for example, which alternate conformably with cretaceous limestone and greensand between Kastri and Damala, in the Morea. They consist in great part of diallage rocks and serpentine, and of an amygdaloid with calcareous kernels, and a base of serpentine. In certain parts of the Morea, the age of these volcanic rocks is established by the following proofs: first, the lithographic limestones of the Cretaceous era are cut through by trap, and then a conglomerate occurs, at Nauplia and other places, containing in its calcareous cement many well-known fossils of the chalk and greensand, together with pebbles formed of rolled pieces of the same serpentinous trap, which appear in the dikes above alluded to.

PERIOD OF OOLITE AND LIAS.

Although the green and serpentinous trap-rocks of the Morea belong chiefly to the Cretaceous era, as before mentioned, yet it seems that some eruptions of similar rocks began during the Oolitic period (Boblaye and Virlet Morea page 23.); and it is probable that a large part of the trappean masses, called ophiolites in the Apennines, and associated with the limestone of that chain, are of corresponding age.

TRAP OF THE NEW RED SANDSTONE PERIOD.

In the southern part of Devonshire, trappean rocks are associated with New Red Sandstone, and, according to Sir H. De la Beche, have not been intruded subsequently into the sandstone, but were produced by contemporaneous volcanic action. Some beds of grit, mingled with ordinary red marl, resemble sands ejected from a crater; and in the stratified conglomerates occurring near Tiverton are many angular fragments of trap porphyry, some of them one or two tons in weight, intermingled with pebbles of other rocks. These angular fragments were probably thrown out from volcanic vents, and fell upon sedimentary matter then in the course of deposition. (De la Beche Geological Proceedings volume 2 page 198.)

TRAP OF THE PERMIAN PERIOD.

The recent investigations of Mr. Archibald Geikie in Ayrshire have shown that some of the volcanic rocks in that county are of Permian age, and it appears highly probable that the uppermost portion of Arthur's Seat in the suburbs of Edinburgh marks the site of an eruption of the same era.

TRAP OF THE CARBONIFEROUS PERIOD.

Two classes of contemporaneous trap-rocks occur in the coal-field of the Forth, in Scotland. The newest of these, connected with the higher series of coal- measures, is well exhibited along the shores of the Forth, in Fifeshire, where they consist of basalt with olivine, amygdaloid, greenstone, wacke, and tuff. They appear to have been erupted while the sedimentary strata were in a horizontal position, and to have suffered the same dislocations which those strata have subsequently undergone. In the volcanic tuffs of this age are found not only fragments of limestone, shale, flinty slate, and sandstone, but also pieces of coal. The other or older class of carboniferous traps are traced along the south margin of Stratheden, and constitute a ridge parallel with the Ochils, and extending from Stirling to near St. Andrews. They consist almost exclusively of greenstone, becoming, in a few instances, earthy and amygdaloidal. They are regularly interstratified with the sandstone, shale, and iron-stone of the lower coal-measures, and, on the East Lomond, with Mountain Limestone. I examined these trap-rocks in 1838, in the cliffs south of St. Andrews, where they consist in great part of stratified tuffs, which are curved, vertical, and contorted, like the associated coal-measures. In the tuff I found fragments of carboniferous shale and limestone, and intersecting veins of greenstone.

FIFE— FLISK DIKE.

A trap dike was pointed out to me by Dr. Fleming, in the parish of Flisk, in the northern part of the county of Fife, which cuts through the grey sandstone and shale, forming the lowest part of the Old Red Sandstone, but which may probably be of carboniferous date. It may be traced for many miles, passing through the amygdaloidal and other traps of the hill called Norman's Law in that parish. In its course it affords a good exemplification of the passage from the trappean into the Plutonic, or highly crystalline texture. Professor Gustavus Rose, to whom I submitted specimens of this dike, found it to be dolerite, and composed of greenish black augite and Labrador feldspar, the latter being the most abundant ingredient. A small quantity of magnetic iron, perhaps titaniferous, is also present. The result of this analysis is interesting, because both the ancient and modern lavas of Etna consist in like manner of augite, Labradorite, and titaniferous iron.

ERECT TREES BURIED IN VOLCANIC ASH AT ARRAN.

An interesting discovery was made in 1867 by Mr. E.A. Wunsch in the carboniferous strata of the north-eastern part of the island of Arran. In the sea-cliff about five miles north of Corrie, near the village of Laggan, strata of volcanic ash occur, forming a solid rock cemented by carbonate of lime and enveloping trunks of trees, determined by Mr. Binney to belong to the genera Sigillaria and Lepidodendron. Some of these trees are at right angles to the planes of stratification, while others are prostrate and accompanied by leaves and fruits of the same genera. I visited the spot in company with Mr. Wunsch in 1870, and saw that the trees with their roots, of which about fourteen had been observed, occur at two distinct levels in volcanic tuffs parallel to each other, and inclined at an angle of about 40 degrees, having between them beds of shale and coaly matter seven feet thick. It is evident that the trees were overwhelmed by a shower of ashes from some neighbouring volcanic vent, as Pompeii was buried by matter ejected from Vesuvius. The trunks, several of them from three to five feet in circumference, remained with their Stigmarian roots spreading through the stratum below, which had served as a soil. The trees must have continued for years in an upright position after they were killed by the shower of burning ashes, giving time for a partial decay of the interior, so as to afford hollow cylinders into which the spores of plants were wafted. These spores germinated and grew, until finally their stems were petrified by carbonate of lime like some of the remaining portions of the wood of the containing Sigillaria. Mr. Carruthers has discovered that sometimes the plants which had thus grown and become fossil in the inside of a single trunk belonged to several distinct genera. The fact that the tree-bearing deposits now dip at an angle of 40 degrees is the more striking, as they must clearly have remained horizontal and undisturbed during a long period of intermittent and contemporaneous volcanic action.

In some of the associated carboniferous shales, ferns and calamites occur, and all the phenomena of the successive buried forests remind us of the sections in Figures 439 and 440 of the Nova Scotia coal-measures, with this difference only, that in the case of the South Joggins the fossilisation of the trees was effected without the eruption of volcanic matter.

TRAP OF THE OLD RED SANDSTONE PERIOD.

By referring to the section explanatory of the structure of Forfarshire, already given (Chapter 5), the reader will perceive that beds of conglomerate, No. 3, occur in the middle of the Old Red Sandstone system, 1, 2, 3, 4. The pebbles in these conglomerates are sometimes composed of granitic and quartzose rocks, sometimes exclusively of different varieties of trap, which last, although purposely omitted in the section referred to, is often found either intruding itself in amorphous masses and dikes into the old fossiliferous tilestones, No. 4, or alternating with them in conformable beds. All the different divisions of the red sandstone, 1, 2, 3, 4, are occasionally intersected by dikes, but they are very rare in Nos. 1 and 2, the upper members of the group consisting of red shale and red sandstone. These phenomena, which occur at the foot of the Grampians, are repeated in the Sidlaw Hills; and it appears that in this part of Scotland volcanic eruptions were most frequent in the earlier part of the Old Red Sandstone period. The trap-rocks alluded to consist chiefly of feldspathic porphyry and amygdaloid, the kernels of the latter being sometimes calcareous, often chalcedonic, and forming beautiful agates. We meet also with claystone, greenstone, compact feldspar, and tuff. Some of these rocks look as if they had flowed as lavas over the bottom of the sea, and enveloped quartz pebbles which were lying there, so as to form conglomerates with a base of greenstone, as is seen in Lumley Den, in the Sidlaw Hills. On either side of the axis of this chain of hills (see Figure 55), the beds of massive trap, and the tuffs composed of volcanic sand and ashes, dip regularly to the south-east or north-west, conformably with the shales and sandstones.

But the geological structure of the Pentland Hills, near Edinburgh, shows that igneous rocks were there formed during the newer part of the Devonian or "Old Red" period. These hills are 1900 feet high above the sea, and consist of conglomerates and sandstones of Upper Devonian age, resting on the inclined edges of grits and slates of Lower Devonian and Upper Silurian date. The contemporaneous volcanic rocks intercalated in this Upper Old Red consist of feldspathic lavas, or feldstones, with associated tuffs or ashy beds. The lavas were some of them originally compact, others vesicular, and these last have been converted into amygdaloids. They consist chiefly of feldstone or compact feldspar. The Pentland Hills, say Messrs. Maclaren and Geikie, afford evidence that at the time of the Upper Old Red Sandstone, the district to the south-west of Edinburgh was for a long while the seat of a powerful volcano, which sent out massive streams of lava and showers of ash, and continued active until well-nigh the dawn of the Carboniferous period. (Maclaren Geology of Fife and Lothians. Geikie Transactions of the Royal Society Edinburgh 1860-1861.)

SILURIAN VOLCANIC ROCKS.

It appears from the investigations of Sir R. Murchison in Shropshire, that when the Lower Silurian strata of that country were accumulating, there were frequent volcanic eruptions beneath the sea; and the ashes and scoriae then ejected gave rise to a peculiar kind of tufaceous sandstone or grit, dissimilar to the other rocks of the Silurian series, and only observable in places where syenitic and other trap-rocks protrude. These tuffs occur on the flanks of the Wrekin and Caer Caradoc, and contain Silurian fossils, such as casts of encrinites, trilobites, and mollusca. Although fossiliferous, the stone resembles a sandy claystone of the trap family. (Murchison Silurian System etc. page 230.)

Thin layers of trap, only a few inches thick, alternate in some parts of Shropshire and Montgomeryshire with sedimentary strata of the Lower Silurian system. This trap consists of slaty porphyry and granular feldspar rock, the beds being traversed by joints like those in the associated sandstone, limestone, and shale, and having the same strike and dip. (Ibid. page 212.)

In Radnorshire there is an example of twelve bands of stratified trap, alternating with Silurian schists and flagstones, in a thickness of 350 feet. The bedded traps consist of feldspar porphyry, and other varieties; and the interposed Llandeilo flags are of sandstone and shale, with trilobites and graptolites. (Murchison Silurian System etc. page 325.)

The Snowdonian hills in Carnarvonshire consist in great part of volcanic tuffs, the oldest of which are interstratified with the Bala and Llandeilo beds. There are some contemporaneous feldspathic lavas of this era, which, says Professor Ramsay, alter the slates on which they repose, having doubtless been poured out over them, in a melted state, whereas the slates which overlie them having been subsequently deposited after the lava had cooled and consolidated, have entirely escaped alteration. But there are greenstones associated with the same formation, which, although they are often conformable to the slates, are in reality intrusive rocks. They alter the stratified deposits both above and below them, and when traced to great distances are sometimes seen to cut through the slates, and to send off branches. Nevertheless, these greenstones appear to belong, like the lavas, to the Lower Silurian period.

CAMBRIAN VOLCANIC ROCKS.

The Lingula beds in North Wales have been described as 5000 feet in thickness. In the upper portion of these deposits volcanic tuffs or ashy materials are interstratified with ordinary muddy sediment, and here and there associated with thick beds of feldspathic lava. These rocks form the mountains called the Arans and the Arenigs; numerous greenstones are associated with them, which are intrusive, although they often run in the lines of bedding for a space. "Much of the ash," says Professor Ramsay, "seems to have been subaerial. Islands, like Graham's Island, may have sometimes raised their craters for various periods above the water, and by the waste of such islands some of the ashy matter became waterworn, whence the ashy conglomerate. Viscous matter seems also to have been shot into the air as volcanic bombs, which fell among the dust and broken crystals (that often form the ashes) before perfect cooling and consolidation had taken place." (Quarterly Geological Journal volume 9 page 170 1852.)

LAURENTIAN VOLCANIC ROCKS.

The Laurentian rocks in Canada, especially in Ottawa and Argenteuil, are the oldest intrusive masses yet known. They form a set of dikes of a fine-grained dark greenstone or dolerite, composed of feldspar and pyroxene, with occasional scales of mica and grains of pyrites. Their width varies from a few feet to a hundred yards, and they have a columnar structure, the columns being truly at right angles to the plane of the dike. Some of the dikes send off branches. These dolerites are cut through by intrusive syenite, and this syenite, in its turn, is again cut and penetrated by feldspar porphyry, the base of which consists of petrosilex, or a mixture of orthoclase and quartz. All these trap- rocks appear to be of Laurentian date, as the Cambrian and Huronian rocks rest unconformably upon them. (Logan Geology of Canada 1863.) Whether some of the various conformable crystalline rocks of the Laurentian series, such as the coarse-grained granitoid and porphyritic varieties of gneiss, exhibiting scarcely any signs of stratification, and some of the serpentines, may not also be of volcanic origin, is a point very difficult to determine in a region which has undergone so much metamorphic action.

CHAPTER XXXI.
PLUTONIC ROCKS.

General Aspect of Plutonic Rocks.
Granite and its Varieties.
Decomposing into Spherical Masses.
Rude columnar Structure.
Graphic Granite.
Mutual Penetration of Crystals of Quartz and Feldspar.
Glass Cavities in Quartz of Granite.
Porphyritic, talcose, and syenitic Granite.
Schorlrock and Eurite.
Syenite.
Connection of the Granites and Syenites with the Volcanic Rocks.
Analogy in Composition of Trachyte and Granite.
Granite Veins in Glen Tilt, Cape of Good Hope, and Cornwall.
Metalliferous Veins in Strata near their Junction with Granite.
Quartz Veins.
Exposure of Plutonic Rocks at the surface due to Denudation.

The Plutonic rocks may be treated of next in order, as they are most nearly allied to the volcanic class already considered. I have described, in the first chapter, these Plutonic rocks as the unstratified division of the crystalline or hypogene formations, and have stated that they differ from the volcanic rocks, not only by their more crystalline texture, but also by the absence of tuffs and breccias, which are the products of eruptions at the earth's surface, whether thrown up into the air or the sea. They differ also by the absence of pores or cellular cavities, to which the expansion of the entangled gases gives rise in ordinary lava, never being scoriaceous or amygdaloidal, and never forming a porphyry with an uncrystalline base, nor alternating with tuffs.

From these and other peculiarities it has been inferred that the granites have been formed at considerable depths in the earth, and have cooled and crystallised slowly under great pressure, where the contained gases could not expand. The volcanic rocks, on the contrary, although they also have risen up from below, have cooled from a melted state more rapidly upon or near the surface. From this hypothesis of the great depth at which the granites originated, has been derived the name of "Plutonic rocks." The beginner will easily conceive that the influence of subterranean heat may extend downward from the crater of every active volcano to a great depth below, perhaps several miles or leagues, and the effects which are produced deep in the bowels of the earth may, or rather must, be distinct; so that volcanic and Plutonic rocks, each different in texture, and sometimes even in composition, may originate simultaneously, the one at the surface, the other far beneath it. The Plutonic formations also agree with the volcanic in having veins or ramifications proceeding from central masses into the adjoining rocks, and causing alterations in these last, which will be presently described. They also resemble trap in containing no organic remains; but they differ in being more uniform in texture, whole mountain masses of indefinite extent appearing to have originated under conditions precisely similar.

The two principal members of the Plutonic family of rocks are Granite and Syenite, each of which, with their varieties, bear very much the same relation to each other as the trachytes bear to the basalts. Granite is a compound of feldspar, quartz, and mica, the feldspars being rich in silica, which forms from 60 to 70 per cent of the whole aggregate. In Syenite quartz is rare or wanting, hornblende taking the place of mica, and the proportion of silica not exceeding 50 to 60 per cent.

(FIGURE 605. Mass of granite near the Sharp Tor, Cornwall.)

(FIGURE 606. Granite having a cuboidal and rude columnar structure, Land's End, Cornwall.)

Granite often preserves a very uniform character throughout a wide range of territory, forming hills of a peculiar rounded form, usually clad with a scanty vegetation. The surface of the rock is for the most part in a crumbling state, and the hills are often surmounted by piles of stones like the remains of a stratified mass, as in Figure 605, and sometimes like heaps of boulders, for which they have been mistaken. The exterior of these stones, originally quadrangular, acquires a rounded form by the action of air and water, for the edges and angles waste away more rapidly than the sides. A similar spherical structure has already been described as characteristic of basalt and other volcanic formations, and it must be referred to analogous causes, as yet but imperfectly understood. Although it is the general peculiarity of granite to assume no definite shapes, it is nevertheless occasionally subdivided by fissures, so as to assume a cuboidal, and even a columnar, structure. Examples of these appearances may be seen near the Land's End, in Cornwall. (See Figure 606.)

(FIGURES 607 and 608. Graphic granite.

(FIGURE 607. Graphic granite. Section parallel to the laminae.)

(FIGURE 608. Graphic granite. Section transverse to the laminae.))

Feldspar, quartz, and mica are usually considered as the minerals essential to granite, the feldspar being most abundant in quantity, and the proportion of quartz exceeding that of mica. These minerals are united in what is termed a confused crystallisation; that is to say, there is no regular arrangement of the crystals in granite, as in gneiss (see Figure 622), except in the variety termed graphic granite, which occurs mostly in granitic veins. This variety is a compound of feldspar and quartz, so arranged as to produce an imperfect laminar structure. The crystals of feldspar appear to have been first formed, leaving between them the space now occupied by the darker-coloured quartz. This mineral, when a section is made at right angles to the alternate plates of feldspar and quartz, presents broken lines, which have been compared to Hebrew characters. (See Figure 608.) The variety of granite called by the French Pegmatite, which is a mixture of quartz and common feldspar, usually with some small admixture of white silvery mica, often passes into graphic granite.

Ordinary granite, as well as syenite and eurite, usually contains two kinds of feldspar: First, the common, or orthoclase, in which potash is the prevailing alkali, and this generally occurs in large crystals of a white or flesh colour; and secondly, feldspar in smaller crystals, in which soda predominates, usually of a dead white or spotted, and striated like albite, but not the same in composition. (Delesse Ann. des Mines 1852 tome 3 page 409 and 1848 tome 13 page 675.)

As a general rule, quartz, in a compact or amorphous state, forms a vitreous mass, serving as the base in which feldspar and mica have crystallised; for although these minerals are much more fusible than silex, they have often imprinted their shapes upon the quartz. This fact, apparently so paradoxical, has given rise to much ingenious speculation. We should naturally have anticipated that, during the cooling of the mass, the flinty portion would be the first to consolidate; and that the different varieties of feldspar, as well as garnets and tourmalines, being more easily liquefied by heat, would be the last. Precisely the reverse has taken place in the passage of most granite aggregates from a fluid to a solid state, crystals of the more fusible minerals being found enveloped in hard, transparent, glassy quartz, which has often taken very faithful casts of each, so as to preserve even the microscopically minute striations on the surface of prisms of tourmaline. Various explanations of this phenomenon have been proposed by MM. de Beaumont, Fournet, and Durocher. They refer to M. Gaudin's experiments on the fusion of quartz, which show that silex, as it cools, has the property of remaining in a viscous state, whereas alumina never does. This "gelatinous flint" is supposed to retain a considerable degree of plasticity long after the granitic mixture has acquired a low temperature. Occasionally we find the quartz and feldspar mutually imprinting their forms on each other, affording evidence of the simultaneous crystallisation of both. (Bulletin 2e serie 4 1304; and d'Archiac Hist. des Progres de la Geol. 1 38.)

According to the experiments and observations of Gustavus Rose, the quartz of granite has the specific gravity of 2.6, which characterises silica when it is precipitated from a liquid solvent, and not that inferior density, namely, 2.3, which belongs to it when it cools in the laboratory from a state of fusion in what is called the dry way. By some it had been rashly inferred that the manner in which the consolidation of granite takes place is exceedingly different from the cooling of lavas, and that the intense heat supposed to be necessary for the production of mountain masses of Plutonic rocks might be dispensed with. But Mr. David Forbes informs me that silica can crystallise in the dry way, and he has found in quartz forming a constituent part of some trachytes, both from Guadeloupe and Iceland, glass cavities quite similar to those met with in genuine volcanic minerals.

These "glass cavities," which with many other kindred phenomena have been carefully studied by Mr. Sorby, are those in which a liquid, on cooling, has become first viscous and then solid without crystallising or undergoing a definite change in its physical structure. Other cavities which, like those just mentioned, are frequently discernible under the microscope in the minerals composing granitic rocks, are filled, some of them with gas or vapour, others with liquid, and by the movements of the bubbles thus included the distinctness of such cavities from those filled with a glassy substance can be tested. Mr. Sorby admits that the frequent occurrence of fluid cavities in the quartz of granite implies that water was almost always present in the formation of this rock; but the same may be said of almost all lavas, and it is now more than forty years since Mr. Scrope insisted on the important part which water plays in volcanic eruptions, being so intimately mixed up with the materials of the lava that he supposed it to aid in giving mobility to the fluid mass. It is well known that steam escapes for months, sometimes for years, from the cavities of lava when it is cooling and consolidating. As to the result of Mr. Sorby's experiments and speculations on this difficult subject, they may be stated in a few words. He concludes that the physical conditions under which the volcanic and granitic rocks originate are so far similar that in both cases they combine igneous fusion, aqueous solution, and gaseous sublimation— the proof, he says, of the operation of water in the formation of granite being quite as strong as of that of heat. (See Quarterly Geological Journal volume 14 pages 465, 488.)

When rocks are melted at great depths water must be present, for two reasons— First, because rainwater and seawater are always descending through fissured and porous rocks, and must at length find their way into the regions of subterranean heat; and secondly, because in a state of combination water enters largely into the composition of some of the most common minerals, especially those of the aluminous class. But the existence of water under great pressure affords no argument against our attributing an excessively high temperature to the mass with which it is mixed up. Bunsen, indeed, imagines that in Iceland water attains a white heat at a very moderate depth. To what extent some of the metamorphic rocks containing the same minerals as the granites may have been formed by hydrothermal action without the intervention of intense heat comparable to that brought into play in a volcanic eruption, will be considered when we treat of the metamorphic rocks in the thirty-third chapter.

PORPHYRITIC GRANITE.

(FIGURE 609. Porphyritic granite. Land's End, Cornwall.)

This name has been sometimes given to that variety in which large crystals of common feldspar, sometimes more than three inches in length, are scattered through an ordinary base of granite. An example of this texture may be seen in the granite of the Land's End, in Cornwall (Figure 609). The two larger prismatic crystals in this drawing represent feldspar, smaller crystals of which are also seen, similar in form, scattered through the base. In this base also appear black specks of mica, the crystals of which have a more or less perfect hexagonal outline. The remainder of the mass is quartz, the translucency of which is strongly contrasted to the opaqueness of the white feldspar and black mica. But neither the transparency of the quartz nor the silvery lustre of the mica can be expressed in the engraving.

The uniform mineral character of large masses of granite seems to indicate that large quantities of the component elements were thoroughly mixed up together, and then crystallised under precisely similar conditions. There are, however, many accidental, or "occasional," minerals, as they are termed, which belong to granite. Among these black schorl or tourmaline, actinolite, zircon, garnet, and fluor spar are not uncommon; but they are too sparingly dispersed to modify the general aspect of the rock. They show, nevertheless, that the ingredients were not everywhere exactly the same; and a still greater difference may be traced in the ever-varying proportions of the feldspar, quartz, and mica.

TALCOSE GRANITE

Talcose Granite, or Protogine of the French, is a mixture of feldspar, quartz, and talc. It abounds in the Alps, and in some parts of Cornwall, producing by its decomposition the kaolin or china clay, more than 12,000 tons of which are annually exported from that country for the potteries.

SCHORL-ROCK, AND SCHORLY GRANITE.

The former of these is an aggregate of schorl, or tourmaline, and quartz. When feldspar and mica are also present, it may be called schorly granite. This kind of granite is comparatively rare.

EURITE, FELDSTONE.

Eurite is a rock in which the ingredients of granite are blended into a finely granular mass, mica being usually absent, and, when present, in such minute flakes as to be invisible to the naked eye. It is sometimes called FELDSTONE, and when the crystals of feldspar are conspicuous it becomes FELDSPAR PORPHYRY. All these and other varieties of granite pass into certain kinds of trap— a circumstance which affords one of many arguments in favour of what is now the prevailing opinion, that the granites are also of igneous origin. The contrast of the most crystalline form of granite to that of the most common and earthy trap is undoubtedly great; but each member of the volcanic class is capable of becoming porphyritic, and the base of the porphyry may be more and more crystalline, until the mass passes to the kind of granite most nearly allied in mineral composition.

SYENITIC GRANITE.

The quadruple compound of quartz, feldspar, mica, and hornblende, may be termed Syenitic Granite, and forms a passage between the granites and the syenites.
This rock occurs in Scotland and in Guernsey.

SYENITE.

We now come to the second division of the Plutonic rocks, or those having less than 60 per cent of silica, and which, as before stated, are usually called syenitic. Syenite originally received its name from the celebrated ancient quarries of Syene, in Egypt. It differs from granite in having hornblende as a substitute for mica, and being without quartz. Werner at least considered syenite as a binary compound of feldspar and hornblende, and regarded quartz as merely one of its occasional minerals.

MIASCITE.

Miascite is one of the varieties of syenite most frequently spoken of; it is composed chiefly of orthoclase and nepheline, with hornblende and quartz as occasional accessory minerals. It derives its name from Miask, in the Ural Mountains, where it was first discovered by Gustavus Rose. ZIRCON-SYENITE is another variety closely allied to Miascite, but containing crystals of Zircon.

CONNECTION OF THE GRANITES AND SYENITES WITH THE VOLCANIC ROCKS.

The minerals which constitute alike the Plutonic and volcanic rocks consist, almost exclusively, of seven elements, namely, silica, alumina, magnesia, lime, soda, potash, and iron (see Table 28.1); and these may sometimes exist in about the same proportions in a porous lava, a compact trap, and a crystalline granite. The same lava, for example, may be glassy, or scoriaceous, or stony, or porphyritic, according to the more or less rapid rate at which it cools.

It would be easy to multiply examples and authorities to prove the gradation of the Plutonic into the trap rocks. On the western side of the Fiord of Christiania, in Norway, there is a large district of trap, chiefly greenstone- porphyry and syenitic-greenstone, resting on fossiliferous strata. To this, on its southern limit, succeeds a region equally extensive of syenite, the passage from the trappean to the crystalline Plutonic rock being so gradual that it is impossible to draw a line of demarkation between them.

"The ordinary granite of Aberdeenshire," says Dr. MacCulloch, "is the usual ternary compound of quartz, feldspar, and mica; though sometimes hornblende is substituted for the mica. But in many places a variety occurs which is composed simply of feldspar and hornblende; and in examining more minutely this duplicate compound, it is observed in some places to assume a fine grain, and at length to become undistinguishable from the greenstones of the trap family. It also passes in the same uninterrupted manner into a basalt, and at length into a soft claystone, with a schistose tendency on exposure, in no respect differing from those of the trap islands of the western coast." The same author mentions, that in Shetland a granite composed of hornblende, mica, feldspar, and quartz graduates in an equally perfect manner into basalt. (System of Geology volume 1 pages 157 and 158.) In Hungary there are varieties of trachyte, which, geologically speaking, are of modern origin, in which crystals, not only of mica, but of quartz, are common, together with feldspar and hornblende. It is easy to conceive how such volcanic masses may, at a certain depth from the surface, pass downward into granite.

GRANITIC VEINS.

(Figures 610 and 611. Junction of granite and argillaceous schist in Glen Tilt. (MacCulloch. (Geological Transactions First Series volume 3 plate 21.))

(FIGURE 610. Junction of granite and argillaceous schist in Glen Tilt.)
(FIGURE 611. Junction of granite and argillaceous schist in Glen Tilt.))

I have already hinted at the close analogy in the forms of certain granitic and trappean veins; and it will be found that strata penetrated by Plutonic rocks have suffered changes very similar to those exhibited near the contact of volcanic dikes. Thus, in Glen Tilt, in Scotland, alternating strata of limestone and argillaceous schist come in contact with a mass of granite. The contact does not take place as might have been looked for if the granite had been formed there before the strata were deposited, in which case the section would have appeared as in Figure 610; but the union is as represented in Figure 611, the undulating outline of the granite intersecting different strata, and occasionally intruding itself in torturous veins into the beds of clay-slate and limestone, from which it differs so remarkably in composition. The limestone is sometimes changed in character by the proximity of the granitic mass or its veins, and acquires a more compact texture, like that of hornstone or chert, with a splintery fracture, and effervescing freely with acids.

The conversion of the limestone and these and many other instances into a siliceous rock, effervescing slowly with acids, would be difficult of explanation, were it not ascertained that such limestones are always impure, containing grains of quartz, mica, or feldspar disseminated through them. The elements of these minerals, when the rock has been subjected to great heat, may have been fused, and so spread more uniformly through the whole mass.

(FIGURE 612. Granite veins traversing clay slate, Table Mountain, Cape of Good Hope. (Captain B. Hall Transactions of the Royal Society of Edinburgh volume 7.))

(FIGURE 613. Granite veins traversing gneiss, Cape Wrath. (MacCulloch (Western Islands plate 31.))

In the Plutonic, as in the volcanic rocks, there is every gradation from a torturous vein to the most regular form of a dike, such as intersect the tuffs and lavas of Vesuvius and Etna. Dikes of granite may be seen, among other places, on the southern flank of Mount Battock, one of the Grampians, the opposite walls sometimes preserving an exact parallelism for a considerable distance. As a general rule, however, granite veins in all quarters of the globe are more sinuous in their course than those of trap. They present similar shapes at the most northern point of Scotland, and the southernmost extremity of Africa, as Figures 612 and 613 will show.

It is not uncommon for one set of granite veins to intersect another; and sometimes there are three sets, as in the environs of Heidelberg, where the granite on the banks of the river Necker is seen to consist of three varieties, differing in colour, grain, and various peculiarities of mineral composition. One of these, which is evidently the second in age, is seen to cut through an older granite; and another, still newer, traverses both the second and the first. In Shetland there are two kinds of granite. One of them, composed of hornblende, mica, feldspar, and quartz, is of a dark colour, and is seen underlying gneiss. The other is a red granite, which penetrates the dark variety everywhere in veins. (MacCulloch System of Geology volume 2 page 58.)

(FIGURE 614. Granite veins passing through hornblende slate, Carnsilver Cove, Cornwall.)

Figure 614 is a sketch of a group of granite veins in Cornwall, given by Messrs. Von Oeynhausen and Von Dechen. (Philosophical Magazine and Annals No. 27 New Series March 1829.) The main body of the granite here is of a porphyritic appearance, with large crystals of feldspar; but in the veins it is fine- grained, and without these large crystals. The general height of the veins is from 16 to 20 feet, but some are much higher.

Granite, syenite, and those porphyries which have a granitiform structure, in short all Plutonic rocks, are frequently observed to contain metals, at or near their junction with stratified formations. On the other hand, the veins which traverse stratified rocks are, as a general law, more metalliferous near such junctions than in other positions. Hence it has been inferred that these metals may have been spread in a gaseous form through the fused mass, and that the contact of another rock, in a different state of temperature, or sometimes the existence of rents in other rocks in the vicinity, may have caused the sublimation of the metals. (Necker Proceedings of the Geological Society No. 26 page 392.)

(FIGURE 615. a, b. Quartz vein passing through gneiss and greenstone. Tronstad Strand, near Christiania.)

Veins of pure quartz are often found in granite as in many stratified rocks, but they are not traceable, like veins of granite or trap, to large bodies of rock of similar composition. They appear to have been cracks, into which siliceous matter was infiltered. Such segregation, as it is called, can sometimes clearly be shown to have taken place long subsequently to the original consolidation of the containing rock. Thus, for example, I observed in the gneiss of Tronstad Strand, near Drammen, in Norway, the section on the beach shown in Figure 615. It appears that the alternating strata of whitish granitiform gneiss and black hornblende-schist were first cut by a greenstone dike, about 2 1/2 feet wide; then the crack a-b passed through all these rocks, and was filled up with quartz. The opposite walls of the vein are in some parts incrusted with transparent crystals of quartz, the middle of the vein being filled up with common opaque white quartz.

(FIGURE 616. Euritic porphyry alternating with primary fossiliferous strata, near Christiania.)

We have seen that the volcanic formations have been called overlying, because they not only penetrate others but spread over them. M. Necker has proposed to call the granites the underlying igneous rocks, and the distinction here indicated is highly characteristic. It was, indeed, supposed by some of the earlier observers that the granite of Christiania, in Norway, was intercalated in mountain masses between the primary or palaeozoic strata of that country, so as to overlie fossiliferous shale and limestone. But although the granite sends veins into these fossiliferous rocks, and is decidedly posterior in origin, its actual superposition in mass has been disproved by Professor Keilhau, whose observations on this controverted point I had opportunities, in 1837, of verifying. There are, however, on a smaller scale, certain beds of euritic porphyry, some a few feet, others many yards in thickness, which pass into granite, and deserve, perhaps, to be classed as Plutonic rather than trappean rocks, which may truly be described as interposed conformably between fossiliferous strata, as the porphyries (a, c, Figure 616) which divide the bituminous shales and argillaceous limestones, f, f. But some of these same porphyries are partially unconformable, as b, and may lead us to suspect that the others also, notwithstanding their appearance of interstratification, have been forcibly injected. Some of the porphyritic rocks above mentioned are highly quartzose, others very feldspathic. In proportion as the masses are more voluminous, they become more granitic in their texture, less conformable, and even begin to send forth veins into contiguous strata. In a word, we have here a beautiful illustration of the intermediate gradations between volcanic and Plutonic rocks, not only in their mineralogical composition and structure, but also in their relations of position to associated formations. If the term "overlying" can in this instance be applied to a Plutonic rock, it is only in proportion as that rock begins to acquire a trappean aspect.

It has been already hinted that the heat which in every active volcano extends downward to indefinite depths must produce simultaneously very different effects near the surface and far below it; and we can not suppose that rocks resulting from the crystallising of fused matter under a pressure of several thousand feet, much less several miles, of the earth's crust can exactly resemble those formed at or near the surface. Hence the production at great depths of a class of rocks analogous to the volcanic, and yet differing in many particulars, might have been predicted, even had we no Plutonic formations to account for. How well these agree, both in their positive and negative characters, with the theory of their deep subterranean origin, the student will be able to judge by considering the descriptions already given.

It has, however, been objected, that if the granitic and volcanic rocks were simply different parts of one great series, we ought to find in mountain chains volcanic dikes passing upward into lava and downward into granite. But we may answer that our vertical sections are usually of small extent; and if we find in certain places a transition from trap to porous lava, and in others a passage from granite to trap, it is as much as could be expected of this evidence.

The prodigious extent of denudation which has been already demonstrated to have occurred at former periods, will reconcile the student to the belief that crystalline rocks of high antiquity, although deep in the earth's crust when originally formed, may have become uncovered and exposed at the surface. Their actual elevation above the sea may be referred to the same causes to which we have attributed the upheaval of marine strata, even to the summits of some mountain chains.

CHAPTER XXXII.
ON THE DIFFERENT AGES OF THE PLUTONIC ROCKS.

Difficulty in ascertaining the precise Age of a Plutonic Rock.
Test of Age by Relative Position.
Test by Intrusion and Alteration.
Test by Mineral Composition.
Test by included Fragments.
Recent and Pliocene Plutonic Rocks, why invisible.
Miocene Syenite of the Isle of Skye.
Eocene Plutonic Rocks in the Andes.
Granite altering Cretaceous Rocks.
Granite altering Lias in the Alps and in Skye.
Granite of Dartmoor altering Carboniferous Strata.
Granite of the Old Red Sandstone Period.
Syenite altering Silurian Strata in Norway.
Blending of the same with Gneiss.
Most ancient Plutonic Rocks.
Granite protruded in a solid Form.

When we adopt the igneous theory of granite, as explained in the last chapter, and believe that different Plutonic rocks have originated at successive periods beneath the surface of the planet, we must be prepared to encounter greater difficulty in ascertaining the precise age of such rocks than in the case of volcanic and fossiliferous formations. We must bear in mind that the evidence of the age of each contemporaneous volcanic rock was derived either from lavas poured out upon the ancient surface, whether in the sea or in the atmosphere, or from tuffs and conglomerates, also deposited at the surface, and either containing organic remains themselves or intercalated between strata containing fossils. But the same tests entirely fail, or are only applicable in a modified degree, when we endeavour to fix the chronology of a rock which has crystallised from a state of fusion in the bowels of the earth. In that case we are reduced to the tests of relative position, intrusion, alteration of the rocks in contact, included fragments, and mineral character; but all these may yield at best a somewhat ambiguous result.

TEST OF AGE BY RELATIVE POSITION.

Unaltered fossiliferous strata of every age are met with reposing immediately on Plutonic rocks; as at Christiania, in Norway, where the Post-pliocene deposits rest on granite; in Auvergne, where the fresh-water Miocene strata, and at Heidelberg, on the Rhine, where the New Red sandstone occupy a similar place. In all these, and similar instances, inferiority in position is connected with the superior antiquity of granite. The crystalline rock was solid before the sedimentary beds were superimposed, and the latter usually contain in them rounded pebbles of the subjacent granite.

TEST BY INTRUSION AND ALTERATION.

But when Plutonic rocks send veins into strata, and alter them near the point of contact, in the manner before described (Chapter 31), it is clear that, like intrusive traps, they are newer than the strata which they invade and alter. Examples of the application of this test will be given in the sequel.

TEST BY MINERAL COMPOSITION.

Notwithstanding a general uniformity in the aspect of Plutonic rocks, we have seen in the last chapter that there are many varieties, such as syenite, talcose granite, and others. One of these varieties is sometimes found exclusively prevailing throughout an extensive region, where it preserves a homogeneous character; so that, having ascertained its relative age in one place, we can recognise its identity in others, and thus determine from a single section the chronological relations of large mountain masses. Having observed, for example, that the syenitic granite of Norway, in which the mineral called zircon abounds, has altered the Silurian strata wherever it is in contact, we do not hesitate to refer other masses of the same zircon-syenite in the south of Norway to a post- Silurian date. Some have imagined that the age of different granites might, to a great extent, be determined by their mineral characters alone; syenite, for instance, or granite with hornblende, being more modern than common or micaceous granite. But modern investigations have proved these generalisations to have been premature.

TEST BY INCLUDED FRAGMENTS.

This criterion can rarely be of much importance, because the fragments involved in granite are usually so much altered that they can not be referred with certainty to the rocks whence they were derived. In the White Mountains, in North America, according to Professor Hubbard, a granite vein, traversing granite, contains fragments of slate and trap which must have fallen into the fissure when the fused materials of the vein were injected from below (Silliman's Journal No. 69 page 123.), and thus the granite is shown to be newer than those slaty and trappean formations from which the fragments were derived.

RECENT AND PLIOCENE PLUTONIC ROCKS, WHY INVISIBLE.

The explanations already given in the 28th and in the last chapter of the probable relation of the Plutonic to the volcanic formations, will naturally lead the reader to infer that rocks of the one class can never be produced at or near the surface without some members of the other being formed below. It is not uncommon for lava-streams to require more than ten years to cool in the open air; and where they are of great depth, a much longer period. The melted matter poured from Jorullo, in Mexico, in the year 1759, which accumulated in some places to the height of 550 feet, was found to retain a high temperature half a century after the eruption. (See Principles Index Jorullo.) We may conceive, therefore, that great masses of subterranean lava may remain in a red-hot or incandescent state in the volcanic foci for immense periods, and the process of refrigeration may be extremely gradual. Sometimes, indeed, this process may be retarded for an indefinite period by the accession of fresh supplies of heat; for we find that the lava in the crater of Stromboli, one of the Lipari Islands, has been in a state of constant ebullition for the last two thousand years; and we may suppose this fluid mass to communicate with some caldron or reservoir of fused matter below. In the Isle of Bourbon, also, where there has been an emission of lava once in every two years for a long period, the lava below can scarcely fail to have been permanently in a state of liquefaction. If then it be a reasonable conjecture, that about 2000 volcanic eruptions occur in the course of every century, either above the waters of the sea or beneath them (Ibid. Volcanic Eruptions.), it will follow that the quantity of Plutonic rock generated or in progress during the Recent epoch must already have been considerable.

But as the Plutonic rocks originate at some depth in the earth's crust, they can only be rendered accessible to human observation by subsequent upheaval and denudation. Between the period when a Plutonic rock crystallises in the subterranean regions and the era of its protrusion at any single point of the surface, one or two geological periods must usually intervene. Hence, we must not expect to find the Recent or even the Pliocene granites laid open to view, unless we are prepared to assume that sufficient time has elapsed since the commencement of the Pliocene period for great upheaval and denudation. A Plutonic rock, therefore, must, in general, be of considerable antiquity relatively to the fossiliferous and volcanic formations, before it becomes extensively visible. As we know that the upheaval of land has been sometimes accompanied in South America by volcanic eruptions and the emission of lava, we may conceive the more ancient Plutonic rocks to be forced upward to the surface by the newer rocks of the same class formed successively below— subterposition in the Plutonic, like superposition in the sedimentary rocks, being usually characteristic of a newer origin.

(FIGURE 617. Diagram showing the relative position which the Plutonic and sedimentary formations of different ages may occupy. I. Primary Plutonic rocks. II. Secondary Plutonic rocks. III. Tertiary Plutonic rocks. IV. Post-tertiary Plutonic rocks. 1. Primary fossiliferous or Palaeozoic strata. 2. Secondary or Mesozoic strata. 3. Tertiary or Cainozoic strata. 4. Post-tertiary strata. The metamorphic rocks are not indicated in this diagram: but the student will infer, from what is said in Chapters 31 and 33, that some portions of the stratified formations, Nos. 1 and 2, invaded by granite, will have become metamorphic.)

In Figure 617 an attempt is made to show the inverted order in which sedimentary and Plutonic formations may occur in the earth's crust. The oldest Plutonic rock, No. I, has been upheaved at successive periods until it has become exposed to view in a mountain-chain. This protrusion of No. I has been caused by the igneous agency which produced the newer Plutonic rocks Nos. II, III and IV. Part of the primary fossiliferous strata, No. I, have also been raised to the surface by the same gradual process. It will be observed that the Recent STRATA No. 4 and the Recent GRANITE or Plutonic rock No. IV are the most remote from each other in position, although of contemporaneous date. According to this hypothesis, the convulsions of many periods will be required before Recent or Post-tertiary granite will be upraised so as to form the highest ridges and central axes of mountain-chains. During that time the RECENT strata No. 4 might be covered by a great many newer sedimentary formations.

MIOCENE PLUTONIC ROCKS.

A considerable mass of syenite, in the Isle of Skye, is described by Dr. MacCulloch as intersecting limestone and shale, which are of the age of the lias. The limestone, which at a greater distance from the granite contains shells, exhibits no traces of them near its junction, where it has been converted into a pure crystalline marble. (Western Islands volume 1 page 330.) MacCulloch pointed out that the syenite here, as in Raasay, was newer than the secondary rocks, and Mr. Geikie has since shown that there is a strong probability that this Plutonic rock may be of Miocene age, because a similar Syenite having a true granitic character in its crystallisation has modified the Tertiary volcanic rocks of Ben More, in Mull, some of which have undergone considerable metamorphism.

EOCENE PLUTONIC ROCKS.

In a former part of this volume (Chapter 16), the great nummulitic formation of the Alps and Pyrenees was referred to the Eocene period, and it follows that vast movements which have raised those fossiliferous rocks from the level of the sea to the height of more than 10,000 feet above its level have taken place since the commencement of the Tertiary epoch. Here, therefore, if anywhere, we might expect to find hypogene formations of Eocene date breaking out in the central axis or most disturbed region of the loftiest chain in Europe. Accordingly, in the Swiss Alps, even the flysch, or upper portion of the nummulitic series, has been occasionally invaded by Plutonic rocks, and converted into crystalline schists of the hypogene class. There can be little doubt that even the talcose granite or gneiss of Mont Blanc itself has been in a fused or pasty state since the flysch was deposited at the bottom of the sea; and the question as to its age is not so much whether it be a secondary or tertiary granite or gneiss, as whether it should be assigned to the Eocene or Miocene epoch.

Great upheaving movements have been experienced in the region of the Andes, during the Post-tertiary period. In some part, therefore, of this chain, we may expect to discover tertiary Plutonic rocks laid open to view; and Mr. Darwin's account of the Chilian Andes, to which the reader may refer, fully realises this expectation: for he shows that we have strong ground to presume that Plutonic rocks there exposed on a large scale are of later date than certain Secondary and Tertiary formations.

But the theory adopted in this work of the subterranean origin of the hypogene formations would be untenable, if the supposed fact here alluded to, of the appearance of tertiary granite at the surface, was not a rare exception to the general rule. A considerable lapse of time must intervene between the formation of Plutonic and metamorphic rocks in the nether regions and their emergence at the surface. For a long series of subterranean movements must occur before such rocks can be uplifted into the atmosphere or the ocean; and, before they can be rendered visible to man, some strata which previously covered them must have been stripped off by denudation.

We know that in the Bay of Baiae in 1538, in Cutch in 1819, and on several occasions in Peru and Chili, since the commencement of the present century, the permanent upheaval or subsidence of land has been accompanied by the simultaneous emission of lava at one or more points in the same volcanic region. From these and other examples it may be inferred that the rising or sinking of the earth's crust, operations by which sea is converted into land, and land into sea, are a part only of the consequences of subterranean igneous action. It can scarcely be doubted that this action consists, in a great degree, of the baking, and occasionally the liquefaction, of rocks, causing them to assume, in some cases a larger, in others a smaller volume than before the application of heat. It consists also in the generation of gases, and their expansion by heat, and the injection of liquid matter into rents formed in superincumbent rocks. The prodigious scale on which these subterranean causes have operated in Sicily since the deposition of the Newer Pliocene strata will be appreciated when we remember that throughout half the surface of that island such strata are met with, raised to the height of from 50 to that of 2000 and even 3000 feet above the level of the sea. In the same island also the older rocks which are contiguous to these marine tertiary strata must have undergone, within the same period, a similar amount of upheaval.

The like observations may be extended to nearly the whole of Europe, for, since the commencement of the Eocene Period, the entire European area, including some of the central and very lofty portions of the Alps themselves, as I have elsewhere shown, has, with the exception of a few districts, emerged from the deep to its present altitude. (See map of Europe, and explanation, in Principles book 1.) There must, therefore, have been at great depths in the earth's crust, within the same period, an amount of subterranean change corresponding to this vast alteration of level affecting a whole continent.

The principal effect of subterranean movements during the Tertiary Period seems to have consisted in the upheaval of hypogene formations of an age anterior to the Carboniferous. The repetition of another series of movements, of equal violence, might upraise the Plutonic and metamorphic rocks of many secondary periods; and, if the same force should still continue to act, the next convulsions might bring up to the day the TERTIARY and RECENT hypogene rocks. In the course of such changes many of the existing sedimentary strata would suffer greatly by denudation, others might assume a metamorphic structure, or become melted down into Plutonic and volcanic rocks. Meanwhile the deposition of a great thickness of new strata would not fail to take place during the upheaval and partial destruction of the older rocks. But I must refer the reader to the last chapter but one of this volume for a fuller explanation of these views.

PLUTONIC ROCKS OF CRETACEOUS PERIOD.

(FIGURE 618. Section through three layers (b, c, d) of the Cretaceous series over granite (A).)

It will be shown in the next chapter that chalk, as well as lias, has been altered by granite in the eastern Pyrenees. Whether such granite be cretaceous or tertiary, can not easily be decided. Suppose b, c, d, Figure 618, to be three members of the Cretaceous series, the lowest of which, b, has been altered by the granite A, the modifying influence not having extended so far as c, or having but slightly affected its lowest beds. Now it can rarely be possible for the geologist to decide whether the beds d existed at the time of the intrusion of A, and alteration of b and c, or whether they were subsequently thrown down upon c. But as some Cretaceous and even Tertiary rocks have been raised to the height of more than 9000 feet in the Pyrenees, we must not assume that plutonic formations of the same periods may not have been brought up and exposed by denudation, at the height of 2000 or 3000 feet on the flanks of that chain.

PLUTONIC ROCKS OF THE OOLITE AND LIAS.

(FIGURE 619. Junction of granite with Jurassic or Oolite strata in the Alps, near Champoleon. (Granite over Altered Rocks over Secondary Schists.))

In the Department of the Hautes Alpes, in France, M. Eliede Beaumont traced a black argillaceous limestone, charged with belemnites, to within a few yards of a mass of granite. Here the limestone begins to put on a granular texture, but is extremely fine-grained. When nearer the junction it becomes grey, and has a saccharoid structure. In another locality, near Champoleon, a granite composed of quartz, black mica, and rose-coloured feldspar is observed partly to overlie the secondary rocks, producing an alteration which extends for about 30 feet downward, diminishing in the beds which lie farthest from the granite. (See Figure 619.) In the altered mass the argillaceous beds are hardened, the limestone is saccharoid, the grits quartzose, and in the midst of them is a thin layer of an imperfect granite. It is also an important circumstance that near the point of contact, both the granite and the secondary rocks become metalliferous, and contain nests and small veins of blende, galena, iron, and copper pyrites. The stratified rocks become harder and more crystalline, but the granite, on the contrary, softer and less perfectly crystallised near the junction. (Elie de Beaumont sur les Montagnes de l'Oisans etc. Mem. de la Soc. d'Hist. Nat. de Paris tome 5.) Although the granite is incumbent in the section (Figure 619), we can not assume that it overflowed the strata, for the disturbances of the rocks are so great in this part of the Alps that their original position is often inverted.

At Predazzo, in the Tyrol, secondary strata, some of which are limestones of the Oolitic period, have been traversed and altered by Plutonic rocks, one portion of which is an augitic porphyry, which passes insensibly into granite. The limestone is changed into granular marble, with a band of serpentine at the junction. (Von Buch Annales de Chimie etc.)

PLUTONIC ROCKS OF CARBONIFEROUS PERIOD.

The granite of Dartmoor, in Devonshire, was formerly supposed to be one of the most ancient of the Plutonic rocks, but is now ascertained to be posterior in date to the culm-measures of that county, which from their position, and, as containing true coal-plants, are now known to be members of the true Carboniferous series. This granite, like the syenitic granite of Christiania, has broken through the stratified formations, on the north-west side of Dartmoor, the successive members of the culm-measures abutting against the granite, and becoming metamorphic as they approach. These strata are also penetrated by granite veins, and Plutonic dikes, called "elvans." (Proceedings of the Geological Society volume 2 page 562 and Transactions second series volume 5 page 686.) The granite of Cornwall is probably of the same date, and, therefore, as modern as the Carboniferous strata, if not newer.

PLUTONIC ROCKS OF SILURIAN PERIOD.

(FIGURE 620. Section through Silurian strata and Granite.)

It has long been known that a very ancient granite near Christiania, in Norway, is posterior in date to the Lower Silurian strata of that region, although its exact position in the Palaeozoic series can not be defined. Von Buch first announced, in 1813, that it was of newer origin than certain limestones containing orthocerata and trilobites. The proofs consist in the penetration of granite veins into the shale and limestone, and the alteration of the strata, for a considerable distance from the point of contact, both of these veins and the central mass from which they emanate. (See Chapter 31.) Von Buch supposed that the Plutonic rock alternated with the fossiliferous strata, and that large masses of granite were sometimes incumbent upon the strata; but this idea was erroneous, and arose from the fact that the beds of shale and limestone often dip towards the granite up to the point of contact, appearing as if they would pass under it in mass, as at a, Figure 620, and then again on the opposite side of the same mountain, as at b, dip away from the same granite. When the junctions, however, are carefully examined, it is found that the Plutonic rock intrudes itself in veins, and nowhere covers the fossiliferous strata in large overlying masses, as is so commonly the case with trappean formations. (See the Gaea Norvegica and other works of Keilhau with whom I examined this country.)

Now this granite, which is more modern than the Silurian strata of Norway, also sends veins in the same country into an ancient formation of gneiss; and the relations of the Plutonic rock and the gneiss, at their junction, are full of interest when we duly consider the wide difference of epoch which must have separated their origin.

(FIGURE 621. Granite sending veins into Silurian strata and gneiss. Christiania, Norway. a. Inclined gneiss. b. Silurian strata.)

The length of this interval of time is attested by the following facts: The fossiliferous, or Silurian, beds rest unconformably upon the truncated edges of the gneiss, the inclined strata of which had been denuded before the sedimentary beds were superimposed (see Figure 621). The signs of denudation are twofold; first, the surface of the gneiss is seen occasionally, on the removal of the newer beds containing organic remains, to be worn and smoothed; secondly, pebbles of gneiss have been found in some of these Silurian strata. Between the origin, therefore, of the gneiss and the granite there intervened, first, the period when the strata of gneiss were denuded; secondly, the period of the deposition of the Silurian deposits upon the denuded and inclined gneiss, a. Yet the granite produced after this long interval is often so intimately blended with the ancient gneiss, at the point of junction, that it is impossible to draw any other than an arbitrary line of separation between them; and where this is not the case, tortuous veins of granite pass freely through gneiss, ending sometimes in threads, as if the older rock had offered no resistance to their passage. These appearances may probably be due to hydrothermal action (see Chapter 33). I shall merely observe in this place that had such junctions alone been visible, and had we not learnt, from other sections, how long a period elapsed between the consolidation of the gneiss and the injection of this granite, we might have suspected that the gneiss was scarcely solidified, or had not yet assumed its complete metamorphic character when invaded by the Plutonic rock. From this example we may learn how impossible it is to conjecture whether certain granites in Scotland, and other countries, which send veins into gneiss and other metamorphic rocks, are primary, or whether they may not belong to some secondary or tertiary period.

OLDEST GRANITES.

It is not half a century since the doctrine was very general that all granitic rocks were PRIMITIVE, that is to say, that they originated before the deposition of the first sedimentary strata, and before the creation of organic beings (see above Chapter 1). But so greatly are our views now changed, that we find it no easy task to point out a single mass of granite demonstrably more ancient than known fossiliferous deposits. Could we discover some Laurentian strata resting immediately on granite, there being no alterations at the point of contact, nor any intersecting granitic veins, we might then affirm the Plutonic rock to have originated before the oldest known fossiliferous strata. Still it would be presumptuous, as we have already pointed out (Chapter 26), to suppose that when a small part only of the globe has been investigated, we are acquainted with the oldest fossiliferous strata in the crust of our planet. Even when these are found, we can not assume that there never were any antecedent strata containing organic remains, which may have become metamorphic. If we find pebbles of granite in a conglomerate of the Lower Laurentian system, we may then feel assured that the parent granite was formed before the Laurentian formation. But if the incumbent strata be merely Cambrian or Silurian, the fundamental granite, although of high antiquity, may be posterior in date to KNOWN fossiliferous formations.

PROTRUSION OF SOLID GRANITE.

In part of Sutherlandshire, near Brora, common granite, composed of feldspar, quartz, and mica is in immediate contact with Oolitic strata, and has clearly been elevated to the surface at a period subsequent to the deposition of those strata. (Murchison Geological Transactions second series volume 2 page 307.) Professor Sedgwick and Sir R. Murchison conceive that this granite has been upheaved in a solid form; and that in breaking through the submarine deposits, with which it was not perhaps originally in contact, it has fractured them so as to form a breccia along the line of junction. This breccia consists of fragments of shale, sandstone, and limestone, with fossils of the oolite, all united together by a calcareous cement. The secondary strata at some distance from the granite are but slightly disturbed, but in proportion to their proximity the amount of dislocation becomes greater.

Mr. T. McKenney Hughes has suggested to me in explanation of these phenomena that they may be the effect of the association of more pliant strata with hard unyielding rocks, the whole of which were subjected simultaneously to great movements, whether of elevation or subsidence, and of lateral pressure, during which the more solid granite, being incapable of compression, was forced through the softer beds of shale, sandstone, and limestone. He remarks that similar breccias with slickensides are observed on a minor scale where rocks of different composition and rigidity are contorted together. Such protrusion may have been brought about by degrees by innumerable shocks of earthquakes repeated after long intervals of time along the same tract of country. The opening of new fissures in the hardest rocks is a frequent accompaniment of such convulsions, and during the consequent vibrations, breccias must often be caused. But these catastrophes, as we well know, do not imply that the land or sea of the disturbed region are rendered uninhabitable by living beings, and by no means indicate a state of things different from that witnessed in the ordinary course of nature.

CHAPTER XXXIII.
METAMORPHIC ROCKS.

General Character of Metamorphic Rocks.
Gneiss.
Hornblende-schist.
Serpentine.
Mica-schist.
Clay-slate.
Quartzite.
Chlorite-schist.
Metamorphic Limestone.
Origin of the metamorphic Strata.
Their Stratification.
Fossiliferous Strata near intrusive Masses of Granite converted into Rocks identical with different Members of the metamorphic Series.
Arguments hence derived as to the Nature of Plutonic Action.
Hydrothermal Action, or the Influence of Steam and Gases in producing Metamorphism.
Objections to the metamorphic Theory considered.

We have now considered three distinct classes of rocks: first, the aqueous, or fossiliferous; secondly, the volcanic; and, thirdly, the Plutonic; and it remains for us to examine those crystalline (or hypogene) strata to which the name of METAMORPHIC has been assigned. The last-mentioned term expresses, as before explained, a theoretical opinion that such strata, after having been deposited from water, acquired, by the influence of heat and other causes, a highly crystalline texture. They who still question this opinion may call the rocks under consideration the stratified hypogene formations or crystalline schists.

These rocks, when in their characteristic or normal state, are wholly devoid of organic remains, and contain no distinct fragments of other rocks, whether rounded or angular. They sometimes break out in the central parts of mountain chains, but in other cases extend over areas of vast dimensions, occupying, for example, nearly the whole of Norway and Sweden, where, as in Brazil, they appear alike in the lower and higher grounds. However crystalline these rocks may become in certain regions, they never, like granite or trap, send veins into contiguous formations. In Great Britain, those members of the series which approach most nearly to granite in their composition, as gneiss, mica-schist, and hornblende-schist, are confined to the country north of the rivers Forth and Clyde.

Many attempts have been made to trace a general order of succession or superposition in the members of this family; clay-slate, for example, having been often supposed to hold invariably a higher geological position than mica-schist, and mica-schist to overlie gneiss. But although such an order may prevail throughout limited districts, it is by no means universal. To this subject, however, I shall again revert, in Chapter 35, where the chronological relations of the metamorphic rocks are pointed out.

PRINCIPAL METAMORPHIC ROCKS.

The following may be enumerated as the principal members of the metamorphic class:— gneiss, mica-schist, hornblende-schist, clay-slate, chlorite-schist, hypogene or metamorphic limestone, and certain kinds of quartz-rock or quartzite.

GNEISS.

(FIGURE 622. Fragment of gneiss, natural size; section made at right angles to the planes of foliation.)

The first of these, gneiss, may be called stratified— or by those who object to that term, foliated— granite, being formed of the same materials as granite, namely, feldspar, quartz, and mica. In the specimen in Figure 622, the white layers consist almost exclusively of granular feldspar, with here and there a speck of mica and grain of quartz. The dark layers are composed of grey quartz and black mica, with occasionally a grain of feldspar intermixed. The rock splits most easily in the plane of these darker layers, and the surface thus exposed is almost entirely covered with shining spangles of mica. The accompanying quartz, however, greatly predominates in quantity, but the most ready cleavage is determined by the abundance of mica in certain parts of the dark layer. Instead of consisting of these thin laminae, gneiss is sometimes simply divided into thick beds, in which the mica has only a slight degree of parallelism to the planes of stratification.

Hand specimens may often be obtained from such gneiss which are undistinguishable from granite, affording an argument to which we shall allude in the concluding part of this chapter, in favour of those who regard all granite and syenite not as igneous rocks, but as aqueous formations so altered as to have lost all signs of their original stratified arrangement. Gneiss in geology is commonly used to designate not merely stratified and foliated rocks having the same component materials as granite or syenite, but also in a wider sense to embrace the formation with which other members of the metamorphic series, such as hornblende-schist, may alternate, and which are then considered subordinate to the true gneiss.

The different varieties of rock allied to gneiss, into which feldspar enters as an essential ingredient, will be understood by referring to what was said of granite. Thus, for example, hornblende may be superadded to mica, quartz, and feldspar, forming a hornblendic or syenitic gneiss; or talc may be substituted for mica, constituting talcose gneiss (called stratified protogine by the French), a rock composed of feldspar, quartz, and talc, in distinct crystals or grains.

EURITE, which has already been mentioned as a Plutonic rock, occurs also with precisely the same composition in beds subordinate to gneiss or mica-slate.

HORNBLENDE-SCHIST is usually black, and composed principally of hornblende, with a variable quantity of feldspar, and sometimes grains of quartz. When the hornblende and feldspar are in nearly equal quantities, and the rock is not slaty, it corresponds in character with the greenstones of the trap family, and has been called "primitive greenstone." It may be termed hornblende rock, or amphibolite. Some of these hornblendic masses may really have been volcanic rocks, which have since assumed a more crystalline or metamorphic texture.

SERPENTINE is a greenish rock, a silicate of magnesia, in which there is sometimes from 30 to 40 per cent of magnesia. It enters largely into the composition of a trap dike cutting through Old Red Sandstone in Forfarshire, and in that case is probably an altered basaltic dike which had contained much olivine. The theory of its having been originally a volcanic product subsequently altered by metamorphism may at first sight seem inconsistent with its occurrence in large and regularly stratified masses in the metamorphic series in Scotland, as in Aberdeenshire. But it has been suggested in explanation that such serpentine may have been originally regularly-bedded trap tuff, and volcanic breccia, with much olivine, which would still retain a stratified appearance after their conversion into a metamorphic rock.

ACTINOLITE SCHIST is a slaty foliated rock, composed chiefly of actinolite, an emerald-green mineral, allied to hornblende, with some admixture of garnet, mica, and quartz.

MICA-SCHIST or MICACEOUS SCHIST is, next to gneiss, one of the most abundant rocks of the metamorphic series. It is slaty, essentially composed of mica and quartz, the mica sometimes appearing to constitute the whole mass. Beds of pure quartz also occur in this formation. In some districts, garnets in regular twelve-sided crystals form an integrant part of mica-schist. This rock passes by insensible gradations into clay-slate.

CLAY-SLATE— ARGILLACEOUS SCHIST— ARGILLITE.

This rock sometimes resembles an indurated clay or shale. It is for the most part extremely fissile, often affording good roofing-slate. Occasionally it derives a shining and silky lustre from the minute particles of mica or talc which it contains. It varies from greenish or bluish-grey to a lead colour; and it may be said of this, more than of any other schist, that it is common to the metamorphic and fossiliferous series, for some clay-slates taken from each division would not be distinguishable by mineral characters alone. It is not uncommon to meet with an argillaceous rock having the same composition, without the slaty cleavage, which may be called argillite.

CHLORITE SCHIST is a green slaty rock, in which chlorite is abundant in foliated plates, usually blended with minute grains of quartz, or sometimes with feldspar or mica; often associated with, and graduating into, gneiss and clay- slate.

QUARTZITE, or QUARTZ ROCK, is an aggregate of grains of quartz which are either in minute crystals, or in many cases slightly rounded, occurring in regular strata, associated with gneiss or other metamorphic rocks. Compact quartz, like that so frequently found in veins, is also found together with granular quartzite. Both of these alternate with gneiss or mica-schist, or pass into those rocks by the addition of mica, or of feldspar and mica.

CRYSTALLINE, OR METAMORPHIC LIMESTONE.

This hypogene rock, called by the earlier geologists PRIMARY LIMESTONE, is sometimes a white crystalline granular marble, which when in thick beds can be used in sculpture; but more frequently it occurs in thin beds, forming a foliated schist much resembling in colour and arrangement certain varieties of gneiss and mica-schist. When it alternates with these rocks, it often contains some crystals of mica, and occasionally quartz, feldspar, hornblende, talc, chlorite, garnet, and other minerals. It enters sparingly into the structure of the hypogene districts of Norway, Sweden, and Scotland, but is largely developed in the Alps.

ORIGIN OF THE METAMORPHIC STRATA.

Having said thus much of the mineral composition of the metamorphic rocks, I may combine what remains to be said of their structure and history with an account of the opinions entertained of their probable origin. At the same time, it may be well to forewarn the reader that we are here entering upon ground of controversy, and soon reach the limits where positive induction ends, and beyond which we can only indulge in speculations. It was once a favourite doctrine, and is still maintained by many, that these rocks owe their crystalline texture, their want of all signs of a mechanical origin, or of fossil contents, to a peculiar and nascent condition of the planet at the period of their formation. The arguments in refutation of this hypothesis will be more fully considered when I show, in Chapter 35, to how many different ages the metamorphic formations are referable, and how gneiss, mica-schist, clay-slate, and hypogene limestone (that of Carrara, for example) have been formed, not only since the first introduction of organic beings into this planet, but even long after many distinct races of plants and animals had flourished and passed away in succession.

The doctrine respecting the crystalline strata implied in the name metamorphic may properly be treated of in this place; and we must first inquire whether these rocks are really entitled to be called stratified in the strict sense of having been originally deposited as sediment from water. The general adoption by geologists of the term stratified, as applied to these rocks, sufficiently attests their division into beds very analogous, at least in form, to ordinary fossiliferous strata. This resemblance is by no means confined to the existence in both occasionally of a laminated structure, but extends to every kind of arrangement which is compatible with the absence of fossils, of sand, pebbles, ripple-mark, and other characters which the metamorphic theory supposes to have been obliterated by Plutonic action. Thus, for example, we behold alike in the crystalline and fossiliferous formations an alternation of beds varying greatly in composition, colour, and thickness. We observe, for instance, gneiss alternating with layers of black hornblende-schist or of green chlorite-schist, or with granular quartz or limestone; and the interchange of these different strata may be repeated for an indefinite number of times. In the like manner, mica-schist alternates with chlorite-schist, and with beds of pure quartz or of granular limestone. We have already seen that, near the immediate contact of granitic veins and volcanic dikes, very extraordinary alterations in rocks have taken place, more especially in the neighbourhood of granite. It will be useful here to add other illustrations, showing that a texture undistinguishable from that which characterises the more crystalline metamorphic formations has actually been superinduced in strata once fossiliferous.

FOSSILIFEROUS STRATA RENDERED METAMORPHIC BY INTRUSIVE MASSES OF GRANITE.

(FIGURE 623. Ground-plan of altered slate and limestone near granite. Christiania. The arrows indicate the dip, and the oblique lines the strike of the beds.)

In the southern extremity of Norway there is a large district, on the west side of the fiord of Christiania, which I visited in 1837 with the late Professor Keilhau, in which syenitic granite protrudes in mountain masses through fossiliferous strata, and usually sends veins into them at the point of contact. The stratified rocks, replete with shells and zoophytes, consist chiefly of shale, limestone, and some sandstone, and all these are invariably altered near the granite for a distance of from 50 to 400 yards. The aluminous shales are hardened, and have become flinty. Sometimes they resemble jasper. Ribboned jasper is produced by the hardening of alternate layers of green and chocolate-coloured schist, each stripe faithfully representing the original lines of stratification. Nearer the granite the schist often contains crystals of hornblende, which are even met with in some places for a distance of several hundred yards from the junction; and this black hornblende is so abundant that eminent geologists, when passing through the country, have confounded it with the ancient hornblende-schist, subordinate to the great gneiss formation of Norway. Frequently, between the granite and the hornblende-slate above-mentioned, grains of mica and crystalline feldspar appear in the schist, so that rocks resembling gneiss and mica-schist are produced. Fossils can rarely be detected in these schists, and they are more completely effaced in proportion to the more crystalline texture of the beds, and their vicinity to the granite. In some places the siliceous matter of the schist becomes a granular quartz; and when hornblende and mica are added, the altered rock loses its stratification, and passes into a kind of granite. The limestone, which at points remote from the granite is of an earthy texture and blue colour, and often abounds in corals, becomes a white granular marble near the granite, sometimes siliceous, the granular structure extending occasionally upward of 400 yards from the junction; the corals being for the most part obliterated, though sometimes preserved, even in the white marble. Both the altered limestone and hardened slate contain garnets in many places, also ores of iron, lead, and copper, with some silver. These alterations occur equally whether the granite invades the strata in a line parallel to the general strike of the fossiliferous beds, or in a line at right angles to their strike, both of which modes of junction will be seen by the ground-plan in Figure 623. (Keilhau Gaea Norvegica pages 61-63.)

The granite of Cornwall sends forth veins into a coarse argillaceous-schist, provincially termed killas. This killas is converted into hornblende-schist near the contact with the veins. These appearances are well seen at the junction of the granite and killas, in St. Michael's Mount, a small island nearly 300 feet high, situated in the bay, at a distance of about three miles from Penzance. The granite of Dartmoor, in Devonshire, says Sir H. De la Beche, has intruded itself into the Carboniferous slate and slaty sandstone, twisting and contorting the strata, and sending veins into them. Hence some of the slate rocks have become "micaceous; others more indurated, and with the characters of mica-slate and gneiss; while others again appear converted into a hard zoned rock strongly impregnated with feldspar." (Geological Manual page 479.)

We learn from the investigation of M. Dufrenoy that in the eastern Pyrenees there are mountain masses of granite posterior in date to the formations called lias and chalk of that district, and that these fossiliferous rocks are greatly altered in texture, and often charged with iron-ore, in the neighbourhood of the granite. Thus in the environs of St. Martin, near St. Paul de Fenouillet, the chalky limestone becomes more crystalline and saccharoid as it approaches the granite, and loses all trace of the fossils which it previously contained in abundance. At some points, also, it becomes dolomitic, and filled with small veins of carbonate of iron, and spots of red iron-ore. At Rancie the lias nearest the granite is not only filled with iron-ore, but charged with pyrites, tremolite, garnet, and a new mineral somewhat allied to feldspar, called, from the place in the Pyrenees where it occurs, "couzeranite."

"Hornblende-schist," says Dr. MacCulloch, "may at first have been mere clay; for clay or shale is found altered by trap into Lydian stone, a substance differing from hornblende-schist almost solely in compactness and uniformity of texture." (System of Geology volume 1 pages 210, 211.) "In Shetland," remarks the same author, "argillaceous-schist (or clay-slate), when in contact with granite, is sometimes converted into hornblende-schist, the schist becoming first siliceous, and ultimately, at the contact, hornblende-schist." In like manner gneiss and mica-schist may be nothing more than altered micaceous and argillaceous sandstones, granular quartz may have been derived from siliceous sandstone, and compact quartz from the same materials. Clay-slate may be altered shale, and granular marble may have originated in the form of ordinary limestone, replete with shells and corals, which have since been obliterated; and, lastly, calcareous sands and marls may have been changed into impure crystalline limestones.

The anthracite and plumbago associated with hypogene rocks may have been coal; for not only is coal converted into anthracite in the vicinity of some trap dikes, but we have seen that a like change has taken place generally even far from the contact of igneous rocks, in the disturbed region of the Appalachians. At Worcester, in the State of Massachusetts, 45 miles due west of Boston, a bed of plumbago and impure anthracite occurs, interstratified with mica-schist. It is about two feet in thickness, and has been made use of both as fuel, and in the manufacture of lead pencils. At the distance of 30 miles from the plumbago, there occurs, on the borders of Rhode Island, an impure anthracite in slates containing impressions of coal-plants of the genera Pecopteris, Neuropteris, Calamites, etc. This anthracite is intermediate in character between that of Pennsylvania and the plumbago of Worcester, in which last the gaseous or volatile matter (hydrogen, oxygen, and nitrogen) is to the carbon only in the proportion of three per cent. After traversing the country in various directions, I came to the conclusion that the carboniferous shales or slates with anthracite and plants, which in Rhode Island often pass into mica-schists, have at Worcester assumed a perfectly crystalline and metamorphic texture; the anthracite having been nearly transmuted into that state of pure carbon which is called plumbago or graphite. (See Lyell Quarterly Geological Journal volume 1 page 199.)

Now the alterations above described as superinduced in rocks by volcanic dikes and granite veins prove incontestably that powers exist in nature capable of transforming fossiliferous into crystalline strata, a very few simple elements constituting the component materials common to both classes of rocks. These elements, which are enumerated in Table 28.1, may be made to form new combinations by what has been termed Plutonic action, or those chemical changes which are no doubt connected with the passage of heat, unusually heated steam and waters, through the strata.

HYDROTHERMAL ACTION, OR THE INFLUENCE OF STEAM AND GASES IN PRODUCING METAMORPHISM.

The experiments of Gregory Watt, in fusing rocks in the laboratory, and allowing them to consolidate by slow cooling, prove distinctly that a rock need not be perfectly melted in order that a re-arrangement of its component particles should take place, and a partial crystallisation ensue. (Philosophical Transactions 1804.) We may easily suppose, therefore, that all traces of shells and other organic remains may be destroyed, and that new chemical combinations may arise, without the mass being so fused as that the lines of stratification should be wholly obliterated. We must not, however, imagine that heat alone, such as may be applied to a stone in the open air, can constitute all that is comprised in Plutonic action. We know that volcanoes in eruption not only emit fluid lava, but give off steam and other heated gases, which rush out in enormous volume, for days, weeks, or years continuously, and are even disengaged from lava during its consolidation.

We also know that long after volcanoes have spent their force, hot springs continue to flow out at various points in the same area. In regions, also, subject to violent earthquakes such springs are frequently observed issuing from rents, usually along lines of fault or displacement of the rocks. These thermal waters are most commonly charged with a variety of mineral ingredients, and they retain a remarkable uniformity of temperature from century to century. A like uniformity is also persistent in the nature of the earthy, metallic, and gaseous substances with which they are impregnated. It is well ascertained that springs, whether hot or cold, charged with carbonic acid, especially with hydrofluoric acid, which is often present in small quantities, are powerful causes of decomposition and chemical reaction in rocks through which they percolate.

The changes which Daubree has shown to have been produced by the alkaline waters of Plombieres in the Vosges, are more especially instructive. (Daubree Sur le Metamorphisme Paris 1860.) These waters have a heat of 160 degrees F., or an excess of 109 degrees above the average temperature of ordinary springs in that district. They were conveyed by the Romans to baths through long conduits or aqueducts. The foundations of some of their works consisted of a bed of concrete made of lime, fragments of brick, and sandstone. Through this and other masonry the hot waters have been percolating for centuries, and have given rise to various zeolites— apophyllite and chabazite among others; also to calcareous spar, arragonite, and fluor spar, together with siliceous minerals, such as opal— all found in the inter-spaces of the bricks and mortar, or constituting part of their re-arranged materials. The quantity of heat brought into action in this instance in the course of 2000 years has, no doubt, been enormous, but the intensity of it developed at any one moment has been always inconsiderable.

From these facts and from the experiments and observations of Senarmont, Daubree, Delesse, Scheerer, Sorby, Sterry Hunt, and others, we are led to infer that when in the bowels of the earth there are large volumes of matter containing water and various acids intensely heated under enormous pressure, these subterranean fluid masses will gradually part with their heat by the escape of steam and various gases through fissures, producing hot springs; or by the passage of the same through the pores of the overlying and injected rocks. Even the most compact rocks may be regarded, before they have been exposed to the air and dried, in the light of sponges filled with water. According to the experiments of Henry, water, under a hydrostatic pressure of 96 feet, will absorb three times as much carbonic acid gas as it can under the ordinary pressure of the atmosphere. There are other gases, as well as the carbonic acid, which water absorbs, and more rapidly in proportion to the amount of pressure. Although the gaseous matter first absorbed would soon be condensed, and part with its heat, yet the continual arrival of fresh supplies from below might, in the course of ages, cause the temperature of the water, and with it that of the containing rock, to be materially raised; the water acts not only as a vehicle of heat, but also by its affinity for various silicates, which, when some of the materials of the invaded rocks are decomposed, form quartz, feldspar, mica, and other minerals. As for quartz, it can be produced under the influence of heat by water holding alkaline silicates in solution, as in the case of the

Plombieres springs. The quantity of water required, according to Daubree, to produce great transformations in the mineral structure of rocks, is very small. As to the heat required, silicates may be produced in the moist way at about incipient red heat, whereas to form the same in the dry way would require a much higher temperature.

M. Fournet, in his description of the metalliferous gneiss near Clermont, in Auvergne, states that all the minute fissures of the rock are quite saturated with free carbonic acid gas; which gas rises plentifully from the soil there and in many parts of the surrounding country. The various elements of the gneiss, with the exception of the quartz, are all softened; and new combinations of the acid with lime, iron, and manganese are continually in progress. (See Principles Index Carbonated Springs etc.)

The power of subterranean gases is well illustrated by the stufas of St. Calogero in the Lipari Islands, where the horizontal strata of tuffs, forming cliffs 200 feet high, have been discoloured in places by the jets of steam often quite above the boiling point, called "stufas," issuing from the fissures; and similar instances are recorded by M. Virlet of corrosion of rocks near Corinth, and by Dr. Daubeny of decomposition of trachytic rocks by sulphureted hydrogen and muriatic acid gases in the Solfatara, near Naples. In all these instances it is clear that the gaseous fluids must have made their way through vast thicknesses of porous or fissured rocks, and their modifying influence may spread through the crust for thousands of yards in thickness.

It has been urged as an argument against the metamorphic theory, that rocks have a small power of conducting heat, and it is true that when dry, and in the air, they differ remarkably from metals in this respect. The syenite of Norway, as we have seen (Chapter 31), has sometimes altered fossiliferous strata both in the direction of their dip and strike for a distance of a quarter of a mile, but the theory of gneiss and mica-schist above proposed requires us to imagine that the same influence has extended through strata miles in thickness. Professor Bischof has shown what changes may be superinduced, on black marble and other rocks, by the steam of a hot spring having a temperature of no more than 133 degrees to 167 degrees Fahrenheit, and we are becoming more and more acquainted with the prominent part which water is playing in distributing the heat of the interior through mountain masses of incumbent strata, and of introducing into them various mineral elements in a fluid or gaseous state. Such facts may induce us to consider whether many granites and other rocks of that class may not sometimes represent merely the extreme of a similar slow metamorphism. But, on the other hand, the heat of lava in a volcanic crater when it is white and glowing like the sun must convince us that the temperature of a column of such a fluid at the depth of many miles exceeds any heat which can ever be witnessed at the surface. That large portions of the Plutonic rocks had been formed under the influence of such intense heat is in perfect accordance with their great volume, uniform composition, and absence of stratification. The forcing also of veins into contiguous stratified or schistose rocks is a natural consequence of the hydrostatic pressure to which columns of molten matter many miles in height must give rise.

OBJECTIONS TO THE METAMORPHIC THEORY CONSIDERED.

It has been objected to the metamorphic theory that the crystalline schists contain a considerable proportion of potash and soda, whilst the sedimentary strata out of which they are supposed to have been formed are usually wanting in alkaline matter. But this reasoning proceeds on mistaken data, for clay, marl, shale, and slate often contain a considerable proportion of alkali, so much so as to make them frequently unfit to be burnt into bricks or pottery, and the Old Red Sandstone in Forfarshire and other parts of Scotland, derived from disintegration of granite, contains much triturated feldspar rich in potash. In the common salt by which strata are often largely impregnated, as in Patagonia, much soda is present, and potash enters largely into the composition of fossil sea-weeds, and recent analysis has also shown that the carboniferous strata in England, the Upper and Lower Silurian in East Canada, and the oldest clay-slates in Norway, all contain as much alkali as is generally present in metamorphic rocks.

Another objection has been derived from the alternation of highly crystalline strata with others less crystalline. The heat, it is said, in its ascent from below, must have traversed the less altered schists before it reached a higher and more crystalline bed. In answer to this, it may be observed, that if a number of strata differing greatly in composition from each other be subjected to equal quantities of heat, or hydrothermal action, there is every probability that some will be much more fusible or soluble than others. Some, for example, will contain soda, potash, lime, or some other ingredient capable of acting as a flux or solvent; while others may be destitute of the same elements, and so refractory as to be very slightly affected by the same causes. Nor should it be forgotten that, as a general rule, the less crystalline rocks do really occur in the upper, and the more crystalline in the lower part of each metamorphic series.

CHAPTER XXXIV.
METAMORPHIC ROCKS CONTINUED.

Definition of slaty Cleavage and Joints.
Supposed Causes of these Structures.
Crystalline Theory of Cleavage.
Mechanical Theory of Cleavage.
Condensation and Elongation of slate Rocks by lateral Pressure.
Lamination of some volcanic Rocks due to Motion.
Whether the Foliation of the crystalline Schists be usually parallel with the original Planes of Stratification.
Examples in Norway and Scotland.
Causes of Irregularity in the Planes of Foliation.

We have already seen that chemical forces of great intensity have frequently acted upon sedimentary and fossiliferous strata long subsequently to their consolidation, and we may next inquire whether the component minerals of the altered rocks usually arrange themselves in planes parallel to the original planes of stratification, or whether, after crystallisation, they more commonly take up a different position.

In order to estimate fairly the merits of this question, we must first define what is meant by the terms cleavage and foliation. There are four distinct forms of structure exhibited in rocks, namely, stratification, joints, slaty cleavage, and foliation; and all these must have different names, even though there be cases where it is impossible, after carefully studying the appearances, to decide upon the class to which they belong.

SLATY CLEAVAGE.

(FIGURE 624. Parallel planes of cleavage intersecting curved strata. (Sedgwick.))

Professor Sedgwick, whose essay "On the Structure of large Mineral Masses" first cleared the way towards a better understanding of this difficult subject, observes, that joints are distinguishable from lines of slaty cleavage in this, that the rock intervening between two joints has no tendency to cleave in a direction parallel to the planes of the joints, whereas a rock is capable of indefinite subdivision in the direction of its slaty cleavage. In cases where the strata are curved, the planes of cleavage are still perfectly parallel. This has been observed in the slate rocks of part of Wales (see Figure 624), which consists of a hard greenish slate. The true bedding is there indicated by a number of parallel stripes, some of a lighter and some of a darker colour than the general mass. Such stripes are found to be parallel to the true planes of stratification, wherever these are manifested by ripple-mark or by beds containing peculiar organic remains. Some of the contorted strata are of a coarse mechanical structure, alternating with fine-grained crystalline chloritic slates, in which case the same slaty cleavage extends through the coarser and finer beds, though it is brought out in greater perfection in proportion as the materials of the rock are fine and homogeneous. It is only when these are very coarse that the cleavage planes entirely vanish. In the Welsh hills these planes are usually inclined at a very considerable angle to the planes of the strata, the average angle being as much as from 30 to 40 degrees. Sometimes the cleavage planes dip towards the same point of the compass as those of stratification, but often to opposite points. (Geological Transactions second series volume 3 page 461.) The cleavage, as represented in Figure 624, is generally constant over the whole of any area affected by one great set of disturbances, as if the same lateral pressure which caused the crumpling up of the rock along parallel, anticlinal, and synclinal axes caused also the cleavage.

(FIGURE 625. Section in Lower Silurian slates of Cardiganshire, showing the cleavage planes bent along the junction of the beds. (T. McK. Hughes.))

Mr. T. McK. Hughes remarks, that where a rough cleavage cuts flag-stones at a considerable angle to the planes of stratification, the rock often splits into large slabs, across which the lines of bedding are frequently seen, but when the cleavage planes approach within about 15 degrees of stratification, the rock is apt to split along the lines of bedding. He has also called my attention to the fact that subsequent movements in a cleaved rock sometimes drag and bend the cleavage planes along the junction of the beds in the manner indicated in Figure 625.

JOINTED STRUCTURE.

In regard to joints, they are natural fissures which often traverse rocks in straight and well-determined lines. They afford to the quarryman, as Sir R. Murchison observes, when speaking of the phenomenon, as exhibited in Shropshire and the neighbouring counties, the greatest aid in the extraction of blocks of stone; and, if a sufficient number cross each other, the whole mass of rock is split into symmetrical blocks. The faces of the joints are for the most part smoother and more regular than the surfaces of true strata. The joints are straight-cut chinks, sometimes slightly open, and often passing, not only through layers of successive deposition, but also through balls of limestone or other matter which have been formed by concretionary action since the original accumulation of the strata. Such joints, therefore, must often have resulted from one of the last changes superinduced upon sedimentary deposits. (Silurian System page 246.)

(FIGURE 626. Stratification, joints, and cleavage (From Murchison's Silurian System page 245.))

In Figure 626 the flat-surfaces of rock, A, B, C, represent exposed faces of joints, to which the walls of other joints, J-J, are parallel. S-S are the lines of stratification; D, D are lines of slaty cleavage, which intersect the rock at a considerable angle to the planes of stratification.

In the Swiss and Savoy Alps, as Mr. Bakewell has remarked, enormous masses of limestone are cut through so regularly by nearly vertical partings, and these joints are often so much more conspicuous than the seams of stratification, that an inexperienced observer will almost inevitably confound them, and suppose the strata to be perpendicular in places where in fact they are almost horizontal. (Introduction to Geology chapter 4.)

Now such joints are supposed to be analogous to the partings which separate volcanic and Plutonic rocks into cuboidal and prismatic masses. On a small scale we see clay and starch when dry split into similar shapes; this is often caused by simple contraction, whether the shrinking be due to the evaporation of water, or to a change of temperature. It is well known that many sandstones and other rocks expand by the application of moderate degrees of heat, and then contract again on cooling; and there can be no doubt that large portions of the earth's crust have, in the course of past ages, been subjected again and again to very different degrees of heat and cold. These alternations of temperature have probably contributed largely to the production of joints in rocks.

In many countries where masses of basalt rest on sandstone, the aqueous rock has, for the distance of several feet from the point of junction, assumed a columnar structure similar to that of the trap. In like manner some hearth-stones, after exposure to the heat of a furnace without being melted, have become prismatic. Certain crystals also acquire by the application of heat a new internal arrangement, so as to break in a new direction, their external form remaining unaltered.

CRYSTALLINE THEORY OF CLEAVAGE.

Professor Sedgwick, speaking of the planes of slaty cleavage, where they are decidedly distinct from those of sedimentary deposition, declared, in the essay before alluded to, his opinion that no retreat of parts, no contraction in the dimensions of rocks in passing to a solid state, can account for the phenomenon. He accordingly referred it to crystalline or polar forces acting simultaneously, and somewhat uniformly, in given directions, on large masses having a homogeneous composition.

Sir John Herschel, in allusion to slaty cleavage, has suggested that "if rocks have been so heated as to allow a commencement of crystallisation— that is to say, if they have been heated to a point at which the particles can begin to move among themselves, or at least on their own axes, some general law must then determine the position in which these particles will rest on cooling. Probably, that position will have some relation to the direction in which the heat escapes. Now, when all, or a majority of particles of the same nature have a general tendency to one position, that must of course determine a cleavage- plane. Thus we see the infinitesimal crystals of fresh-precipitated sulphate of barytes, and some other such bodies, arrange themselves alike in the fluid in which they float; so as, when stirred, all to glance with one light, and give the appearance of silky filaments. Some sorts of soap, in which insoluble margarates exist (Margaric acid is an oleaginous acid, formed from different animal and vegetable fatty substances. A margarate is a compound of this acid with soda, potash, or some other base, and is so named from its pearly lustre.), exhibit the same phenomenon when mixed with water; and what occurs in our experiments on a minute scale may occur in nature on a great one." (Letter to the author dated Cape of Good Hope February 20, 1836.)

MECHANICAL THEORY OF CLEAVAGE.

Professor Phillips has remarked that in some slaty rocks the form of the outline of fossil shells and trilobites has been much changed by distortion, which has taken place in a longitudinal, transverse, or oblique direction. This change, he adds, seems to be the result of a "creeping movement" of the particles of the rock along the planes of cleavage, its direction being always uniform over the same tract of country, and its amount in space being sometimes measurable, and being as much as a quarter or even half an inch. The hard shells are not affected, but only those which are thin. (Report British Association Cork 1843 Section page 60.) Mr. D. Sharpe, following up the same line of inquiry, came to the conclusion that the present distorted forms of the shells in certain British slate rocks may be accounted for by supposing that the rocks in which they are imbedded have undergone compression in a direction perpendicular to the planes of cleavage, and a corresponding expansion in the direction of the dip of the cleavage. (Quarterly Geological Journal volume 3 page 87 1847.)

(FIGURE 627. Vertical section of slate rock in the cliffs near Ilfracombe, North Devon. Scale one inch to one foot. (Drawn by H.C. Sorby.) a, b, c, e. Fine-grained slates, the stratification being shown partly by lighter or darker colours, and partly by different degrees of fineness in the grain. d, f. A coarser grained light-coloured sandy slate with less perfect cleavage.)

Subsequently (1853) Mr. Sorby demonstrated the great extent to which this mechanical theory is applicable to the slate rocks of North Wales and Devonshire (On the Origin of Slaty Cleavage by H.C. Sorby Edinburgh New Philosophical Journal 1853 volume 55 page 137.), districts where the amount of change in dimensions can be tested and measured by comparing the different effects exerted by lateral pressure on alternating beds of finer and coarser materials. Thus, for example, in Figure 627 it will be seen that the sandy bed d-f, which has offered greater resistance, has been sharply contorted, while the fine-grained strata, a, b, c, have remained comparatively unbent. The points d and f in the stratum d-f must have been originally four times as far apart as they are now. They have been forced so much nearer to each other, partly by bending, and partly by becoming elongated in the direction of what may be called the longer axes of their contortions, and lastly, to a certain small amount, by condensation. The chief result has obviously been due to the bending; but, in proof of elongation, it will be observed that the thickness of the bed d-f is now about four times greater in those parts lying in the main direction of the flexures than in a plane perpendicular to them; and the same bed exhibits cleavage planes in the direction of the greatest movement, although they are much fewer than in the slaty strata above and below.

Above the sandy bed d-f, the stratum c is somewhat disturbed, while the next bed, b, is much less so, and a not at all; yet all these beds, c, b, and a, must have undergone an equal amount of pressure with d, the points a and g having approximated as much towards each other as have d and f. The same phenomena are also repeated in the beds below d, and might have been shown, had the section been extended downward. Hence it appears that the finer beds have been squeezed into a fourth of the space they previously occupied, partly by condensation, or the closer packing of their ultimate particles (which has given rise to the great specific gravity of such slates), and partly by elongation in the line of the dip of the cleavage, of which the general direction is perpendicular to that of the pressure. "These and numerous other cases in North Devon are analogous," says Mr. Sorby, "to what would occur if a strip of paper were included in a mass of some soft plastic material which would readily change its dimensions. If the whole were then compressed in the direction of the length of the strip of paper, it would be bent and puckered up into contortions, while the plastic material would readily change its dimensions without undergoing such contortions; and the difference in distance of the ends of the paper, as measured in a direct line or along it, would indicate the change in the dimensions of the plastic material."

By microscopic examination of minute crystals, and by other observations, Mr. Sorby has come to the conclusion that the absolute condensation of the slate rocks amounts upon an average to about one half their original volume. Most of the scales of mica occurring in certain slates examined by Mr. Sorby lie in the plane of cleavage; whereas in a similar rock not exhibiting cleavage they lie with their longer axes in all directions. May not their position in the slates have been determined by the movement of elongation before alluded to? To illustrate this theory some scales of oxide of iron were mixed with soft pipe-clay in such a manner that they inclined in all directions. The dimensions of the mass were then changed artificially to a similar extent to what has occurred in slate rocks, and the pipe-clay was then dried and baked. When it was afterwards rubbed to a flat surface perpendicular to the pressure and in the line of elongation, or in a plane corresponding to that of the dip of cleavage, the particles were found to have become arranged in the same manner as in natural slates, and the mass admitted of easy fracture into thin flat pieces in the plane alluded to, whereas it would not yield in that perpendicular to the cleavage. (Sorby as cited above page 741 note.)

Dr. Tyndall, when commenting in 1856 on Mr. Sorby's experiments, observed that pressure alone is sufficient to produce cleavage, and that the intervention of plates of mica or scales of oxide of iron, or any other substances having flat surfaces, is quite unnecessary. In proof of this he showed experimentally that a mass of "pure white wax, after having been submitted to great pressure, exhibited a cleavage more clean than that of any slate-rock, splitting into laminae of surpassing tenuity." (Tyndall View of the Cleavage of Crystals and Slate rocks.) He remarks that every mass of clay or mud is divided and subdivided by surfaces among which the cohesion is comparatively small. On being subjected to pressure, such masses yield and spread out in the direction of least resistance, small nodules become converted into laminae separated from each other by surfaces of weak cohesion, and the result is that the mass cleaves at right angles to the line in which the pressure is exerted. In further illustration of this, Mr. Hughes remarks that "concretions which in the undisturbed beds have their longer axes parallel to the bedding are, where the rock is much cleaved, frequently found flattened laterally, so as to have their longer axes parallel to the cleavage planes, and at a considerable angle, even right angles, to their former position."

Mr. Darwin attributes the lamination and fissile structure of volcanic rocks of the trachytic series, including some obsidians in Ascension, Mexico, and elsewhere, to their having moved when liquid in the direction of the laminae. The zones consist sometimes of layers of air-cells drawn out and lengthened in the supposed direction of the moving mass. (Darwin Volcanic Islands pages 69, 70.)

FOLIATION OF CRYSTALLINE SCHISTS.

After studying, in 1835, the crystalline rocks of South America, Mr. Darwin proposed the term FOLIATION for the laminae or plates into which gneiss, mica-schist, and other crystalline rocks are divided. Cleavage, he observes, may be applied to those divisional planes which render a rock fissile, although it may appear to the eye quite or nearly homogeneous. Foliation may be used for those alternating layers or plates of different mineralogical nature of which gneiss and other metamorphic schists are composed.

That the planes of foliation of the crystalline schists in Norway accord very generally with those of original stratification is a conclusion long since espoused by Keilhau. (Norske Mag. Naturvidsk. volume 1 page 71.) Numerous observations made by Mr. David Forbes in the same country (the best probably in Europe for studying such phenomena on a grand scale) confirm Keilhau's opinion. In Scotland, also, Mr. D. Forbes has pointed out a striking case where the foliation is identical with the lines of stratification in rocks well seen near

Crianlorich on the road to Tyndrum, about eight miles from Inverarnon, in Perthshire. There is in that locality a blue limestone foliated by the intercalation of small plates of white mica, so that the rock is often scarcely distinguishable in aspect from gneiss or mica-schist. The stratification is shown by the large beds and coloured bands of limestone all dipping, like the folia, at an angle of 32 degrees N.E. (Memoir read before the Geological Society London January 31, 1855.) In stratified formations of every age we see layers of siliceous sand with or without mica, alternating with clay, with fragments of shells or corals, or with seams of vegetable matter, and we should expect the mutual attraction of like particles to favour the crystallisation of the quartz, or mica, or feldspar, or carbonate of lime, along the planes of original deposition, rather than in planes placed at angles of 20 or 40 degrees to those of stratification.

We have seen how much the original planes of stratification may be interfered with or even obliterated by concretionary action in deposits still retaining their fossils, as in the case of the magnesian limestone (see Chapter 4). Hence we must expect to be frequently baffled when we attempt to decide whether the foliation does or does not accord with that arrangement which gravitation, combined with current-action, imparted to a deposit from water. Moreover, when we look for stratification in crystalline rocks, we must be on our guard not to expect too much regularity. The occurrence of wedge-shaped masses, such as belong to coarse sand and pebbles— diagonal lamination (Chapter 2)— ripple- marked, unconformable stratification,— the fantastic folds produced by lateral pressure— faults of various width— intrusive dikes of trap— organic bodies of diversified shapes, and other causes of unevenness in the planes of deposition, both on the small and on the large scale, will interfere with parallelism. If complex and enigmatical appearances did not present themselves, it would be a serious objection to the metamorphic theory. Mr. Sorby has shown that the peculiar structure belonging to ripple-marked sands, or that which is generated when ripples are formed during the deposition of the materials, is distinctly recognisable in many varieties of mica-schists in Scotland. (H.C. Sorby Quarterly Geological Journal volume 19 page 401.)

(FIGURE 628. Lamination of clay-stone. Montagne de Seguinat, near Gavarnie, in the Pyrenees.)

In Figure 628 I have represented carefully the lamination of a coarse argillaceous schist which I examined in 1830 in the Pyrenees. In part it approaches in character to a green and blue roofing-slate, while part is extremely quartzose, the whole mass passing downward into micaceous schist. The vertical section here exhibited is about three feet in height, and the layers are sometimes so thin that fifty may be counted in the thickness of an inch. Some of them consist of pure quartz. There is a resemblance in such cases to the diagonal lamination which we see in sedimentary rocks, even though the layers of quartz and of mica, or of feldspar and other minerals, may be more distinct in alternating folia than they were originally.

CHAPTER XXXV.
ON THE DIFFERENT AGES OF THE METAMORPHIC ROCKS.

Difficulty of ascertaining the Age of metamorphic Strata.
Metamorphic Strata of Eocene date in the Alps of Switzerland and Savoy.
Limestone and Shale of Carrara.
Metamorphic Strata of older date than the Silurian and Cambrian Rocks.
Order of Succession in metamorphic Rocks.
Uniformity of mineral Character.
Supposed Azoic Period.
Connection between the Absence of Organic Remains and the Scarcity of calcareous Matter in metamorphic Rocks.

According to the theory adopted in the last chapter, the metamorphic strata have been deposited at one period, and have become crystalline at another. We can rarely hope to define with exactness the date of both these periods, the fossils having been destroyed by Plutonic action, and the mineral characters being the same, whatever the age. Superposition itself is an ambiguous test, especially when we desire to determine the period of crystallisation. Suppose, for example, we are convinced that certain metamorphic strata in the Alps, which are covered by cretaceous beds, are altered lias; this lias may have assumed its crystalline texture in the cretaceous or in some tertiary period, the Eocene for example.

When discussing the ages of the Plutonic rocks, we have seen that examples occur of various primary, secondary, and tertiary deposits converted into metamorphic strata near their contact with granite. There can be no doubt in these cases that strata once composed of mud, sand, and gravel, or of clay, marl, and shelly limestone, have for the distance of several yards, and in some instances several hundred feet, been turned into gneiss, mica-schist, hornblende-schist, chlorite- schist, quartz rock, statuary marble, and the rest. (See Chapters 33 and 34.) It may be easy to prove the identity of two different parts of the same stratum; one, where the rock has been in contact with a volcanic or Plutonic mass, and has been changed into marble or hornblende-schist, and another not far distant, where the same bed remains unaltered and fossiliferous; but when hydrothermal action, as described in Chapter 33, has operated gradually on a more extensive scale, it may have finally destroyed all monuments of the date of its development throughout a whole mountain chain, and all the labour and skill of the most practised observers are required, and may sometimes be at fault. I shall mention one or two examples of alteration on a grand scale, in order to explain to the student the kind of reasoning by which we are led to infer that dense masses of fossiliferous strata have been converted into crystalline rocks.

EOCENE STRATA RENDERED METAMORPHIC IN THE ALPS.

In the eastern part of the Alps, some of the Palaeozoic strata, as well as the older Mesozoic formations, including the oolitic and cretaceous rocks, are distinctly recognisable. Tertiary deposits also appear in a less elevated position on the flanks of the Eastern Alps; but in the Central or Swiss Alps, the Palaeozoic and older Mesozoic formations disappear, and the Cretaceous, Oolitic, Liassic, and at some points even the Eocene strata, graduate insensibly into metamorphic rocks, consisting of granular limestone, talc-schist, talcose-gneiss, micaceous schist, and other varieties.

As an illustration of the partial conversion into gneiss of portions of a highly inclined set of beds, I may cite Sir R. Murchison's memoir on the structure of the Alps. Slates provincially termed "flysch" (see Chapter 16), overlying the nummulite limestone of Eocene date, and comprising some arenaceous and some calcareous layers, are seen to alternate several times with bands of granitoid rock, answering in character to gneiss. In this case heat, vapour, or water at a high temperature may have traversed the more permeable beds, and altered them so far as to admit of an internal movement and re-arrangement of the molecules, while the adjoining strata did not give passage to the same heated gases or water, or, if so, remained unchanged because they were composed of less fusible or decomposable materials. Whatever hypothesis we adopt, the phenomena establish beyond a doubt the possibility of the development of the metamorphic structure in a tertiary deposit in planes parallel to those of stratification. The strata appear clearly to have been affected, though in a less intense degree, by that same Plutonic action which has entirely altered and rendered metamorphic so many of the subjacent formations; for in the Alps this action has by no means been confined to the immediate vicinity of granite. Granite, indeed, and other Plutonic rocks, rarely make their appearance at the surface, notwithstanding the deep ravines which lay open to view the internal structure of these mountains. That they exist below at no great depth we can not doubt, for at some points, as in the Valorsine, near Mont Blanc, granite and granitic veins are observable, piercing through talcose gneiss, which passes insensibly upward into secondary strata.

It is certainly in the Alps of Switzerland and Savoy, more than in any other district in Europe, that the geologist is prepared to meet with the signs of an intense development of Plutonic action; for here strata thousands of feet thick have been bent, folded, and overturned, and marine secondary formations of a comparatively modern date, such as the Oolitic and Cretaceous, have been upheaved to the height of 12,000, and some Eocene strata to elevations of 10,000 feet above the level of the sea; and even deposits of the Miocene era have been raised 4000 or 5000 feet, so as to rival in height the loftiest mountains in Great Britain. In one of the sections described by M. Studer in the highest of the Bernese Alps, namely in the Roththal, a valley bordering the line of perpetual snow on the northern side of the Jungfrau, there occurs a mass of gneiss 1000 feet thick, and 15,000 feet long, which I examined, not only resting upon, but also again covered by strata containing oolitic fossils. These anomalous appearances may partly be explained by supposing great solid wedges of intrusive gneiss to have been forced in laterally between strata to which I found them to be in many sections unconformable. The superposition, also, of the gneiss to the oolite may, in some cases, be due to a reversal of the original position of the beds in a region where the convulsions have been on so stupendous a scale.

NORTHERN APENNINES.— CARRARA.

The celebrated marble of Carrara, used in sculpture, was once regarded as a type of primitive limestone. It abounds in the mountains of Massa Carrara, or the "Apuan Alps," as they have been called, the highest peaks of which are nearly 6000 feet high. Its great antiquity was inferred from its mineral texture, from the absence of fossils, and its passage downward into talc-schist and garnetiferous mica-schist; these rocks again graduating downward into gneiss, which is penetrated, at Forno, by granite veins. But the researches of MM. Savi, Boue, Pareto, Guidoni, De la Beche, Hoffman, and Pilla demonstrated that this marble, once supposed to be formed before the existence of organic beings, is, in fact, an altered limestone of the Oolitic period, and the underlying crystalline schists are secondary sandstones and shales, modified by Plutonic action. In order to establish these conclusions it was first pointed out that the calcareous rocks bordering the Gulf of Spezia, and abounding in Oolitic fossils, assume a texture like that of Carrara marble, in proportion as they are more and more invaded by certain trappean and Plutonic rocks, such as diorite, serpentine, and granite, occurring in the same country.

It was then observed that, in places where the secondary formations are unaltered, the uppermost consist of common Apennine limestone with nodules of flint, below which are shales, and at the base of all, argillaceous and siliceous sandstones. In the limestone fossils are frequent, but very rare in the underlying shale and sandstone. Then a gradation was traced laterally from these rocks into another and corresponding series, which is completely metamorphic; for at the top of this we find a white granular marble, wholly devoid of fossils, and almost without stratification, in which there are no nodules of flint, but in its place siliceous matter disseminated through the mass in the form of prisms of quartz. Below this, and in place of the shales, are talc-schists, jasper, and hornstone; and at the bottom, instead of the siliceous and argillaceous sandstones, are quartzite and gneiss. (See notices of Savi, Hoffman, and others, referred to by Boue, Bull. de la Soc. Geol. de France tome 5 page 317 and tome 3 page 44; also Pilla, cited by Murchison Quarterly Geological Journal volume 5 page 266.) Had these secondary strata of the Apennines undergone universally as great an amount of transmutation, it would have been impossible to form a conjecture respecting their true age; and then, according to the method of classification adopted by the earlier geologists, they would have ranked as primary rocks. In that case the date of their origin would have been thrown back to an era antecedent to the deposition of the Lower Silurian or Cambrian strata, although in reality they were formed in the Oolitic period, and altered at some subsequent and perhaps much later epoch.

METAMORPHIC STRATA OF OLDER DATE THAN THE SILURIAN AND CAMBRIAN ROCKS.

It was remarked (Figure 617) that as the hypogene rocks, both stratified and unstratified, crystallise originally at a certain depth beneath the surface, they must always, before they are upraised and exposed at the surface, be of considerable antiquity, relatively to a large portion of the fossiliferous and volcanic rocks. They may be forming at all periods; but before any of them can become visible, they must be raised above the level of the sea, and some of the rocks which previously concealed them must have been removed by denudation.

In Canada, as we have seen (Chapter 27), the Lower Laurentian gneiss, quartzite, and limestone may be regarded as metamorphic, because, among other reasons, organic remains (Eozoon Canadense) have been detected in a part of one of the calcareous masses. The Upper Laurentian or Labrador series lies unconformably upon the Lower, and differs from it chiefly in having as yet yielded no fossils. It consists of gneiss with Labrador-feldspar and feldstones, in all 10,000 feet thick, and both its composition and structure lead us to suppose that, like the Lower Laurentian, it was originally of sedimentary origin and owes its crystalline condition to metamorphic action. The remote date of the period when some of these old Laurentian strata of Canada were converted into gneiss may be inferred from the fact that pebbles of that rock are found in the overlying Huronian formation, which is probably of Cambrian age (Chapter 27).

The oldest stratified rock of Scotland is the hornblendic gneiss of Lewis, in the Hebrides, and that of the north-west coast of Ross-shire, represented at the base of the section given at Figure 82. It is the same as that intersected by numerous granite veins which forms the cliffs of Cape Wrath, in Sutherlandshire (see Figure 613), and is conjectured to be of Laurentian age. Above it, as shown in the section (Figure 82), lie unconformable beds of a reddish or purple sandstone and conglomerate, nearly horizontal, and between 3000 and 4000 feet thick. In these ancient grits no fossils have been found, but they are supposed to be of Cambrian date, for Sir R. Murchison found Lower Silurian strata resting unconformably upon them. These strata consist of quartzite with annelid burrows already alluded to (Chapter 7), and limestone in which Mr. Charles Peach was the first to find, in 1854, three or four species of Orthoceras, also the genera Cyrtoceras and Lituites, two species of Murchisonia, a Pleurotomaria, a species of Maclurea, one of Euomphalus, and an Orthis. Several of the species are believed by Mr. Salter to be identical with Lower Silurian fossils of Canada and the United States.

The discovery of the true age of these fossiliferous rocks was one of the most important steps made of late years in the progress of British Geology, for it led to the unexpected conclusion that all the Scotch crystalline strata to the eastward, once called primitive, which overlie the limestone and quartzite in question, are referable to some part of the Silurian series.

These Scotch metamorphic strata are of gneiss, mica-schist, and clay-slate of vast thickness, and having a strike from north-east to south-west almost at right angles to that of the older Laurentian gneiss before mentioned. The newer crystalline series, comprising the crystalline rocks of Aberdeenshire, Perthshire, and Forfarshire, were inferred by Sir R. Murchison to be altered Silurian strata; and his opinion has been since confirmed by the observations of three able geologists, Messrs. Ramsay, Harkness, and Geikie. The newest of the series is a clay-slate, on which, along the southern borders of the Grampians, the Lower Old Red, containing Cephalaspis Lyelli, Pterygotus Anglicus, and Parka decipiens, rests unconformably.

ORDER OF SUCCESSION IN METAMORPHIC ROCKS.

There is no universal and invariable order of superposition in metamorphic rocks, although a particular arrangement may prevail throughout countries of great extent, for the same reason that it is traceable in those sedimentary formations from which crystalline strata are derived. Thus, for example, we have seen that in the Apennines, near Carrara, the descending series, where it is metamorphic, consists of, first, saccharine marble; secondly, talcose-schist; and thirdly, of quartz-rock and gneiss: where unaltered, of, first, fossiliferous limestone; secondly, shale; and thirdly, sandstone.

But if we investigate different mountain chains, we find gneiss, mica-schist, hornblende-schist, chlorite-schist, hypogene limestone, and other rocks, succeeding each other, and alternating with each other in every possible order. It is, indeed, more common to meet with some variety of clay-slate forming the uppermost member of a metamorphic series than any other rock; but this fact by no means implies, as some have imagined, that all clay-slates were formed at the close of an imaginary period when the deposition of the crystalline strata gave way to that of ordinary sedimentary deposits. Such clay-slates, in fact, are variable in composition, and sometimes alternate with fossiliferous strata, so that they may be said to belong almost equally to the sedimentary and metamorphic order of rocks. It is probable that, had they been subjected to more intense Plutonic action, they would have been transformed into hornblende- schist, foliated chlorite-schist, scaly talcose-schist, mica-schist, or other more perfectly crystalline rocks, such as are usually associated with gneiss.

UNIFORMITY OF MINERAL CHARACTER IN HYPOGENE ROCKS.

It is true, as Humboldt has happily remarked, that when we pass to another hemisphere, we see new forms of animals and plants, and even new constellations in the heavens; but in the rocks we still recognise our old acquaintances— the same granite, the same gneiss, the same micaceous schist, quartz-rock, and the rest. There is certainly a great and striking general resemblance in the principal kinds of hypogene rocks in all countries, however different their ages; but each of them, as we have seen, must be regarded as geological families of rocks, and not as definite mineral compounds. They are more uniform in aspect than sedimentary strata, because these last are often composed of fragments varying greatly in form, size, and colour, and contain fossils of different shapes and mineral composition, and acquire a variety of tints from the mixture of various kinds of sediment. The materials of such strata, if they underwent metamorphism, would be subject to chemical laws, simple and uniform in their action, the same in every climate, and wholly undisturbed by mechanical and organic causes. It would, however, be a great error to assume, as some have done, that the hypogene rocks, considered as aggregates of simple minerals, are really more homogeneous in their composition than the several members of the sedimentary series. Not only do the proportional quantities of feldspar, quartz, mica, hornblende, and other minerals, vary in hypogene rocks bearing the same name; but what is still more important, the ingredients, as we have seen, of the same simple mineral are not always constant (Chapter 28 and Table 28.1).

SUPPOSED AZOIC PERIOD.

The total absence of any trace of fossils has inclined many geologists to attribute the origin of the most ancient strata to an azoic period, or one antecedent to the existence of organic beings. Admitting, they say, the obliteration, in some cases, of fossils by Plutonic action, we might still expect that traces of them would oftener be found in certain ancient systems of slate which can scarcely be said to have assumed a crystalline structure. But in urging this argument it seems to have been forgotten that there are stratified formations of enormous thickness, and of various ages, some of them even of Tertiary date, and which we know were formed after the earth had become the abode of living creatures, which are, nevertheless, in some districts, entirely destitute of all vestiges of organic bodies. In some, the traces of fossils may have been effaced by water and acids, at many successive periods; indeed the removal of the calcareous matter of fossil shells is proved by the fact of such organic remains being often replaced by silex or other minerals, and sometimes by the space once occupied by the fossil being left empty, or only marked by a faint impression.

Those who believed the hypogene rocks to have originated antecedently to the creation of organic beings, imputed the absence of lime, so remarkable in metamorphic strata, to the non-existence of those mollusca and zoophytes by which shells and corals are secreted; but when we ascribe the crystalline formations to Plutonic action, it is natural to inquire whether this action itself may not tend to expel carbonic acid and lime from the materials which it reduces to fusion or semi-fusion. Not only carbonate of lime, but also free carbonic acid gas, is given off plentifully from the soil and crevices of rocks in regions of active and spent volcanoes, as near Naples and in Auvergne. By this process, fossil shells or corals may often lose their carbonic acid, and the residual lime may enter into the composition of augite, hornblende, garnet, and other hypogene minerals. Although we can not descend into the subterranean regions where volcanic heat is developed, we can observe in regions of extinct volcanoes, such as Auvergne and Tuscany, hundreds of springs, both cold and thermal, flowing out from granite and other rocks, and having their waters plentifully charged with carbonate of lime.

If all the calcareous matter transferred in the course of ages by these and thousands of other springs from the lower part of the earth's crust to the atmosphere could be presented to us in a solid form, we should find that its volume was comparable to that of many a chain of hills. Calcareous matter is poured into lakes and the ocean by a thousand springs and rivers; so that part of almost every new calcareous rock chemically precipitated, and of many reefs of shelly and coralline stone, must be derived from mineral matter subtracted by Plutonic agency, and driven up by gas and steam from fused and heated rocks in the bowels of the earth.

The scarcity of limestone in many extensive regions of metamorphic rocks, as in the Eastern and Southern Grampians of Scotland, may have been the result of some action of this kind; and if the limestones of the Lower Laurentian in Canada afford a remarkable exception to the general rule, we must not forget that it is precisely in this most ancient formation that the Eozoon Canadense has been found. The fact that some distinct bands of limestone from 700 to 1500 feet thick occur here, may be connected with the escape from destruction of some few traces of organic life, even in a rock in which metamorphic action has gone so far as to produce serpentine, augite, and other minerals found largely intermixed with the carbonate of lime.

CHAPTER XXXVI.
MINERAL VEINS.

Different Kinds of mineral Veins.
Ordinary metalliferous Veins or Lodes.
Their frequent Coincidence with Faults.
Proofs that they originated in Fissures in solid Rock.
Veins shifting other Veins.
Polishing of their Walls or "Slicken sides."
Shells and Pebbles in Lodes.
Evidence of the successive Enlargement and Reopening of veins.
Examples in Cornwall and in Auvergne.
Dimensions of Veins.
Why some alternately swell out and contract.
Filling of Lodes by Sublimation from below.
Supposed relative Age of the precious Metals.
Copper and lead Veins in Ireland older than Cornish Tin.
Lead Vein in Lias, Glamorganshire.
Gold in Russia, California, and Australia.
Connection of hot Springs and mineral Veins.

The manner in which metallic substances are distributed through the earth's crust, and more especially the phenomena of those more or less connected masses of ore called mineral veins, from which the larger part of the precious metals used by man are obtained, are subjects of the highest practical importance to the miner, and of no less theoretical interest to the geologist.

ON DIFFERENT KINDS OF MINERAL VEINS.

The mineral veins with which we are most familiarly acquainted are those of quartz and carbonate of lime, which are often observed to form lenticular masses of limited extent traversing both hypogene strata and fossiliferous rocks. Such veins appear to have once been chinks or small cavities, caused, like cracks in clay, by the shrinking of the mass, during desiccation, or in passing from a higher to a lower temperature. Siliceous, calcareous, and occasionally metallic matters have sometimes found their way simultaneously into such empty spaces, by infiltration from the surrounding rocks. Mixed with hot water and steam, metallic ores may have permeated the mass until they reached those receptacles formed by shrinkage, and thus gave rise to that irregular assemblage of veins, called by the Germans a "stockwerk," in allusion to the different floors on which the mining operations are in such cases carried on.

The more ordinary or regular veins are usually worked in vertical shafts, and have evidently been fissures produced by mechanical violence. They traverse all kinds of rocks, both hypogene and fossiliferous, and extend downward to indefinite or unknown depths. We may assume that they correspond with such rents as we see caused from time to time by the shock of an earthquake. Metalliferous veins referable to such agency are occasionally a few inches wide, but more commonly three or four feet. They hold their course continuously in a certain prevailing direction for miles or leagues, passing through rocks varying in mineral composition.

THAT METALLIFEROUS VEINS WERE FISSURES.

(FIGURES 629, 630 and 631. Vertical sections of the mine of Huel Peever, Redruth, Cornwall.

(Figure 629. Vertical section of the mine of Huel Peever, Redruth, Cornwall. Tin.)

(FIGURE 630. Vertical section of the mine of Huel Peever, Redruth, Cornwall. Copper.)

(FIGURE 631. Vertical section of the mine of Huel Peever, Redruth, Cornwall. Clay and copper.))

As some intelligent miners, after an attentive study of metalliferous veins, have been unable to reconcile many of their characteristics with the hypothesis of fissures, I shall begin by stating the evidence in its favour. The most striking fact, perhaps, which can be adduced in its support is, the coincidence of a considerable proportion of mineral veins with FAULTS, or those dislocations of rocks which are indisputably due to mechanical force, as above explained (Chapter 5). There are even proofs in almost every mining district of a succession of faults, by which the opposite walls of rents, now the receptacles of metallic substances, have suffered displacement. Thus, for example, suppose a-a, Figure 629, to be a tin lode in Cornwall, the term LODE being applied to veins containing metallic ores. This lode, running east and west, is a yard wide, and is shifted by a copper lode (b-b) of similar width. The first fissure (a-a) has been filled with various materials, partly of chemical origin, such as quartz, fluor-spar, peroxide of tin, sulphuret of copper, arsenical pyrites, bismuth, and sulphuret of nickel, and partly of mechanical origin, comprising clay and angular fragments or detritus of the intersected rocks. The plates of quartz and the ores are, in some places, parallel to the vertical sides or walls of the vein, being divided from each other by alternating layers of clay or other earthy matter. Occasionally the metallic ores are disseminated in detached masses among the vein-stones.

It is clear that, after the gradual introduction of the tin and other substances, the second rent (b-b) was produced by another fracture accompanied by a displacement of the rocks along the plane of b-b. This new opening was then filled with minerals, some of them resembling those in a-a, as fluor-spar (or fluate of lime) and quartz; others different, the copper being plentiful and the tin wanting or very scarce. We must next suppose a third movement to occur, breaking asunder all the rocks along the line c-c, Figure 630; the fissure, in this instance, being only six inches wide, and simply filled with clay, derived, probably, from the friction of the walls of the rent, or partly, perhaps, washed in from above. This new movement has displaced the rock in such a manner as to interrupt the continuity of the copper vein (b-b), and, at the same time, to shift or heave laterally in the same direction a portion of the tin vein which had not previously been broken.

Again, in Figure 631 we see evidence of a fourth fissure (d-d), also filled with clay, which has cut through the tin vein (a-a), and has lifted it slightly upward towards the south. The various changes here represented are not ideal, but are exhibited in a section obtained in working an old Cornish mine, long since abandoned, in the parish of Redruth, called Huel Peever, and described both by Mr. Williams and Mr. Carne. (Geological Transactions volume 4 page 139; Transactions of the Royal Geological Society Cornwall volume 2 page 90.) The principal movement here referred to, or that of c-c, Figure 631, extends through a space of no less than 84 feet; but in this, as in the case of the other three, it will be seen that the outline of the country above, d, c, b, a, etc., or the geographical features of Cornwall, are not affected by any of the dislocations, a powerful denuding force having clearly been exerted subsequently to all the faults. (See Chapter 5.) It is commonly said in Cornwall, that there are eight distinct systems of veins, which can in like manner be referred to as many successive movements or fractures; and the German miners of the Hartz Mountains speak also of eight systems of veins, referable to as many periods.

Besides the proofs of mechanical action already explained, the opposite walls of veins are often beautifully polished, as if glazed, and are not unfrequently striated or scored with parallel furrows and ridges, such as would be produced by the continued rubbing together of surfaces of unequal hardness. These smoothed surfaces resemble the rocky floor over which a glacier has passed (see Figure 106). They are common even in cases where there has been no shift, and occur equally in non-metalliferous fissures. They are called by miners "slicken-sides," from the German schlichten, to plane, and seite, side. It is supposed that the lines of the striae indicate the direction in which the rocks were moved.

In some of the veins in the mountain limestone of Derbyshire, containing lead, the vein-stuff, which is nearly compact, is occasionally traversed by what may be called a vertical crack passing down the middle of the vein. The two faces in contact are slicken-sides, well polished and fluted, and sometimes covered by a thin coating of lead-ore. When one side of the vein-stuff is removed, the other side cracks, especially if small holes be made in it, and fragments fly off with loud explosions, and continue to do so for some days. The miner, availing himself of this circumstance, makes with his pick small holes about six inches apart, and four inches deep, and on his return in a few hours finds every part ready broken to his hand. (Conybeare and Phil. Geol. page 401 and Farey's Derbyshire page 243.)

That a great many veins communicated originally with the surface of the country above, or with the bed of the sea, is proved by the occurrence in them of well- rounded pebbles, agreeing with those in superficial alluviums, as in Auvergne and Saxony. Marine fossil shells, also, have been found at great depths, having probably been ingulfed during submarine earthquakes. Thus, a gryphaea is stated by M. Virlet to have been met with in a lead-mine near Semur, in France, and a madrepore in a compact vein of cinnabar in Hungary. (Fournet Etudes sur les Depots Metalliferes.) In Bohemia, similar pebbles have been met with at the depth of 180 fathoms; and in Cornwall, Mr. Carne mentions true pebbles of quartz and slate in a tin lode of the Relistran Mine, at the depth of 600 feet below the surface. They were cemented by oxide of tin and bisulphuret of copper, and were traced over a space more than twelve feet long and as many wide. (carne Transactions of the Geological Society Cornwall volume 3 page 238.) When different sets or systems of veins occur in the same country, those which are supposed to be of contemporaneous origin, and which are filled with the same kind of metals, often maintain a general parallelism of direction. Thus, for example, both the tin and copper veins in Cornwall run nearly east and west, while the lead veins run north and south; but there is no general law of direction common to different mining districts. The parallelism of the veins is another reason for regarding them as ordinary fissures, for we observe that faults and trap dikes, admitted by all to be masses of melted matter which have filled rents, are often parallel.

FRACTURE, RE-OPENING AND SUCCESSIVE FORMATION OF VEINS.

Assuming, then, that veins are simply fissures in which chemical and mechanical deposits have accumulated, we may next consider the proofs of their having been filled gradually and often during successive enlargements.

Werner observed, in a vein near Gersdorff, in Saxony, no less than thirteen beds of different minerals, arranged with the utmost regularity on each side of the central layer. This layer was formed of two plates of calcareous spar, which had evidently lined the opposite walls of a vertical cavity. The thirteen beds followed each other in corresponding order, consisting of fluor-spar, heavy spar, galena, etc. In these cases the central mass has been last formed, and the two plates which coat the walls of the rent on each side are the oldest of all. If they consist of crystalline precipitates, they may be explained by supposing the fissure to have remained unaltered in its dimensions, while a series of changes occurred in the nature of the solutions which rose up from below: but such a mode of deposition, in the case of many successive and parallel layers, appears to be exceptional.

(FIGURE 632. Copper lode, near Redruth, enlarged at six successive periods.)

If a vein-stone consist of crystalline matter, the points of the crystals are always turned inward, or towards the centre of the vein; in other words, they point in the direction where there was space for the development of the crystals. Thus each new layer receives the impression of the crystals of the preceding layer, and imprints its crystals on the one which follows, until at length the whole of the vein is filled: the two layers which meet dovetail the points of their crystals the one into the other. But in Cornwall, some lodes occur where the vertical plates, or COMBS, as they are there called, exhibit crystals so dovetailed as to prove that the same fissure has been often enlarged. Sir H. De la Beche gives the following curious and instructive example (Figure 632), from a copper-mine in granite, near Redruth. (Geological Report on Cornwall page 340.) Each of the plates or combs (a, b, c, d, e, f) is double, having the points of their crystals turned inward along the axis of the comb. The sides or walls (2, 3, 4, 5 and 6) are parted by a thin covering of ochreous clay, so that each comb is readily separable from another by a moderate blow of the hammer. The breadth of each represents the whole width of the fissure at six successive periods, and the outer walls of the vein, where the first narrow rent was formed, consisted of the granitic surfaces 1 and 7.

A somewhat analogous interpretation is applicable to many other cases, where clay, sand, or angular detritus, alternate with ores and vein-stones. Thus, we may imagine the sides of a fissure to be incrusted with siliceous matter, as Von Buch observed, in Lancerote, the walls of a volcanic crater formed in 1731 to be traversed by an open rent in which hot vapours had deposited hydrate of silica, the incrustation nearly extending to the middle. (Principles chapter 27 8th edition page 422.) Such a vein may then be filled with clay or sand, and afterwards re-opened, the new rent dividing the argillaceous deposit, and allowing a quantity of rubbish to fall down. Various metals and spars may then be precipitated from aqueous solutions among the interstices of this heterogeneous mass.

That such changes have repeatedly occurred, is demonstrated by occasional cross-veins, implying the oblique fracture of previously formed chemical and mechanical deposits. Thus, for example, M. Fournet, in his description of some mines in Auvergne worked under his superintendence, observes that the granite of that country was first penetrated by veins of granite, and then dislocated, so that open rents crossed both the granite and the granitic veins. Into such openings, quartz, accompanied by sulphurets of iron and arsenical pyrites, was introduced. Another convulsion then burst open the rocks along the old line of fracture, and the first set of deposits were cracked and often shattered, so that the new rent was filled, not only with angular fragments of the adjoining rocks, but with pieces of the older vein-stones. Polished and striated surfaces on the sides or in the contents of the vein also attest the reality of these movements. A new period of repose then ensued, during which various sulphurets were introduced, together with hornstone quartz, by which angular fragments of the older quartz before mentioned were cemented into a breccia. This period was followed by other dilatations of the same veins, and the introduction of other sets of mineral deposits, as well as of pebbles of the basaltic lavas of Auvergne, derived from superficial alluviums, probably of Miocene or even Older Pliocene date. Such repeated enlargement and re-opening of veins might have been anticipated, if we adopt the theory of fissures, and reflect how few of them have ever been sealed up entirely, and that a country with fissures only partially filled must naturally offer much feebler resistance along the old lines of fracture than anywhere else.

CAUSE OF ALTERNATE CONTRACTION AND SWELLING OF VEINS.

(FIGURES 633 to 635. Irregular fissures.

(FIGURE 633.)
(FIGURE 634.)
(FIGURE 635.))

A large proportion of metalliferous veins have their opposite walls nearly parallel, and sometimes over a wide extent of country. There is a fine example of this in the celebrated vein of Andreasburg in the Hartz, which has been worked for a depth of 500 yards perpendicularly, and 200 horizontally, retaining almost everywhere a width of three feet. But many lodes in Cornwall and elsewhere are extremely variable in size, being one or two inches in one part, and then eight or ten feet in another, at the distance of a few fathoms, and then again narrowing as before. Such alternate swelling and contraction is so often characteristic as to require explanation. The walls of fissures in general, observes Sir H. De la Beche, are rarely perfect planes throughout their entire course, nor could we well expect them to be so, since they commonly pass through rocks of unequal hardness and different mineral composition. If, therefore, the opposite sides of such irregular fissures slide upon each other, that is to say, if there be a fault, as in the case of so many mineral veins, the parallelism of the opposite walls is at once entirely destroyed, as will be readily seen by studying Figures 633 to 635.

Let a-b, Figure 633, be a line of fracture traversing a rock, and let a-b, Figure 634, represent the same line. Now, if we cut in two a piece of paper representing this line, and then move the lower portion of this cut paper sideways from a to a', taking care that the two pieces of paper still touch each other at the points 1, 2, 3, 4, 5, we obtain an irregular aperture at c, and isolated cavities at d, d, d, and when we compare such figures with nature we find that, with certain modifications, they represent the interior of faults and mineral veins. If, instead of sliding the cut paper to the right hand, we move the lower part towards the left, about the same distance that it was previously slid to the right, we obtain considerable variation in the cavities so produced, two long irregular open spaces, f, f, Figure 635, being then formed. This will serve to show to what slight circumstances considerable variations in the character of the openings between unevenly fractured surfaces may be due, such surfaces being moved upon each other, so as to have numerous points of contact.

(FIGURE 636. Nipped ores where the course of a vein departs from verticality.)

Most lodes are perpendicular to the horizon, or nearly so; but some of them have a considerable inclination or "hade," as it is termed, the angles of dip being very various. The course of a vein is frequently very straight; but if tortuous, it is found to be choked up with clay, stones, and pebbles, at points where it departs most widely from verticality. Hence at places, such as a, Figure 636, the miner complains that the ores are "nipped," or greatly reduced in quantity, the space for their free deposition having been interfered with in consequence of the pre-occupancy of the lode by earthy materials. When lodes are many fathoms wide, they are usually filled for the most part with earthy matter, and fragments of rock, through which the ores are disseminated. The metallic substances frequently coat or encircle detached pieces of rock, which our miners call "horses" or "riders." That we should find some mineral veins which split into branches is also natural, for we observe the same in regard to open fissures.

CHEMICAL DEPOSITS IN VEINS.

If we now turn from the mechanical to the chemical agencies which have been instrumental in the production of mineral veins, it may be remarked that those parts of fissures which were choked up with the ruins of fractured rocks must always have been filled with water; and almost every vein has probably been the channel by which hot springs, so common in countries of volcanoes and earthquakes, have made their way to the surface. For we know that the rents in which ores abound extend downward to vast depths, where the temperature of the interior of the earth is more elevated. We also know that mineral veins are most metalliferous near the contact of Plutonic and stratified formations, especially where the former send veins into the latter, a circumstance which indicates an original proximity of veins at their inferior extremity to igneous and heated rocks. It is moreover acknowledged that even those mineral and thermal springs which, in the present state of the globe, are far from volcanoes, are nevertheless observed to burst out along great lines of upheaval and dislocation of rocks. (See Dr. Daubeny's Volcanoes.) It is also ascertained that all the substances with which hot springs are impregnated agree with those discharged in a gaseous form from volcanoes. Many of these bodies occur as vein-stones; such as silex, carbonate of lime, sulphur, fluor-spar, sulphate of barytes, magnesia, oxide of iron, and others. I may add that, if veins have been filled with gaseous emanations from masses of melted matter, slowly cooling in the subterranean regions, the contraction of such masses as they pass from a plastic to a solid state would, according to the experiments of Deville on granite (a rock which may be taken as a standard), produce a reduction in volume amounting to 10 per cent. The slow crystallisation, therefore, of such Plutonic rocks supplies us with a force not only capable of rending open the incumbent rocks by causing a failure of support, but also of giving rise to faults whenever one portion of the earth's crust subsides slowly while another contiguous to it happens to rest on a different foundation, so as to remain unmoved.

Although we are led to infer, from the foregoing reasoning, that there has often been an intimate connection between metalliferous veins and hot springs holding mineral matter in solution, yet we must not on that account expect that the contents of hot springs and mineral veins would be identical. On the contrary, M. E. de Beaumont has judiciously observed that we ought to find in veins those substances which, being least soluble, are not discharged by hot springs— or that class of simple and compound bodies which the thermal waters ascending from below would first precipitate on the walls of a fissure, as soon as their temperature began slightly to diminish. The higher they mount towards the surface, the more will they cool, till they acquire the average temperature of springs, being in that case chiefly charged with the most soluble substances, such as the alkalies, soda and potash. These are not met with in veins, although they enter so largely into the composition of granitic rocks. (Bulletin 4 page 1278.)

To a certain extent, therefore, the arrangement and distribution of metallic matter in veins may be referred to ordinary chemical action, or to those variations in temperature which waters holding the ores in solution must undergo, as they rise upward from great depths in the earth. But there are other phenomena which do not admit of the same simple explanation. Thus, for example, in Derbyshire, veins containing ores of lead, zinc, and copper, but chiefly lead, traverse alternate beds of limestone and greenstone. The ore is plentiful where the walls of the rent consist of limestone, but is reduced to a mere string when they are formed of greenstone, or "toad-stone," as it is called provincially. Not that the original fissure is narrower where the greenstone occurs, but because more of the space is there filled with vein-stones, and the waters at such points have not parted so freely with their metallic contents.

"Lodes in Cornwall," says Mr. Robert W. Fox, "are very much influenced in their metallic riches by the nature of the rock which they traverse, and they often change in this respect very suddenly, in passing from one rock to another. Thus many lodes which yield abundance of ore in granite, are unproductive in clay- slate, or killas and vice versa.

SUPPOSED RELATIVE AGE OF THE DIFFERENT METALS.

After duly reflecting on the facts above described, we can not doubt that mineral veins, like eruptions of granite or trap, are referable to many distinct periods of the earth's history, although it may be more difficult to determine the precise age of veins; because they have often remained open for ages, and because, as we have seen, the same fissure, after having been once filled, has frequently been re-opened or enlarged. But besides this diversity of age, it has been supposed by some geologists that certain metals have been produced exclusively in earlier, others in more modern times; that tin, for example, is of higher antiquity than copper, copper than lead or silver, and all of them more ancient than gold. I shall first point out that the facts once relied upon in support of some of these views are contradicted by later experience, and then consider how far any chronological order of arrangement can be recognised in the position of the precious and other metals in the earth's crust.

In the first place, it is not true that veins in which tin abounds are the oldest lodes worked in Great Britain. The government survey of Ireland has demonstrated that in Wexford veins of copper and lead (the latter as usual being argentiferous) are much older than the tin of Cornwall. In each of the two countries a very similar series of geological changes has occurred at two distinct epochs— in Wexford, before the Devonian strata were deposited; in Cornwall, after the Carboniferous epoch. To begin with the Irish mining district: We have granite in Wexford traversed by granite veins, which veins also intrude themselves into the Silurian strata, the same Silurian rocks as well as the veins having been denuded before the Devonian beds were superimposed. Next we find, in the same county, that elvans, or straight dikes of porphyritic granite, have cut through the granite and the veins before mentioned, but have not penetrated the Devonian rocks. Subsequently to these elvans, veins of copper and lead were produced, being of a date certainly posterior to the Silurian, and anterior to the Devonian; for they do not enter the latter, and, what is still more decisive, streaks or layers of derivative copper have been found near Wexford in the Devonian, not far from points where mines of copper are worked in the Silurian strata.

Although the precise age of such copper lodes can not be defined, we may safely affirm that they were either filled at the close of the Silurian or commencement of the Devonian period. Besides copper, lead, and silver, there is some gold in these ancient or primary metalliferous veins. A few fragments also of tin found in Wicklow in the drift are supposed to have been derived from veins of the same age. (Sir H. De la Beche MS. Notes on Irish Survey.)

Next, if we turn to Cornwall, we find there also the monuments of a very analogous sequence of events. First, the granite was formed; then, about the same period, veins of fine-grained granite, often tortuous (see Figure 614), penetrating both the outer crust of granite and the adjoining fossiliferous or primary rocks, including the coal-measures; thirdly, elvans, holding their course straight through granite, granitic veins, and fossiliferous slates; fourthly, veins of tin also containing copper, the first of those eight systems of fissures of different ages already alluded to. Here, then, the tin lodes are newer than the elvans. It has, indeed, been stated by some Cornish miners that the elvans are in some instances posterior to the oldest tin-bearing lodes, but the observations of Sir H. de la Beche during the survey led him to an opposite conclusion, and he has shown how the cases referred to in corroboration can be otherwise interpreted. (Report on the Geology of Cornwall page 310.) We may, therefore, assert that the most ancient Cornish lodes are younger than the coal- measures of that part of England, and it follows that they are of a much later date than the Irish copper and lead of Wexford and some adjoining counties. How much later, it is not so easy to declare, although probably they are not newer than the beginning of the Permian period, as no tin lodes have been discovered in any red sandstone which overlies the coal in the south-west of England.

There are lead veins in Glamorganshire which enter the lias, and others near Frome, in Somersetshire, which have been traced into the Inferior Oolite. In Bohemia, the rich veins of silver of Joachimsthal cut through basalt containing olivine, which overlies tertiary lignite, in which are leaves of dicotyledonous trees. This silver, therefore, is decidedly a tertiary formation. In regard to the age of the gold of the Ural mountains, in Russia, which, like that of California, is obtained chiefly from auriferous alluvium, it occurs in veins of quartz in the schistose and granitic rocks of that chain, and is supposed by Sir R. Murchison, MM. Deverneuil and Keyserling to be newer than the syenitic granite of the Ural— perhaps of tertiary date. They observe that no gold has yet been found in the Permian conglomerates which lie at the base of the Ural Mountains, although large quantities of iron and copper detritus are mixed with the pebbles of those Permian strata. Hence it seems that the Uralian quartz veins, containing gold and platinum, were not formed, or certainly not exposed to aqueous denudation, during the Permian era.

In the auriferous alluvium of Russia, California, and Australia, the bones of extinct land-quadrupeds have been met with, those of the mammoth being common in the gravel at the foot of the Ural Mountains, while in Australia they consist of huge marsupials, some of them of the size of the rhinoceros and allied to the living wombat. They belong to the genera Diprotodon and Nototherium of Professor Owen. The gold of Northern Chili is associated in the mines of Los Hornos with copper pyrites, in veins traversing the cretaceo-oolitic formations, so-called because its fossils have the character partly of the cretaceous and partly of the oolitic fauna of Europe. (Darwin's South America page 209 etc.) The gold found in the United States, in the mountainous parts of Virginia, North and South Carolina, and Georgia, occurs in metamorphic Silurian strata, as well as in auriferous gravel derived from the same.

Gold has now been detected in almost every kind of rock, in slate, quartzite, sandstone, limestone, granite, and serpentine, both in veins and in the rocks themselves at short distances from the veins. In Australia it has been worked successfully not only in alluvium, but in vein-stones in the native rock, generally consisting of Silurian shales and slates. It has been traced on that continent over more than nine degrees of latitude (between the parallels of 30 degrees and 39 degrees S.), and over twelve of longitude, and yielded in 1853 an annual supply equal, if not superior, to that of California; nor is there any apparent prospect of this supply diminishing, still less of the exhaustion of the gold-fields.

ORIGIN OF GOLD IN CALIFORNIA.

Mr. J. Arthur Phillips, in his treatise "On the Gold Fields of California," has shown that the ore in the gold workings is derived from drifts, or gravel clay, and sand, of two distinct geological ages, both comparatively modern, but belonging to different river-systems, the older of which is so ancient as to be capped by a thick sheet of lava divided by basaltic columns. (Proceedings of the Royal Society 1868 page 294.) The auriferous quartz of these drifts is derived from veins apparently due to hydrothermal agency, proceeding from granite and penetrating strata supposed to be of Jurassic and Triassic date. The fossil wood of the drift is sometimes beautifully silicified, and occasionally the trunks of trees are replaced by iron pyrites, but gold seems not to have been found as in the pyrites of similarly petrified trees in the drift of Australia.

The formation of recent metalliferous veins is now going on, according to Mr. Phillips, in various parts of the Pacific coast. Thus, for example, there are fissures at the foot of the eastern declivity of the Sierra Nevada in the state of that name, from which boiling water and steam escape, forming siliceous incrustations on the sides of the fissures. In one case, where the fissure is partially filled up with silica inclosing iron and copper pyrites, gold has also been found in the vein-stone.

It has been remarked by M. de Beaumont, that lead and some other metals are found in dikes of basalt and greenstone, as well as in mineral veins connected with trap-rock, whereas tin is met with in granite and in veins associated with the Plutonic series. If this rule hold true generally, the geological position of tin accessible to the miner will belong, for the most part, to rocks older than those bearing lead. The tin veins will be of higher relative antiquity for the same reason that the "underlying" igneous formations or granites which are visible to man are older, on the whole, than the overlying or trappean formations.

If different sets of fissures, originating simultaneously at different levels in the earth's crust, and communicating, some of them with volcanic, others with heated Plutonic masses, be filled with different metals, it will follow that those formed farthest from the surface will usually require the longest time before they can be exposed superficially. In order to bring them into view, or within reach of the miner, a greater amount of upheaval and denudation must take place in proportion as they have lain deeper when first formed and filled. A considerable series of geological revolutions must intervene before any part of the fissure which has been for ages in the proximity of the Plutonic rock, so as to receive the gases discharged from it when it was cooling, can emerge into the atmosphere. But I need not enlarge on this subject, as the reader will remember what was said in the 30th, 32d, and 35th chapters on the chronology of the volcanic and hypogene formations.

Printed in Great Britain
by Amazon